Preface

The present volume and its companion Volume 2 document the proceedings of the Symposium on Surface Contamination: Its Genesis, Detection and Control held in Washington, D.C., September 10-13, 1978. This Symposium was a part of the 4th International Symposium on Contamination Control held under the auspices of the International Committee of Contamination Control Societies, and the Institute of Environmental Sciences (U.S.A.) was the official host.

The ubiquitous nature of surface contamination causes concern to everyone dealing with surfaces, and the world of surfaces is wide and open-ended. The technological areas where surface cleaning is of cardinal importance are too many and very diversified. To people working in areas such as adhesion, composites, adsorption, friction, lubrication, soldering, device fabrication, printed circuit boards, etc., surface contamination has always been a _bete noire_. In short, people dealing with surfaces are afflicted with molysmophobia†, and rightfully so.

In the past, the subject of surface contamination had been discussed in various meetings, but this symposium was hailed as the most comprehensive symposium ever held on this important topic, as the technical program comprised 70 papers by more than 100 authors from 10 countries. The symposium was truly international in scope and spirits and was very well attended. The attendees represented a broad spectrum of backgrounds, interests, and professional affiliations, but all had a common interest and concern about surface contamination and cleaning.

This Symposium was organized with the following objectives in mind: to bring together scientists, technologists, and engineers

†Molysmophobia means fear of dirt or contamination, from Mrs. Byrne's Dictionary of Unusual, Obscure, and Preposterous Words, University Books, Secaucus, N.J. 1974.

interested in all aspects of surface contamination, to review and assess the current state of the knowledge, to provide a forum for exchange and cross-fertilization of ideas, and to define problem areas which need intensified efforts; the comments from the audience confirm that these objectives were definitely fulfilled. For example, there were brisk and enlightening - not exothermic - discussions and information exchanges between the presenters and the audience. It should be added that the purpose of a symposium is to present the state of the knowledge of the topic under consideration, and this is best accomplished by inviting the leading workers to present overview papers covering topics of their special interest; these should be complemented and augmented by original research papers. This is exactly what was done for this symposium as there were 20 invited overviews covering a wide range of subtopics within the purview of surface contamination. May it be added that there are three components of the subject of surface contamination: (i) cleaning of surfaces, (ii) characterization of the degree of surface cleanliness, and (iii) storage of clean surfaces or the kinetics of recontamination. All these components were covered in this symposium.

The present proceedings volumes contain a total of 64 papers (some papers from the original program are not included for a variety of reasons, and a few papers which were not presented but are included) which were reviewed by at least two qualified reviewers, and most of these were modified and revised in light of the reviewers' comments before inclusion in these volumes. Only about half of these papers (unreviewed) in their abbreviated form (long abstract) were included in the document distributed to the registrants at the time of the meeting. So the present volumes contain a great deal more material. The papers included are in their full form and are divided into four sections as follows: General Papers; Cleaning of Surfaces; Surface Contamination Detection, Identification, Characterization, and Control; and Implications of Surface Contamination. Sections I & II are embodied in Volume 1, and Volume 2 contains Sections III & IV. Broadly speaking, all kinds of surfaces (metal, oxide, ceramic, glass, alloy, polymer,and liquid) and contaminants (organic film, inorganic, particulate, microbial,and radioactive) are covered in these volumes as the intent of the Symposium was to cover the topic of surface contamination in general and broad terms rather than to concentrate on a particular surface/contamination combination.

The topics covered include: sources, forms, and mechanism(s) of surface contamination; various techniques (including solvent, plasma, UV/ozone, chemical, mechanical, ion milling, and surface chemical) for cleaning surfaces; radioactive and microbial contamination of surfaces; atomically-clean surfaces; preparation of clean mineral surfaces; particle adhesion; preparation of clean

THE UNIVERSITY OF STRATHCLYDE
ANDERSONIAN
LIBRARY
Chemical Division Library

Surface Contamination

Genesis, Detection, and Control

Volume 1

Surface Contamination

Genesis, Detection, and Control

Volume 1

Edited by

K. L. Mittal

IBM Corporation
East Fishkill Facility
Hopewell Junction, New York

PLENUM PRESS · NEW YORK AND LONDON

Library of Congress Cataloging in Publication Data

Main entry under title:

Surface contamination.

Proceedings of a symposium on Surface Contamination: Its Genesis, Detection, and Control held at the fourth International Symposium on Contamination Control in Washington, D.C., on September 10–14, 1978, sponsored by the Institute of Environmental Sciences and the International Committee of Contamination Control Societies

Includes index.

1. Surface contamination–Congresses I. Mittal K. L., 1945- II. International Symposium on Contamination Control, 4th, Washington, D.C., 1978. III. Institute of Environmental Sciences. IV. International Committee of Contamination Control Societies.

TA418.7.S8 620.1'129 79-15433

ISBN 0-306-40176-2 (v. 1)

Proceedings of a Symposium on Surface Contamination: Its Genesis, Detection, and Control, held at the Fourth International Symposium on Contamination Control in Washington, D.C., on September 10–14, 1978, published in two volumes, of which this is volume one.

A Division of Plenum Publishing Corporation
227 West 17th Street, New York, N.Y. 10011

Printed in the United States of America

water surfaces; cleaning of polymeric surfaces; various techniques [including contact angle or wettability, evaporative rate analysis, indium adhesion test, surface potential difference, spectroscopy (Auger, ESCA, ISS, SIMS), ellipsometry, plasma chromatography, ion chromatography, and microfluorescence] for characterizing the degree of surface cleanliness; ionic contamination detection and characterization; and implications of surface contamination.

It should be added that the availability of sophisticated surface analytical tools has been a boon in the area of surface contamination and a series of papers deal with the utility of such techniques for monitoring and characterizing microamounts of surface contaminants. The papers dealing with implications of surface contamination cover topics ranging from microelectronics to the public health sector.

In essence, these proceedings volumes present a very comprehensive coverage of the latest state of the knowledge relative to the important subject of surface contamination. A special feature of these volumes is the inclusion of 20 invited overviews which should provide a veritable gold mine of valuable information, and these overviews coupled with the 44 contributed research papers should serve as a *vade mecum* for anyone interested in surface contamination and cleaning. It should be added that originally it was intended to include a Discussion at the end of each section, but in spite of the continuous exhortation, the number of questions received in written form (although there was a brisk verbal discussion at the end of each paper) did not warrant inclusion of written Discussions in these volumes.

Acknowledgements: First of all, I am thankful to the management of IBM Corporation for permitting me to organize this symposium and to edit these volumes. Particularly, I would like to acknowledge the understanding and patience of my manager, Dr. H. R. Anderson, Jr., during this activity. Special thanks are due to Mr. R. W. Martin for his help and cooperation during the various stages of putting together the technical program, to Mrs. Betty Peterson of the Institute of Environmental Sciences for her ready and willing help whenever it was needed. The reviewers should be thanked for their sacrifice of time and many valuable comments. The cooperation and enthusiasm of the authors is a must for any proceedings volume and I would like to thank them for their efforts and cooperation in submitting the manuscripts. Thanks are due to all the secretaries who helped with the correspondence typing. I am thankful to my wife, Usha, for letting me work late hours during the tenure of editing these volumes and also for helping with the subject index; to my daughter, Anita, and son, Rajesh, for letting me spend those hours which rightfully belonged to them.

Last, but not least, I would like to express my thanks to Mrs. Edith Oakley (Forbes Services) for meeting, without complaint, various deadlines for typing of the manuscripts.

K. L. Mittal
Symposium Organizer and Chairman

IBM Corporation
East Fishkill Facility
Hopewell Junction, N.Y. 12533

Contents of Volume 1

PART I. GENERAL PAPERS

PART II. CLEANING OF SURFACES

Contents of Volume 2

PART III. SURFACE CONTAMINATION DETECTION, IDENTIFICATION, CHARACTERIZATION, AND CONTROL

Part I
General Papers

SURFACE CONTAMINATION: AN OVERVIEW

K. L. Mittal

IBM Corporation, East Fishkill Facility

Hopewell Junction, New York 12533

Surface contamination is of great concern wherever surfaces are used, and there is a legion of products, processes and technologies where surfaces play a very vital role. There is no universally acceptable or applicable definition of a "clean" surface, as a clean surface can be defined in a variety of manners depending upon the purpose of and the requirements imposed on the surface. Generally speaking, a clean surface is one which is free from contaminants, and any unwanted matter or energy is regarded as a contaminant.

There are four components of a surface contamination program: (i) causes and mechanism(s) of surface contamination, (ii) techniques for removing contamination, i.e., for cleaning surfaces, (iii) techniques for monitoring the degree of surface cleanliness, and (iv) kinetics of recontamination or the storage of clean surfaces. In this review, all these aspects are considered with emphasis on recent developments. The availability of sophisticated surface analytical techniques has been a boon for identifying, characterizing, and quantifying microamounts of surface contaminants, and examples are cited where such techniques have proved of immense value. Although the emphasis is on the film-type contaminants on solid surfaces, but the particulate and other contaminations of such surfaces and the contamination of liquid surfaces are also accorded due coverage.

INTRODUCTION

Surface contamination is of concern to everyone dealing with surfaces and the world of surfaces is wide and open-ended. The technological areas where surface cleaning is of cardinal importance are too many and diversified. To people working in areas such as adhesion, adsorption, friction, lubrication, device fabrication, composites, printed circuits, etc., the surface contamination has always been a bete noire. The issue of surface contamination has important implications in many human endeavors ranging from spacecrafts to microelectronics to public health sector.[1-17] Here a few eclectic examples should suffice to underscore the importance of surface contamination or cleaning. Trace amounts of surface contaminants can play havoc with the reliability of a sensitive electronic device, and particulate contamination can be a significant yield detractor particularly in light of the fact that line patterns for large scale integrated circuits are getting closer and closer with the result that even very small particles can not be tolerated. Ionic contamination (e.g. left from fluxing operation) can lead to electrical circuit leakage and unacceptably low circuit reliability. Ionic contamination can also provide the electrolyte necessary for corrosion to ensue.

Before one can bond surfaces, these must be clean; if it is not clean, it will not stick. Even a very minute level of undesirable material on the surface can lead to gross bond failures. Very minute contamination can interfere with the functioning of optical devices. The consequences of opaque film formation on reflecting mirrors in space can be quite serious. Presence of radioactive contamination on surfaces has serious implications. Or just imagine what would happen if the open wound were contaminated with microorganisms. A very interesting paper[18] has recently appeared dealing with redispersion of surface contamination and its consequences. Also the interest in and importance of surface contamination is well attested by the large number (64) of papers included in these proceedings volumes. I do not mean to be facetious when I say that people dealing with surfaces are afflicted with molysmophobia*, and rightfully so.

There are four components of the surface contamination program: (i) causes and mechanism(s) of surface contamination, (ii) techniques for removing surface contamination, i.e., for cleaning

*Molysmophbia means fear of dirt or contamination, from Mrs. Byrne's Dictionary of Unsual, Obscure, and Preposterous Words, University Books, Secaucus, N.J. 1974.

surfaces, (iii) techniques for monitoring the degree of surface cleanliness, and (iv) kinetics of recontamination or the storage of clean surfaces. All these four aspects are discussed in this review with emphasis on recent developments.

The topic of surface contamination and cleaning has been in the past the subject of various symposia,[19-26] books, monographs, reports and review papers,[27-50] and also a great deal of material has been discussed by other authors in these proceedings volumes, so there is no need to expatiate on certain topics here and only references to these should suffice. Although the emphasis in this review is on the film-type contaminants on solid surfaces, but the particulate and other contaminations of such surfaces and the contamination of liquid surfaces are also accorded due consideration. Clean surfaces can be categorized as: (i) Atomically clean surfaces and (ii) Technologically or practically clean surfaces; the emphasis in this review is on the latter type, and the surfaces considered include metals, metal oxides, glass, polymers, and liquids.

SOURCES, CAUSES AND FORMS OF SURFACE CONTAMINATION

Surface contamination can originate from a number of sources: (i) Adsorption of impurities (e.g. hydrocarbons) from the ambient which culminates into adsorbed layer type contaminants, (ii) Reaction of the surface with the reactive species (e.g. oxygen, sulfur) which results in reaction layers, and (iii) Preferential diffusion of one component (in case of multicomponent) can give rise to variable composition type of surface contaminant. The contaminants caused by these sources are film or layer type in nature, but a surface could also be contaminated by particles adsorbed from the ambient. In the case of polymeric surfaces, there is an additional source of surface contamination due to the leaching of low molecular weight fractions or plasticizers.

In many cases, the cleaning itself may be the source of contamination, e.g., a detergent layer left after cleaning if the surface is not thoroughly rinsed. In solvent cleaning technique, some solvents may oxidize leaving films behind on the surface. Human breath is a common source of contamination and the following remark was overheard: 'If you want a clean environment, do not let anybody breathe.' The choice is between a dead person or a clean ambient.

Surface contamination could be gaseous, liquid or solid in physical state and may be present in film or particulate form. Furthermore, it could be ionic or nonionic in its chemical makeup and organic or inorganic in character. In addition to chemical type, the surface contamination could be microbial, radioactive, or radiation.

The most common organic contaminants are shown in Table I, and Table II presents some of the common sources of particulate contamination.

Table I. Common Organic Contaminants.

Motor Oils
Pump Oils
Bearing Lubricants
Fluxes
Solvent Vapors
Plasticizers
Dioctylphthalate From Filter Testing
Skin Oils
Cosmetics
Vacuum Pump Exhausts

Table II. Sources of Particulate Contamination.

- Dust
- Street Clothing
- Human Hair
- Perspiration
- Smoking
- Cosmetics - Flaking
- Metal Chips and Burrs from Assembly Process
- Solid Film Lubricants
- Metal Oxides - Nonadherent or Flakey
- Fretting Corrosion
- Air Circulating Systems - Suspended Solids
- Air

WHAT IS A CLEAN SURFACE?

There is no single definition of a clean surface which will have ecumenical appeal and universal validity. "How clean is clean" is a very subjective matter as it depends on the individual's requirements. Cleanliness, like beauty, is in the eyes of the beholder. A particular surface may be considered as a highly contaminated surface by someone interested in doing fundamental adsorption studies, whereas the same surface could be totally acceptable as a clean surface for other applications or investigations. A good example would be that the oxide on a metal surface will be considered a contaminant by someone who wants to study adsorption on the metal surface, whereas the same oxide is not only acceptable but highly desirable for bonding purposes. Figure 1 delineates the real life situation on a metal surface and raises the question: How many of these layers should one remove before this metal sample can be called a "clean" surface? The answer depends on the requirements imposed on this metal surface, i.e., the intended usage of this surface, as pointed out earlier.

Clean surfaces can be broadly divided into two categories: Atomically clean surfaces, and technologically or practically clean surfaces. Atomically clean surfaces are required for special purposes and these can only be realized in high vacuum ($<10^{-9}$ Torr), see below; whereas we are more concerned on a day-to-day basis with technologically clean surfaces, i.e., surfaces which are clean enough to be processed and used for subsequent operations.

Below are given certain definitions and conditions as gleaned from the literature for a contaminant and a clean surface from pragmatic point of view.

Contaminant:

A contaminant is any substance or energy which is unwanted or adversely affects a product or process of interest. Or a contaminant is a material in the wrong place; this is analogous to the definition of dirt as used in detergency. Hydrocarbons on a surface are always considered as contaminants and undesirable;

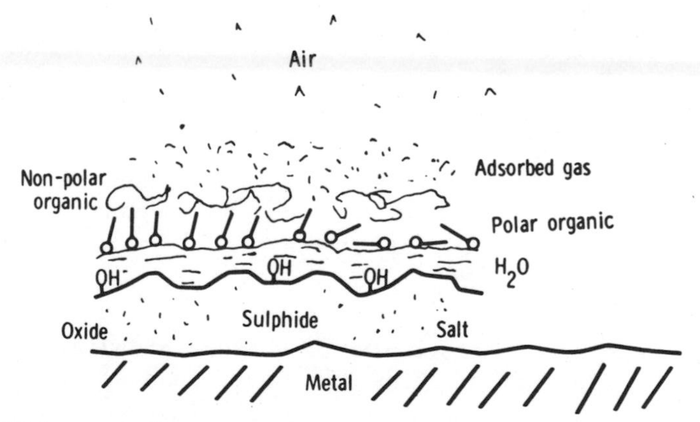

Figure 1. "Hierarchy" of spontaneously adsorbed layers on a "metal" surface (from reference 51).

however, with the present energy situation one would love these hydrocarbons. The point is that these organics become contaminants when they happen to be in the wrong place.

Clean Surface:

1. A clean surface is one that contains no significant amounts of undesired material.[50]

2. A clean surface is that which contains no material that will affect measurements on or processing of the surface.

3. A clean surface is one that is free of all but a few percent of a single monolayer of foreign atoms, either adsorbed on or substantially replacing surface atoms of the parent lattice. This kind of surface is called atomically clean surface.

Or putting it differently, for a surface to be called clean, the degree of surface cleanliness must meet the following two criteria: it must be sufficient for subsequent processing and (ii) it must be sufficient to ensure the future reliability of the product for which that surface will be used.

TECHNIQUES FOR CLEANING SURFACES

In light of the discussion above, it is quite clear that the choice of a technique for cleaning a surface will depend on the following: (i) Nature of the surface; techniques which are applicable to metal surfaces may be totally unacceptable for polymeric surfaces, (ii) Nature, form, and amount of the contaminant to be removed; certain techniques which are suitable for removing ionic contamination are not useful for the removal of nonionics, and (iii) Extent to which the contaminants should be removed, i.e., the degree of surface cleanliness required.

Atomically Clean Surfaces

As far as atomically clean surfaces are concerned, there are special techniques to obtain these and such techniques have been reviewed and discussed in detail elsewhere[52-54] which obviates the need to cover these here.

Cleaning of Metals, Metal Oxides, and Glass (Removal of Film-Type Contaminants)

There is available in the literature a cornucopia of techniques,[27-47,50] ranging from very simple to very sophisticated, for cleaning metals, metal oxides and glasses. In addition to these references, pertinent references will be cited later for specific cleaning techniques, surfaces, and contaminants.

A detailed discussion of each technique will render this review prohibitively long so only selected techniques (with emphasis on recent developments) will be discussed, and references to others, where germane, will be given.

Solvent or Chemical Cleaning. Solvent cleaning has been used from times immemorial and even today it is commonly used for cleaning surfaces, and a number of solvents are effectively used to remove a variety of contaminats.[55-64] It should be noted that solvent cleaning is always included whenever cleaning of surfaces is discussed (see, e.g., references 27-47, 50.) However, recently the use of solubility parameter(δ) principle for selecting the most suitable organic solvent or mixture of organic solvents for a particular contaminant has been discussed and it has been effectively utilized.[55-57] Figure 2 shows that for the removal of solder flux, a solvent with δ of 8.29 should be the most effective; but the same solvent may be totally inadequate for removal of grease or other contaminants (see Figure 3). The ordinate in these figures is a measure of the contaminant

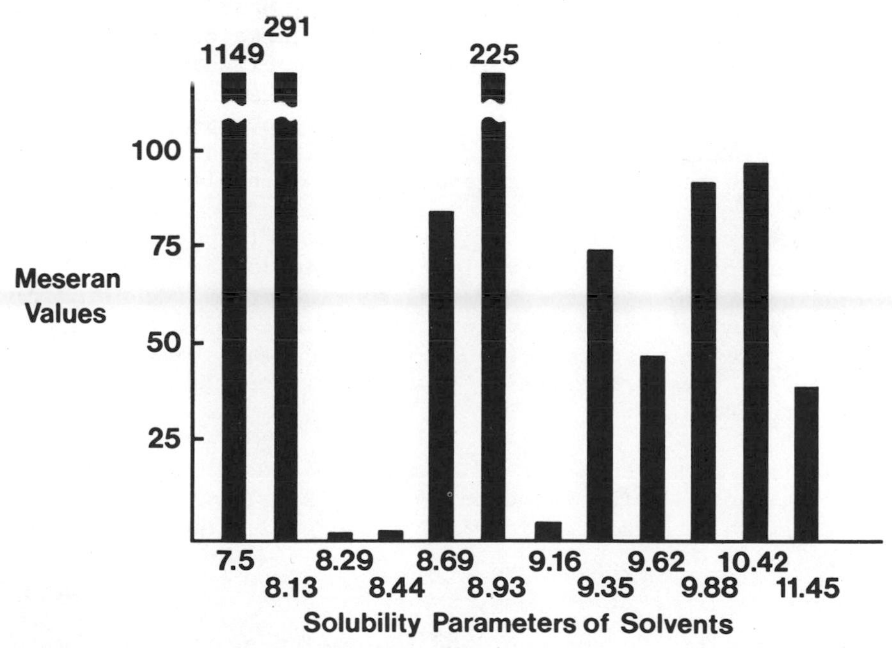

Figure 2. Effectiveness of various solvents for the removal of a solder flux (from reference 57).

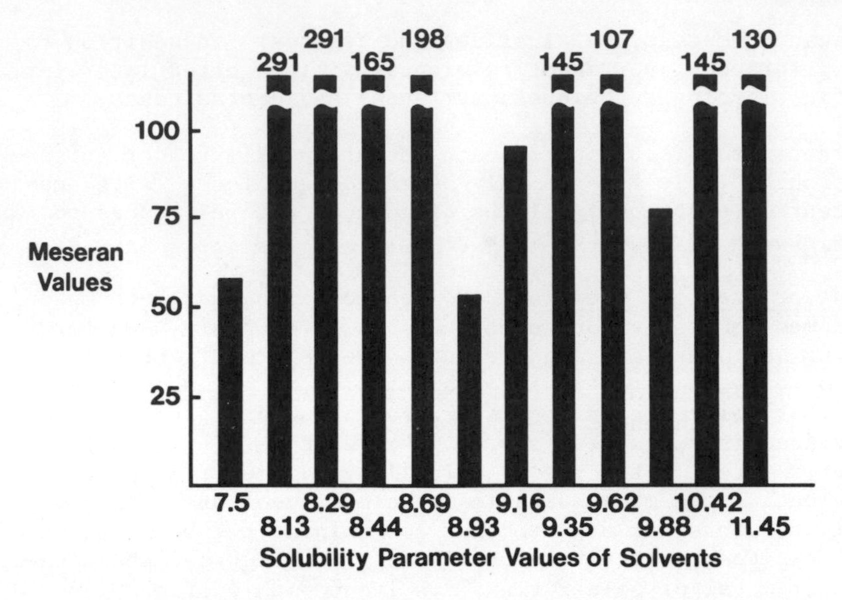

Figure 3. Effectiveness of various solvents for the removal of grease (from reference 57).

remaining on the surface as determined by the MESERAN* technique, the details of which are given later in the section on techniques for characterization of the degree of surface cleanliness. The lower the MESERAN value, the cleaner the surface. Solvents are available with varying δ's[65-66] and furthermore these can be blended to obtain the desired δ. For detailed application and utility of this technique, the reader should consult the original references.

Solvents are also used for cleaning ionic contaminants.[62-64] For cleaning of glasses, solvents, along with other techniques, are commonly used.[67-72] In addition to organic solvents other chemicals are also used for cleaning metals, metal oxides and glasses.[73-75] Use of chemicals based on surface-chemical principle is discussed later.

Mayer and Shwartzman[75] have recently discussed the use of MEGASONIC cleaning using standard chemical solution for the removal of particles and organic films (see details under Removal of Particles).

---

*MESERAN stands for Measurement and Evaluation of Surfaces by Evaporative Rate ANalysis. It is a trademark of ERA Systems Inc., Ooltewah, TN.

Glow Discharge Plasma or Sputter Cleaning. Glow discharge plasmas of both inert gas (e.g. argon) and oxygen are commonly and effectively utilized for cleaning of surfaces of metals, metal oxides, and glasses. The effect of plasma treatment on polymers is different from that on inorganic solids as discussed later.

In the case of oxygen plasma, there are present ozone and UV radiation. Ozone is a strong oxidizing agent and UV is very effective in causing bond scission which allows effective oxidation, with the result that hydrocarbon contamination is converted into volatile products. Plasma oxidation is akin to UV/ozone cleaning process (disucssed later) but the UV/ozone cleaning rates are lower than the oxygen plasma cleaning rates since the concentration and intensities are lower. However, in the case of inert gas discharge (glow discharge cleaning) the cleaning mechanism is poorly defined since the surface is being bombarded with ions, electrons, and high-energy neutrals as well as radiation from the plasma. In the case of glow discharge cleaning the substrate itself is not a part of the glow-discharge circuit as it is in sputter cleaning.

Glow discharge plasma cleaning has been used for cleaning both metals and oxides; however, the plasma oxidation can not be used for metal surfaces which are sensitive to oxidation. For example, O'Kane and Mittal[76] used this technique for cleaning Fe-Co and rhodium surfaces as shown in Table III. The results clearly show the superiority of the plasma technique over that of solvent cleaning. The degree of surface cleanliness (as monitored in terms of carbon from hydrocarbon contaminant) was monitored both by Auger spectroscopy and by contact angle measurements. It should be noted that different surfaces require different amounts of gas flow, time of plasma treatment, power, etc.; so one has to optimize these variables for the surface of interest. For further details and latest developments anent glow discharge plasma cleaning see references 50, 77-79.

In addition to inert gas or oxygen plasma, reactive plasmas (plasmas containing species which can chemically react with the material to be removed) have also been effectively used for cleaning purposes. Kominiak and Mattox[80] have discussed the reactive plasma cleaning of metals and have shown that reactive plasma cleaning in Ar-HCl was as effective as r.f. sputter cleaning and could be accomplished at much lower power levels. For example, reactive plasma cleaning of titanium was accomplished in 90 min. using 500 V r.f. and 30 W in an Ar-5% HCl plasma. Similar r.f. sputter cleaning required 90 min at 2KV r.f. and 600 W in air. Their results are shown in Figure 4.

Sputter cleaning is another commonly used technique to clean both metals and oxides. This technique involves bombardment of

Table III. Change In Surface Wettability As a Function Of Surface Cleaning Conditions (from reference 76).

Type of Surface	Method of Cleaning	Water Contact angle Θ_{H_2O}
Fe-Co	No cleaning of contaminated surface	96°
	Organic Solvents	80°
	Organic Solvents and helium-oxygen plasma[a]	<5°
	Freshly evaporated Fe-Co surfaces	<5°
Rhodium	Vapor degreasing	38°
	Vapor degreasing + argon plasma[b]	19°
	Vapor degreasing + argon plasma[c]	13°
	Vapor degreasing + argon plasma[d]	8°

a 10 W for 1 min.
b 40 W for 1 min.
c 40 W for 3 min.
d 40 W for 6 min.

the substrate with high energy particles which results in the removal of the material by a momentum transfer process called sputtering. In sputter cleaning a conductive surface, the applied sputtering potential may be dc or r.f.; whereas in the case of an insulator surface, an r.f. potential is impressed on the surface. This is a very effective technique but problems may arise because of back scattering of sputtered material to the surface, particularly at high pressures. Also the substrate may possibly be roughened and foreign material from the counterelectrode may be deposited on the substrate. For further details and its applications in cleaning, see, for example references 45a,50.

It should be noted that recently, Bouwman et al.[81] have used hydrogen at a pressure of 5×10^{-5} Torr and at room temperature in an ion-etching mode for cleaning copper and Fe/Cr/Ni steel surfaces. Copper is almost completely cleaned of carbon, sulfur and oxygen within 30 min (as analyzed by Auger spectroscopy). This cleaning method seems to be very surface specific and nondestructive to subsurface layers. Their results for copper are shown in Figure 5.

UV/Ozone Cleaning. Recently the ability of ultraviolet (UV) radiation to decompose hydrocarbons has been used to clean a number of surfaces. Sowell et al.[82] have reported that prolonged (~ 15 hours) exposure to UV radiation produced clean surfaces on gold contacts and glass slides in ambient air, but the time of exposure was reduced in a vacuum system when a pressure

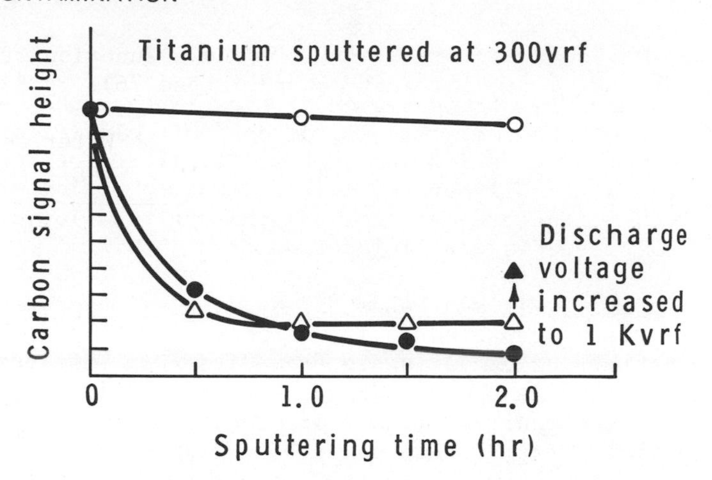

Figure 4. The cleaning of a titanium disk as measured by the carbon peak height, for low power r.f. discharges in Ar, Ar - 1% CCl_4 and Ar - 5% HCl:●, Ar - 5% HCl, $P_{in} \approx 30W$; Δ,Ar - 1% CCl_4, $P_{in} \approx 50W$;o, 100% Ar, $P_{in} \approx 30W$; total pressure 0.8 Pa (5mTorr) (from reference 80).

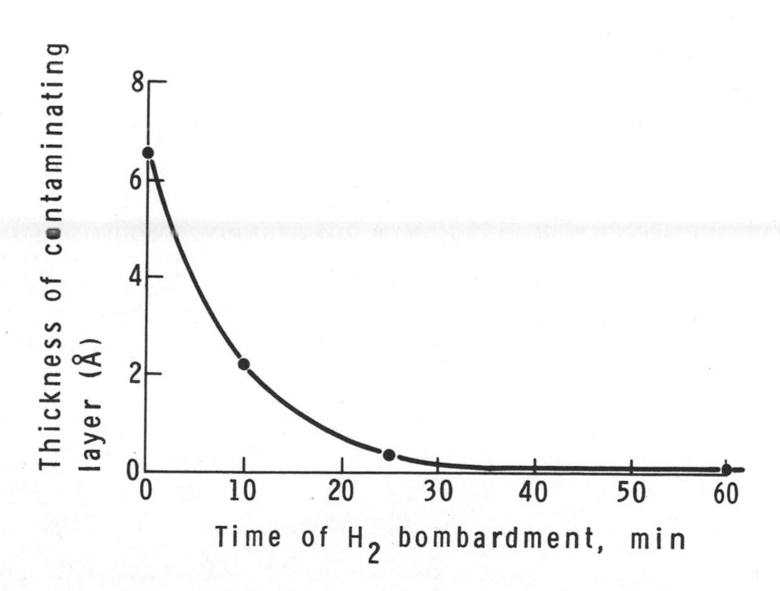

Figure 5. Decrease of thickness of contamination layer with time of hydrogen bombardment (from reference 81).

of 10^{-4} Torr of oxygen was present. Their results are shown in Figures 6 and 7. In Figure 6, the contact angle of water, Θ_{H_2O}, was used as a measure of the degree of cleanliness; the lower the Θ_{H_2O}, the cleaner the surface. In the case of gold (Figure 7) the co-efficient of adhesion was used as the criterion for surface cleanliness; the higher the co-efficient, the cleaner the surface. Both of these techniques for characterization of cleanliness are discussed in detail later.

More recently, Vig and LeBus,[83a] and Vig[83b] have shown that the UV/ozone cleaning procedure is a highly effective method for removing a variety of contaminants from different surfaces. The experimental set-up is shown in Figure 8. It should be noted that the combination of short UV light and ozone produces a clean surface substantially faster than either short-wave UV light without ozone or ozone without UV light. The virtues of this technique and the variables which control the cleaning efficiency are summarized in Table IV.

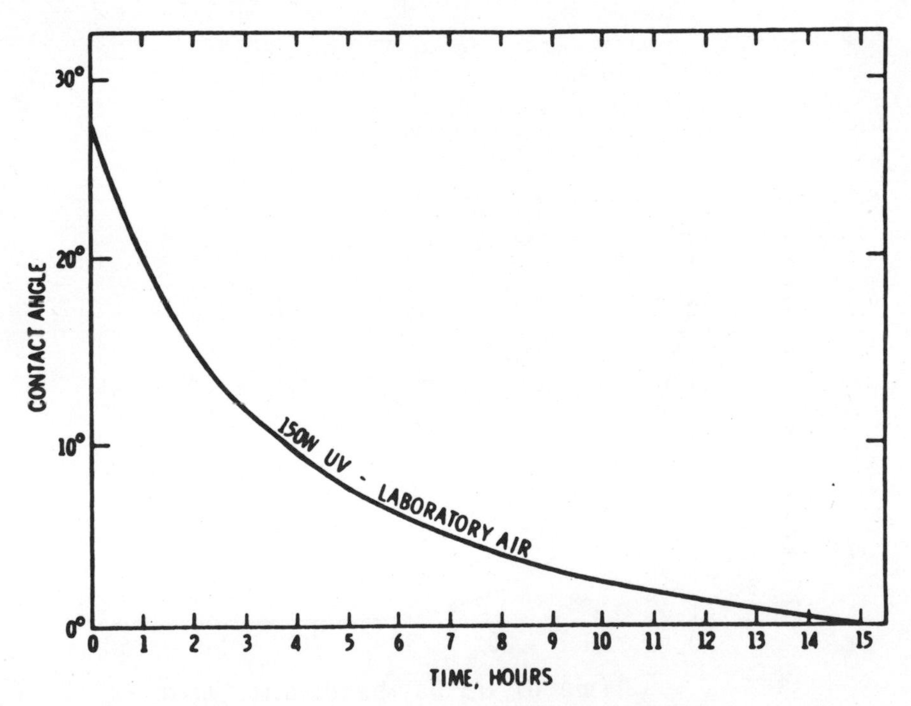

Figure 6. Cleaning of glass surface by UV irradiation in normal laboratory air (from reference 82).

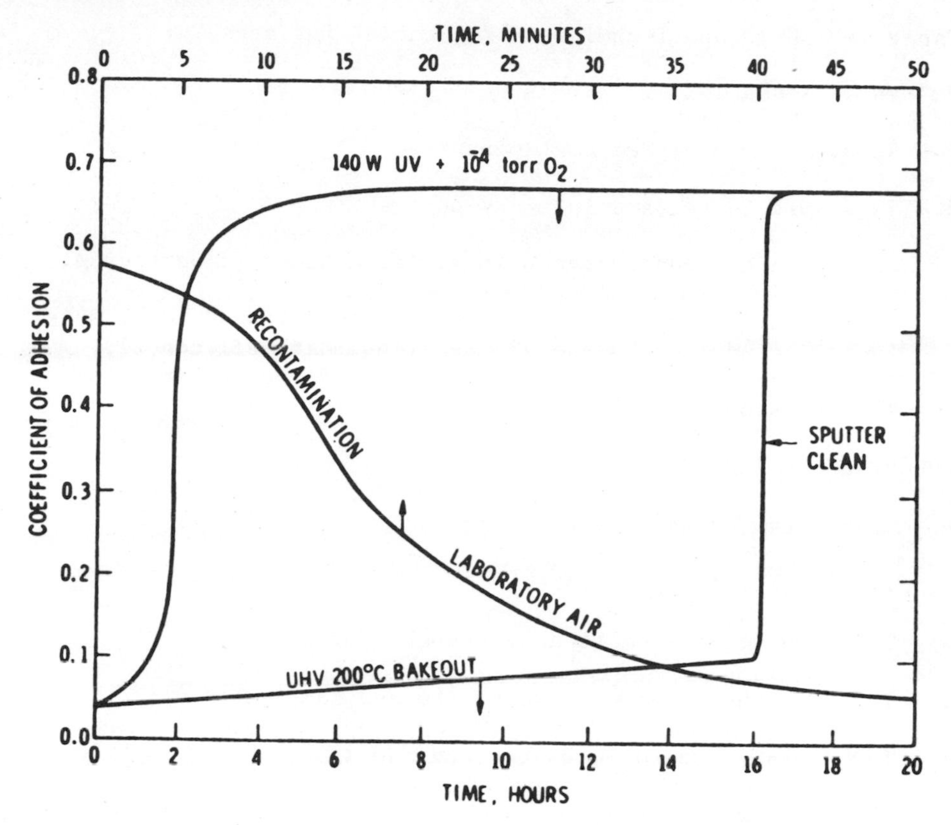

Figure 7. Cleaning of a gold surface by UHV 200°C bakeout, argon sputter cleaning, and UV irradiation at 10^{-4} Torr (lower scale). Recontamination rate in normal laboratory air is also shown (upper scale) (from reference 82).

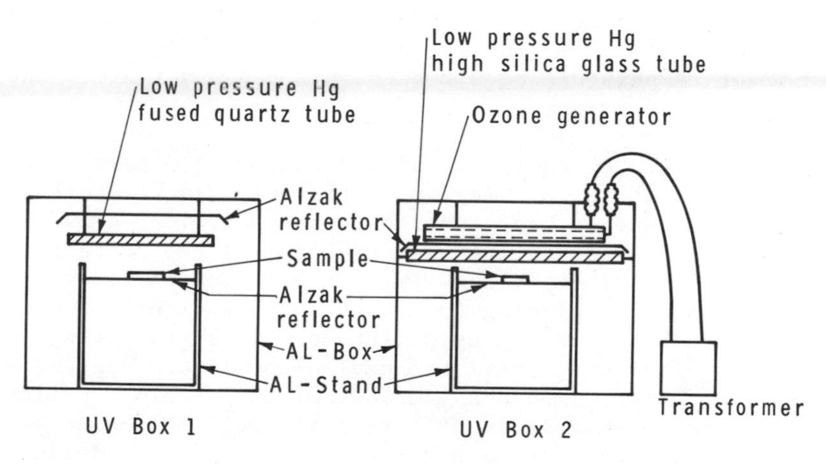

Figure 8. Apparatus for UV cleaning experiments (from reference 83a).

Table IV. UV/Ozone Technique For Cleaning Surfaces

POINTS IN FAVOR

1. Removes a variety of contaminants.

2. Is simple to use and is a dry process.

3. Can produce clean surfaces in air at ambient temperatures.

4. Successfully removed contaminants from quartz and metal surfaces include: oils and greases (including silicones), fluxes, skin oils, and contamination adsorbed during prolonged exposure to air.

IMPORTANT VARIABLES ARE:

- The contaminants initially present.
- The precleaning procedure.
- The wave lengths emitted by the UV source.
- The atmosphere between the source and sample.
- The distance between source and sample.
- The time of exposure.

One example of the effectiveness of UV/O_3 cleaning is given in Figure 9. This figure shows the thermocompression ball bonding of gold to gold of surfaces which have been UV/O_3 cleaned and the ones that have been sputter cleaned then exposed to laboratory ambient.[84] Cleaning efficiency of the UV/O_3 is shown by the lower temperature required to make bonds vis-a-vis the non-UV/O_3 cleaned surfaces.

Other Techniques. Recently Petvai and Schnitzel[85] have discussed the application of ion-milling in surface cleaning. Another interesting technique is based on the surface-chemical principle[86] in which suitable liquids are used to displace the contaminants on the surface. For example, water and oils can be removed from solid surfaces by displacing them with another liquid, the displacing action being driven by differences in surface tension. For water displacement, butyl and amyl alcohols are most effective. Certain fluorinated compounds and low molecular weight silicones have been found particularly effective in the displacement of organic fluids. This approach has been successfully applied to the salvage of electronic and electrical equipment contaminated

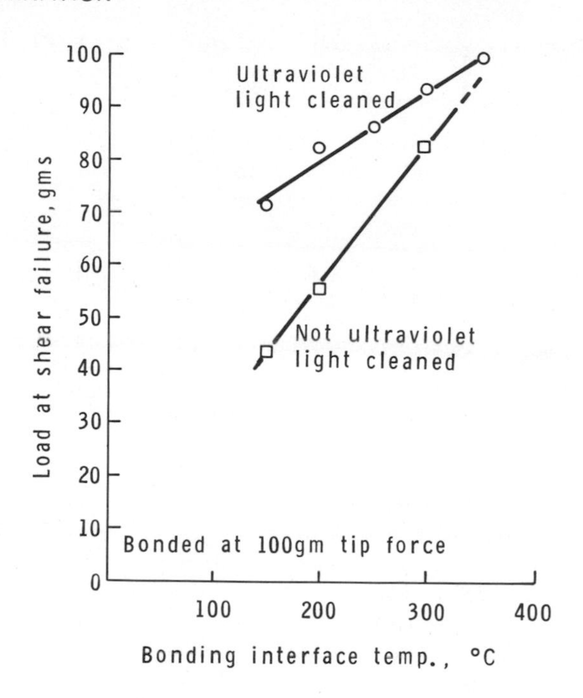

Figure 9. Effect of UV/O_3 cleaning on bondability of gold-on-gold using thermocompression bonding (from reference 84).

by water, seawater, fuels and lubricants, and smoke deposits. The displacing liquids should have certain desirable characteristics and these are discussed in detail by these authors.

Cleaning of Metals, Metal Oxides, and Glass (Removal of Particulates)

So far we have discussed the removal of film-type contaminants and the removal of particulates requires special techniques. A number of authors[87-94] have discussed the forces of adhesion and removal of particles, mechanism(s) of removal of particulates, and the factors which influence particle removal. Kroeck et al.[87] have discussed the removal of particles by various means. Particles are removed by applying an air jet (blow dusting) and recently it has been shown that[91] solvent spray is very effective for the removal of particulates; his findings are presented in Table V.

Mayer and Shwartzman[75] have discussed the MEGASONIC cleaning which uses high frequency ultrasonic energy (0.8 MHz) and a standard chemical solution which is not heated. This process effec-

Table V. Removal Efficiency of 5 μm and Greater Al_2O_3 Particles From Optically Polished Glass and Polished Metal Surfaces (from reference 91).

Removal Method	From Glass Surface	From Metal Surface
Vapor Degreaser Freon TF Solvent	11%-28%	....
Photon Irradiation (Flashing) 10 Flashings Each 20 J/cm^2, 800 μs	32%-34%	33%
Compressed Gas Jet, 10 s. 690 kPa(100 psi)	....	52%-61%
Ultrasonic Cleaning in Freon TF Solvent, 1-2 Min	69%-92%	85%-91% <1% for ≥ 1μm Particles
Liquid Spray Freon TF Solvent 30 s, 345 kPa(50 psi)	97%	92%-95% 3% for ≥ 1μm Particles
Liquid Spray Freon TF Solvent 5-10 s, 6.9 MPa(1000 psi)	99.8%-99.95%	99.5%-99.7% 81% for ≥ 1μm Particles

tively removes particles down to about 0.3 μm diameter simultaneously from the front and back surfaces, thin organic films and many ionic impurities.

An interesting approach to particle removal has recently been discussed in the literature. It is based on the use of hydrophilic polymers[89] to trap the particles. For details, the reader should refer to the original article.

Cleaning of Polymeric Surfaces

Many of the techniques commonly used for cleaning metals, metal oxides, and glasses are not suitable for cleaning of polymeric surfaces. Cleaning of polymeric surfaces poses some special concerns and problems which require special handling of polymers. It should be emphasized that quite frenquently cleaning of a polymer surface may mean simply the modification of the surface so that the contaminant initially present (e.g. low molecular weight fractions, plasticizers, etc.) is no longer considered a contaminant. For example, the effect of glow discharge plasma on the polymer is not as simple as that on the metal, because plasma treatment of a polymer modifies its surface considerably. The details of polymer cleaning or etching and modification have been discussed elsewhere.[50,95]

Cleaning of Liquid Surfaces

So far we have discussed the cleaning of solid surfaces, but the cleaning of liquid surfaces is also very important for many interfacial studies. Scott,[96-97] Mysels and Florence[49] have discussed the techniques for cleaning liquid (particularly water) surfaces.

Removal of Radioactive and Microbial Contamination

The removal of radioactive and microbial type of surface contaminants is extremely important as, in addition to their harmful effect on the surface on which these are present, these can redisperse and deposit on the neighboring surfaces. Allen et al.[98] have discussed the use of electropolishing as a large-scale rapid and effective technique for removal of plutonium and other radionuclide contamination from a variety of metal surfaces. Schaeffer and Brewer[99] have discussed the surface contamination and cleaning characteristic of two radioisotopic heat sources. Radioisotopic thermoelectric generators (RTG's) employ the self-heating characteristic of decaying radioactive material (in this case plutonium 238) to generate electricity. The cleaning of the exterior surfaces of the containers is quite important.

Recently Costerton et al.[100] have discussed the mechanism by which bacteria present in natural water systems adhere to surfaces which come in contact with such water systems. An understanding of the adherent mode of growth of these bacteria offers the prospect of controlling their deposition on surfaces.

TECHNIQUES FOR MONITORING THE DEGREE OF SURFACE CLEANLINESS

In order to monitor the effectiveness of a particular surface cleaning technique, the need for quantitative methods of assessing the degree of surface cleanliness is quite patent. The literature is replete with techniques for characterizing surfaces in general[101-105] and surface cleanliness in particular.[28,30,39,41-43,45,50,106-111] Techniques for monitoring the degree of surface cleanliness range from very simple to very sophisticated. It should be pointed out that the availability of modern surface analytical tools has been a boon in the area of surface contamination. Table VI lists virtues of an ideal test for monitoring the degree of surface cleanliness. The whole gamut of techniques availabile for monitoring surface cleanliness can be broadly divided into two categories: Direct methods and indirect methods. In the case of direct methods,

Table VI. Virtues of An Ideal Test For Characterizing Cleanliness.

1. Objective
2. Reproducible
3. Reliable
4. Quick
5. Independent of operator's experience
6. Nondestructive to the surface
7. Applicable to all materials
8. Capable of identifying all types of contamination (organic and inorganic)
9. Quantitative
10. Capable of yielding some "number" on contamination scale

one can directly identify the nature of the contaminant; whereas the indirect methods utilize certain property of the surface as an indicator of the presence of a contaminant. Certain direct methods can provide information as to the state, chemical composition, physical nature, and quantity of the contaminant.

As there is a voluminous literature dealing with surface characterization methods and techniques for monitoring cleanliness, so here I will concentrate on a few selected methods and references to others, where pertinent, should suffice.

Spectroscopic Methods

Among the spectroscopic techniques for analyzing and characterizing surfaces, the commonly used for monitoring surface cleanliness are: AES,[112-113] ESCA,[114-116] ISS and SIMS*[112b,117-121] Table VII presents a comparative summary of these techniques. The spectroscopic techniques fall into the category of direct methods for monitoring surface cleanliness. As a number of papers in these proceedings volumes deal with the application of such techniques for monitoring surface contamination, so here I will cite a few selected examples to illustrate their usefulness. Figure 10 is a good example of the utility of Auger spectroscopy in analyzing the presence of surface contaminants, thereby monitoring the effectiveness of various surface cleaning techniques. Figure 11 shows clearly that ISS/SIMS can be very effective in monitoring trace amounts of surface contaminants. For application of ESCA, see references 114-116; ESCA is particularly useful for analyzing and characterizing organic polymeric surfaces.

*AES → Auger Electron Spectroscopy
ESCA → Electron Spectroscopy for Chemical Analysis
ISS → Ion Scattering Spectrometry
SIMS → Secondary Ion Mass Spectroscopy

Table VII. Comparison of ESCA, AES, ISS/SIMS Techniques (from reference 105).

Characteristic	ESCA	AES	ISS/SIMS
Minimum Detectability Limit (one part in ________) of elements	10^4	10^3	10^6
Topography	Examines entire surface	Can sweep sample with spots a few hundred microns wide. Can ultimately produce electron spot size of 5μ	Examine entire surface
In depth profiling	Possible but slow	Sputtering rate to order of few hundred Å/min	Natural consequence of the technique (5Å/min)
Destructive ? (What materials can be examined?)	Most	Solids organics only very special conditions	Solids only
Depth to which information is produced	50 Å	20 Å	1 monolayer
Chemistry	Sensitive to oxidation state and crystal structure	Some sensitivity to oxidation state	SIMS gives molecular composition

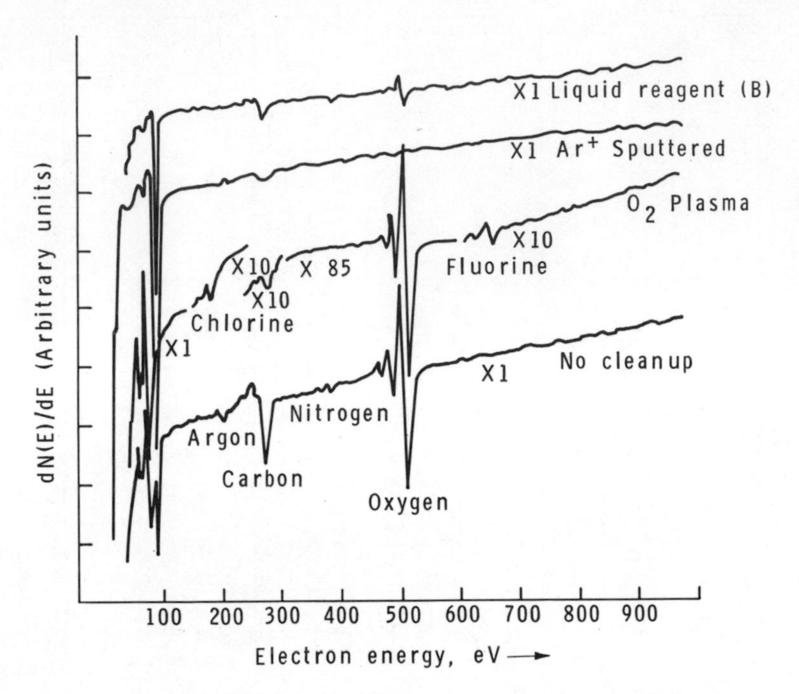

Figure 10. Typical Auger spectra after various surface cleaning procedures (from reference 113).

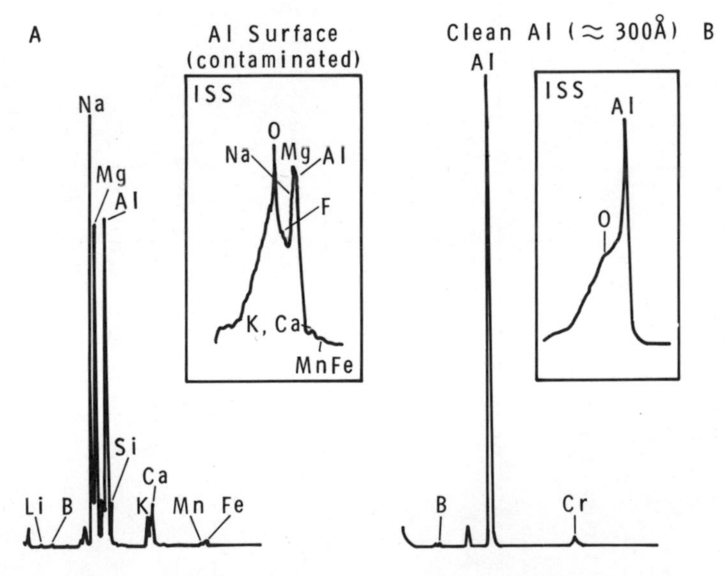

Figure 11. ISS and SIMS spectra of 1024 Al sheeting prior to and after chemical cleaning procedures (from reference 119).

It should be added that such techniques have been extremely helpful in many cases for unraveling (by analyzing trace amounts of undesirable material present on the surfaces) the poor bondability behavior of a variety of substrates.

Evaporative Rate Analysis

Evaporative rate analysis (ERA) technique has been shown to be quite effective for monitoring the degree of surface cleanliness.[122-124] For example, Garrett and Good[124] used ERA to monitor the cleanliess (removal of cutting oil) of steel and aluminum surfaces. The principle of this technique is as follows: a radioactive compound (dissolved in a suitable solvent) is deposited on the surface to be analyzed. The rate of evaporation of a volatile material from a surface is an inverse function of the amount of pre-existing contamination, i.e., $R \propto 1/C$ where R is the rate of evaporation and C is the concentration of pre-existing contamination. So it is clear that the greater the R, the cleaner the surface. Or one can monitor the radioactivity remaining on the surface; the contaminated surface exhibits a greater amount of radioactivity than the corresponding clean surface. This is given in terms of MESERAN numbers which signify the amount of radioactivity remaining on the surface; the lower the MESERAN numbers, the cleaner the surface (see Figure 12.) For details of this technique and its applications, see reference 123.

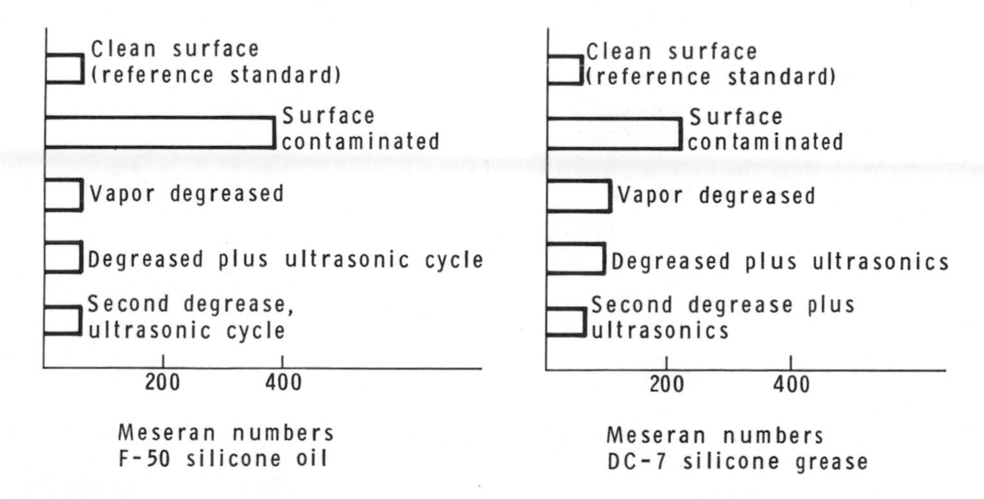

Figure 12. Cleaning process efficiency in the removal of silicone contamination from copper (trichloroethylene solvent); MESERAN numbers = radioactivity counts in 112 seconds. (from reference 55).

Surface Potential Difference

Another technique which has recently shown a great deal of interest is the surface potential difference (SPD) or contact potential difference (CPD) method.[125-128] Guttenplan[125] has described the use of SPD monitor as a low-cost, nondestructive, in-line test instrument for monitoring surface cleanliness or surface condition. The experimental set-up is shown in Figure 13. SPD represents the difference in potential between the test surface and a reference electrode (probe), and any change in the test surface condition is reflected in changes in SPD. SPD is useful for detecting both hydrophilic contaminants, but the surfaces must be dried for measurements. According to the author, SPD measurements show good correlation with the effectiveness of the cleaning process. Figure 14 shows SPD results with the water-break test (discussed below). In this figure, except for cleaning process No. 2, there seems to be a good correlation between the SPD findings and the water-break test. Figure 15 shows a good correlation of SPD and particle counts as a function of cleaning method. He has also used the SPD method for monitoring the effectiveness of flux removal as shown in Table VIII. For further applications of and viewpoints regarding this technique, the reader should consult references 126-128.

Bijlmer[128] used SPD to monitor the condition of metal surfaces. He found that abraded surfaces of different metals (e.g. aluminum) showed SPD values in the same sequence as is given in the electromotive force series. He monitored the quality of surface film formed after the chemical and electrochemical treatment. Furthermore, he observed that cracks in anodic layers increased the SPD in relation to the crack density. Smith[127] used SPD on phosphoric acid anodized Al 7075-T6 samples deliberately contaminated (with about 30 different contaminants) to varying controlled levels for purposes of their detection and found that SPD could detect only about half of these.

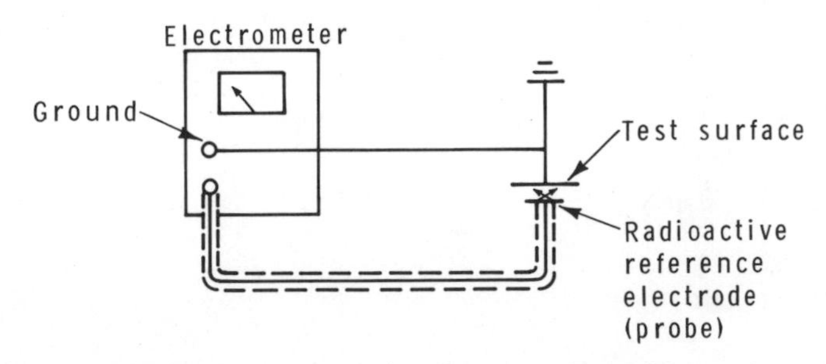

Figure 13. Schematic diagram for measurement of SPD by ionization method (from reference 125).

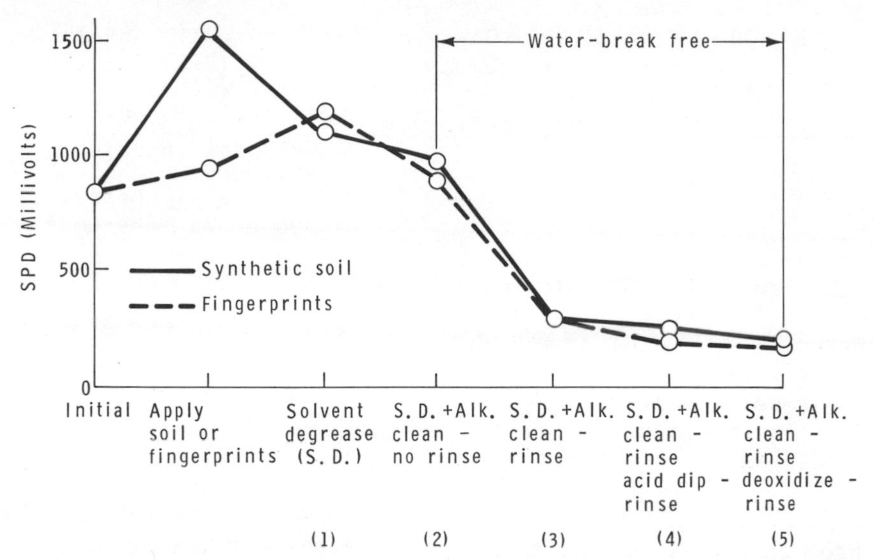

Figure 14. Surface cleanliness -- SPD vs. water-break test (from reference 125).

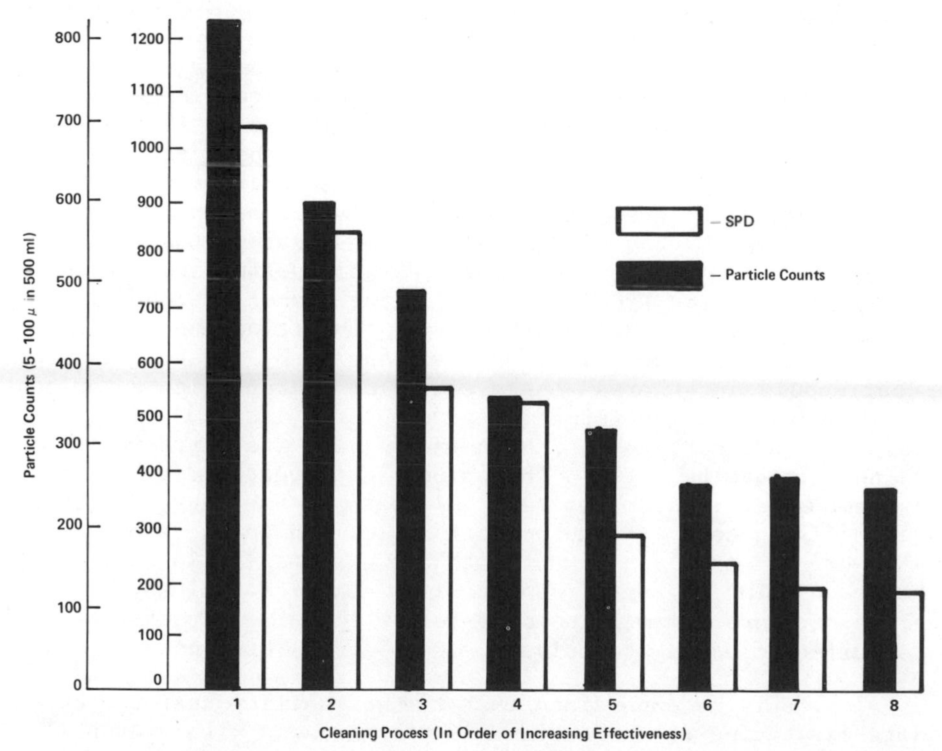

Figure 15. Correlation of SPD and particle counts with cleaning method (from reference 125).

Table VIII. Flux Removal From Ceramic Printed Circuits - Relative Effectiveness of Different Solvents (from Reference 125).

Solvent	Relative Ranking	
	SPD Meter	Referee Method*
TMS (Freon TS-Methanol)	1	1
TE 35 (65 Freon TF - 35 Ethanol)	1	2
Alpha 1003 (Fluorocarbon 122 - Propanol)	3	3
TES (Freon TF - Ethanol Azeotrope)	4	4
TMC (Freon TF - Methylene chloride)	5	5
Toluene (Present solvent)	6	6

*Based on water extraction of ionic residues and conductivity measurement (1-Best; 6=Least effective).

From this discussion it appears that the SPD method should be an useful nondestructive technique for detection of surface contaminants and for monitoring the quality of surfaces.

Wettability or Contact Angle

The concept of wettability of a surface is commonly employed to monitor the cleanliness of a surface.[39,42,44,45,106-108,129-133] In addition to these references, it should be noted that this technique is so prevalent for monitoring cleanliness that it is always included in every list or discussion of surface cleanliness test methods. The basic idea behind this test is that if the surface (metal, metal oxide, glass; but not polymeric, see comment below) is free from hydrophobic (the common hydrophobic contamination is of the hydrocarbon type) impurities then the water should wet the surface (i.e., contact angle, Θ, is zero) and form a continuous film, whereas the presence of hydrophobic contaminants will culminate in the formation of droplets (i.e., $\Theta > 0^{o}$). Table IX lists surfaces which in their clean state are wettable and nonwettable by water. The principle of wettability is used in various ways: atomizer test,[131a] water-break test,[131b] black breath figure test,[132] and contact angle. While the atomizer, water-break, and black breath tests are qualitative in nature, the measurement of contact angle gives numerical values. The values of Θ are commonly used to follow the effectiveness of various surface cleaning techniques, (see, e.g., references 44 and 76.)

It should be emphasized that the wettability test can be quite misleading if the surface is contaminated with hydrophilic impurities. For example, if the surface is covered with a layer of detergent or an inorganic salt the water will spread indicating

Table IX. Wettability of Surfaces (from Reference 42).

I. Surfaces that Wet

1. Oxides.
2. Metal plus metal oxide film.
3. Metal plus foreign oxides (alumina, silicates).
4. Metal plus adsorbed chemical films (wetting agents).
5. Hydrated silica (Si-OH surface bonds).

II. Surfaces that Do Not Wet

1. Gold*
2. Metals free of oxides
3. Organic polymers or surfaces with organic films.
4. Surfaces with organics in the surface.
5. Metals plus certain adsorbed ions (F^-).
6. HF etched silicon (Si-F bonds).
7. Strongly heated silica (Si-O-Si surface bonds).

*The issue of wettability of gold by water has been a polemical one and both wettability and nonwettability of gold has been reported in the literature. However, Schrader[129] has recently shown that the pure gold (gold surface prepared in ultra-high vacuum) is wettable by water.

that the surface is clean. Also wettability studies on rough surfaces are very difficult, as roughness interferes with Θ measurements and one can get spurious results. Furthermore, it should be noted that this test can not be applied to test the cleanliness of low energy surfaces, i.e., surfaces with surface free energy less than 72 dynes/cm (all polymers fall in the category of low energy surfaces), as these even in their absolutely clean state are not completely wetted by water. However, if a polymer surface is contaminated with high surface energy contaminants then the Θ_{H_2O} will be lowered as compared with that on the clean polymer.

This test offers the virtue of being quick and simple to carry out to test for hydrophobic contaminants, and there are available two ASTM Standards[131] utilizing the concept of wettability.

Indium Adhesion Test[134-135]

In this test the surface to be tested is brought in contact with indium (a soft metal) and force is applied to make a bond, and subsequently tensile force is applied to disrupt the bond.

The co-efficient of adhesion, σ, is defined as the ratio of the tensile force for adhesive failure to the joining force; the higher the σ, the cleaner the surface. Figure 16 shows some results obtained using this test. A few comments are in order, (i) this test is applicable to both rough and polished surfaces, (ii) both hydrophobic and hydrophilic contaminants can be detected, (iii) σ of unity indicates a surface cleanliness barely applicable for critical applications. Freshly broken semiconductor-grade silicon and vigorously cleaned glass have σ values as high as 2, and (iv) wide variations of results on contaminated surfaces may result partly from poor definition of the contaminated area and irregular distribution of contaminants on this part of the surface.

Plasma Chromatography

Recently a new technique known as plasma chromatography has been effectively utilized as an ultra-sensitive analytical technique which permits the characterization of trace contaminants of the order of parts per billion or less.[136] The technique is capable of detecting both positive and negative ions which are formed as a result of ion-molecule interactions. Figure 17 & 18 show typical spectra of clean SiO_2 and SiO_2 contaminated with ethyl cellosolve acetate.

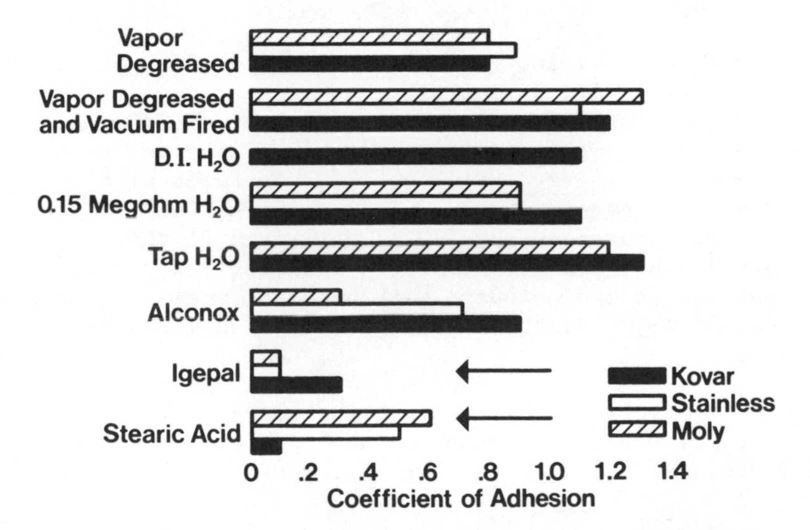

Figure 16. Effect of cleaning and subsequent contamination on the coefficient of adhesion of metals. An arrow indicates that no adhesion or very little adhesion was obtained for one or more of the tests in that series, indicating that there were areas of greater contamination than the average value reflects (from reference 134).

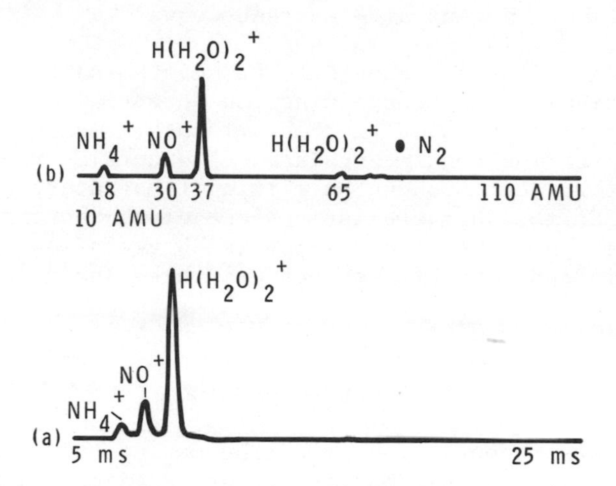

Figure 17. The positive-mode spectra of a clean thermal oxide surface: (a) ion mobility spectrum; (b) mass spectrum (from reference 136).

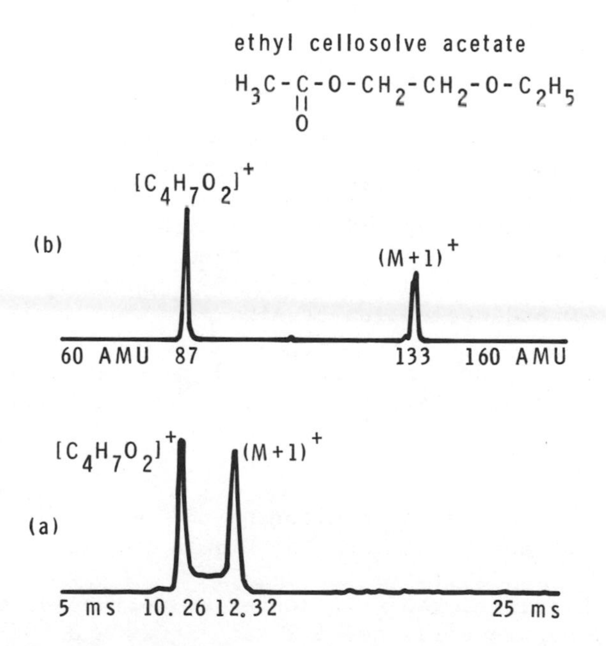

Figure 18. The positive-mode spectra of SiO_2 contaminated with ethyl cellosolve acetate: (a) ion mobility spectrum; (b) mass spectrum (from reference 136).

Potential - Current Curve

Cuthrell and Jones[137] have detected contaminants on the surface of electrical contacts by using the potential - current curves as characteristics of the contaminants. A leveling of the potential (decreasing contact resistance) occurs as the junction temperature is raised through the melting temperature of the contaminant. So the characteristic shape of the potential - current curves can be used to distinguish between low and high melting organic contaminants and between these and oxides or refractory particles.

Microfluorescence Method

Froot[138] has recently described the use of this method for identifying complex organic particulates and films. This technique can rapidly and nondestructively detect and identify organic particulates without the necessity of removing them from or altering the surface in question. It can also identify particles embedded in transparent layers and detect the presence of thin organic films. Detection of films is particularly sensitive on either textured surfaces or those containing some form of pattern. Using this method, films have been detected that were as thin as 0.3 nm as determined by ellipsometry.

Other Techniques

In addition to some of the techniques discussed above the following have also been used to monitor the degree of surface cleanliness. Ellipsometry,[139] microscopic,[140] electrochemical,[48] electrode potential,[141-142] radioisotope tracer,[143,144] exoelectron emission,[130] remission photometry,[130] photoelectron emission,[127] co-efficient of friction,[145] fluorescent dyes,[146] edge lifting,[45a] and nucleation of zinc.[147]

The techniques for monitoring the degree of radioactive surface contamination have been discussed by Anzai and Kikuchi[148] and Gale.[149]

Techniques for Monitoring the Degree of Surface Cleanliness of Liquid Surfaces

The characterization of the degree of surface cleanliness of liquid surfaces requires special techniques and these have been discussed by Scott,[96-97] and Mysels and Florence.[49] Such techniques are primarily based on the principle that the contamination of a liquid surface changes its surface or interfacial properties. These techniques are indirect in nature as these use some property of the surface to monitor its cleanliness.

Measurement of Ionic Contamination

The measurement or characterization of ionic species on a surface requires special techniques and there is a considerable amount of literature dealing with this topic.[150-156] The common technique for measurement of ionic surface contamination is to measure the electrical conductivity of solutions of aqueous extracts of the surface. Brous[155] has recently discussed the theory and practice of dynamic and static measurement of ionic residues. In addition, two interesting papers have recently appeared. Wargotz[156] has described a new analytical technique known as ion chromatography (IC) which has been applied for the identification and quantification of various ionic species in samples of water digests of printing wiring. This analytical technique offers the advantage of providing a rapid analysis (30 minutes) of conductive and corrosive contaminants encountered on printed wiring boards, and he has demonstrated the selectivity, sensitivity, calibration and repeatability of this technique.

Rickabaugh[155] has described an ionic contamination detection system (ICDS) with improved performance for quantizing residue ionic species on printed wiring board circuits. Also it has been used to determine the effectiveness of several cleaning procedures for flux removal. The completed system is capable of automatically detecting ionic contamination at a level as low as the calibrated equivalence of 0.35 μg NaCl with a repeatability within 2%. Manual override can be used to lower the detectable limit to the calibrated equivalence of 0.10 μg NaCl.

KINETICS OF RECONTAMINATION OR STORAGE OF CLEAN SURFACES

Kinetics of Recontamination

It is important to clean a surface and assure its degree of cleanliness, but equally important is to understand its kinetics of recontamination so that proper methods can be devised to store them clean. White[44] showed, using contact angle technique, that different surfaces (all so called "clean" to start with) exhibited different contact angles with water after exposure to a particular hydrocarbon environment for a given time, see Figure 19. Figure 20 shows the kinetics of recontamination of different surfaces when exposed to laboratory air. These results show that different

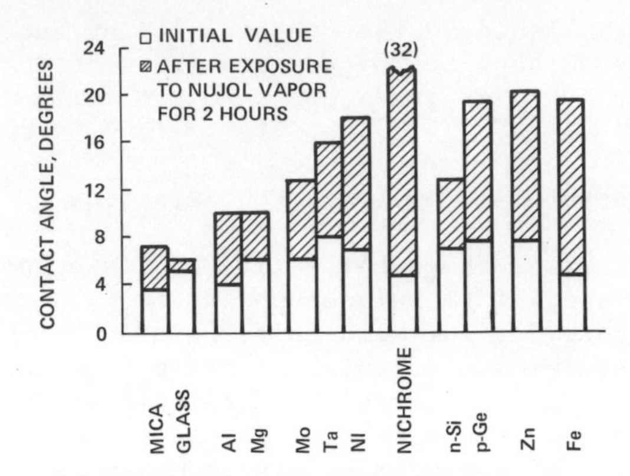

Figure 19. Contact angles on oxidized metal surfaces (from reference 44).

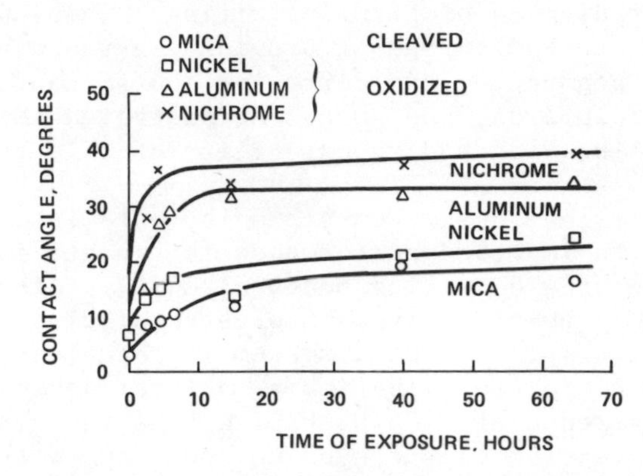

Figure 20. Contamination of surfaces exposed to laboratory air (from reference 44).

surfaces get contaminated to different extent and brings out the importance of specific surface chemical interactions involved in the contamination process. The results of O'Kane and Mittal,[76] shown in Figure 21 agree with the findings of White. Both samples of Fe-Co were clean ($\Theta_{H_2O} \sim 5^o$) at time zero but the rate of recontamination is appreciably higher in the case of plasma cleaned sample vis-a-vis a freshly evaporated surface. These results show clearly that the kinetics of recontamination depends on the mode of cleaning the surface or, in other words, on the history of the surface.

Sowell et al.[82] have also reported on the kinetics of recontamination of clean gold surface when exposed to laboratory air,

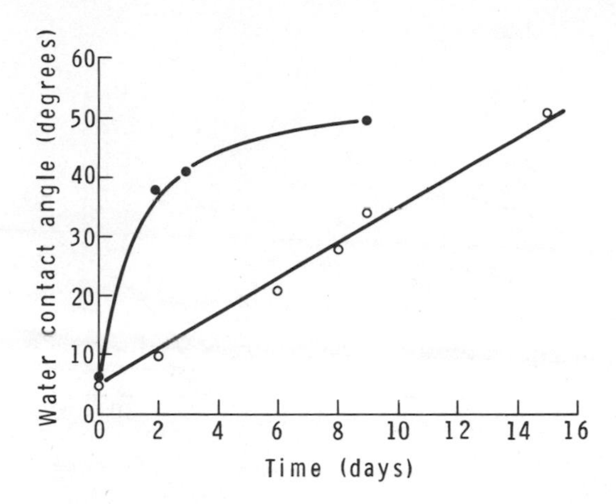

Figure 21. Rate of adsorption of hydrophobic contaminants on Fe-Co alloy surfaces; • plasma cleaned surface, o freshly evaporated surfaces (from reference 76).

as shown in Figure 7. More recently, Tamai et al.[157] have studied the effect of water vapor on the contamination of metallic oxide surfaces, again using contact angle of water as the technique for monitoring the degree of contamination as was done by the other investigators. Their findings are shown in Figure 22. Their results show that the contamination rates of the clean surfaces of metallic oxides (TiO_2, $SrTiO_3$, NiO, Al_2O_3, SiO_2) when exposed to liquid paraffin vapor were largely reduced by the coexisting water vapor, and such reduction is related to the hydrophilicity of oxides.

It is interesting to note that if the contact angle of water (Θ_{H_2O}) is taken as the criterion for surface contamination or cleanliness then one would construe from the results of White, O'Kane and Mittal that it takes hours or days for a surface to be totally contaminated with hydrocarbons. However, every textbook teaches that it takes less than a second for a surface to be covered with a monolayer of impurities when a surface is exposed to the ambient. One possible explanation is that the contamination is adsorbed in patches (due to the heterogeneous nature of the surface) and it takes a long time for the formation of a continuous film of contamination.

Storage of Clean Surfaces

The issue of storage of surfaces in their clean condition is very important because if the surfaces can be kept clean then these need not be used immediately after cleaning them. To store

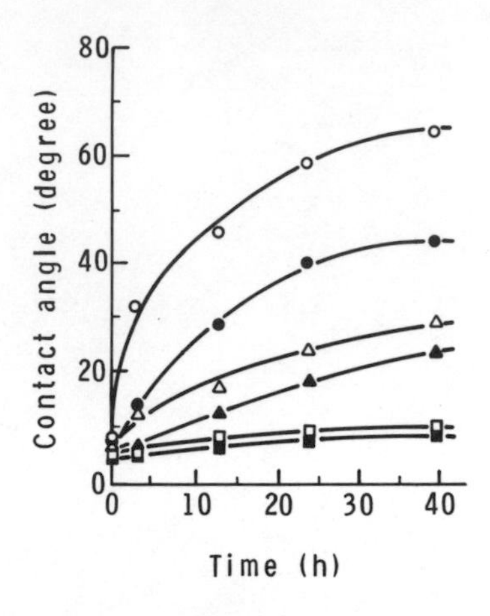

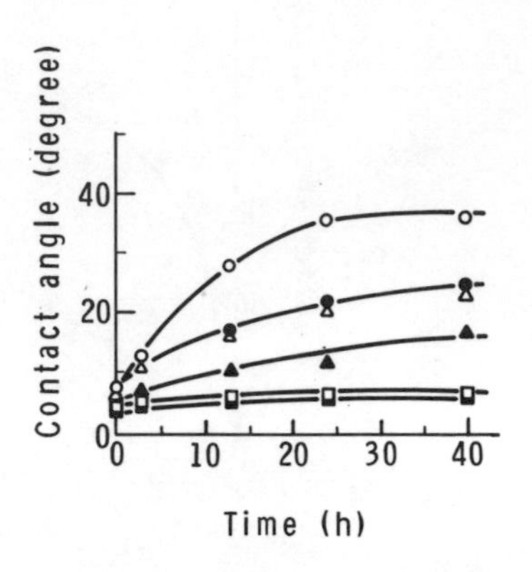

Figure 22. a
Variation of contact angles under the dried atmosphere.
o: TiO_2, •: $SrTiO_3$,
Δ: Cr_2O_3, ▲: NiO,
□: Al_2O_3, ■: SiO_2.

b
Variations of contact angles under the wet atmosphere. (Symbols are the same as in Figure a) (from reference 158).

a clean surface in a plastic bag is a cardinal sin as most plastics give off plasticizers and other low molecular weight material, with the result that these materials will adsorb and form a film on the clean surface. One possible approach to keep surfaces clean is to store them under the influence of UV/ozone, as it has been claimed that UV/ozone not only cleans a surface but also keeps it clean.[83]

Another approach is to store them in containers containing clean and oxidized nichrome or aluminum shots; aluminum shots are preferred as these are easier to prepare. The rationale behind this approach is that the hydrocarbons present inside the container will selectively adsorb on the aluminum shots and the test surface will stay clean. Here the aluminum shots function as "getters" for the organic contaminants, which is analogous to the use of water getters such as silica gel. White carried out an interesting study regarding the effectiveness of various containers for maintaining nichrome in a clean condition. The reason for the selection of nichrome as a test surface was that it was found to be most sensitive to contamination. The results are shown in Figure 23. The best (cleanest) condition is obtained when cleaned and oxidized aluminum shot (HPS-10, -8 mesh Reynolds Metal Company) is used, for there is no change in contact angle from the value of 6° obtained immediately after cleaning of nichrome. It should be noted that for the aluminum shots to stay effective, these must be periodically cleaned of adsorbed material, and this can be done by degreasing and refiring in air at 500°C.

Also it is interesting to note that the findings of Tamai et al.[158] imply that the contamination of oxide surfaces by hydrocarbons can be reduced if these are stored under wet atmosphere.

SUMMARY

This overview brings out the state of the knowledge with respect to understanding, monitoring and controlling surface contamination. A number of techniques for both cleaning and characterization of cleanliness of surfaces are discussed, and the importance of the storage of surfaces in their clean condition is emphasized and a few promising approaches in this direction are outlined.

It is obvious from this overview that there is a brisk R & D activity and interest in the various ramifications of surface contamination control, and all signals indicate there is room and a need for developing better techniques for controlling surface contamination. As the implications of surface contamination are quite serious, so let us all "think clean."

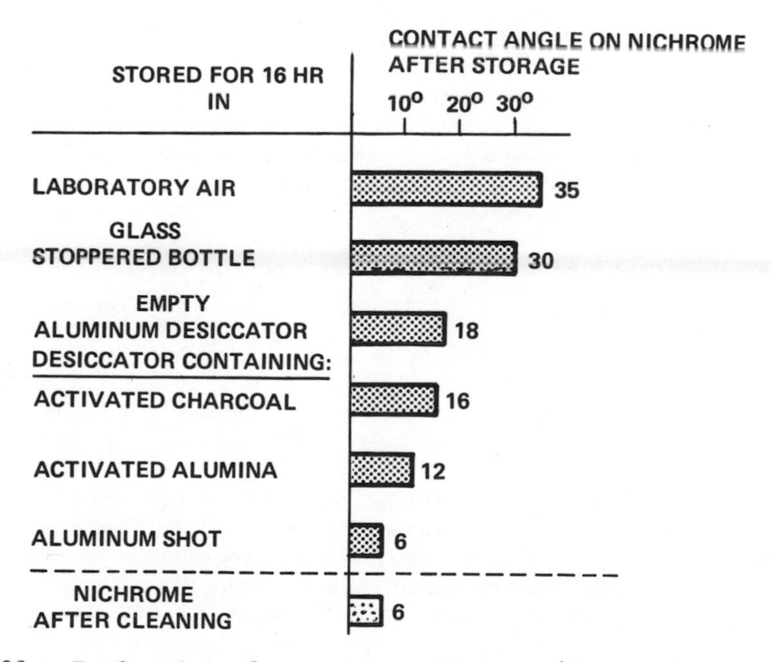

Figure 23. Evaluation of storage conditions (from reference 44).

REFERENCES

Note: It should be noted that the format for citation of references here is different from the rest of the proceedings volumes in the sense that in these references titles of papers are also given. When this paper was being written it was felt that the inclusion of paper titles would be helpful but at this stage it was too late to ask other authors to include the required titles.

1. S. V. Caruso, "Some Aspects of Contamination Detection, Analysis and Control in Microcircuits for the NASA Space Shuttle Program," paper presented at the 1978 Electronic Components Conference.
2. R. P. Young, "Low-Scatter Mirror Degradation by Particle Contamination," Optical Eng. 15, 516-520 (1976).
3. J. L. Jellison, "Effect of Surface Contamination on Solid Phase Welding - An Overview," in Surface Contamination: Its Genesis, Detection and Control, K. L. Mittal, Editor, Vol. 2 pp. 899-923, Plenum Press, New York, 1979.
4. A. V. Ferris-Prabhu, "Contamination and Reliability Concerns in Microelectronics," in Surface Contamination: Its Genesis, Detection and Control, K. L. Mittal, Editor, Vol. 2, pp. 925-944, Plenum Press, New York, 1979.
5. A. A. Bergh and G. H. Schneer, "The Effect of Ionic Contaminants on Silicon Transistor Stability," IEEE Trans. Reliability R-18, No. 2. 34-38 (May 1969).
6. D. L. Cannon and O. D. Trapp, "The Effect of Cleanliness on Integrated Circuit Reliability," in Proc. 6th Annual Reliability Physics Symposium, pp. 68-79, 1968
7. J. L. Jellison, "Effect of Surface Contamination on the Thermocompression Bonding of Gold," in Proceedings of the Electronic Components Conference, 1975.
8. J. M. Chen, T. S. Sun, J. D. Venables and R. Hopping, "Effects of Fluorine Contamination on the Microstructure and Bondability of Aluminum Surfaces." SAMPE J. pp. 22-28 (July/Aug. 1978).
9. M. A. Danforth and R. J. Sunderland, "Contamination of Adhesive Bonding Surface Treatment," J. Polymer Sci. Appl. Polymer Symp. 32, 201-215 (1977).
10. E. L. Koehler, "The Influence of Contaminants on the Failure of Protective Organic Coatings on Steel," Corrosion 33, No. 6, 209-217 (1977).
11. K. Fujiwara and Y. Yamaguchi, "Effects of Surface Contamination on Adhesion and Transfer of Gold," in Proceedings of the 8th International Conference on Electrical Contact Phenomena, held August 22-26, 1976, Tokyo, Japan.
12. J. S. Vermaak, L. W. Snyman and F. D. Auret, "On the Growth of Au on Clean and Contaminated GaAs (001) Surfaces," J. Crystal Growth, 42, 132-135 (1977).

13. G. J. Engelland and W. G. Kenyon, "Non-ionic Surface Contamination Degrades Insulation Resistance," Circuits Mfg. pp. 34-40 (June 1977).
14. G. J. Engelland and W. G. Kenyon, "Insulation Resistance Degradation by Non-ionic Surface Contamination - Its Causes and Cure," 1PC Technical Review, Feb. 1977.
15. G. Karady, "Effect of Surface Contamination on High Voltage Insulator Performance," in Surface Contamination: Its Genesis, Detection and Control, K. L. Mittal, Editor, Vol. 2, pp. 945-965, Plenum Press, New York, 1979.
16. T. Tamai, "Electrical Conduction Mechanisms of Electric Contacts Covered with Contaminant Films," in Surface Contamination: Its Genesis, Detection and Control, K. L. Mittal, Editor, Vol. 2, pp. 967-981, Plenum Press, New York, 1979.
17. J. D. Landsberg, F. W. Bodyfelt and M. E. Morgan, "Retention of Chemical Contaminants by Glass, Polyethylene, and Polycarbonate Multiuse Milk Containers," J. Food Protection, 40(11), 772-777 (1977).
18. E. B. Sansone and M. W. Slein, "Redispersion of Indoor Surface Contamination: A Review," J. Hazardous Materials, 2, 347-361 (1977/78).
19. American Society for Metals Conference on Cleaning, Finishing and Coating of Metal Surfaces, Louisville, KY, May 23-24, 1978.
20. USAF/NASA International Spacecraft Contamination Conference, U.S. Air Force Academy, Colorado, March 7-8, 1978.
21. Fourth Contamination Control Seminar, General Electric Ordnance Dept. Pittsfield, MA, May 16-18, 1978.
22. Third Annual Contamination Seminar, The Charles Stark Draper Lab., Cambridge, MA, Oct. 27, 1976.
23. Contamination: Its Effect, Detection, and Control, Rockwell International, Annaheim, CA, Oct. 8-9, 1975.
24. Contamination: Its Effect, Detection and Control, Honeywell Aerospace, St. Petersburg, FL, Jan. 22-24, 1975.
25. American Ceramic Society/ASTM Joint Symposium on Glass Cleaning, Washington, D.C. May 7, 1975.
26. AEC/NASA Symposium on Contamination Control, Held in Albuquerque, NM, Sept. 12-14, 1967.
27. R. C. Snorgen, Handbook of Surface Preparation, Palmerton Publishing Co., Atlanta, GA, 1974
28. C. Tautscher, The Contamination of Printed Wiring Boards and Assemblies, Omega Scientific Services, Bothell, WA.
29. P. W. Morrison, Editor, Environmental Control in Electronic Manufacturing, Van Nostrand Reinhold Co., New York, 1973
30. G. Goldfinger, Editor, Clean Surfaces: Their Preparation and Characterization for Interfacial Studies, Marcel Dekker, New York, 1970.
31. C. P. Marsden, Editor, Silicon Device Processing, NBS Special Publication No. 337, November 1970 (Available from Superintendent of Documents, U.S. Govt. Printing Office, Washington, D.C. 20402).

32. F. L. Dwyer, Contamination Analysis and Control, Reinhold Publishing Corp. New York, N.Y. 1966.
33. B. R. Fish, Editor, Surface Contamination, Pergamon Press, New York, N.Y., 1966.
34. Cold Cleaning With Halogenated Solvents, ASTM STP No. 403, American Society for Testing and Materials, Philadelphia, PA, July, 1966.
35. S. Spring, Metal Cleaning, Rheinhold Publishing Co., New York, N.Y., 1963.
36. Cleaning and Materials Processing for Electronics and Space Apparatus, ASTM STP No. 342, American Society for Testing and Materials, Philadelphia, PA, November, 1963.
37a. Handbook of Vapor Degreasing, ASTM STP No. 310, American Society for Testing and Materials, Philadelphia, PA, April, 1962.
b. Materials and Electron Device Processing, ASTM STP No. 300, American Society for Testing and Materials, Philadelphia, PA, 1961
38. Cleaning of Electronic Device Components and Materials, ASTM STP No. 246, American Society for Testing and Materials, Philadelphia, PA, 1959.
39. A Survey of Contamination (Its Nature, Detection and Control), prepared by Spectroscopy Division 5525, Sandia Laboratories, Albuquerque, SC-M-70-303, 64 pp. March, 1971.
40. Proceedings of the NASA Cleaning Conference, January, 1966.
41. A. H. Szkudlapski, "Surface Cleaning Practice," in Environmental Control in Electronic Manufacturing, P.W. Morrison, Editor, pp. 146-178, Von Nostrand Reinhold Co., New York, N. Y. 1973
42. J. F. Pudvin, "Surface Cleaning Theory," ibid., pp. 146-178.
43. M. L. White, "Clean Surface Technology," in Proceedings of the 27th Annual Frequency Control Symposium, held June, 1977, Cherry Hill, N.J., pp. 79-88, Electronics Industries Associates, Washington, D.C.
44. M. L. White, "The Detection and Control of Organic Contaminants on Surfaces," in Clean Surfaces, G. Goldfinger, Editor, pp. 361-373, Marcel Dekker, New York, 1970
45a. R. Brown, in Handbook of Thin Film Technology, L. I. Maissel and R. Glang, Editors, Ch. 6, pp. 37-42, McGraw-Hill, New York, 1970.
b. NASA Contamination Control Handbook, SP 5076, 1969.
46. J. Shields, Adhesives Handbook, Chapter 6, CRC Press, 1970
47. J. W. Faust, Jr., "Surface Contamination," ref. 31, pp. 436-441.
48. S. Srinivasan and P. N. Sawyer, "Electrochemical Techniques for the Characterization of Clean Surfaces in Solution," ref. 30, pp. 195-218.
49. K. J. Mysels and A. T. Florence, "Techniques and Criteria in the Purification of Aqueous Surfaces, ref. 30, pp. 227-265.

50. D. M. Mattox, "Surface Cleaning in Thin Film Technology," Thin Solid Films, 53, 81-96 (1978).
51. F. R. Eirich, "Factors in Interface Conversion for Polymer Coatings," in Interface Conversion for Polymer Coatings, P. Weiss and G. Dale Cheever, Editors, pp. , Elsevier, New York, 1968.
52a. J. Verhoeven, "Techniques to Obtain Atomically Clean Surfaces," this proceedings volume, pp. 499-512.
b. J. Verhoeven, "Techniques to Obtain Atomically Clean Surfaces," Nederlands Tijdschrift voor Vacuumtechnick 13, 55-59 (1975).
53. H. E. Farnsworth, "Atomically Clean Surfaces," DECHEMA - Monograph 78 (1537-1548), 9-22 (1975).
54. H. E. Farnsworth, "Atomically Clean Solid Surfaces - Preparation and Evaluation" in The Solid-Gas Interface, F. Allison Flood, Editor, Chapter 13, Marcel Dekker, New York, 1967.
55. L. C. Jackson, "Removal of Silicone Grease and Oil Contaminants," Adhesives Age, pp. 29-32 (April 1977).
56. L. C. Jackson, "Solvent Cleaning Process Efficiency, Contaminant Removal in the Electronic Industry," Adhesives Age, pp. 31-34 (July, 1976).
57. L. C. Jackson, "How to Select a Substrate Cleaning Solvent," Adhesives Age, pp. 23-31 (Dec. 1974).
58. P. G. Creter, "A New Cleaning Method for Ceramic Microelectronic Substrates," in Proceedings of the International Microelectronics Symposium, held Oct. 24-26, 1977.
59. R. Monahan, "Vapor Degreasing with Chlorinated Solvents," Metal Finishing, pp. 26-31 (Nov. 1977).
60. R. B. Ramsey, Jr. "The Niche for Fluorinated Solvents," Metal Progress, pp. 71-74 (April 1975).
61. W. A. Hume, "Proper Cleaning of Electronic Assemblies," Assembly Eng. (March 1968).
62. R. A. Geckle, "Cleaning Procedures and Solvents for Semiconductors and P.C. Boards," Electron. Packaging Production, pp. 127-138 (July 1975).
63. W. G. Kenyon, "Part 2, Vapor Defluxing Systems that Meet Today's PCB Cleaning Needs," Insulation/Circuits, pp. 35-40 (March 1978).
64a. J. W. Dennison, Jr., "Cleaning of Printed Circuit Boards to Remove Ionic Soils," Materials Performance 14(3), 36-40 (March 1975).
b. H. E. Phillips, "Ionic Residue Removal: Which Solvent is Best," Electron. Packaging and Production 13(9), 177-180 (1973).
65. A. F. M. Barton, "Solubility Parameters," Chem. Rev. 75, 731-753 (1975).
66. K. L. Hoy, Tables of Solubility Parameters, Union Carbide Corp, South Charleston, W. VA, July 21, 1969.
67. L. L. Hench and E. C. Ethridge, "Glass Cleaning and Characterization of Cleanliness," this proceedings volume, pp. 313-326.

68. P. B. Adams, "A Systematic Approach to Glass Cleaning," this proceedings volume, pp.
69. S. Tsuchihashi, "How to Clean Glass Surfaces," Kagaku, 33(7), 545-547 (1978).
70. P.B. Adams, "Bibliography on Clean Glass," J. Testing Evaluation 5, No. 1, 53-57 (1977).
71. C. G. Pantano, Jr. and L. L. Hench, "Cleaning Borosilicate Glass for Biological Application," J. Testing Evaluation 5, 66-69 (1977).
72. L. Holland, The Properties of Glass Surfaces, Wiley, New York, 1964.
73. W. Kern and D. A. Puotinen, "Cleaning Solutions Based on Hydrogen Peroxide for Use in Silicon Semiconductor Technology," RCA Reviews 31, 187-206 (1970).
74. P. H. Holloway and R. L. Long, Jr., "On Chemical Cleaning for Thermocompression Bonding," IEEE Trans. PHP, PHP-11, No. 2, 83-88 (June, 1975).
75. A. Mayer and S. Shwartzman, "Megasonic Cleaning - A New Cleaning and Drying System for Use in Semiconductor Processing," abstracts of the 1978 Electronic Materials Conference.
76. D. F. O'Kane and K. L. Mittal, "Plasma Cleaning of Metal Surfaces," J. Vac. Sci. Technol. 11, 567-569 (1974).
77. H. B. Bonham and P. V. Plunkett, "Plasma Cleaning for Improved Wire Bonding on Thin-Film Hybrids," Electron. Packaging and Production, pp. 42- (Feb. 1979).
78. H. B. Bonham and P. V. Plunkett, "Surface Contamination Removal from Solid State Devices by Dry Chemical Processing," this proceedings volume, pp. 271-285.
79. W. Balwanz, "Plasma Cleaning of Surface," this proceedings volume, pp. 255-269.
80. G. J. Kominiak and D. M. Mattox, "Reactive Plasma Cleaning of Metals," Thin Solid Films, 40, 141-148 (1977).
81. R. Bouwman, J. B. van Mechelen and A. A. Holscher, "Surface Cleaning by Low Temperature Bombardment with Hydrogen Particles: An AES Investigation on Copper and Fe-Cr-Ni Steel Surfaces," J. Vac. Sci. Technol. 15, 91-94 (1978).
82. R. R. Sowell, R. E. Cuthrell, D. M. Mattox and R. D. Bland, "Surface Cleaning by Ultraviolet Radiation," J. Vac. Sci. Technol. 11, 474-475 (1974).
83a. J. R. Vig and J. W. LeBus, "UV/Ozone Cleaning of Surfaces," IEEE Trans. on PHP, PHP-12, No. 4, pp. 365-370 (December 1976).
b. J. Vig, "UV/Ozone Cleaning of Surfaces: A Review," this proceedings volume, pp. 235-254.
84. J. L. Jellison, private communication, quoted in D. M. Mattox, "Surface Cleaning in Thin Film Technology," Report from Sandia Laboratories, Albuquerque, NM, SAND 74-0344, January, 1975.

85. S. Petvai and R. H. Schnitzel, "Some Promising Applications of Ion-Milling in Surface Cleaning," this proceedings volume, pp. 297-311.

86a. R. Bolster, "Removal of Fluid Contaminats by Surface Chemical Displacement," this proceedings volume, pp. 359-368.

b. H. R. Baker, P. B. Leach, C. R. Singleterry, and W. A. Zisman, "Cleaning by Surface Displacement of Water and Oils," Ind. Eng. Chem. 59(6), 29-40 (1967).

87. W. H. Kroeck, R. F. Doll and E. L. Stokes, "Removal of Particulate Contamination from Hard Surface Photomasks by Various Cleaning Techniques," in Proceedings Kodak Microelectronics Seminar, pp. 26-32, October, 1977.

88a. A. D. Zimon, Adhesion of Dust and Powders, 2nd edition, Khimia, Moscow, 1976; Plenum Press, New York, 1969.

b. M. Corn, "Adhesion of Particles," in Aerosol Science, C. N. Davies, Editor, Ch. 11, Academic Press, New York, 1966.

c. H. Krupp, "Particle Adhesion: Theory and Experiment," Adv. Colloid Interface Sci., 1, 111 - (1967).

89. "Polymer Removes Submicron Particles," Circuit Manufacturing, pp. 65-67, (June 1978).

90. H. K. Raaschou Nielsen, L. U. Kornum, and O. Saberg, "Dirt Retention on Painted Surfaces," J. Coating Technol. 50, 69-80 (Jan. 1978).

91. I. F. Stowers, "Advances in Cleaning Metal and Glass Surfaces to Micro-level Cleanliness," J. Vac. Sci. Technol., 15(2), 751-754 (1978).

92. S. Bhattacharya and K. L. Mittal, "Mechanics of Removing Glass Particulates from a Solid Surface," Surface Technol. 7, 413-425 (1978).

93. W. J. Whitfield, "A Study of the Effects of Relative Humidity on Small Particle Adhesion to Surfaces," this proceedings volume, pp. 73-81.

94. D. A. Brandreth and R. E. Johnson, Jr. "Particulate Removal in Microelectronics Manufacturing," this proceedings volume, pp. 83-88.

95. J. S. Mijovic and J. A. Koutsky, "Etching of Polymeric Surfaces: A Review," Polym.-Plast. Technol. Eng. 9(2), 139-179 (1977).

96. J. C. Scott, "The Preparation of Clean Water Surfaces for Fluid Mechanics," this proceedings volume, pp. 477-497.

97. J. C. Scott, "The Preparation of Water for Surface-Clean Fluid Mechanics," J. Fluid Mechanics 69, Part 2, 339-351 (1975).

98. R. P. Allen, H. W. Arrowsmith and W. C. Budke, Electropolishing as a Large-Scale Decontamination Technique, 1977, Report from Battelle Pacific Northwest Laboratories, Richland, WA.

99. D. R. Schaeffer and C. O. Brewer, "$^{238}PuO_2$ Surface Contamination of Radioisotopic Heat Sources," this proceedings volume, pp. 195-210.
100a. J. W. Costerton, G. G. Geesey and K. J. Cheng, "How Bacteria Stick," Scientific American 238, No. 1, 86-95 (Jan. 1978).
b. J. W. Costerton and G. G. Geesey, "Microbial Contamination of Surfaces," this proceedings volume, pp. 211-221.
101. N. S. McIntyre, Editor, Quantitative Surfaces Analysis Techniques, American Society for Testing and Materials, Philadelphia, PA, 1978.
102. L. H. Lee, Editor, Characterization of Metals and Polymers, Vol. 1 and Vol. 2, Academic Press, New York, 1977
103. A. W. Czanderna, Editor, Methods of Surface Analysis, Elsevier, Amsterdam, 1975.
104. P. F. Kane and G. B. Larrabee, Characterization of Solid Surfaces, Plenum Press, New York, 1974.
105. W. E. Moddeman, C. R. Cothern and J. N. Black, "Characterization of Solid Surfaces," J. Environmental Sci. 18, 27-33 (Sept./Oct. 1975).
106. G. J. Hof, Present Practices in The Verification of Cleanliness, Report from Sandia Labs., Albuquerque, NM, SCTM 147-63 (25), Sept. 1963.
107. R. N. Miller, "Rapid Method for Determining the Degree of Cleanliness of Metal Surfaces," Materials Protection & Performance, pp. 31-36 (May 1973).
108. L. K. Jones, Verification of Cleanliness - What Does it Mean, Report from Sandia Labs., SC-R-65-903, April 1965.
109. "Detecting and Measuring Contamination on PWB Surfaces," Circuits Mfg. pp. 42-44 (May 1976).
110. T. O. Duyck, "Testing Large Printed Circuit Boards for Cleanliness," Insulation/Circuits, pp. 38-41 (Oct. 1978).
111. C. J. Tautscher, "Printed Wiring Board Cleanliness Testing," Circuit World 4, No. 2, 30-32 (Jan. 1978).
112a. P. A. Lindfors, "Application of Auger Electron Spectroscopy to Characterize Contamination," in Surface Contamination: Its Genesis, Detection and Control, K. L. Mittal, Editor, Vol. 2, pp. 587-594, Plenum Press, New York, 1979.
b. J.S. Solomon and W.L. Baun, "Surface Characterization of Contamination on Adhesive Bonding Materials," ibid., pp. 609-634.
113. M. G. Yang, K. M. Koliwad and G. E. McGuire, "Auger Electron Spectroscopy of Clean-up Related Contamination on Silicon Surfaces," J. Electrochem. Soc. 122, 675-678 (1975).
114. D. W. Dwight and J. P. Wightman, "Identification of Contaminants with Energetic Beam Techniques," in Surface Contamination: Its Genesis, Detection and Control, K. L. Mittal, Editor, Vol. 2, pp. 569-586, Plenum Press, New York, 1979.
115. B. D. Ratner, J. J. Rosen, A. S. Hoffman and L. H. Scharpen, "An ESCA Study of Surface Contaminants on Glass Substrates for Cell Adhesion," ibid., pp. 669-686.

116. C. E. Bryson, III, L. H. Scharpen and P. L. Zajicek, "An ESCA Analysis of Several Surface Cleaning Techniques," ibid., pp. 687-696.

117. G. R. Sparrow and S. R. Smith, "Application of ISS/SIMS in Characterizing Thin Layers (~ 10nm) of Surface Contaminants," ibid., pp. 635-654.

118. A. C. Miller and A. W. Czanderna, "Ion Scattering Analysis of Contaminated Copper Oxide Surfaces Before and After Cleaning," Vacuum 28, 9-10 (1978).

119. G. R. Sparrow, "Characterization of Cleaned and Prepared Bonding Surfaces by ISS/SIMS," paper presented at the 2nd International Conference on Deburring and Surface Conditioning, Society of Manufacturing Engineers, Chicago, June 7-9, 1977.

120. M. G. Dowsett, R. M. King and E. H. C. Parker, "SIMS Evaluation of Contamination on Ion-Cleaned (100) InP Substrates," Appl. Phys. Lett. 31(8), 529-531 (1977).

121. G. R. Sparrow, "Surface Analysis of Polymers and Glass by Combined ISS/SIMS," paper presented at the 1977 Pittsburgh Conference on Analytical Chemistry and Applied Fields, Cleveland, OH, Feb. 28-March 4, 1977.

122. J. L. Anderson, "Quantitative Detection of Surface Contaminants," J. Amer. Assoc. Contamination Control II, No. 6, pp. 9- (June 1963).

123. J. L. Anderson, "Evaporative Rate Analysis: Its First Decade," in Characterization of Metals and Polymer Surfaces, L. H. Lee, Editor, Vol. 2, Academic Press, New York, 1977.

124. C. E. Garrett and E. F. Good, "Characterization of Bonding Surfaces Using Surface Analytical Equipment," in Surface Contamination: Its Genesis, Detection and Control, K. L. Mittal, Editor, Vol. 2. pp. 857-875, Plenum Press, New York, 1979.

125. S. D. Guttenplan, "SPD - A NDT Method for Detecting Surface Contamination," Plating & Surface Finishing, pp. 54-58 (June 1978).

126. D. H. Kim, "Contact Potential Difference (CPD) Measurement Method for Pre-Bond Non-Destructive Surface Inspection," SAMPE Q. 9, No. 4, 59-63 (July 1978).

127. T. Smith, "Quantitative Techniques for Monitoring Surface Contamination," in Surface Contamination: Its Genesis, Detection and Control, K. L. Mittal, Editor, Vol. 2, pp. 697-712, Plenum Press, New York, 1979.

128. P. F. A. Bijlmer, "Characterization of the Surface Quality by Means of Surface Potential Difference," ibid., pp. 723-748.

129. M. E. Schrader, "Surface Contamination Detection Through Wettability Measurements," ibid., pp. 541-555.

130. H. G. Kollek and W. Brockmann,"Detection of Surface Contamination in Metal Bonding by Simple Methods," ibid., pp. 713-721.
131a. "Standard Method of Test for Hydrophobic Surface Films by the Atomizer Test," Book of ASTM Standards Part 8, F 21-65, pp. 437-441, 1969; previously F 21-62T.
b. "Standard Method of Test for Hydrophobic Surface Films by the Water-Break Test," Book of ASTM Standards Part 8, F 22-65, pp. 442-443, 1969; previously F 22-62T.
132. Ref. 72, Chapter 5.
133. H. B. Lindford and E. B. Saubestre, "A New Degreasing Evaluation Test: The Atomizer Test," ASTM Bulletin No. 190, 47-50 (May 1953).
134. G. L. Krieger and G. J. Wilson, "Measuring Surface Cleanliness by Indium Adhesion," Materials Res.,pp. 341-349 (July 1965).
135. R. E. Cuthrell, "The Quantitative Detection of Molecular Layers with the Indium Adhesion Tester," Sandia Labs. Albuquerque, NM, Report. SC-DR-66-300, July, 1966.
136. T. W. Carr, "Analysis of Surface Contaminants by Plasma Chromatography - Mass Spectroscopy," Thin Solid Films, 45, 115-122 (1977).
137. R. E. Cuthrell and L. K. Jones, "Surface Contaminant Characterization Using Potential Current Curves," IEEE Trans CHMT, CHMT - 1, No. 2, 167-171 (1978).
138. H. Froot, "Microfluorescence Technique for Detection and Identifying Organic Contaminants on a Variety of Surfaces," in Surface Contamination: Its Genesis, Detection and Control, K. L. Mittal, Editor, Vol. 2, pp. 805-817, Plenum Press, New York, 1979.
139. W. E. J. Neal, "Application of Ellipsometry for Monitoring Surface Contamination and Degree of Surface Cleanliness," ibid., pp. 749-767.
140. W. C. McCrone, "Microscopic Identification of Surface Contaminants," ibid., pp. 557-567.
141. D. O. Feder and E. S. Jacob, "Electrode Potential: A Tool for the Control of Materials and Processes in Electron Device Fabrication, I. EMF - Time Studies of Clean and Contaminated Platinum Electrodes," in ref. 37b, pp. 53-66.
142. D. G. Schimmel, " Detection of Inorganic Contamination on Surfaces by an EMF Measurement," in ref. 37b, pp. 46-52.
143. T. J. Bulat, "Use of Radioisotope Tracer Techniques in Contamination Studies," in ref. 37b, pp. 77-81.
144. M. N. Slater and D. Joseph Donahue, "Radiotracers in Evaluating Parts Cleaning," in ref. 38, pp. 146-154.
145. Ref. 72, Chapter 7.
146. L. Missel, D. R. Torgeson and H. M Wagner "Surface Cleanliness Testing," Electron Packaging and Production, pp. 70- (Dec. 1966).

147. S. Nielsen, " "
in 7th National Symposium on Vacuum Technology Transactions, C. Robert Meissner, Editor, p. 293, Pergaman Press, New York, 1961.
148. I Anzai and T. Kikuchi, "A New Monitoring Technique of Surface Contamination - The Test Surface Method," Health Physics 34, 271-273 (March 1978).
149. H. J. Gale, "Monitoring Surface Contamination," Phys. Med. Biol. 20(4), 656-657 (1975).
150. T. F. Egan, "Determination of Plating Salt Residues," Plating, pp. 350-354 (April 1973).
151. "Detecting and Removing Ionic Contaminants" Circuit Manufacturing, 4 pp. (Feb. 1973).
152. P. Altavilla, "Analysis of Chloride Contaminants on PCB's Reveals Causes, Suggests Preventive Action," Insulation/Circuits, 20(10), 46-48 (1974).
153. J. Brous and H. Cole, "Measurement of Residual Ionic Contamination on Printed Circuit Boards," in Proceedings National Electronic Packaging and Production Conference pp. 167-174, 1974.
154. L. J. Rickabaugh, "An Ionic Contamination Detection System (ICDS) with Inproved Performance for Quantizing Residue Ionic Species," IEEE Trans. CHMT, CHMT-2, No. 1, 134-139 (Feb. 1979).
155. J. Brous, "Extraction Methods for Measurements of Ionic Surface Contamination," in Surface Contamination: Its Genesis, Detection and Control, K. L. Mittal, Editor, Vol. 2, pp. Plenum Press, 1979.
156. W. B. Wargotz, "Ion Chromatography - Quantification of Contaminant Ions in Water Extracted of Printed Wiring," ibid. pp.
157. Y. Tamai, T. Matsunaga, and K. Suzuki, "The Effect of Water Vapor on Contamination of Metallic Oxide Surfaces," Bull. Chem. Soc. Japan 50, 1881-1882 (1977).

Note: It might be interesting to add here that, to my knowledge, there are two journals[1,2] and one newsletter[3] which contain the word "contamination" in their titles.

1. Revista de Contaminacion Y Prevencion
   Londres 41
   Madrid 28, Spain
2. Surface Contamination & Cleaning Design
   Kindaihenshusha Co.
   5-23-5-501 Hirai
   Edogawa-ku, Tokyo, Japan
3. Contamination Newsletter, published by
   Contamination Control Labs., Inc.
   13324 Farmington Road
   Livonia, MI 48150

RELATIONSHIP BETWEEN SURFACE ENERGY AND SURFACE CONTAMINATION

Toshiaki Matsunaga

Department of Fuel Chemistry, Mining College
Akita University
Akita 010, Japan

As a characterizing method of engineering materials, surface measurement of contact angles is an effective one. The relation of surface energies determined by the two-liquid-contact-angle method to surface contamination is described. By measuring the contact angles of water drops on solid specimens in hydrocarbon liquid, not only the dispersion component of surface tension (γ_s^d) but also the nondispersive interaction energy between surface and water (I_{sw}^n) (mainly due to hydrogen bonding) can be determined.

For metallic oxides such as rutile, nickel oxide, alumina etc., the I_{sw}^n value, a characteristic parameter expressing the affinity of the surface to water, was observed to have a clear correlation to the contamination rates of these surfaces by organic vapor. Moreover, the coexisting water vapor with organic vapor largely suppressed the contamination rates, which may be suggestive for storage of clean surfaces.

Using the results of the contact angle method applied to some organic polymers, the method for estimating I_{sw}^n values from the concentration of organic surface groups is discussed.

INTRODUCTION

In order to obtain clean surfaces, the nature of interaction between surfaces and contaminants would be one of the main problems to be solved. Since the interaction energy determines the adsorption phenomena of contaminants on surface, regardless to the origin of contaminants, for example, from the atmosphere or from the bulk body, surface energy should have a relation to surface contamination.

As a characterizing method of surfaces of engineering materials, measurement of contact angles is an effective one. Although accurate determination of surface energies of solids is very difficult, measurement of contact angles of appropriate liquids can give us some valuable informations about the surface force components of solids, as proposed by Fowkes[1] and other researchers, such as Kaelble[2]. Tamai[3] has extended the Fowkes' method as the two-liquid-contact-angle method. He could obtain not only the London dispersion force component (γ_S^d) but also the nondispersive interaction energy between surface and water (I_{SW}^n) by measuring contact angles of water drops on solid specimens in hydrocarbon liquid.

Contact angle measurement is also a good method to detect contamination of solid surface, especially for organic contaminants, as reported by many researchers such as White[4] and Mittal[5].

In this report, the relation of surface energies of some metallic oxides determined by the two-liquid method to their surface contamination, and some results on organic polymers will be described.

SURFACE FREE ENERGY ANALYSIS BY CONTACT ANGLE METHOD

As Fowkes proposed[1] and its theoretical basis was discussed by Tamai[6], the surface tension can be expanded into its several terms according to the intermolecular (or bonding) forces.

$$\gamma = \gamma^d + \gamma^e + \cdots \qquad (1)$$

γ^d : the dispersion force component

γ^e : the electrostatic term

etc.

For contact angle θ, the well known Young's equation holds as

$$\gamma_{SV} = \gamma_S - \pi_L = \gamma_{SL} + \gamma_L \cos\theta \qquad (2)$$

where the subscripts S, L and V mean solid, liquid and vapor phase respectively, and π is the surface pressure.

Here, the interfacial tension γ_{SL} can be related to the work of adhesion I_{SL} as

$$\gamma_{SL} = \gamma_S + \gamma_L - I_{SL} \qquad (3)$$

Assuming the geometrical mean for the dispersive interaction energy at the solid - liquid interface,

$$\gamma_{SL} = \gamma_S + \gamma_L - 2\,(\gamma_S^d\,\gamma_L)^{1/2} \qquad (4)$$

if the $\gamma_L = \gamma_L^d$ liquid such as saturated hydrocarbon is used.

Combining Equations (2), (3), and (4),

$$\gamma_S^d = \gamma_L\,(1 + \cos\theta)^2/4 \qquad (5)$$

if π be negligible. This method, denoted the one-liquid method, can be applied to many so-called low energy solid surfaces, since on these surfaces many liquids can give some definite angles.

However, high energy solids may be spreadingly wettable with the usually employable liquids in the air. To obtain finite contact angles for these surfaces, a system of two liquids and solid, for example, water, hydrocarbon, and metallic oxide, can satisfactorily be employed. The Young's equation in this system is

$$\gamma_{SH} = \gamma_{SW} + \gamma_{HW}\cos\theta \qquad (6)$$

Here, the interaction energy at the solid and liquid interfaces may be given as

$$I_{SH} = 2\,(\gamma_S^d\,\gamma_H)^{1/2} \qquad (7)$$

$$I_{SW} = 2\,(\gamma_S^d\,\gamma_W^d)^{1/2} + I_{SW}^n \qquad (8)$$

assuming geometrical mean for the dispersive interaction energy and introducing an interaction due to nondispersive forces as I_{sw}^n in Equation (8).

From Equations (6), (7), and (8), the basic equation for the two-liquid method

$$\gamma_H - 2\,(\gamma_S^d\,\gamma_H)^{1/2} = \gamma_W - 2\,(\gamma_S^d\,\gamma_W^d)^{1/2} - I_{SW}^n + \gamma_{HW}\cos\theta \qquad (9)$$

is obtained. If contact angles are measured using two different hydrocarbons, the two unknown quantities γ_S^d and I_{sw}^n in Equation (9) can be solved as followings.

$$\gamma_S^d = \frac{(\gamma_{H_1} - \gamma_{H_2}) - (\gamma_{H_1W}\cos\theta_1 - \gamma_{H_2W}\cos\theta_2)}{2\,(\gamma_{H_1}^{1/2} - \gamma_{H_2}^{1/2})} \qquad (10)$$

$$I_{SW}^{n} = \frac{1}{(\gamma_{H_1}^{1/2} - \gamma_{H_2}^{1/2})} \Big[\{\gamma_{H_2}^{1/2} - (\gamma_{W}^{d})^{1/2}\}\{\gamma_{H_1} - \gamma_{W} - \gamma_{H_1W} \cos\theta_1\} - \{\gamma_{H_1}^{1/2} - (\gamma_{W}^{d})^{1/2}\}\{\gamma_{H_2} - \gamma_{W} - \gamma_{H_2W} \cos\theta_2\}\Big] \quad (11)$$

Here, the subscripts 1 and 2 mean the two hydrocarbon liquids. Although I_{sw}^{n} values are obtained with a sufficient accuracy as will be seen later, γ_s^d values from Equation (10) are especially sensitive to the small difference of contact angles so that their accuracy is usually only around 20%.

In obtaining γ_s^d from Equation (5) of the one-liquid method, the surface pressure π had to be neglected as described above, but this assumption seems to remain, sometimes, an open question. On the contrary, even though some effects of surface pressures of two liquids due to adsorption at the liquid-solid interfaces would possibly exist, they may slightly influence the γ_s^d values cancelling each other, as discussed elsewhere[7].

Thus, by the two-liquid method, the γ_s^d as well as I_{sw}^{n} can be obtained for high energy and low energy solids. The I_{sw}^{n} value, considered to be due to mainly hydrogen bonding force between solid and water, presents the information about the nondispersive surface forces, and consequently, about the reactivity of solid surfaces.

SURFACE CONTAMINATION AND I_{SW}^{n} FOR METALLIC OXIDES

High energy solid surfaces such as those of metallic oxides or metals are liable to be contaminated by organic vapors in the atmospheric air. The surface contaminated by organics shows a large change of wettability by water, which presents a method to detect organic contamination.

We examined the relation of contamination rates to the surface energy characteristics for six metallic oxides[8]:

quartz glass SiO_2, rutile TiO_2 (001), Chromia Cr_2O_3 (0001), alumina Al_2O_3 (0001), nickel oxide NiO (100), and strontium titanate $SrTiO_3$.

Single crystal plates were used except quartz glass. The numbers in parentheses indicate the crystal planes investigated.

The specimens, polished with emery paper and diamond powder were cleaned before each measurement in a detergent solution with an ultrasonic cleaner, followed by thorough washing with redistil-

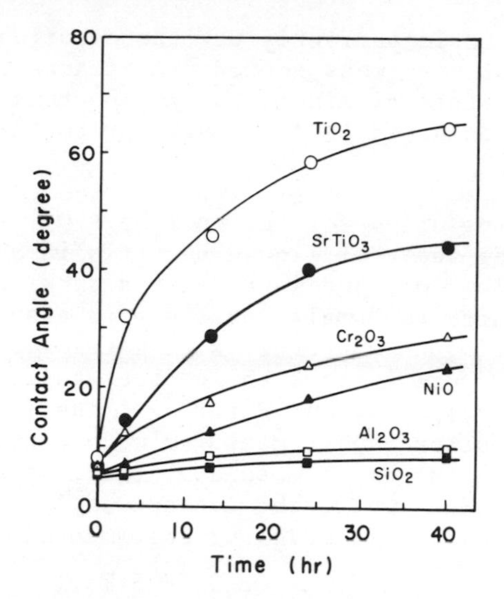

Figure 1. Contamination rates on metallic oxide surfaces under the dried atmosphere.[8]

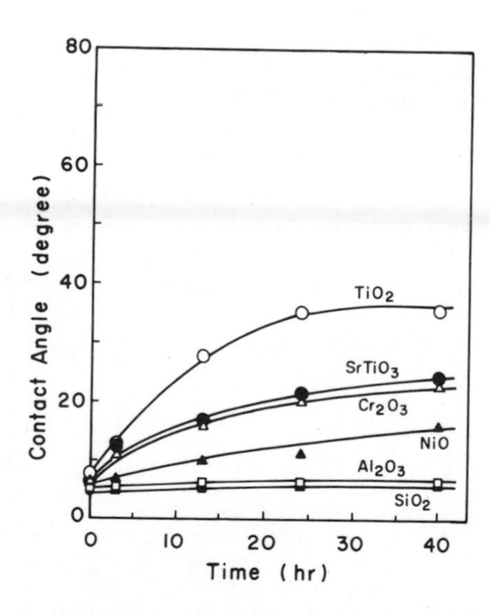

Figure 2. Contamination rates on metallic oxide surfaces under the wet atmosphere.[8]

led water. Water could perfectly wet these surfaces. The Auger spectra for alumina specimens showed only traces of phosphorus and carbon atoms other than aluminum and oxygen atoms, and therefore, the cleaning is considered to be enough for contamination measurement.

These specimens dried for one hour by P_2O_5 were allowed to be contaminated by the vapor of liquid paraffin in a glass vessel, i.e. a cleaned oxide plate was put in a Petri dish with a cover, 15 cm in diameter, in which two small vessels were also placed, one filled with liquid paraffin, and the other with water for the case of the wet atmosphere. Figure 1 shows the contact angles plotted against time in the dry atmosphere, and Figure 2 those in the wet atmosphere saturated with water vapor, measured with a goniometer-telescope system at 20°C. In the dry atmosphere, the most quickly contaminated sample was rutile, while the contact angles on silica and alumina increased slowly. The order of contamination rates was :

$$SiO_2 < Al_2O_3 < NiO < Cr_2O_3 < SrTiO_3 < TiO_2$$

In the wet atmosphere, on the other hand, the contamination was largely depressed for all specimens, and especially silica and alumina showed only a small increment of contact angles even after forty hours. However, the order of contamination rates was just the same as that in the dry atmosphere.

Additionally, when a clean rutile plate was placed in a vessel with a contaminated nickel oxide plate which had been in contact with liquid paraffin vapor for previous one hundred hours, a rapid increase of contact angles on rutile was observed as shown in Figure 3, although the decrease of contact angles on nickel oxide was slow. As for organic contamination of high energy hydrophilic surfaces such as metallic oxides, the contaminant organic molecules should adsorb onto the solid surface competing with water molecules.

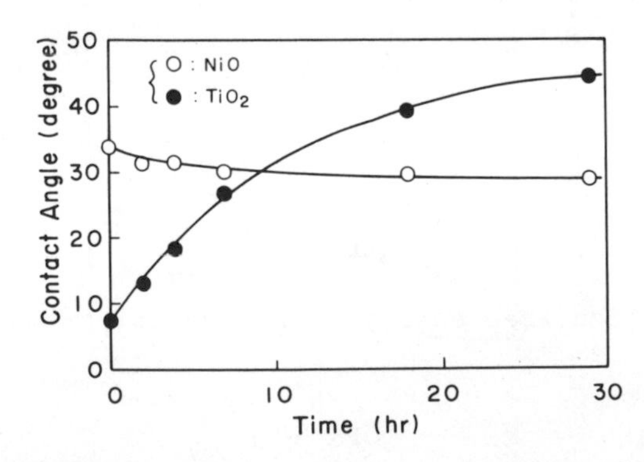

Figure 3. Changes of contact angles on a cleaned rutile plate placed in a vessel with a contaminated nickel oxide.

Therefore, if the surface has a larger affinity to water, it will be contaminated more slowly by organics. Then, the value I^n_{sw} from the two-liquid method may be used as a prameter to evaluate contamination rates.

The analysis of surface energies of these oxide specimens by the two-liquid method was conducted measuring the contact angles of water drops in cyclohexane and isooctane at 20°C. The I^n_{sw} values were as follows[9] :

TiO_2 91.4, $SrTiO_3$ 93.2, Cr_2O_3 93.6

NiO 94.0, Al_2O_3 97.6, SiO_2 99.2 (in mJ m^{-2})

The average error of these values is ± 0.3 mJ m^{-2}.

The order of I^n_{sw} values clearly corresponds to the reverse order of the contamination rates, which shows the surface energetical parameter I^n_{sw} to be a good one for evaluation of surface contamination.

The oxide surface thus measured may be said to be similar to the water surface from the surface energetical viewpoint, if we compare the I^n_{sw} values with the corresponding value for an imaginary wate-water interface I^n_{ww}, calculated as followingly. When Equation (3) is applied to the imaginary water-water interface, the next relation will be obtained,

$$\gamma_{WW} = 0 = \gamma_W + \gamma_W - I^n_{WW}$$

or

$$I_{WW} = 2\,\gamma_W \qquad (12)$$

Here, the adhesion work, I_{ww}, may be divided into the two parts, the dispersion force component I^d_{ww} and the nondispersive one, I^n_{ww}. Then,

$$I^n_{WW} = I_{WW} - I^d_{WW} = 2\,\gamma_W - 2\,\gamma^d_W = 100.2 \text{ mJ m}^{-2} \qquad (13)$$

where 72.8 and 22.7 mJ m^{-2} are used as γ_W and γ^d_W for the water surface, respectively. Comparing the I^n_{sw} values of oxides with the above I^n_{ww} value, silica seems to have a surface nearly equal to water.

SURFACE FREE ENERGY ANALYSIS FOR ORGANIC POLYMERS

The interaction free energy or the adhesion work due to the nondispersive force I^n_{sw} should have an intimate relation to adhesive properties of polymers, and it will be valuable if the relation between the interaction energies and the surface composition will be obtained. The experimental work to correlate the I^n_{sw} values for some commercial organic polymers with their surface compo-

Table I. Estimation of I^n_{SW} from Polymer Structure.[7, 10, 11]

Polymer	x	$I^n_{SW}(x)$ [a]	s(x)	$I^n_{SW}(x) \cdot s(x)$	I^n_{SW}, mJ m^{-2} Calcd.	I^n_{SW}, mJ m^{-2} Obsd.
Nylon 6	-CONH-	17	2.7	45.9	45.9	37.8
Nylon 6,6	-CONH-	17	2.7	45.9	45.9	47.8
Nylon 6,12	-CONH-	17	1.8	30.6	30.6	30.7
Poly(methylmethacrylate)	-COOR	8.4[c]	3.7	31.1	31.1	27.4
Poly(vinyl acetate)	-CCOR	8.4[c]	4.1	34.4	34.4	34.4
Poly(styrene)	-Benz.	1	3.3	3.3	3.3	3.3
Poly(vinyl chloride)	-Cl	2	5.7	11.4	11.4	12.5
Poly(vinylidene fluoride)	$-F_2$	3	6.5	19.5	19.5	21.1
Polyethylene terephthalate	-COO- -Benz.	4.5 1	5.3 2.6	23.9 2.6 }	26.5	26.7
Polycarbonate	-COO- -Benz.	4.5 1	2.1 4.2	9.5 4.2 }	13.7	10.6
Poly(oxymethylene)	-O-	2.5	9.2	23.0	23.0	23.4
Polyphenol	-OH -Benz.	10 1	3.7 3.7	37.0 3.7 }	40.7	40.9

a) In the unit of 10^{-21} J/group.
b) In the unit of 10^{10} m^{-2}.
c) See the reference (11).

sition was published recently[10, 11].

Since I^n_{sw} of polymers is considered to be due to mainly hydrogen bonding between functional groups on a given polymer surface and water molecules, and further, the hydrogen bonding is of short range characteristics, effective interactions may be restricted to those between the functional groups and their surrounding nearest water molecules. Then, I^n_{sw} may be approximately calculated as the sum of the contribution of each group as

$$I^n_{SW} = \sum_x I^n_{SW}(x) \cdot s(x) \qquad (14)$$

where x denotes a functional group, $I^n_{sw}(x)$ is the contribution of this x group to I^n_{sw} value, and s(x) is the surface density of the x group. The results are shown in Table I, where s(x) was calculated assuming that the bulk structure continues upto its surface. I^n_{sw} values were determined so that the observed I^n_{sw} values agree with the calculated ones by Equation (14) as well as possible. Although there are no comparable data with these $I^n_{sw}(x)$ in Table I, some of them apparently work rather well for different polymers at first, and secondly, the relative amount of $I^n_{sw}(x)$ for -OH, -COO-, and -O- groups can be reasonably compared with the solubilities of some organic compounds having those functional groups, i.e.,

$C_2H_4 \cdot OH \cdot C_3H_7$ 0.54, $CH_3 \cdot COO \cdot C_4H_9$ 0.37, $C_2H_5 \cdot O \cdot C_3H_7$ 0.21 in mol/kg H_2O at 25°C[12].

SUMMARY AND CONCLUSIONS

Our recent results of the analytical method with a two-liquids -solid system, the two-liquid-contact-angle method, including their relation to surface contamination are described.

Contamination rates of several metallic oxides by organic vapor correspond to their nondispersive components of the interaction free energy between solid and water, I^n_{sw}, determined by the two-liquid method. The I^n_{sw} values for organic polymers, possibly a valuable parameter of the surface reactivity or adhesive properties, were approximately estimated from the composition of polymer surface structure.

In spite of some deffects, e.g., a low accuracy of γ^d_s, the two-liquid method may present some useful informations in relation to surface contamination as well as an estimation of the surface reactivity.

ACKNOWLEDGEMENTS

The author wishes to express his sincere thanks to Prof. Yasukatsu Tamai, Tohoku University, for his encouragement and valuable discussions.

REFERENCES

1. F. M. Fowkes, Ind. Eng. Chem., 56 (No. 12), 40 (1964).
2. D. H. Kaelble, "Physical Chemistry of Adhesion," Wiley Intersci., New York (1971).
3. Y. Tamai, K. Makuuchi, and M. Suzuki, J. Phys. Chem., 71, 4176 (1967).
4. M. L. White, in "Clean Surfaces," G. Goldfinger, Editor, p. 361, Marcel Dekker, New York 1970.
5. D. F. O'Kane and K. L. Mittal, J. Vac. Sci. Technol., 11, 567 (1974).
6. Y. Tamai, J. Phys. Chem., 79, 965 (1975).
7. Y. Tamai, T. Matsunaga, and K. Horiuchi, J. Colloid Interface Sci., 60, 112 (1977).
8. Y. Tamai, T. Matsunaga, and K. Suzuki, Bull. Chem. Soc. Jpn., 50, 1881 (1977).
9. Y. Tamai, T. Matsunaga, and K. Suzuki, in "Proc. VIIth Intern. Cong. on Surface Active Substances" held in Moscow in 1976, in press.
10. T. Matsunaga, J. Appl. Polym. Sci., 21, 2847 (1977).
11. T. Matsunaga and Y. Tamai, J. Appl. Polym. Sci., 22, 3525 (1978).
12. "Chemistry Handbook," p. 816, Chem. Soc. Japan, Maruzen (1972).

INVESTIGATION OF THE INTERACTION OF CERTAIN LOW ENERGY LIQUIDS WITH POLYTETRAFLUOROETHYLENE AND POLYETHYLENE AND ITS IMPLICATIONS IN THE CONTAMINATION OF POLYMERIC SURFACES

F. Galembeck*, S. E. Galembeck*, H. Vargas+, C. A. Ribeiro+, L. C. M. Miranda+, and C. C. Ghizoni++

*University of Sao Paulo, Brazil; +State University of Campinas, Brazil and ++INPE, Sao Jose dos Campos, Brazil

The interaction of some liquids with polytetrafluoethylene Teflon is examined, showing that low cohesive energy density liquids, characterized by low solubility parameters (Hildebrand's δ), are absorbed in significant amounts by this polymer, as observed in the case of other polymers, including polyethylene. More stable contaminants can be formed if an absorbed liquid undergoes any chemical reaction before desorption, yielding solid, non diffusive products. The case of the interaction of Teflon and polyethylene with the organometallic, low δ liquid, iron pentacarbonyl is examined in detail and it is found that this liquid can be absorbed by the polymers and react *in situ* to give iron salts and oxides. Surface properties of the plastics are drastically altered, as evidenced by liquid contact angle measurements. Samples of iron oxide-contaminated Teflon were examined by electron microscopy and photoacoustical and ESR spectra were obtained. These experiments show that the contaminant is ultrafine, superparamagnetic α- Fe_2O_3. It is suggested that a number of other chemicals can have an important role in polymer surface and bulk contamination, not only for their ability to be sorbed but also for their derivatives which, once formed, are practically impossible to remove.

INTRODUCTION

Many synthetic polymers are very well known as solids having very low surface energies. This characteristic is usually related to the absence of chemically active groups in the macromolecule as well as to the fact that their permanent dipole moments are very close or equal to zero. Polytetrafluoroethylene[1] (PTFE) and polyethylene are remarkable for their peculiar surface properties which originate in the fact that they are among the solids having the lowest surface energies known[2].

PTFE macromolecules contain polar C-F bonds in a symmetrical arrangement which is such that the repeat unit has a zero dipole moment.Studies on the interaction of liquids with PTFE surfaces have revealed that quadrupole and higher-order momenta are effective[3] but another possible factor, namely, the presence of polar chain-ends, does not appear to affect PTFE surfaces to any measurable extent. Some procedures have been developed in an effort to increase PTFE surface energies. Since this substance is highly inert to chemical attack, most of the successful methods which are known are very harsh, relying on the use of metallic sodium, strongly alkaline solutions or ionizing radiation[4-6]. PTFE and polyethylene surfaces may also be rendered wettable by immersion in aqueous dispersions of colloidal tin and iron hydrous oxides, due to the deposition of the high surface energy oxides on the polymer surfaces[7].

Polyethylene is not as difficult to surface-modify. It is particularly amenable to oxidation and the formation of polar derivatives is well known[8,9]. It is also affected by a number of organic solvents. Both PTFE and polyethylene are extensively used as inert materials in the fabrication of equipment for the laboratory and for the chemical plant, primarily because of their inertness and also because they do not tend to contaminate most substances with which they are brought into contact. PTFE is considered to be the ultimate solution for a number of problems in critical laboratory equipment construction notwithstanding its high cost and the difficulties found in putting this material in the desired shapes.

Some years ago, during a previous work[10] developed in the Institute of Chemistry at USP, it has been possible to notice that every piece of labware made of PTFE, polyethylene and polypropylene which was exposed to the organometallic liquid, iron pentacarbonyl, or to the related hydride, $H_2Fe(CO)_4$, was found to end up more or less heavily stained, displaying an orange-brick color rather similar to the color of hydrated iron (III) oxide. This contamination was removed with considerable difficulty only, and in many cases it was impossible to obtain clean surfaces, as judged from their discoloration. These were puzzling observations because the oxide in any of its solid forms or any other iron derivative which could have been originated under these conditions are very unlikely to be miscible with either PTFE or polyethylene.

More recently we have realized that this phenomenon might be interesting from two points of view: first, it could open the way to new approaches for polymer surface modification. Second, this system could be considered an useful model in a study of contamination of polymeric surfaces. The conclusions which have emerged from this work have allowed us to suggest a general mechanism for some types of contamination of low surface energy, polymeric solids.

The suggested mechanism involves two major steps: in the first, a low cohesive energy density substance is absorbed in the polymer[11]; then, this substance reacts giving a more stable product which is formed in situ, already mixed with the polymeric matrix. The miscibility of this newly formed substance is not relevant to the onset of polymer contamination and it should be considered only if one is interested in spontaneous decontamination by phase separation.

In this work we describe experiments on a) the sorption of some liquids by PTFE, b) surface modification of PTFE and of polyethlene by iron and manganese oxides and c) physical characterization of iron oxide incorporated in a PTFE matrix by $Fe(CO)_5$ sorption followed by its in situ oxidation. We describe the results of the examination of PTFE-Fe_2O_3 samples by electron microscopy, photoacoustic spectroscopy (PAS) and electron paramagnetic resonance, which give a rather detailed picture of the physical arrangement and properties of the iron oxide dispersed in the polymer.

EXPERIMENTAL

PTFE samples were: 1) sheets made of Dupont Teflon and fabricated by Incoflon (São Paulo) or 2) films prepared in our laboratory by sedimenting "Fluon" PTFE (unstabilized latex) on glass slides, drying and heating at 360°C. Except if indicated otherwise, experiments were performed with the former samples.

The identity of the samples was checked by IR absorption spectroscopy, melting point measurement and ESCA. Their degree of crystallinity was determined by X-Ray diffraction in a Norelco instrument. Thermal treatment was done as follows: samples were annealed by heating at 206°C for two hours; to obtain samples with a decreased degree of crystallinity, they were heated to 360°C for 15 minutes and rapidly immersed in ethanol.

Low-density polyethylene and polypropylene samples were pieces cut from 3 mm thick sheets, bought from local manufacturers. Chemicals used were pro analysis or of the best available grade. All the liquids used in the sorption measurements were redistilled prior to use.

The instrument for photoacoustic spectroscopy consisted of a 200 watt tungsten filament lamp, a variable-speed light chopper, monochromator, an air-filled aluminium sample cell with a condenser microphone connected to a low-noise preamplifier and a lock-in amplifier. The sample compartment is a cylindrical chamber with a diameter of 1.0 cm. Samples were weighed to correct for mass variation.

Magnetic resonance spectra were measured from liquid nitrogen to room temperature using an X-band Varian E-12 ESR spectrometer with a 100 KHz rectangular cavity.As usual, the first derivative of the absorbed power was recorded as a function of the applied magnetic field. The magnetic field was measured using the proton NMR signal.

RESULTS AND DISCUSSION

Sorption of Liquids in PTFE

We have determined the amounts of some liquids which are sorbed in PTFE, gravimetrically. The liquids used were: acetone, acetic acid, n-butanol, acetic anhydride, ethyl acetate, acetaldehyde, n-octanol, carbon tetrachloride, chloroform, benzene, n-pentane, acetonitrile, n-pentanol, pyridine, pentacarbonyl iron, trifluoroacetic acid, nitromethane and dioxane. Swelling coefficients, defined by Equation (1), were calculated and they are presented in Figure 1 as a function of the solubility parameters of the liquids, Hildebrand's δ [12]:

$$Q = \frac{m - m_o}{m_o} \times \frac{1}{d} \qquad (1)$$

m= polymer mass after sorption
m_o= initial polymer mass
d= density of the liquid.

These results show that the liquids having the lower cohesive energy densities are sorbed in higher amounts, in PTFE. On the other side, only part of the right-side branch of the Gaussian curve predicted by Gee and cols.[13] is observed, because the δ of PTFE is lower than δ of any of the liquids examined in this work. Using the tables in Reference 12, PTFE's δ can be estimated to be 1.2×10^4 $J^{1/2}$ $m^{-3/2}$. $Fe(CO)_5$ is outstanding for displaying a swelling coefficient in PTFE which is higher than the Qs obtained for most other substances examined, as expected considering that it is a liquid of lower cohesive energy density than most other existing liquids, except hydrocarbons and fluorocarbons.

The sorption of the liquids examined in this work is a relatively fast process. For instance, the amount of acetic acid sorbed in 0.21 mm thick PTFE reaches a plateau after 6 hours at 100°C ($Q=4.0$ m^3/kg)

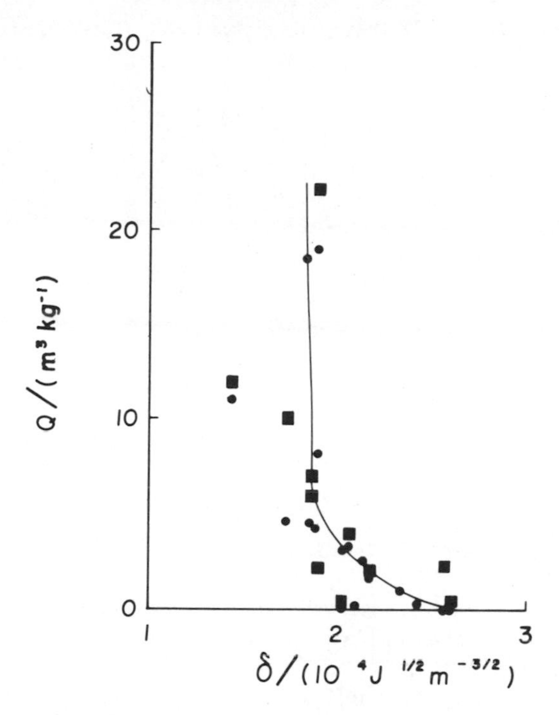

Figure 1. Swelling coefficients of liquids sorbed in PTFE Teflon as a function of their solubility parameter. t = 30°C, 48 hours immersion. Degree of crystallinity of PTFE: 70% (●) and 55% (■).

and after 25 hours at 30°C (Q=4.6 m^3/kg). Data published by Wade[14] shows that CCl_4 vapor sorption by finely divided PTFE occurs in a much faster time-scale, at 20°C.

It is indicated in the previous paragraph that the swelling coefficient decreases with temperature increase. This is not the usual behavior for most liquids, according to the examples given in Table I.

An important factor in the amounts of liquid which are sorbed by PTFE is the degree of crystallinity of the polymer sample. We have already observed[11] that iron pentacarbonyl is sorbed to a higher extent in PTFE of a lower degree of crystallinity. This observation has been extended to some other liquids, which are included in Figure 1. In this respect, it should be mentioned that the data in Table I indicates noticeable differences in the amounts of acetone and chloroform sorbed at 0°C and at 30°C. Since PTFE undergoes a phase transition around 20°C, these observations suggest that not only the degree of crystallinity but also the changes induced in both the amorphous and crystalline regions of the polymer by this

Table I. Temperature Effect on the Swelling Coefficients of Some Liquids in Polytetrafluoroethylene.

	t/°C	$\delta/(10^4 J^{1/2} m^{-3/2})$	$Q/(m^3 Kg^{-1})$
chloroform	0		1.2
	30	1.90	8.2
	55		8.3
carbon tetrachloride	0		0.5
	30	1.75	4.7
	55		13.8
acetone	0		1.0
	30	2.02	3.6
	55		3.8

phase transition can affect the amounts of liquids sorbed.

Reactions of $Fe(CO)_5$ Sorbed in PTFE

Iron carbonyl sorbed in PTFE displays some of its usual chemical reactions: a) it is thermally decomposed at ca. 110°C; b) it is oxidized by air exposure; in these two cases the characteristic CO-stretching bands disappear but the IR spectra of these samples do not offer any evidence for chemical changes in the polymer; it is photochemically transformed to $Fe_2(CO)_9$[11]; irradiation of co-sorbed isoprene and $Fe(CO)_5$ with ultra-violet light yields the known derivative, isoprenyltricarbonyliron[15]. $Fe(CO)_5$ can thus be considered as a general precursor of iron compounds obtained in situ, already mixed with the polymeric matrix. The iron oxide obtained by the procedures outlined above can be located in the bulk or in the surface of the polymer solid, depending on the details of the oxidation procedure utilized. Polyethylene[16] and polypropylene are also able to sorb low-δ liquids and we have found that their behavior in the presence of $Fe(CO)_5$ parallels the behavior of PTFE[17].

Surface Modification of PTFE, Polyethylene and Polypropylene with Iron and Manganese Oxides

A number of oxidizing reagents can react with $Fe(CO)_5$, giving a host of products. If the iron carbonyl is previously sorbed in a polymer which is then exposed to an oxidant unable to cross the polymer surface, a preferred site for reaction is the oxidant-polymer interface. This assumption allowed us to formulate simple, effective procedures for the incorporation of iron (III) and manganese oxides on plastic surfaces, as follows: PTFE films (0.21 mm thick) are immersed in iron carbonyl for 24 hours, at room temperature. The films are then immersed in an alkaline solution of potassium permanganate (0.04% $KMnO_4$, 0.1N NaOH) or in a 3% solution of hydrogen

Table II. Water-Polymer Contact Angles (degrees).

	advancing	receding
PTFE[c]		
pure	114	76
surface-modified[a]	44	zero
Polyethylene		
pure	90	
surface-modified[b]	40	
Polypropylene		
pure	90	
surface-modified[b]	42	

a) 1 hour in $Fe(CO)_5$ followed by 2 hours in alkaline permanganate solution; b) 12 hours in $Fe(CO)_5$, 5 hours in alkaline permanganate; c) films made of Fluon latex.

peroxide in 0.1N NaOH. After these treatments the surfaces appear heavily colored and the films show weight increases of 1.2% (permanganate oxidation) and 0.3% (peroxide). The deposits of oxides in PTFE are adherent and they cannot be removed except by abrasion with carborundum paper, which removes the oxides together with PTFE. The presence of Mn and Fe in the modified surfaces was further confirmed by X-Ray fluorescence spectrometry and the state of oxidation of Fe as Fe (III) was verified by electron spin resonance, as described later in this paper.

It may be possible to obtain deposits of MnO_2 in PTFE by sorption of some other reducing liquid, followed by oxidation with $KMnO_4$ solutions; so far, we did not yet obtain good results from such experiments.

Iron and manganese oxides can be incorporated to surfaces of polyethylene and polypropylene also, by using procedures similar to those described for PTFE. The incorporation of the oxides to plastic surfaces[18] leads to considerable changes in water-polymer contact angles, as shown in Table II.

Reactions of Fe_2O_3 Obtained _in situ_ in a PTFE Matrix

The contamination of polymeric surfaces with the reactive and adsorbent[7,19] oxide, Fe_2O_3 results in an increase in surface energies which opens the possibility to further contamination of the plastics by other chemicals. This is demonstrated by the behavior of Fe_2O_3-contaminated PTFE in the presence of some acids: hydrochloric, acetic, mercaptoacetic and phosphoric.

Hydrochloric Acid: 0.15 mm thick PTFE was immersed in $Fe(CO)_5$ and air-oxidized to a content of 1.1% Fe_2O_3 (8.3 mg). The film was

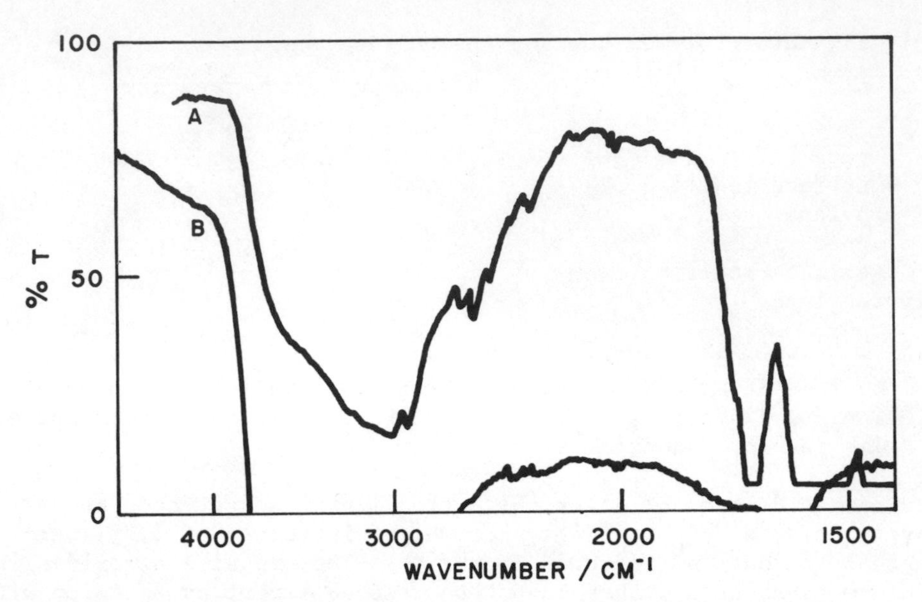

Figure 2. Infrared absorption spectra of PTFE films after incorporation of Fe_2O_3 and exposure to: A) acetic acid and B) azeotropic hydrochloric acid.

then immersed in azeotropic HCl during 48 hours, in a boiling-water bath. The dissolution of some of the oxide was visible but the film showed a further mass increase of 23.3 mg. The IR spectrum of this film is shown in Figure 2; it it clear that a considerable amount of water has been incorporated in the film, concurrent with the observed color change from brick red to pale yellow.

We conclude that some of the oxide in the plastic was unable to dissolve in the acidic solution, although it was completely converted to hydrated iron (III) chloride. This is a further example of the possibility to contaminate PTFE with a substance which is highly immiscible with this polymer. Another point to be noted is that the oxide-impregnated PTFE becames permeable to HCl although the unmodified PTFE does not sorb azeotropic HCl to any measurable extent.

Acetic Acid: The immersion of 1.1% Fe_2O_3 - PTFE film in acetic acid, in a boiling-water bath, leads to a definite color change to a salmon-pink, the same as the color of iron (III) acetate in acetic acid. The IR absorption spectrum of the film thus obtained shows intense bands, which can be assigned to the presence of water and acetate/acetic acid (Figure 2).

Phosphoric Acid: 3 mm thick pieces of PTFE having a superficial deposit of iron oxide (0.5 mg cm^{-2}), obtained by immersion in $Fe(CO)_5$

Table III. Liquid Contact Angles in PTFE, in Fe_2O_3-Impregnated PTFE and in Fe_2O_3-PTFE after Immersion in H_3PO_4 (degrees).

		PTFE	$PTFE/Fe_2O_3$ [17]	$PTFE/Fe_2O_3/H_3PO_4$
Water	a)	95	65	91
Benzene	a)	20	zero	17
Dimethyl-	a)	58	20	57
sulfoxide	r)	45	zero	35
Dimethyl-	a)	59	28	56
formamide	r)	48	zero	24

a) advancing, r) receding

followed by oxidation in an alkaline solution of hydrogen peroxide, were immersed in 5% phosphoric acid overnight. Their color changed to a brilliant red and liquid contact angles increased back to values close to those obtained in pure PTFE, as shown in Table III.

Mercaptoacetic Acid: Iron oxide-impregnated PTFE films are blackened by immersion in mercaptoacetic acid, probably due to its partial reduction to Fe_3O_4. This reaction is observed with the most superficial layers of Fe_2O_3, only. Exposure of the black films to air gives back the usual brick red color.

Microscopical and Spectroscopical Examination of Fe_2O_3-Contaminated Polytetrafluoroethylene

Attempts to establish the identity of the iron oxide obtained *in situ* in PTFE by X-Ray diffraction were unsuccessful because no diffraction lines were observed other than those assigned to the polymer itself.

Electron microscopical, photoacoustical (PAS) and ESR experiments were performed on a set of samples of PTFE (0.2 mm thick films) subjected to immersion in $Fe(CO)_5$ for varying times, followed by oxidation under air and heating to 105°C. The concentration of contaminants in the samples ranged from 0.34 to 1.50% (w/w) and the absence of $Fe(CO)_5$ residues was checked by IR spectroscopy. This work is described in detail elsewhere[20,21] and some conclusions are given here to substantiate our picture of Fe_2O_3-contaminated PTFE.

A typical electron microphotograph is given in Figure 3 and some iron oxide particle size distribution histograms are in Figure 4. It is clear that the oxide particles are rather small and uniform. Less contaminated samples (0.34 - 0.94%) contained particles smaller (average diameter, 36 angstrom) than the more contaminated ones (average diameter, 44 angstrom).

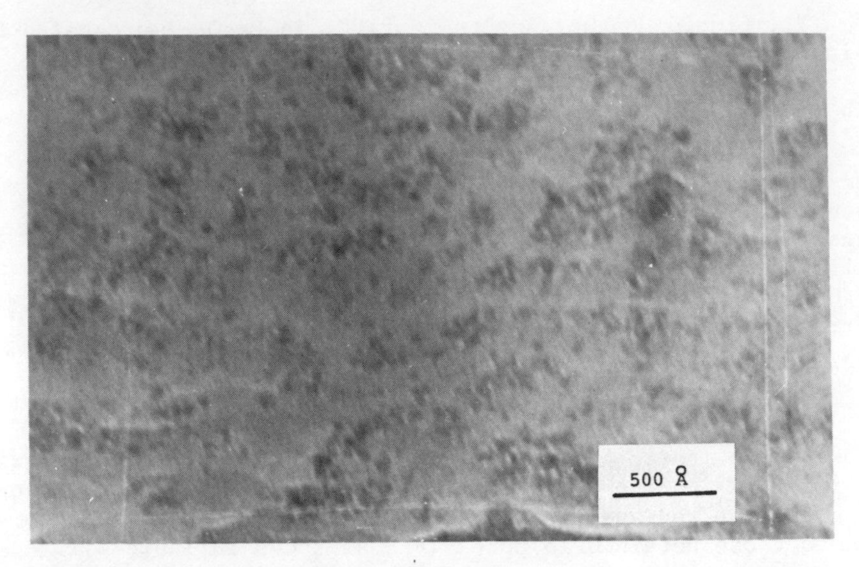

Figure 3. Electron Micrograph of Fe_2O_3-impregnated PTFE.

This observation was confirmed by PAS measurements. A photoacoustical spectrum is in Figure 5, displaying strong absorption bands assigned to charge transfer from oxygen to a central Fe(III) in an octaedral configuration. In these spectra, one might expect absorption peak intensity to be a monotonic function of iron oxide concentration. Figure 6 shows that this is not the case and it has been interpreted as due to the existence of two characteristic oxide

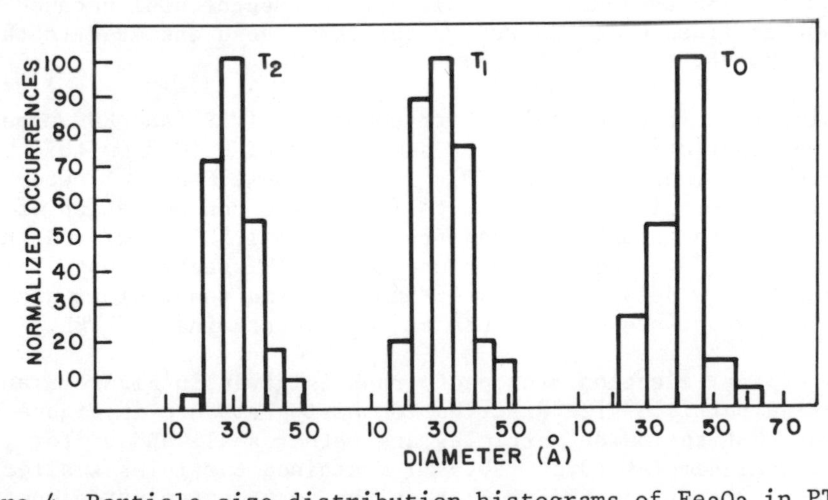

Figure 4. Particle size distribution histograms of Fe_2O_3 in PTFE. Fe_2O_3 concentration in the samples: T_o, 0.34; T_1, 0.58; T_2, 0.62%.

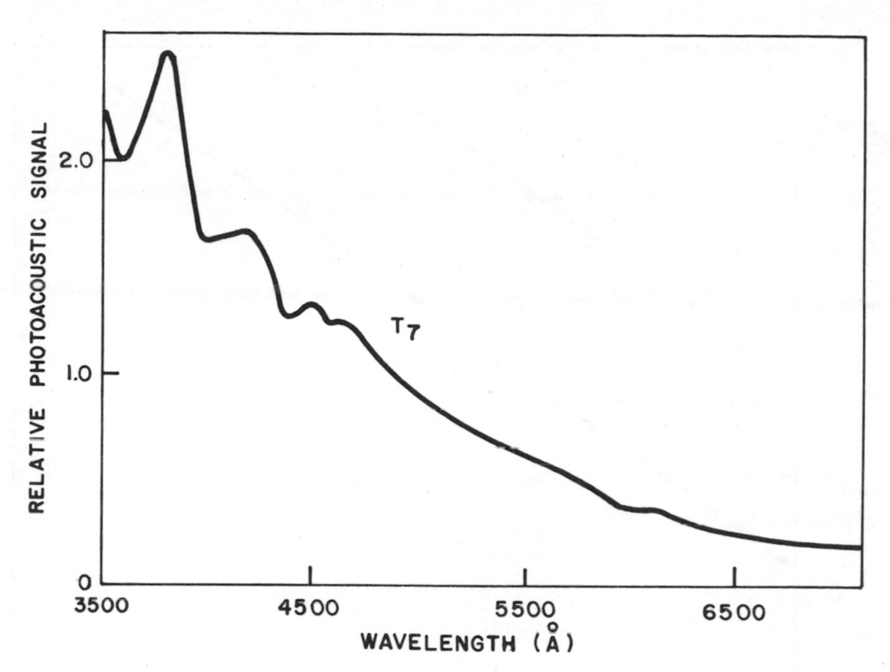

Figure 5. PAS spectrum of Fe_2O_3-contaminated PTFE, 1.39% w/w Fe_2O_3 concentration. Signal intensity relative to the intensity of a carbon black standard.

particle volume distributions. As it is now well established[22], the photoacoustical signal, for an optically opaque and thermally thick sample, is proportional to the product of the optical absorption coefficient β and to the intensity of light which is converted into heat within the sample. The proportionality constant involves the thermal parameters of the cell and sample. In the present case, however, the sample consists of fine, absorbing particles in a non-absorbing matrix and the scattering of light by the particles reduces the actual light intensity available for the energy-conversion process. Taking this effect into account, the photoacoustical signal can be written as[20]:

$$Q \propto (\beta - bd^6) \qquad (2)$$

where Q is the signal intensity, d is the average diameter of Fe_2O_3 particles and b is a constant. From the observed dependence of the PAS signal intensity on concentration (Figure 6), we have used this proportionality to estimate the average particle diameter ratio of the two populations (sixth root of intercept ratio for data of Figure 6). The resultant diameter ratio for the two populations agrees

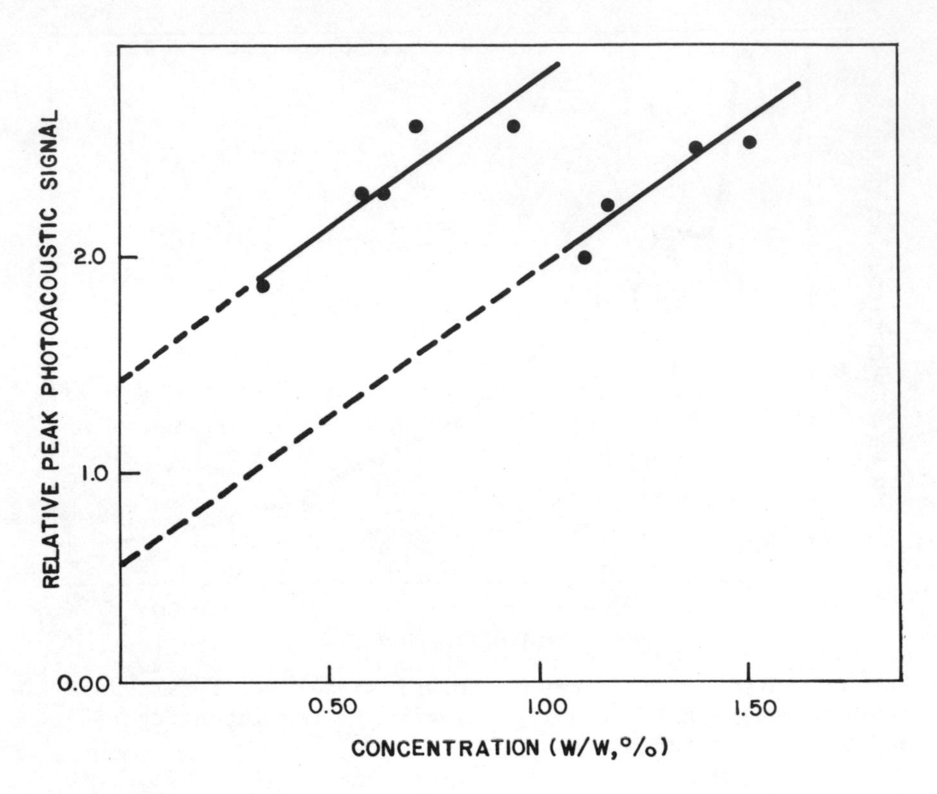

Figure 6. Photoacoustical signal intensity at maximum absorption wavelength as a function of iron oxide concentration in PTFE.

with the corresponding electron microscopic determination within experimental error.

ESR spectra (Figure 7) of Fe_2O_3-impregnated PTFE gave the following results: i) g-value in all cases was found to be g=2.00, which is typical of iron (III); the peak-to-peak resonance linewidth ΔH shows little change in the temperature range used (liquid nitrogen to room temperature) and it is larger in the low concentration samples; ii) the linewidth varies with the inverse of particle volume and iii) the linewidth is independent of the external magnetic field, as verified by measuring spectra in the Q-band (12 KOe), also. The last observation supports the previous conclusion that these oxide particles are very small[23], so that they retain superparamagnetic[24] behavior in the higher field used. Other points above were explained[21] taking Neel's description of superparamagnetism and assuming that the anisotropy energy density K increases with the inverse of particle[25] volume. Calculation of K gave values ranging from 1.5×10^5 to

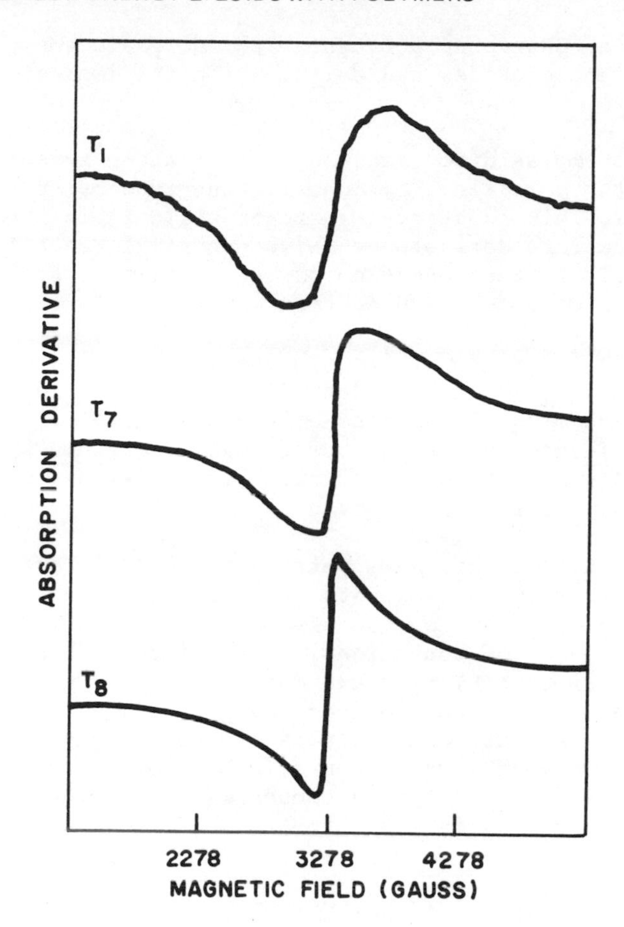

Figure 7. ESR spectra of PTFE-Fe_2O_3. Oxide concentration (w/w): T_1, 0.58; T_7, 1.39; T_8, 1.50%.

4.2×10^4 erg cm^{-3}, for different samples, which can be compared favorably with 4.7×10^4 erg cm^{-3}, determined for 150 angstrom diameter, α-Fe_2O_3 particles by Mossbauer spectroscopy.

To sum up, the evidence given in this section shows that the product of $Fe(CO)_5$ sorption in PTFE followed by air oxidation is finely divided (20-80 angstrom diameter), α-Fe_2O_3.

SUMMARY AND CONCLUSIONS

We have shown that inert, polymeric substances such as PTFE Teflon, polyethylene and polypropylene can be surface and bulk contaminated to a considerable extent with inorganic substances such as iron and manganese oxides and salts, which are immiscible with the polymers.

In the examples discussed the contamination arises due to the sorption of the volatile, low cohesive energy density liquid, iron pentacarbonyl. This substance can react while it is absorbed in the polymer to generate derivatives which can still react further. A striking result is the obtention of Teflon films containing appreciable amounts of water as hydration water in a ferric salt.

Polymer surface properties are drastically altered, due to the incorporation of the metal oxides. It may be expected that many other physical and chemical properties of the polymers will be altered to some extent. These changes may be desirable for some applications of the plastics but if they occurr accidentally they will probably raise serious practical problems.

We suggest that this same pattern of contamination can be induced by a number of other substances which have some analogy to iron pentacarbonyl. Any low cohesive energy density, reactive compound can start a cascade of contaminating reactions such as described in this work. Metal carbonyls, alkyl metals, other organometallics, covalent metal halides and some organic substances such as vinyl monomers and unsaturated halocarbons are to be considered as possible precursors of polymer surface contamination by their stable, non-volatile derivatives and by substances which can adsorb on the newly created high-energy surfaces.

ACKNOWLEDGEMENTS

This work was supported by FAPESP (São Paulo, Brazil), Grant 77/0355. SEG received a fellowship from FAPESP. FG thanks E. J. Vicente and E. H. Galembeck for help in the sorption measurements and the support of the Brazilian Academy of Sciences and Gessy-Lever S/A to his laboratory is gratefully acknowledged.

REFERENCES

1. R. W. Moncrieff, "Man-Made Fibres," pp. 571-573, Wiley, New York, 1970.
2. W. A. Zisman in "Contact Angle, Wettability and Adhesion," Advances in Chemistry Series No. 43, pp. 9-22, American Chemical Society, Washington, D.C., 1964.
3. R. J. Good in "Contact Angle, Wettability and Adhesion," Advances in Chemistry Series No. 43, pp. 74-87, American Chemical Society, Washington, D.C., 1964.

4. D. W. Dwight and W. M. Riggs, J. Colloid Interface Sci. 47,650 (1974).
5. H. Fitz and F. Mayer, G. Pat. 2,354,210; cf. Chem. Abs. 83, P148351z (1975).
6. N. I. Egorenkov, D. A. Rodchenko, A. I. Barki and V. V. Kumaneva, USSR Pat. 458,567; cf. Chem. Abs. 83, 60935g (1975).
7. J. T. Kenney, W. P. Townsend and J. A. Emerson, J. Colloid Interface Sci. 42, 589 (1973).
8. D. Sage, P. Berticat and G. Vallet, Angew, Makromol. Chem. 54, 151 (1976).
9. A. Baszkin, M. Nishino and L. Ter-Minassian Saraga, J. Colloid Interface Sci. 54, 317 (1976).
10. F. Galembeck and P. Krumholz, J. Am. Chem. Soc. 93, 1909 (1971).
11. F. Galembeck, J. Polymer Sci. Polym. Chem. Ed. (in press).
12. J. H. Hildebrand and R. L. Scott, "The Solubility of Non-Electrolytes," 3rd ed., Reinhold, New York, 1950.
13. G. Gee, G. Allen and G. Wilson, Polymer 1, 456 (1960).
14. W. H. Wade, J. Colloid Interface Sci. 47, 676 (1974).
15. M. A. DePaoli, I. Tanashiro and F. Galembeck (1979), unpublished data.
16. M. L. Santos, N. F. Corrêa and D. M. Leitão, J. Colloid Interface Sci. 47, 621 (1974).
17. F. Galembeck, "Modificacão Superficial e Impregnacão de Polímeros com Óxido de Ferro III," Tese de Livre-Docência, São Paulo, 1977.
18. F. Galembeck, J. Polymer Sci. Polym. Lett. Ed. 15, 107 (1977).
19. A. Breeuwsma and J. Lyklema, J. Colloid Interface Sci. 43, 437 (1972).
20. F. Galembeck, C. C. Ghizoni, C. A. Ribeiro, H. Vargas and L. C. M. Miranda, (1978), submitted for publication.
21. F. Galembeck, N. F. Leite, H. Vargas and L. C. M. Miranda, (1978), submitted for publication.
22. A. Rosencwaig and A. Gersho, J. Appl. Phys. 47, 64 (1976).
23. M. Eibshutz and S. Shtrikman, J. Appl. Phys. 39, 997 (1968).
24. L. Neél, Ann. Geophys. 5, 99 (1949).
25. C. Kittel, Phys. Rev. 73, 810 (1948).

A STUDY OF THE EFFECTS OF RELATIVE HUMIDITY ON SMALL PARTICLE ADHESION TO SURFACES

W. J. Whitfield

Sandia Laboratories

Albuquerque, New Mexico 87185

This paper describes a study of relative humidity effects on the adhesion of small particles to surfaces. Ambient dust ranging in size from less than one micron up to 140 microns was used as test particles. A 20 psi nitrogen blowoff was used as the removal mechanism to test for particle adhesion. Particles were counted before and after blowoff to determine retention characteristics. Particle adhesion increased drastically as relative humidity increased above 50%. The greatest adhesion changes occurred within the first hour of conditioning time. Data are presented for total particle adhesion, for particles 10 micron and larger, and 50 microns and larger.

INTRODUCTION

A series of experiments was conducted to examine the effects of relative humidity on particle contamination adhesion to surfaces. Almost all naturally occurring particles are affected in some manner by water vapor in the surrounding air, particularly as the water content of air increases. Of special interest are those particles that pick up moisture from the air which partially or totally dissolve and then form a very strong bond with the surface on which the particles are located. At elevated humidities this occurs quickly - in a very few minutes. These particles cannot be effectively removed by dry wipes, vacuum cleaning or other cleaning methods that are permitted on many critical surfaces.[1] Thus, the final particle burden of a surface can be affected by the relative humidity of the air in which it was assembled.

EQUIPMENT

These experiments were set up to study relative humidity effects from thirty-three percent to one hundred percent. Glass dessicators containing the following saturated salt solutions and water were used for conditioning chambers.

H_2O	Water	100%
$NH_4H_2PO_4$	Ammonium Phosphate	93%
KBr	Potassium Bromide	84%
NaCl	Sodium Chloride	76%
$NaNO_2$	Sodium Nitrite	66%
$Na_2Cr_2O_7 - 2H_2O$	Sodium Dichromate	52%
$MgCl_2 - 6H_2O$	Magnesium Chloride	33%

Test particles were used that simulated as nearly as possible the type particles expected in a clean room environment. Test particles were obtained by sieving building vacuum cleaner dust to exclude particles larger than 140 microns. After sieving, the test particles were stored in dry air over a dessicant bed until use.

Test surfaces were 1" x 1" highly polished metal foils cemented to 1 x 3 inch glass microscope slides. Test surfaces were etched to permit photographing the exact same area before and after "blowoff."

A 3.3 cu. ft. particle loading chamber was used to load the test surfaces prior to conditioning at the various relative humidity levels. An agitator fan was located near the bottom of the chamber and a glass tube was used to feed test particles into the fan inlet during loading. A horizontal rack was positioned in the upper half of the loading chamber to hold test slides during loading.

A blowoff fixture was used to retain and position test slides during "blowoff." The fixture consisted of a 1/8" diameter jet located 1/2" above the test strip. Dry nitrogen was used as the blowoff gas which was controlled by a solenoid valve and timer. Nitrogen pressure was regulated to 20 psi during blowoff.

The microscope-camera system is a Leitz Ortholux equipped with Leitz Ultra-Pack vertical illumination equipment. This combination with the highly polished foils provided an excellent high contrast dark field illumination system. The system will resolve particles less than one micron size. A 4" x 5" Polaroid camera back was used for photographing test slides. A magnification of 90x was used for photographs to be counted and higher magnifications for individual particle analysis.

EXPERIMENTAL PROCEDURE

The following sequence of steps was followed during the experiment.

1. Twelve clean slides were placed on the loading chamber rack.

2. The timer was set for 60 seconds to start blower.

3. Two ml (approximately 1.2 gm) test particles were released into the loading tube during first 30 seconds of the loading cycle.

4. After the load cycle, the 12 test slides were carefully removed and placed in the humidity control chamber. Every effort was made to avoid air currents, drafts, vibration, and rapid movement of the test slides during handling. One slide was removed from the humidity control chamber after each of the following conditioning periods - 84%, 93%, and 100% - 5, 10, 15, 30 minutes; 1, 2, 4, 8, 24, 48, 72 hours. 33%, 52%, 66%, 76% - 30 minutes; 1, 2, 4, 8, 24, 72, 200, 720 hours.

5. After removal from the humidity control chamber, each slide was immediately photographed. Then it was exposed to a "blowoff" treatment for 10 seconds at 20 psi. (Following "blowoff" it was immediately rephotographed. Two separate defined areas were photographed before and after "blowoff").

6. Particles, as recorded on the photographs, were sized and counted in ranges of (a) less than 10 microns, (b) 10 microns and larger, and (c) 50 microns and larger. The lower limit of particle size count was approximately 1 micron. The area photographed from each slide was approximately 1.6 mm^2. Approximately 120 particles were counted per 1.6 mm^2 before blowoff (initial load).

7. Four test slides were loaded, photographed, then subjected to "blowoff" procedure, rephotographed, and counted for reference or control.

RESULTS

Particle count data were converted to "Percent Particles Remaining After Blowoff," designated as "Retention %" and defined as

$$\text{Retention \%} = \frac{\text{Count after blowoff}}{\text{Count before blowoff}} \times 100.$$

Data are presented in graph form in Figures 1 through 7 for individual relative humidity levels. Results are plotted as "Retention %" versus the time (in hours) of conditioning at the various humidity levels. At relative humidity levels of 76%, 84%, 93%, and 100%, "Retention %" is plotted for the particle size categories "total particles," "10 microns and larger," and "50 microns and larger," marked as curves A, B, and C. The "50 microns and larger" is not shown for relative humidity levels below 76% since there was very little retention of large particles at these humidities. Figure 8 shows a comparison plot of total particles for all relative humidity levels.

This series of experiments indicates that the major effect of relative humidity on particle retention occurs within one hour at any of the relative humidity levels investigated. At higher humidity levels, particles become firmly attached to the surface in a few minutes. Disassociation of the particles occurred at all humidity levels; (occurring much more rapidly at the higher humidity levels). The dissociation or "breakup" of particles left large numbers of small particles adhering to the test surface that were much harder to remove than the original, or parent, particles.

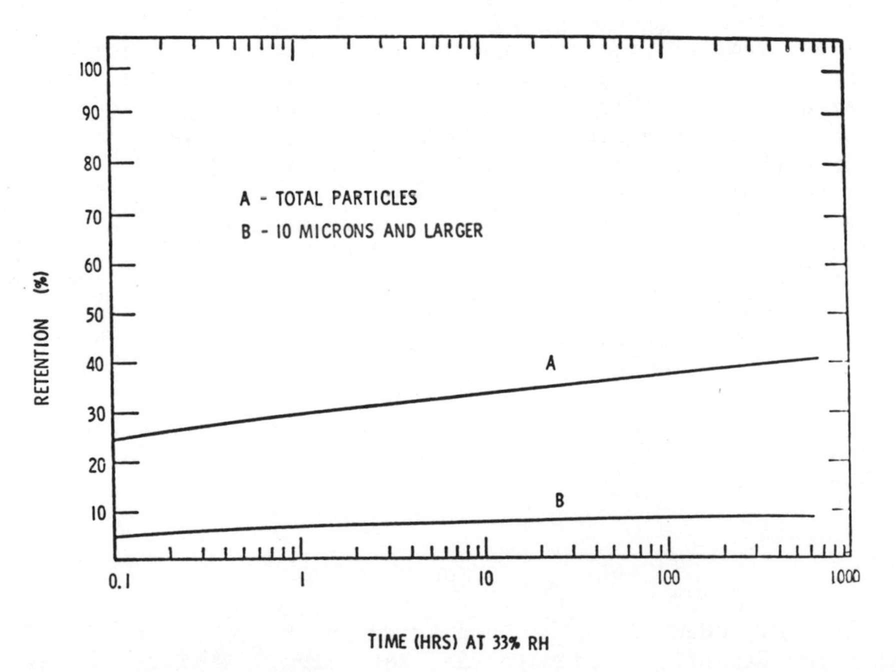

Figure 1. Effects of relative humidity on surface particle retention.

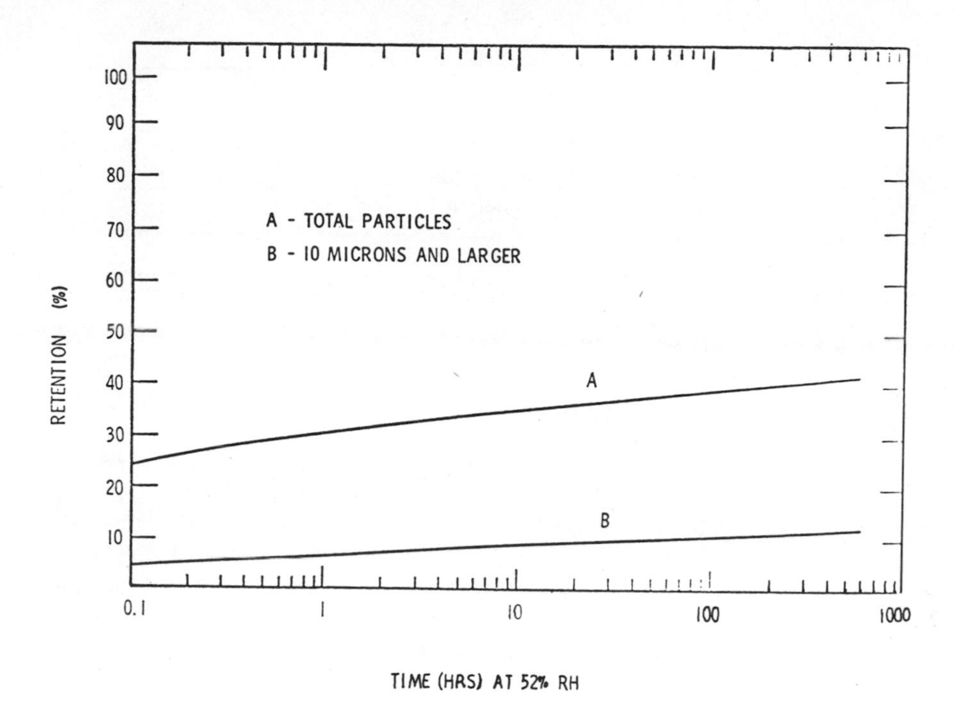

Figure 2. Effects of relative humidity on surface particle retention.

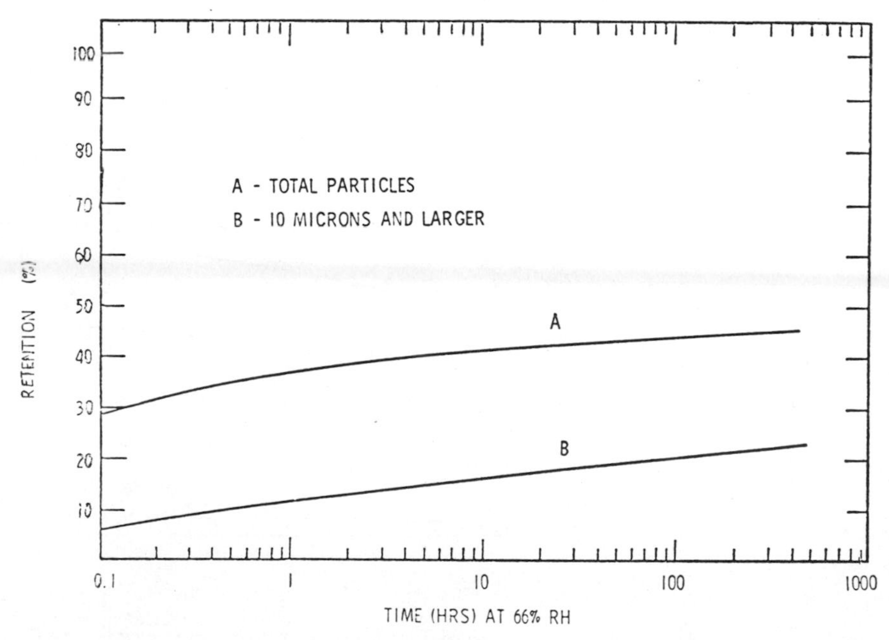

Figure 3. Effects of relative humidity on surface particle retention.

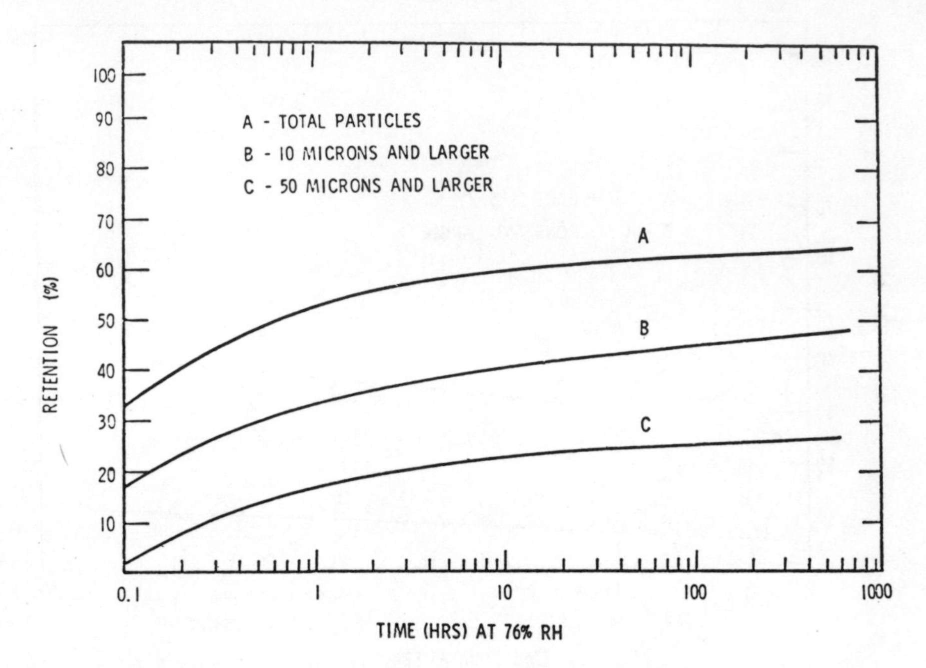

Figure 4. Effects of relative humidity on surface particle retention.

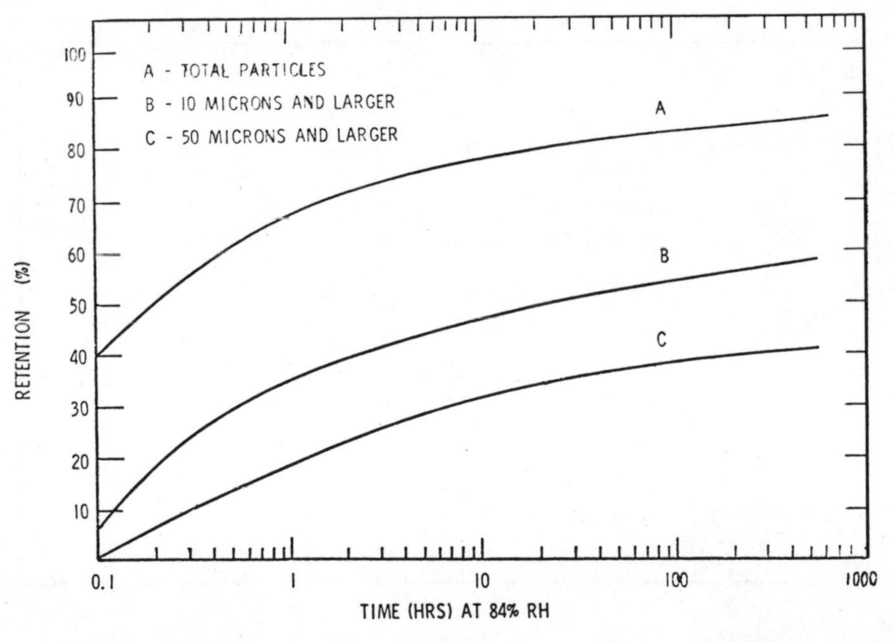

Figure 5. Effects of relative humidity on surface particle retention.

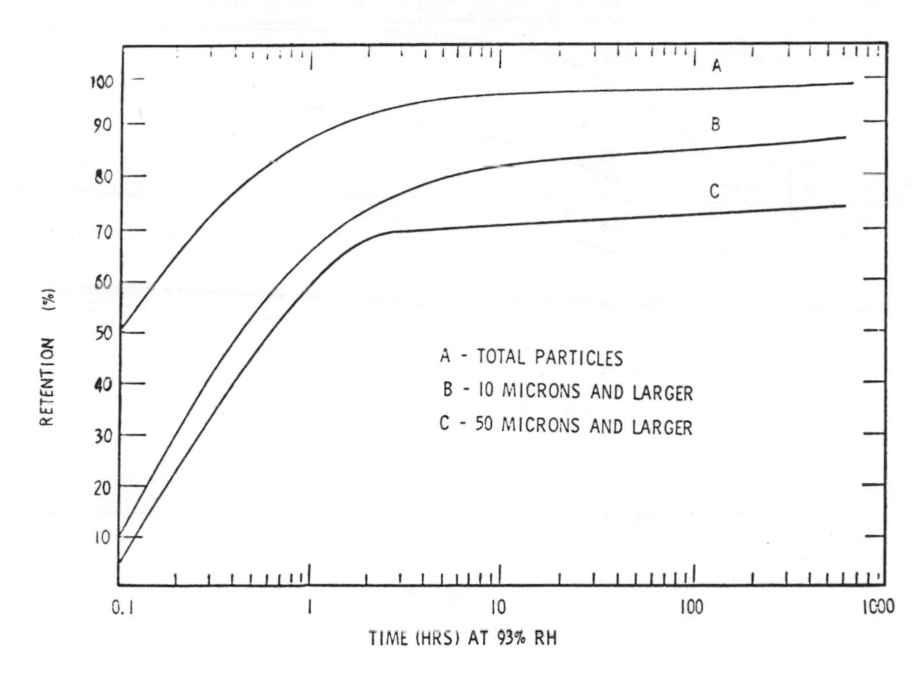

Figure 6. Effects of relative humidity on surface particle retention.

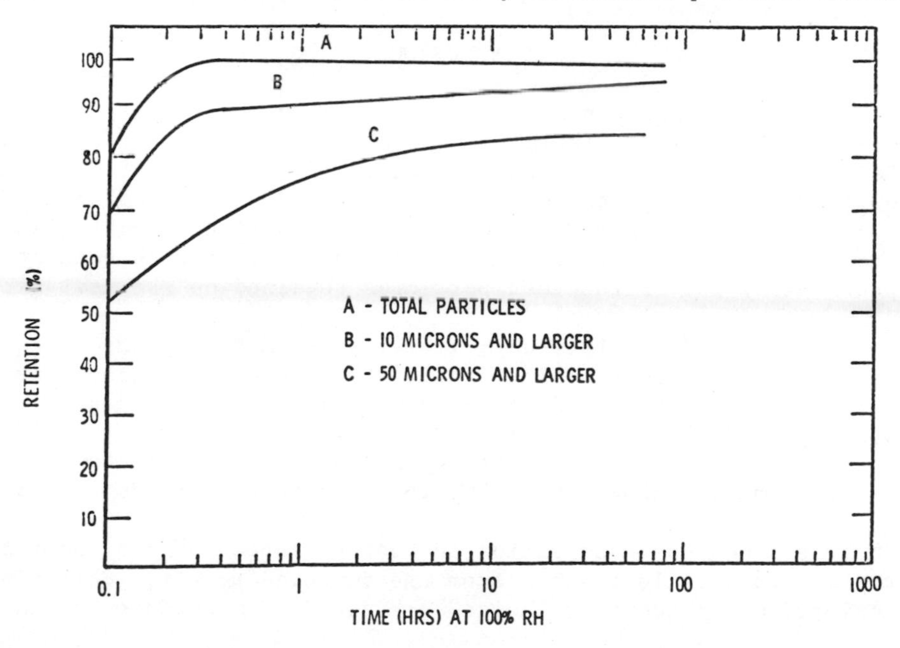

Figure 7. Effects of relative humidity on surface particle retention.

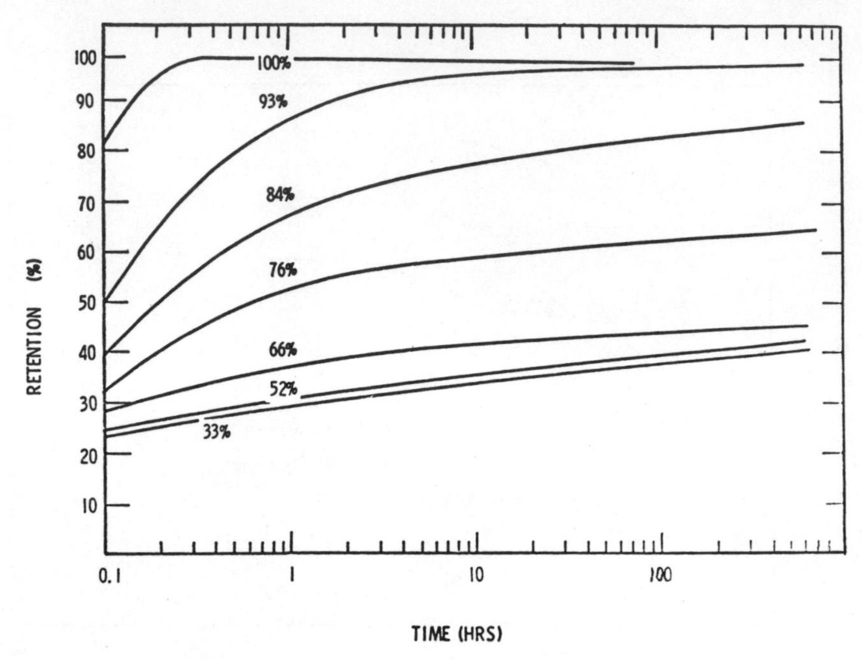

Figure 8. Effects of relative humidity on surface particle retention.

CONCLUSIONS

These experiments show relative humidity under these conditions to be a very important factor in surface particle retention which relates directly to total loading factors for a surface. The hygroscopic property of particles - the ability to pick up moisture from air - appears to be a major factor in the adhesion of particles to surfaces.[2] Molecular, electrostatic and other forces account for approximately 20% of the test particles retained on the test strips.

As a result of this study, it may be seen that particle removal from a surface is most easily facilitated either by cleaning or natural environmental removal factors when the surface has not been exposed to high humidity environments. In particular, it would appear that surfaces should not be exposed to environments with relative humidity above about 50% for even short periods of time.

This study calls attention to another aspect of hygroscopic particles that collect water from the surrounding air. Corrosion or oxidation can occur as "holes" or "pits" in a surface as a result of collected water. Should the hygroscopic particle cause a change in pH on absorption of water, then accelerated corrosion of the surface can be expected.

REFERENCES

1. S. Bhattacharya and K. L. Mittal, Surface Technol. 7, 413 (1978).
2. A. D. Zimon, "Adhesion of Dusts and Powder," pp 63-113, Plenum Press, New York, 1969.

PARTICULATE REMOVAL IN MICROELECTRONICS MANUFACTURING

D. A. Brandreth and R. E. Johnson, Jr.

E. I. du Pont de Nemours & Company

Wilmington, Delaware 19898

The ever-decreasing dimensions of microelectronic components emphasize the need for better cleaning solvents, equipment, and procedures. The fundamentals of particle adhesion to surfaces are discussed, and it is shown that solvents such as mixtures of ethanol or i-propanol with fluorocarbons such as trichlorotrifluoroethane significantly diminish particle-surface adhesion enough to allow removal of particles to the micron-level when used with intense agitation.

INTRODUCTION

The removal of microscopic amounts of contamination in microelectronics manufacturing operations is well established and, with ever-increasing miniaturization of circuitry components, this problem has become even more important. Whereas proper cleaning solvents can remove even monomolecular levels of soluble soils readily, particulate contaminants are very difficult to remove. The fundamentals of particle-surface adhesion and how particulate removal can be facilitated by the proper choice of cleaning solvent are presented here.

Microcircuit production costs depend strongly on the yield of usable circuits--that is the fraction of the total circuits free of defects. Defects are caused by several sources, but a major source is particulate contamination. Dust can cause photomasks to have pinholes in dark areas or opaque specks in areas supposed to be clear. Particulates in the photoprinting operation or other processing steps can affect a critical spot in the circuit that will cause failures. Whereas the minimum average dimension in integrated circuits was on the order of 40μ in 1960, today it is about 6μ, so it is obvious that it is necessary to remove much smaller particles than previously.

PARTICLE-SURFACE ADHESION

It is convenient to divide particulates into three size regimes to discuss adhesion of particles to surfaces. These demarcation lines are roughly: over 1000μ, 0.1μ - 1000μ, and below 0.1μ. Gravitational, intermolecular, and electrostatic forces attract particles independently of their size, but it is the ratio of these forces which is relevant. Thus, for a 1000μ (1 mm) particle resting on a surface, the ratio of the gravitational attraction to the intermolecular forces (always attractive) is very large. Gravity is a long-range force wherein the entire bulk of the particle helps determine the strength of the interaction,* whereas the intermolecular forces are of such short range that, in effect, only the tiny fraction of the particle in immediate contact with the surface enters into the calculation. As the particle size decreases, the relative amount of particle mass within the effective range of the intermolecular forces increases. Finally, at a particle diameter of 0.1μ, the intermolecular forces are far larger than gravitational forces. Between these two extremes, there is an intermediate range in which both forces are important.

Of course, the total force of adhesion for a large particle is much greater than for a small particle, but in practical terms the removal of the larger particle is much easier because larger inertial forces can readily be applied to large particles.

Forces of attraction due to static electricity are also of importance in some cases; but, with the application of the best cleaning solvents available today, rapid charge dissipation occurs and these static forces are thereby overcome.

By way of illustration, Table I shows the relative adhesive force in air for various particle sizes in terms of the number of g's. The forces can vary by orders of magnitude, depending on molecular adhesion, softness, etc.

It is apparent that particles in the 1μ range are held so tenaciously that mechanical cleaning cannot dislodge them. When

---

* Gravitational and static electrical forces are "inverse square" forces, i.e. for point particles at rest, $F = K/r^2$, where K is a constant and r is the distance between particles; whereas intermolecular forces are inverse seventh power forces, $F = C/r^7$, where C is also a constant, when calculated for individual molecules. For the case of spherical particle very close to a flat surface of infinite extent, $F \approx C'/r^2$.

Table I. Relative Adhesive Force (Intermolecular Forces) in Air for Various Sizes of Particles.

Particle Size (micrometers)	Relative Adhesive Force in Air (g's)
500	2
50	200
5	20,000
0.5	2,000,000

a solvent replaces air as the medium, the attractive force is diminished. The resultant attraction depends very much on the chemical nature of the solvent.

Three principal types of interactions are important in physical adhesion:

- dipole-dipole interaction
- dipole-induced dipole interaction
- dispersion forces (London forces)

All molecules exhibit dispersion forces and molecules with permanent dipoles such as HCl, H_2O interact by dipole-dipole interaction. Dipole-induced dipole interactions occur when a polar molecule such as HCl induces a dipole in a nonpolar molecule such as CCl_4. Dipole-dipole interactions are particularly important when a strong specific type of dipole-dipole interaction, known as hydrogen bonding, occurs. Water and alcohols are the most common molecules which form strong hydrogon bonds, and it is these relatively simple, inexpensive substances which are of great importance to free surfaces of particulates. These solvents interact strongly with the substrate surfaces and, thus, diminish the forces which hold the particulates to the surface. This adhesion force is reduced by several orders of magnitude.

THE ROLE OF SOLVENTS

Since water, by itself, often must be excluded from consideration as a precision cleaning agent because it promotes corrosion, alcohols and ketones come to the fore and, indeed, these solvents are capable of doing an excellent cleaning job. However, such pure solvents are highly flammable and, in some cases, they are too aggressive toward plastic components and cannot be used.

A common misconception is that solubility, per se, is the determining criterion for a good cleaning solvent. In cases which involve gross amounts of soluble soils, high solvency for the soil is important, but such cases are not of particular importance in cleaning microelectronics components. In removing particulates, the solvent serves two functions in varying degrees: (1) it adsorbs on the substrate by physical adsorption or chemical adsorption, and (2) it provides a different medium in which the interaction of the particle and the surface is changed.

When solvent molecules adsorb on both substrate and particulate, they create a different outermost layer which has its own (different) attraction for the soil particulate. But, in addition to this change in the attractive force between particle and surface, there is another change. The solvent acts as a different medium, not air or vacuum. To fully understand this "medium effect", one must resort to the physics of adhesion, which demands considerable development of definitions of terms and notation.[1,2] To capture the essence of the idea, however, is to understand that the ultimate attraction of the substrate for the particle is influenced by the nature of the surrounding medium which changes the relative values of the interactions between particle and surface, particle and solvent, and solvent and surface. A good solvent cleans by both adsorption on the surfaces and by the "medium effect" which diminishes the attraction between the surface and the particulates. With the addition of a solute, such as a surface-active agent, the adsorption role can be enhanced still further.

It should also be pointed out that whereas a particle not completely surrounded by liquid experiences forces due to capillarity which tend to cause the particle to adhere to the surface,[4] the disappearance of the gas/solid interface erases the capillary forces and yields the case of the totally immersed particle.

To illustrate some of these points, we present data taken from experiments carried out as part of an extensive research program on precision cleaning agents in the Freon® Products section of the Du Pont Company.

In one type of experiment, which measures directly the effects of solvents on the adhesion of particles to surfaces, diamond dust is deposited on a clean specimen. The particles are counted and sized after deposition. This system is then centrifuged while immersed in the test solvent, and the particles are again counted and sized. In some experiments where only low forces are studied, the system is merely inverted. The better solvents decrease the adhesive forces enough to enable the centrifugal force to release many of the particles. Although simple in concept, in practice meticulous care is required.

The ability of the hydroxyl group to form hydrogen bonds is illustrated when cyclohexane and cyclohexanol are compared. Figure 1 shows the particle size distribution for diamond particles which adhere to glass after inverting the system in the presence of the solvent. There is a pronounced superior particle removal by the cyclohexanol due to the stronger interaction of the hydroxyl group with the glass surface.

As noted before, low molecular weight alcohols such as methanol and ethanol are effective cleaning agents, but they are subject to the severe drawback of being flammable. When a small amount of alcohol (4-6%) is mixed with FC-113, an azeotropic

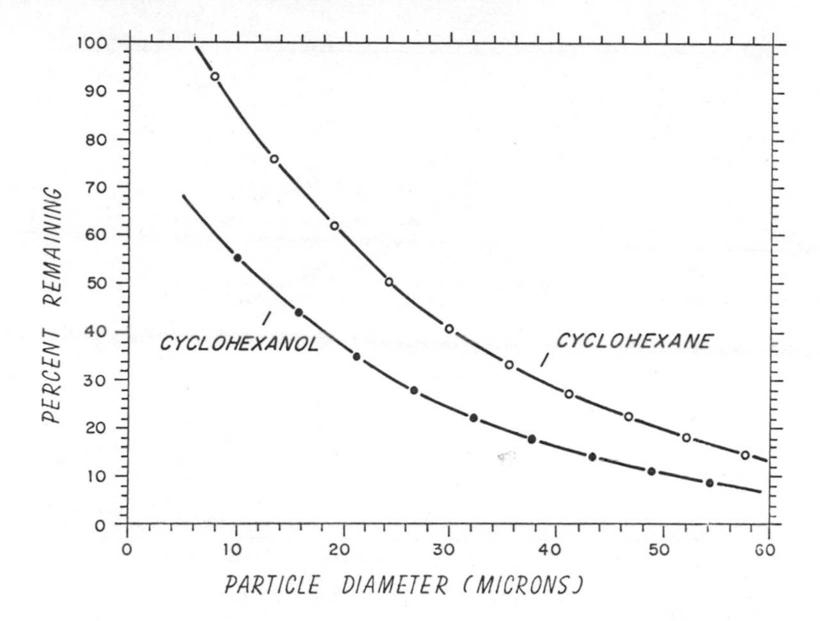

Figure 1. Adhesion of diamond particles to glass in cyclohexane and in cyclohexanol at 1g force.

composition results which exhibits a constant boiling point; that is, the composition does not change as the mixture evaporates.

Figure 2 presents data for diamond particle adhesion to glass after inverting the system (1g) in the presence of several individual solvents: FC-113 (Freon® TF), Freon® TE, Freon® TES, Freon® TM, and Freon® TMS. Freon® TMS and Freon® TM (an unstabilized methanol-FC-113 azeotrope) clearly exhibit the best particle removal of any of these solvents. Freon® TMS is a nonflammable mixture and can be boiled and recovered at a low energy input due to the low latent heat of vaporization of the principal component, FC-113. Furthermore, the Freon® TMS azeotrope is a rather mild solvent and does not attack most polymeric materials nearly so readily as pure methanol does.

It must be emphasized that much better removal can be effected if higher g-forces are employed. For our purposes, it was easier to use lower g-values to demonstrate effects for a particle size range in which size determinations are easier and more accurate and where discrimination among solvents could be more easily discerned. Of course, many practical cleaning problems have the requirement that particles in the 1-10 μ -range be virtually completely removed; and, to accomplish that objective, strong removal forces are needed even when the adhesion has been greatly diminished by proper solvents. Stowers[3] has

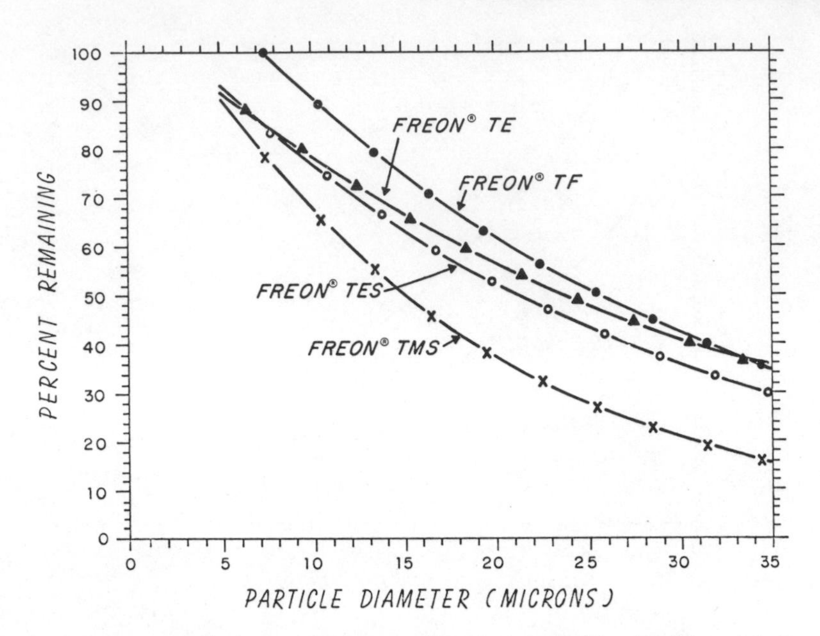

Figure 2. Effect of solvents on adhesion of diamond particles to glass at 1g force.

pointed out the effectiveness of high pressure trichlorotrifluoroethane liquid jets in removing 5-10μ Al_2O_3 particles from glass and metal surfaces.

There are also other significant advantages in the use of the Freon® TMS azeotrope as opposed to pure FC-113. Static electricity charges do not readily dissipate when FC-113 is the medium; whereas, with the polar methanol component and concomitant trace water in Freon® TMS, dissipation is rapid. Another advantage of Freon® TMS over FC-113 is that particles tend to stay suspended in Freon® TMS, so that redeposition is greatly lessened.

In summary, it is evident that very small particles are not subject to removal by mechanical action unless the adhesive forces are diminished by the action of properly chosen solvents. Practical solvents which are safe and effective are now available.

REFERENCES

1. H. C. Hamaker, Physica, IV, 1058 (1937).
2. H. Krupp, Adv. Colloid Interface Sci. 1, 111 (1967).
3. I. F. Stowers, J.Vac. Sci. Technol. 15, 751 (1978).
4. S. Bhattacharya and K. L. Mittal, Surface Technol., 7, 413 (1978).

ELECTROSTATIC EFFECTS IN THE ADHESION OF POWDER LAYERS

J. A. Cross

Wolfson Applied Electrostatics Advisory Unit
Department of Electrical Engineering
University of Southampton
Southampton, England

The adhesion of powder layers deposited onto metallic and non-metallic substrates was measured using a centrifuge. The powder was applied by a number of different techniques which charged the powder in different ways. The effect of the electrostatic force on adhesion could therefore be investigated. A summary of the forces of adhesion which could act on the particles is given. It is shown that the forces which hold the layer together are complicated and that electrostatic effects play an important part.

INTRODUCTION

The work carried out at Southampton University on the adhesion of powder layers was initially performed to further understanding of the electrostatic powder coating process. This technique is applicable only to insulating materials but the results obtained were unexpected and measurements were extended to a wide range of different powders in an attempt to gain insight into the fundamental adhesion mechanism. The results are therefore relevant to many industrial activities in addition to electrostatic coating and precipitation e.g. filling operations often have a problem with frictionally charged powder sticking to the outside of containers and powder transport can be affected by layer build-up on the pipe walls.

MECHANISMS OF ADHESION

Assuming that the powders were dry and the relative humidity low, there are three primary mechanisms which could supply the force of adhesion between particles in a powder layer. The maximum probable magnitude of these forces can be estimated.

Coulomb Forces

A single charged particle in a charged powder layer is held onto the surface by the combined action of the repulsive forces of particles deposited on top of it and the attraction of the image charges in the substrate. The combined action of these other charges provides an electric field, E, which cannot exceed the breakdown limit of air (3.10^6V/m)[1]. In fact in some cases during deposition this limit is reached and discharges occur to hold the field to the air breakdown limit.

The maximum Coulomb force F_c on a particle of charge, q is therefore given by the equation

$$F_c = qE = 3.10^6 q$$

The theoretical maximum charge on a dielectric particle of radius a charged in a corona discharge is $4\pi a^2 EC\varepsilon_o$ where C is a constant depending on the dielectric constant ε_r

$$C = 2\,\frac{\varepsilon_r - 1}{\varepsilon_r + 2} + 1$$

For a 40μ particle this is 6.10^{-14}C/particle, see Pauthenier[2]. Practical values for frictional or corona charging are normally an order of magnitude lower than this. Using the theoretical maximum charge, the greatest possible force on a 40μ particle which can be explained by Coulomb adhesion is therefore 1.8 10^{-7}N.

Dipole Forces

If the particles are nonuniformly charged so as to have a dipole moment, p, there is a force in a nonuniform electric field. This force is given by $F_d = p\frac{dE}{ds}$ where $\frac{dE}{ds}$ is the field gradient[3]. The field at the surface of the layer is close to zero and at the substrate reaches 3.10^6V/m; hence, across a 100μ layer there is a field gradient of 3.10^{10}V/m^2. The induced dipole moment on a particle of radius, a, in an electric field, E, is

$$\frac{4}{3}\pi a^3 (\varepsilon - 1)\varepsilon_o E$$

where ε is the dielectric constant and ε_o the permittivity of free space. For a maximum field of $3 \cdot 10^6$V/m the induced dipole moment is 1.10^{-18}Cm. The dipole force F_d is thus 3.10^{-8}N. If the field gradient is nonuniform, this could be increased.

Van der Waals Forces

Molecular forces act between any two molecules and are due to the dipole moment induced in one molecule by the instantaneous dipole moment of the other[2]. The force varies with distance according to a d^{-7} law at distances less than 20 nm and d^{-8} law at larger distances, see Israelachvilli[4]. The force between two particles is the integrated effect of all the intermolecular forces and depends critically on the true contact area between the two particles. Bradley[5] has shown that for two perfectly smooth spheres of radius, a, the adhesion is given by

$$F_m = \frac{Aa}{12{z_o}^2}$$

where A, is the Hamaker constant ($\sim 10^{-19}$J) and z_o is the separation constant ($\sim$4nm). For two perfectly smooth spheres of 40μ diameter, substitution gives

$$F_m \simeq 10^{-6}\mathrm{N}$$

The electrostatic attraction and van der Waals force together could cause some flattening which would increase this value. It is impossible to make an accurate estimate of the strength of molecular forces because of the difficulty of estimating the contact area. However, adhesion of up to a few 10^{-6}N would be possible for smooth particles. In practice particles will not be perfectly smooth and surface asperities are likely to increase to $\sim$10nm considerably reducing F_m. It can be seen that for irregular particles it is not possible to be certain which of the three forces will be dominant from theoretical considerations alone.

ADHESION MEASUREMENT

There are a large number of different techniques for applying a force to give a measure of adhesion. At Southampton a centrifugal technique was chosen as this applies a normal force which can be directly interpreted as an adhesion. The apparatus is shown in Figure 1.

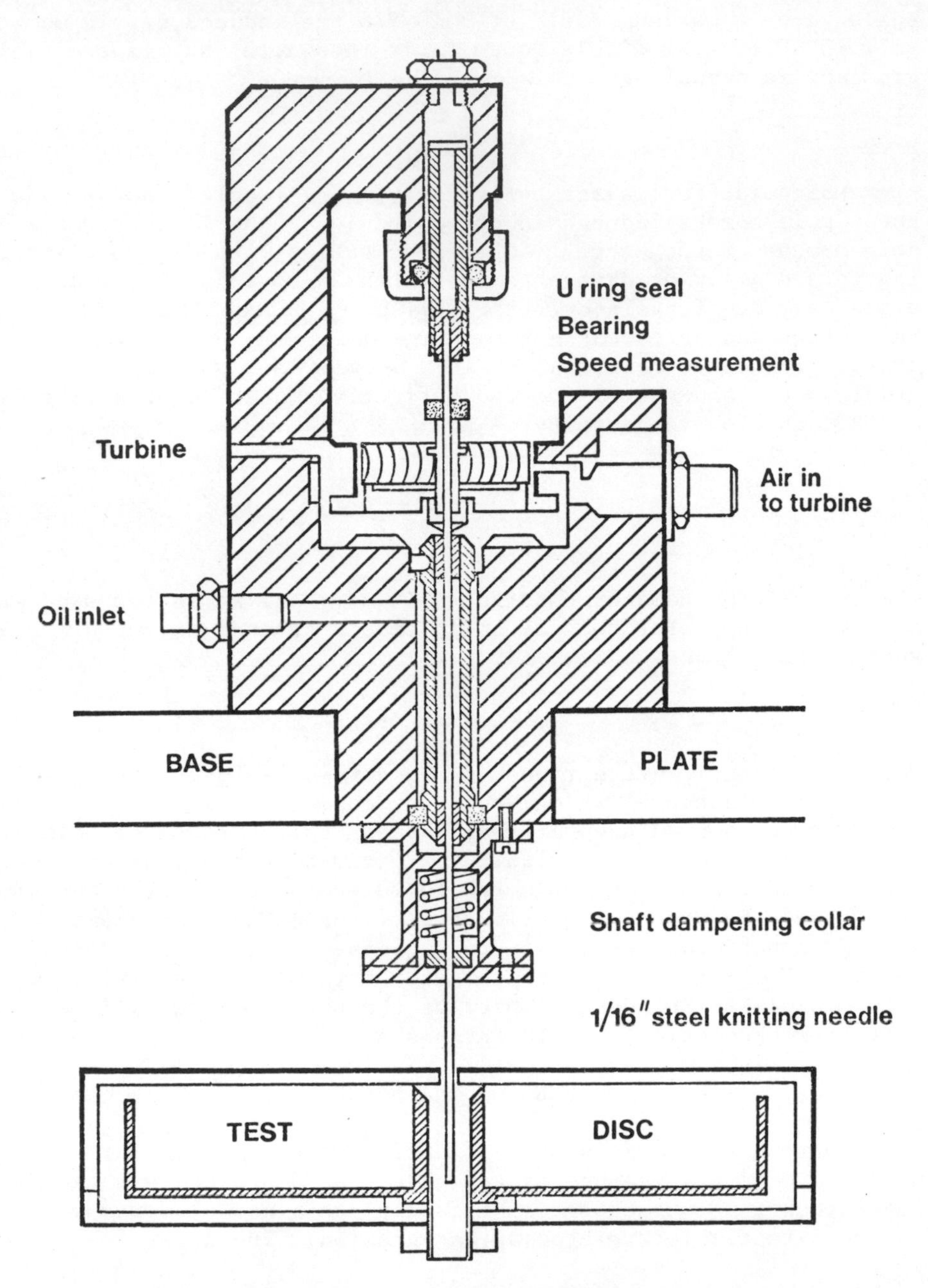

Figure 1. Ultracentrifuge

A disc 13cm in diameter and 2.5cm. deep was coated with powder on its outer edge. The disc was then spun on the centrifuge and weighed to obtain a measure of the amount of powder removed. The disc was usually metallic but glass and painted surfaces were also tested. The apparatus was built from drawings given by the National Gas Turbine Establishment where a similar piece of equipment is used for testing rotor blades to destruction. The shaft is a $^1/_{16}$ inch flexible needle. The disc is able to find its own centre of gravity and no accurate balancing is required. Speeds up to 9000 r.p.m. were obtained in air and up to 20,000 r.p.m. could be achieved when the chamber was evacuated. Evacuation of the test chamber did not affect the adhesion measurement. The rather large disc was chosen so it reasonably represented a plane when electrostatic deposition was studied.

It was found that both thick layers and mono-layers of powder were removed progressively as the velocity increased. There were no sudden steps and no unique adhesion could be defined. The results are therefore presented as a graph of mass of powder removed against removal force applied. The removal force is given as an acceleration as multiples of G (the acceleration due to gravity) so that no assumptions need to be made about the size and density of the powder particles. In the majority of the work presented here, powder-powder adhesion was studied with more than a monolayer deposited on the surface. The relative humidity was always less than 30% so water surface tension forces could be ignored[6].

POWDER DEPOSITION

A wide range of powdered materials was deposited using different techniques, some giving neutral and some charged particles.

Corona Charging

Powder was blown through a pipe at the exit of which there was an array of corona points to which a high voltage was applied. These created a current of about 10μA of ions of one polarity which charged the powder by attachment. The powder was charged to a mean level which could be measured and was accompanied by a stream of free ions. This method of charging can be applied to conducting or insulating powders.

Induction Charging

Induced charge can be applied only to conducting powders. Two electrodes 10cm in diameter were placed one above the other at an angle so the spacing varied between 2 and 3cm. Powder on the lower electrode (at +5kV) acquired a positive charge from it and was then attracted to the other electrode held at the opposite polarity (-5kV). Here the particles discharged and recharged negative so they were repelled back to the first electrode. The powder bounced between the two electrodes moving to the lower field region until it was deposited on the disc. The charge could be measured by catching the powder in a faraday cup.

Friction Charging

High speed rubbing was used to give powder a frictional charge. This was applied by a fan devide shown in Figure 2.

RESULTS

Insulating Powders

The charge decay time of a powder can be shown to be given by $\varepsilon\varepsilon_o\rho$ where ε is the dielectric constant, ε_o the permittivity of free space and ρ the powder resistivity. The first class of mater-

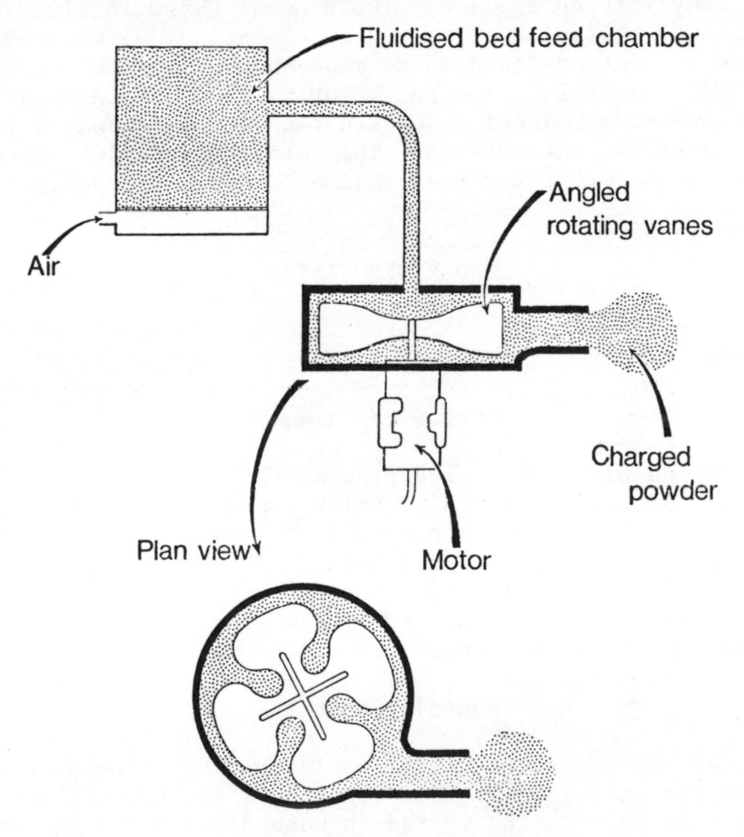

Figure 2. Schematic presentation of assembly used for charging powder

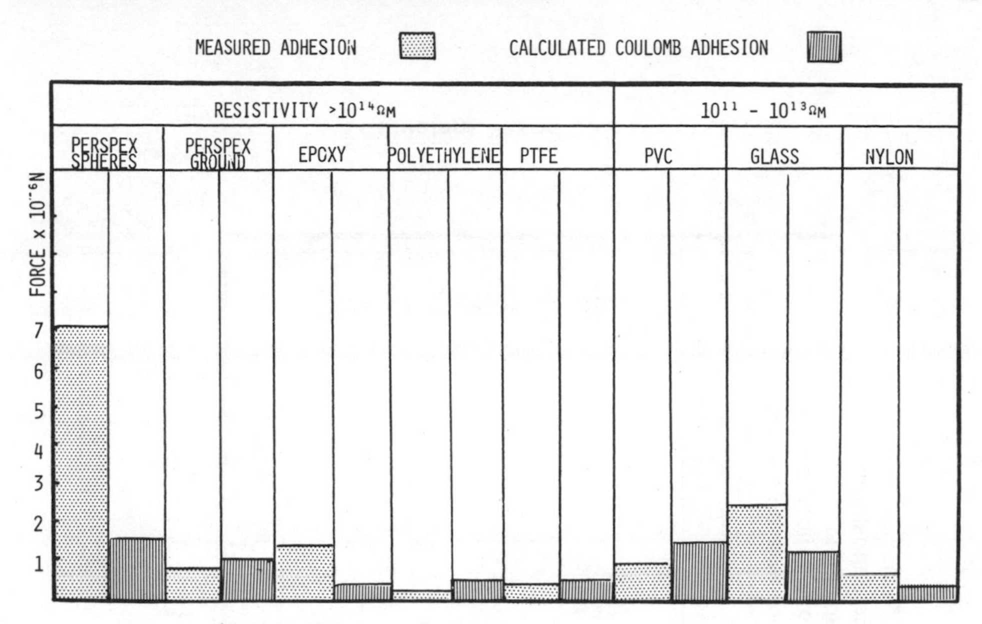

Figure 3. Adhesion of resistive powders.

ials tested were those with resistivities exceeding $10^{14}\Omega m$ which retained their charge throughout the duration of the adhesion measurement. Figure 3 shows the mean adhesion force in Newtons for several powders compared with the maximum possible Coulomb force. As discussed previously the particles were not of one size and there was no unique adhesion value. The mean adhesion force is defined as the force/particle required to remove 50% by weight of the layer. It can be seen that shape and material are both important. Low values of adhesion were obtained for PTFE and polyethylene where low molecular forces would be expected. The figure also shows that in some cases the measured adhesion force exceeded the maximum Coulomb force calculated according to the equation given above.

Figure 4 shows the effect of different charging methods. Induction charging cannot be used for insulating powders so only tribo and corona charging are shown. The charge applied by friction exceeded the corona charge but adhesion was considerably lower. A test was carried out with a high field but no ions to see whether the free ions or high electric field present during corona deposition affected adhesion and it was found that there was very little increase of adhesion when a frictionally charged layer was treated with a high field with no ions but that an ion current considerably improved adhesion.

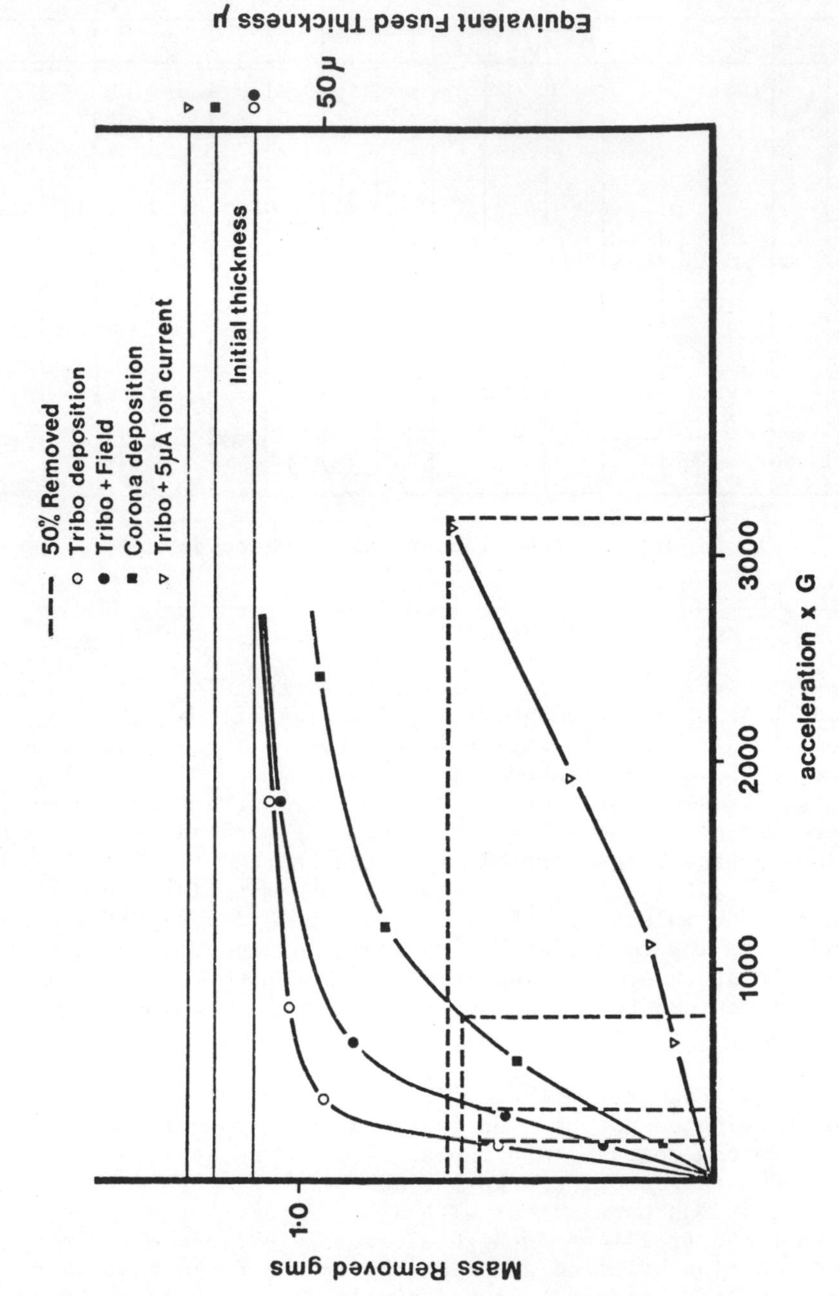

Figure 4. Adhesion as a function of deposition method.

Intermediate Resistivities

Powders between 10^{11} and $10^{13}\Omega m$ are fairly resistive. The electric field builds up to high values in the deposited layers so the powder charge decays at least partly by means of discharges but all charge had been lost by the time the adhesion was measured. The adhesion of corona deposited particles is shown in Figure 3. It was not significantly lower than the particles which retained their charge. It can be seen that the highest adhesion values were for the smooth spheres.

Low Resistivity Powders

When the powder resistivity was between $10^{10}\Omega m$ and $10^{7}\Omega m$ it was found that the charge decayed very rapidly without discrete discharges. The adhesion of all powders in this class was an order of magnitude lower than that of the higher resistance materials. Adhesion curves are shown for calcium tungstate on different substrates in Figure 5. The adhesion was highest on the paint substrate where charge could not decay away. For the metal substrate corona deposition gave considerably more adhesion than charging by induction even though the charge decayed instantaneously. Tribo charging could not be tested as the adhesion was not sufficient to withstand the wind produced by the depositon device. A powder with resistivity below $10^{6}\Omega m$ (iron) would not charge effectivly by induction. The charge was found to be lower by 3 orders of magnitude and the powder would not adhere sufficiently for the disc to be carried to the centrifuge for measurement. Corona charging again produced a considerable improvement in adhesion but it did not reach the level of the more resistive copper or calcium tungstate.

Polarity Dependence

All the powders tested except the highly conducting iron powder adhered better when charged positive than negative, e. g. Figure 6. This applied to all charging methods. It was found that the magnitude of the charge on the particles did not depend significantly on polarity, see J. Sells[7].

Since it appeared that the nature of the charge carriers was important, tests were carried out with powder charged by corona in a nitrogen atmosphere. With positive polarity no difference would be expected from the results in air but with negative polarity the charge produced by the corona would be electrons rather than negative ions. The current increased by nearly three orders of magnitude because of the high mobility of electrons compared with ions. The adhesion curves for an insulating powder are shown in Figure 7. The use of nitrogen improved the adhesion at negative polarities

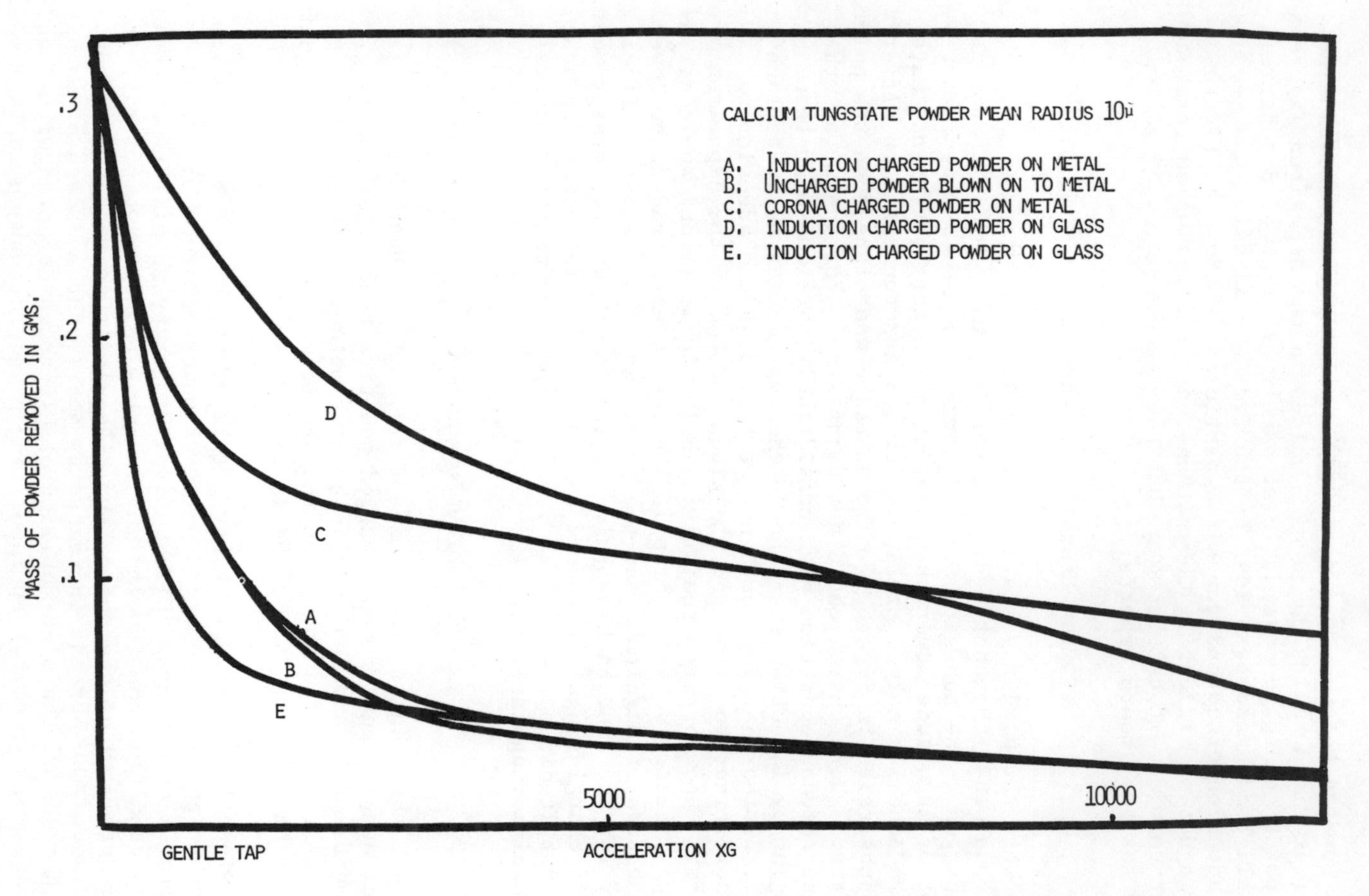

Figure 5. Adhesion curves for calcium tungstate powder.

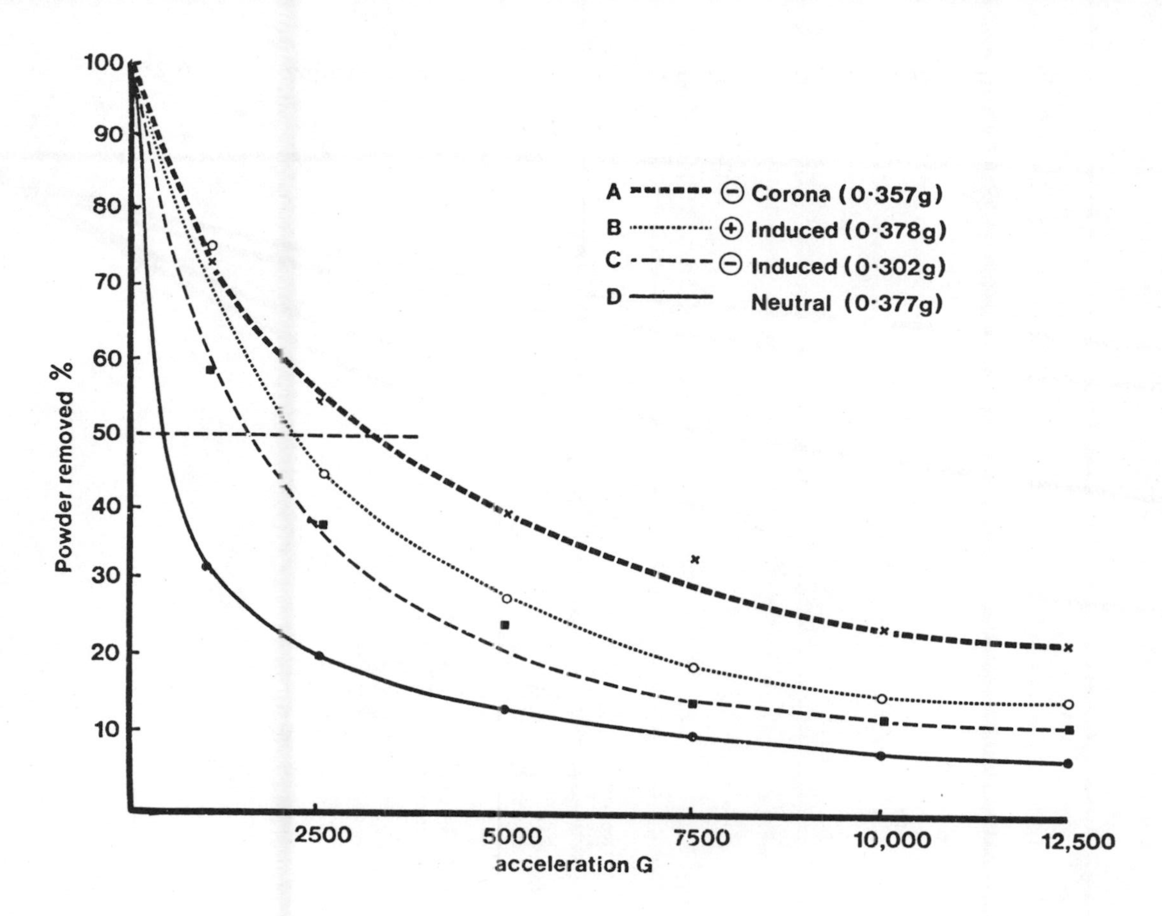

Figure 6. Adhesion characteristics of copper powder on cellulose for corona and induction charge and neutral powder.

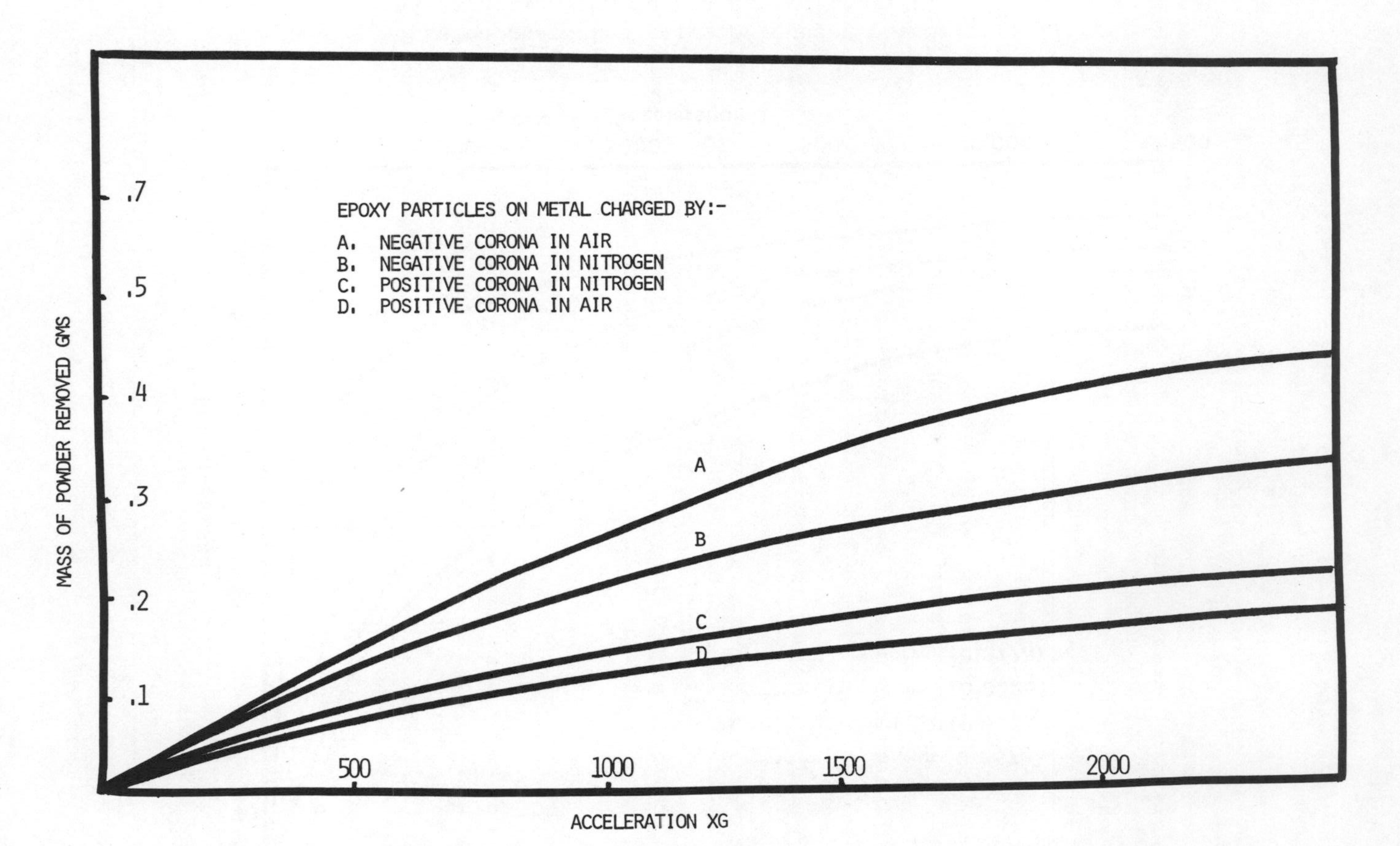

Figure 7. Adhesion of powder deposited in nitrogen environment.

but not so it exceeded the adhesion with positive polarities in air.

DISCUSSION

In summary the tests showed:

(i) The measured adhesion exceeded the maximum Coulomb attraction calculated from the equation given above. It was also higher than would be expected from van der Waals forces for rough particles.

(ii) The adhesion of spheres was higher than the adhesion of uneven shaped particles.

(iii) Better adhesion was obtained if the particles were charged, even when the charge decayed to earth before the adhesion measurement was made.

(iv) Positively charged particles were more adherent than negative except for the highly conducting iron powder where the tendency was slightly the other way.

(v) Corona charged powders were more adherent than induction or frictionally charged powders and this was shown to be due to the presence of ions rather than of the high electric field.

(vi) In order of strength of adhesion the charge carriers should be positive ions, electrons, negative ions.

(vii) A very highly conductive material (iron powder) showed lower adhesion than slightly less conducting materials (e.g. copper).

(viii) There was a clear indication that the adhesion force was directly related to the extent to which the powder discharged by means of discharges (back ionization). Cross[1] found an increase in adhesion as back ionization current increased.

It is extremely difficult to postulate an adhesion mechanism which fits all these results. Since high adhesion levels could be obtained with some materials after the charge had decayed, Coulomb attraction could not be the dominant force. However, the charge initially on the powder played an extremely important part. After the early work with low melting point plastics, showing a direct dependence of adhesion on back ionization current, a localized welding mechanism was postulated. However, the highest current deposition (corona charging in nitrogen) did not give the highest adhesion and the adhesion did not decrease for the high melting point metals. It is therefore felt that this mechanism is unlikely.

It has been suggested that the differences in adhesion are caused by differences in packing under the influence of charge but the strong increase in adhesion when a frictionally charged layer of powder was treated with corona ions contradicts this. The amount of charge on the particle and the manner in which it discharges obviously plays an extremely important part. It is possible that when the powder discharges the process involves addition of charges of the opposite polarity rather than true neutralization. This is certainly probable for insulating powders where it is known that charge frequently becomes injected below the surface. The discharge process can involve a complete polarity change, see Cross[8], with powder particles being thrown off the surface and on this basis partial polarity change must be considered to be a possibility. A strong dipole moment could result which with local field nonuniformities could increase the adhesion due to dipole forces. This mechanism would not apply to highly conducting powders which were found to have very low adhesion.

CONCLUSIONS

The mechanism of adhesion in charged powder layers is obviously complicated and it is extremely difficult to postulate a mechanism which fits all the observations.

Measured adhesions are higher than would be expected. Although Coulomb attraction cannot be the dominant force of adhesion, the amount and polarity of charge and the manner in which the particle discharges at the surface plays a strong role. It is postulated that the adhesion is increased by inter-particle effects with particles acquiring a dipole moment during discharge and localized non-uniformities in the layer. Alternatively, one must look for a considerably increased true contact area due to the discharge process.

REFERENCES

1. J. A. Cross, in "Proceedings of Conference on Static Electrification", A. R. Blythe, Editor, p.202, Institute of Physics, London, 1975.
2. M. Pauthenier, J. Phys. Radium Ser., 7,590 (1932).
3. B.I. Bleaney, "Electricity and Magnetism", Clarendon Press, 1959.
4. J.N. Israelachvilli, Contemporary Phys., 15 159 (1974).
5. R.S. Bradley, Trans Faraday Soc., 32, 1088 (1976).
6. N.L. Cross and R.G. Picknett, Trans Faraday Soc., No. 484 59 (4) 846 (1963).
7. J. Sells, 3rd year project report, Electrical Engineering Dept. Southampton University (1978).
8. J.A. Cross and J.D. Bassett, Trans Inst. Metal Finishing 52, 112 (1974).

CAUSES AND CURES OF DIRTY WINDOWS

P. B. Adams

Corning Glass Works

Corning, New York 14830

Glass surfaces that are exposed to the atmosphere degrade by two processes usually acting in combination with one another: weathering and particulate deposition. Both processes are influenced by the glass surface composition which may be quite different from the bulk composition. There are a variety of options available for cleaning. A self-cleaning surface is most desirable. Optimum cleaning choices as well as "self-cleaning" characteristics are also a function of glass composition.

INTRODUCTION

The degradation in performance of glass products that are used outdoors is of considerable practical importance. Loss in the clarity or transmission of ordinary window glass is only one example. Lighting fixtures that decline in output as a result of weathering mean lost energy. Glass panels that lose their gloss are no longer aesthetically appealing; worse yet if they fail mechanically because of environmentally induced changes. The expanding field of optical waveguides brings into focus the question of how long such devices will perform when buried in the ground. Solar reflectors and collectors that employ glass components must continue to perform at high levels of efficiency if they are to be practical energy collecting systems which justify the enormous capital outlay in construction. The necessity to understand the complex phenomena of degradation of glass products because of environmental factors is clear.

"Dirty" glass surfaces result from two factors usually acting

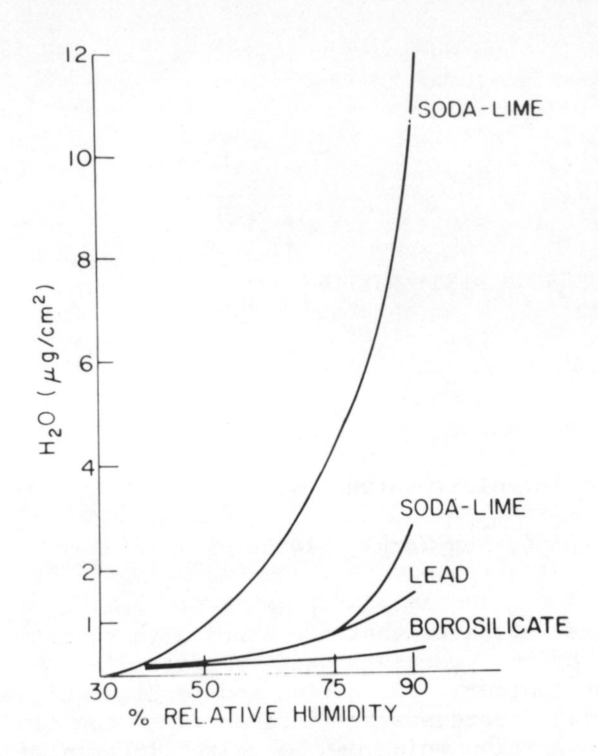

Figure 1. Water adsorbed on several glasses as a function of relative humidity after 7 days (after Walters and Adams, ref. 1).

in combination with one another: glass can be weathered, i.e., chemically attacked by environmental agents or airborne contaminants may adhere to the glass surface. If these two processes are understood, it is easier to design useful products and design optimum cleaning methods.

This paper will describe weathering - what it is and how it depends on glass composition -, define types of "dirt" and adhesion mechanisms, discuss the interactions between dirt and weathering processes, refer the reader to cleaning principles and illustrate why glass composition is important.

WEATHERING

Weathering is the interaction of a glass surface with chemical agents in the atmosphere. The most prevalent agent in this process is water. Others include a variety of chemicals ranging from alkaline dust to salt spray to "acid" rain.

Consider weathering by water.[1] Water is adsorbed onto the

glass surface, the quantity being a function of relative humidity and glass type as illustrated in Figure 1. Water then reacts with the glass surface by a sequence of steps involving leaching (ion exchange) and etching (first order reaction), both of which have been described in another paper in these proceedings volumes.[2] The process is shown schematically in Figure 2.

Leaching of the glass surface may actually enhance weathering resistance. Since leaching involves ion exchange, alkali is removed leaving a silica-rich surface layer. If this layer is mechanically stable, it acts as a barrier to further chemical attack. This plus the fact that reaction products are washed away accounts for the practical utility of ordinary window glass. However, if leaching is severe, i.e., too deep or too little silica remains, the surface will be visibly dulled and iridescent or may craze and spall.

When alkaline reaction products are not removed, the next and more serious stage of weathering occurs. The alkali reacts with the silica network to totally disrupt and destroy the glass structure.[3] In the initial stage, this will be evidenced by a cloudy surface that eventually becomes roughened and covered with tenacious reaction products. Only mechanical polishing can now restore the surface. For a further discussion of the role of water in the weathering process as well as for a description of methods of measuring weathering, see reference 1.

Outdoor weathering exhibits three other factors. The outdoor weathering process is often cyclic with respect to humidity and temperature, chemical agents in the atmosphere may be involved, and mechanical processes such as the shedding effect of melting ice can be present.

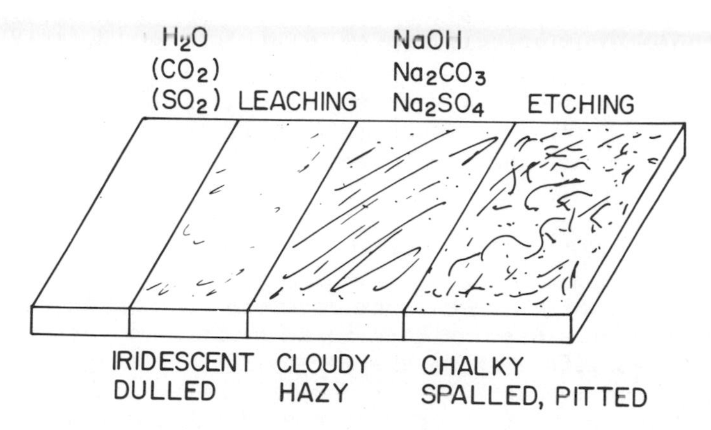

Figure 2. Schematic representation of weathering process.

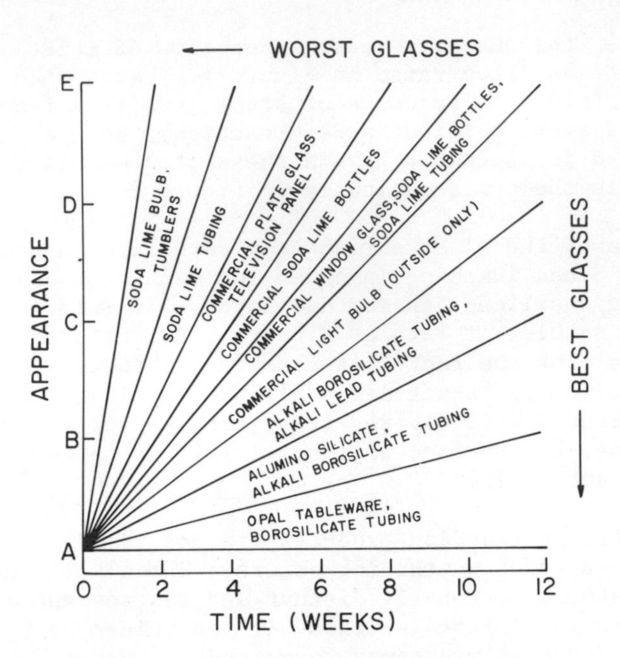

Figure 3. Average weatherability of various glasses at 98% relative humidity and 50°C (after Walters and Adams, ref. 1).

Cycling relative humidity, especially when coupled with periods of water runoff, can result in a surface that is more stable than the original surface. Conversely, cyclic conditions can result in a layer that cracks and sheds on drying if the silica content of the glass is too low to provide mechanical stability.

The action of other chemical agents can be considered in the context of the prior discussion of leaching and etching. A more complete discussion is available in other references.[3,4,5]

Glass composition is a most important factor. Figure 3 shows the range of weathering resistances that is observed for a variety of commerical glasses when evaluated in a laboratory test.

DIRT

A discussion of the mechanical factors involved in dirt accumulation has been presented by Berg.[6] He points out that three factors, convective diffusion at the air-surface boundary layer, sedimentation and impact all play a role in the sticking of dirt to glass surfaces. He further states that adhesion mechanisms exhibit increasing force in the order: gravity, electrostatics, double layer charge, surface energy, capillary force and chemical bonding.

Particle size plays an important role in deposition and bonding of dirt to a glass surface. Figure 4 is a superposition of data from Brandreth[7] and Berg.[6] The plot of Stokes velocity versus size shows that the speed necessary to prevent a given particle from falling out of the wind stream increases with particle size. The plot of relative adhesive force versus size shows that smaller particles adhere more tenaciously. To illustrate -- a one micron dust particle will stay suspended at 1/100th the air speed required for a 10 micron particle. However, if that particle lands on a surface it will adhere with 100 times the force of a 10 micron particle. Thus, a ten-fold decrease in size results in a 10,000 fold increase in the probability of finding a particle on the surface. Obviously, this applies to an idealized totally dry system.

The glass surface provides many active sites. The silanol radicals with their dangling OH groups offer an opportunity for a variety of chemical reactions to occur. Figure 5 shows the effect of pH on adhesion of clay to a soda-lime window glass. Glue-like processes may also often be involved - as will the surface tension effects of water.[8]

INTERACTIONS BETWEEN WEATHERING AND DIRT

The chemical corrosion processes associated with weathering produce reaction products as well as a surface that is changed

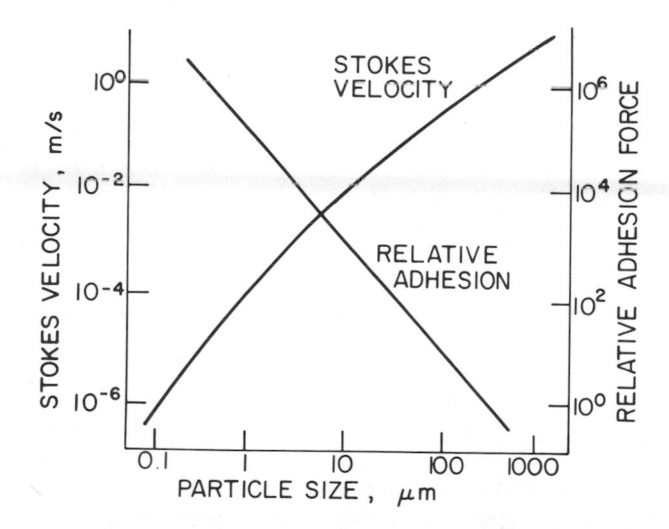

Figure 4. Particle size versus Stokes velocity and relative adhesive force (after Berg, ref. 6 and Brandreth and R. E. Johnson, Jr., ref. 7).

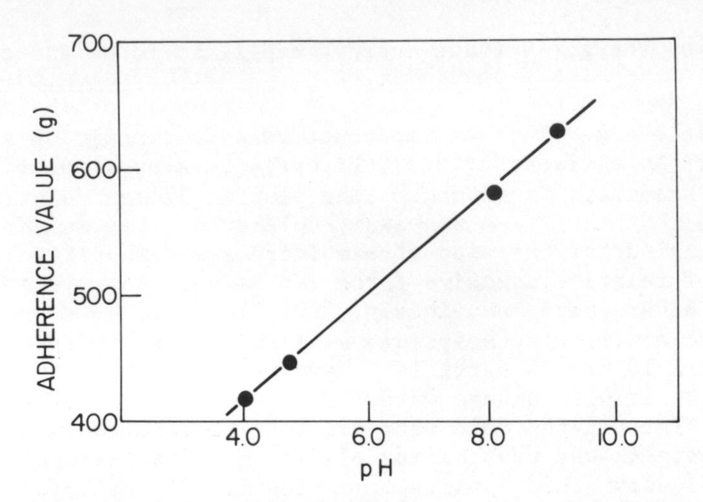

Figure 5. Strength of adherence between clay and window glass as function of slurry pH. (after Anderson et al, ref. 9).

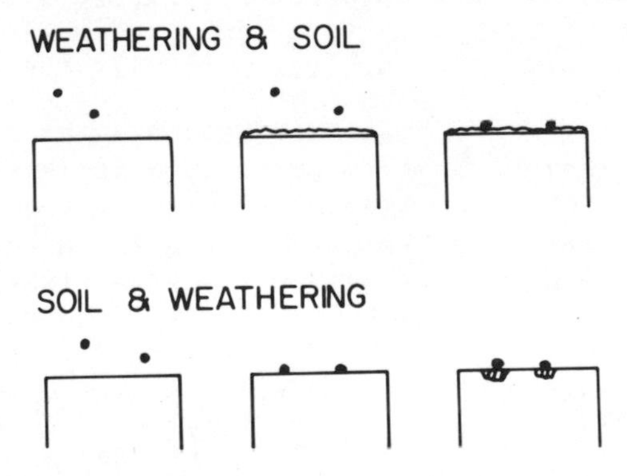

Figure 6. Schematic of interaction processes between weathering and soil.

physically and chemically. Dirt from the atmosphere is thus more likely to adhere. This implies that a glass which tends to weather less will also tend to accumulate less dirt.

The two processes are by no means independent in time. As weathering influences the collection of dirt, so dirt influences the progress of weathering; i.e., the two processes interact chemically, it is impossible to sort them out in the practical case. Two extreme views of the interaction are presented schematically in Figure 6.

The effect of relative humidity on weathering has been previously discussed. It should be noted that there is an additional effect. The capillary condensation of water between a particle and the surface increases the physical bonding as shown in Figure 7 and as discussed by Bhattacharya and Mittal.[8] The net effect is also to increase the total water available for reaction.

CLEANING

Design of a cleaning process should be done with sound technical input. An approach to this problem has been proposed by an ASTM committee.[10] If the surface has been mechanically damaged, i.e., weathering has proceeded to the point of etching, chalking, spalling, etc. then no amount of cleaning will restore it. Only a mechanical process such as buffing or polishing will succeed.

Self-cleaning ability is a desirable property. It is obvious that a glass with poor chemical durability will leach, craze, weather and will accumulate more dirt. A glass with good durability which retains a smooth pristine surface will tend to shed dirt.

Environmental factors, be they chemical, electrostatic or mechanical, will also interact with the glass surface to affect "self-cleaning". Some environmental chemical agents will be good chemical solvents that do not alter the glass; others will produce insoluble reaction products or will react with the glass. Electrostatic charges present will vary depending on the mode of installation and the conditions of use. The mechanical action of ice and snow as it sheds from the glass surface may be a very positive factor in removing dirt.

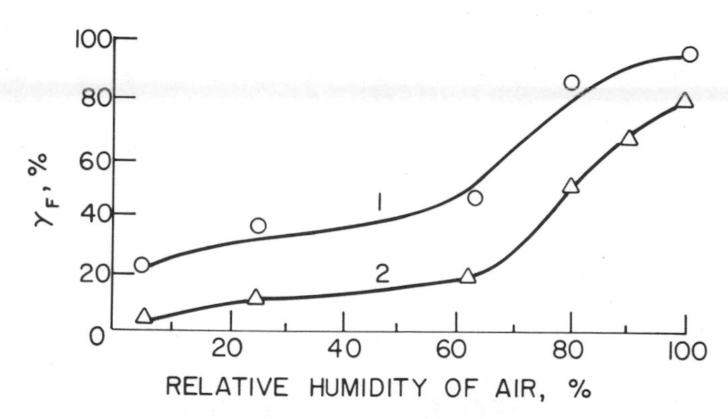

Figure 7. Influence of atmospheric humidity on the forces of particle adhesion: 1) particles 40 - 60μ, removal force $2.25 \cdot 10^{-1}$ dyne; 2) particles 20-30μ, removal force $2.81 \cdot 10^{-2}$ dyne. (after Zimon, ref. 11).

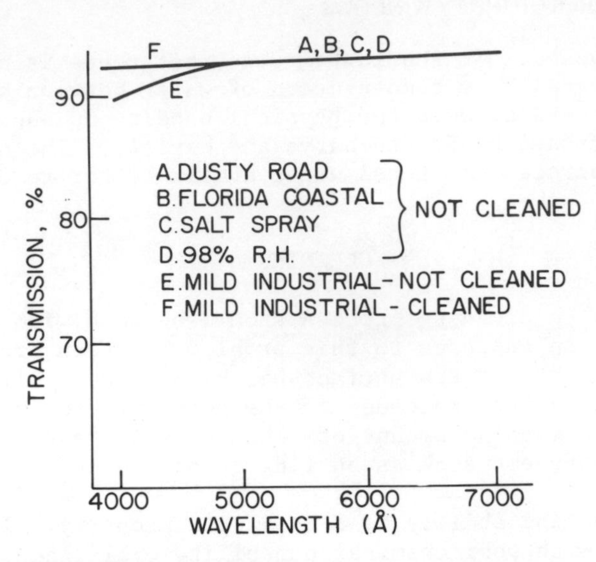

Figure 8. Effects on the transmission of a borosilicate glass caused by outdoor and laboratory exposure tests.

GLASS EXPOSURE EXPERIMENTS

The role of glass composition in dirt accumulation, weathering and self-cleaning can be seen from the experimental results shown in Figures 8 and 9. Flat plates of two glasses, one a chemical borosilicate glass and the other a conventional soda-lime glass were used. The relative weathering resistances of these glasses when tested in the laboratory at 98% RH and 50°C are shown in Figure 3. Typical chemical resistance results obtained by the US Pharmacopoeia powder extraction test in dilute acid[3] are 0.005% for the borosilicate and 0.03% for the soda-lime. In strong alkali (5% NaOH) both lose one mg/cm^2 at 95°C in 6 hours. These results demonstrate that the soda-lime is more readily leached than the borosilicate but that both glasses suffer damage in alkaline media.

Samples of both glasses were exposed in several locations and laboratory devices:

1. "Mild Industrial", on a factory rooftop in Corning, New York for 2 years.

2. "Dusty Road", about one foot above ground level along a dirt road in central New York State for 3 years.

3. "Florida Coastal", 100 yards inland on east coast rooftop for 3 years.

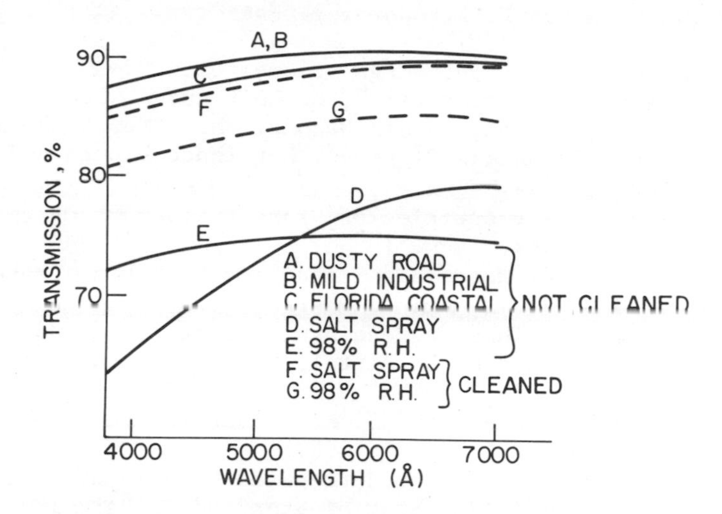

Figure 9. Effects on the transmission of a soda-lime glass caused by outdoor and laboratory exposure tests.

4. "Weathering Chamber", at 98% RH and 50°C for 1 month.

5. "Salt Spray Chamber", 5% salt fog at 35°C for 1 month.

The borosilicate glass, Figure 8, showed virtually no change during the tests. It can be characterized as very weathering resistant, with a low tendency to accumulate dirt and a high "self-cleaning" ability.

The soda-lime glass showed no change during some tests. But, the salt spray test left an accumulation of material that dropped transmission by about 15% although it could be removed. The high relative humidity test damaged the glass beyond restoration. The soda-lime glass exhibits a greater tendency to weather and a lesser degree of self cleaning ability than the borosilicate.

These experiments are not comprehensive. They do substantiate the premise that chemical durability is associated with the causes and cures of dirty windows.

CONCLUSION

"Dirty" glass surfaces that impair functional properties of glass products used in outdoor exposures are caused both by environment and the inherent characteristics of the glass. They can be minimized by proper choice of glass.

1. H. V. Walters and P. B. Adams, J. Non-Cryst. Solids, 19, 183 (1975)
2. P. B. Adams, This proceedings volume, pp. 327-339.
3. P. B. Adams, in "Ultrapurity", M. Zeif and R. Speights, Editors, pp. 293-351, Marcel Dekker, Inc., New York 1972.
4. M. B. Volf, "Technical Glasses", Sir Isaac Pitman and Sons Ltd., London, 1964.
5. L. Holland, "The Properties of Glass Surfaces", Chapman and Hall, London, 1964.
6. R. S. Berg, Solar Materials Technology Workshop, Denver, CO, March 28-30, 1978.
7. D. A. Brandreth, This proceedings volume, pp. 83-88.
8. S. Bhattacharya and K. L. Mittal, Surface Technology, 7, 413 (1978)
9. S. Anderson, D. Tandon, L. B. Kohlenberger and F. G. Blair, J. Am. Ceramic Soc., p. 521 (1969)
10. Proposed New Standard Method prepared by ASTM Committee C14.03 for "Annual ASTM Book of Standards", Part 17, American Society for Testing and Materials, Philadelphia, PA, 1978
11. A.D. Zimon, Colloid J. USSR., 25, 317 (1963)

SURFACE CONTAMINATION BY ION BOMBARDMENT

K. Shimizu and H. Kawakatsu

Electrotechnical Laboratory

Tanashi, Tokyo, Japan 188

In electron and ion beam systems, a surface bombarded with the beam is often contaminated by growth of a thin organic layer. This phenomenon is due to the polymerization of organic molecules adsorbed on the surface by electron or ion bombardment, and the sources of the organic molecules are vapors of diffusion pump oil, grease and other chemicals adsorbed in the inner wall of the vacuum chamber. In this paper, the contamination growth by ion bombardment as compared with electron bombardment is described. The experimental data are compared with a simple theory which takes sputtering into account and good qualitative agreement is obtained. According to this theory, the higher contamination growth by ion bombardment as compared with electron bombardment can be explained by the larger cross section for polymerization of organic molecules adsorbed on the surface. Furthermore, it is ascertained that heating the substrate or using a cold trap surrounding the substrate can effectively reduce the contamination growth.

INTRODUCTION

In electron and ion beam systems, a surface bombarded with the beam is often contaminated by growth of a thin organic layer. This phenomenon is due to the polymerization of organic molecules adsorbed on the surface by electron or ion bombardment, and the sources of the organic molecules are vapors of diffusion pump oil, grease and other chemicals adsorbed in the inner wall of the vacuum chamber.[1] Thus improvement of the vacuum conditions is the best method to reduce the contamination, but it is expensive and would lead to many practical difficulties in designing and operating usual electron and ion beam apparatuses. The contamination layers formed by electron bombardment have been studied by many workers,[1-11] but only a few workers have studied the contamination by ion bombardment.[12,13]

A remarkable feature of contamination growth by ion bombardment as compared with electron bombardment is that a much smaller dose is required. For example, a contamination growth rate of about 0.2 Å s^{-1} for an electron beam of 0.1 mA cm^{-2} has been reported.[6] A similar growth rate is observed by argon ion bombardment with the beam current density of a few μA cm^{-2}.[13] Thus, contamination growth often becomes a serious problem for practical ion beam applications. On the other hand, the contamination phenomenon is often used for a conventional method to obtain the ion beam spot size, emittance, current distribution and other ion beam images, because the contamination thickness depends on the ion density.[12,14] Möllenstedt and Speidel[15] have reported a method of reproducing grids by the use of the contamination shadow pattern produced by ion bombardment and subsequent chemical etching. Another different feature of contamination growth by ion (as opposed to electron) bombardment is that sputtering acts simultaneously, so that the phenomenon is more complicated. In fact if the ion beam is strong enough, no contamination is left on the bombarded surface.[16] However, contamination growth is serious even when the ion beam is strongly focused as in ion-beam processing, because the beam current density is low in the region surrounding the spot.

In this paper, the contamination growth by ion bombardment as compared with electron bombardment is described. As there are only a few experimental results in the literature on the surface contamination by ion bombardment, the data descibed in this paper are mainly the author's results that have been obtained by the bombardment of argon ions with energies varying from a few keV to a few tens keV.[13] In the experiment, the partial pressure of organic molecules was not measured. It is desirable for a theoretical treatment that the experiment be carried out under conditions where the organic atmosphere was precisely known, monitored and capable of control. However, so far as we know, such experiment has not been carried out probably due to the many experimental difficulties. Furthermore, for a practical interest, the knowledge under the usual conditions, where the column or chamber is evacuated by a conventional diffusion pump system, is necessary. Under the conditions, multi-source of the organic molecules exist and the quantities of some parameters that may affect the contamination growth are not exactly known. Thus, the rigorous analysis of the experimental results is difficult, but the quantitative explanation will be given by a simple phenomenological theory.

EXPERIMENTAL METHOD

The experiments were carried out by using an ion-beam apparatus described elsewhere,[17,18] in which Lion-A (a mixture of β-hexadecyl- and octadecyl-naphthalene), as the diffusion pump oil, and silicone grease of Dow Corning Corporation were used. The evacuating system had no liquid nitrogen trap, and the total pressure in the vacuum chamber was about 3×10^{-5} Torr during the experiments.

The contamination layer was formed on an aluminum film about 500-1000 Å thick which had been evaporated on to a flat glass plate. The contamination growth rate can be greatly increased by covering the surface of the substrate with organic material such as vacuum grease and collodion film,[17,19] but this was not done in these experiments. The spot size of the bombarding ion beam was about 7 mm in diameter. The ion-beam current intensity was uniform except for the edge region of the spot. One half of the beam was masked by a flat aluminum plate, which touched the substrate directly so as to obtain a clear boundary line for the thickness measurement. The metal substrate and mask were electrically grounded through a micro-ammeter which monitored the ion-beam current onto the substrate. The ion-beam current was measured by using a Faraday cage before and after the bombardment, and it was 0.5-20 μA. The thickness of the contamination layer was measured interferometrically,[20] and the accuracy of the measurement is estimated to be about ± 50 Å.

EXPERIMENTAL RESULTS

The contamination layer formed on a metal substrate is usually observed as a dark-brown deposit. However the color of the contamination layer is greatly due to the interference of visible light which is reflected from the surfaces of the contamination layer and the substrate. Of course, the contamination layer have its own absorption color which depends on the kind of organic vapor source, but the thin contamination layer formed on a glass plate in our experiment was almost transparent.

Figure 1 shows typical growth curves of the contamination layer with time produced by constant argon ion bombardments on the aluminum substrate at room temperature. The contamination grew almost linearly with time and the growth rates were of the order of 0.1 Å s^{-1}. However, the contamination seemed to grow faster in the early stage of ion bombardment (see the broken curves). There was a remarkably increased growth when the substrate had been left in the vacuum chamber for a long period before the bombardment.

Figure 2 shows a typical variation of contamination thickness against ion current density for a constant bombardment time of 30 min. at room temperature. The contamination thickness increases with low ion current densities but decreases at higher current densities. When the aluminum film was thin enough and the ion current density was high, the aluminum substrate was completely sputtered away. The sputtering of the substrate was particularly noticeable when a silver film was used as a substrate instead of an aluminum film: in this case, more than 1000 Å was completely sputtered away by ion bombardment with a beam current density of only a few μA cm^{-2}. However, the contamination layer remained on the glass plate even when the metallic substrate was completely sputtered away, and the contamination layer on the glass plate was almost transparent. This shows that the sputtering of the target material and the contamination growth occur simultaneously at an early stage of the ion bombardment, which is important from a practical viewpoint becuase it suggests that contamination may remain on the bombarded area even

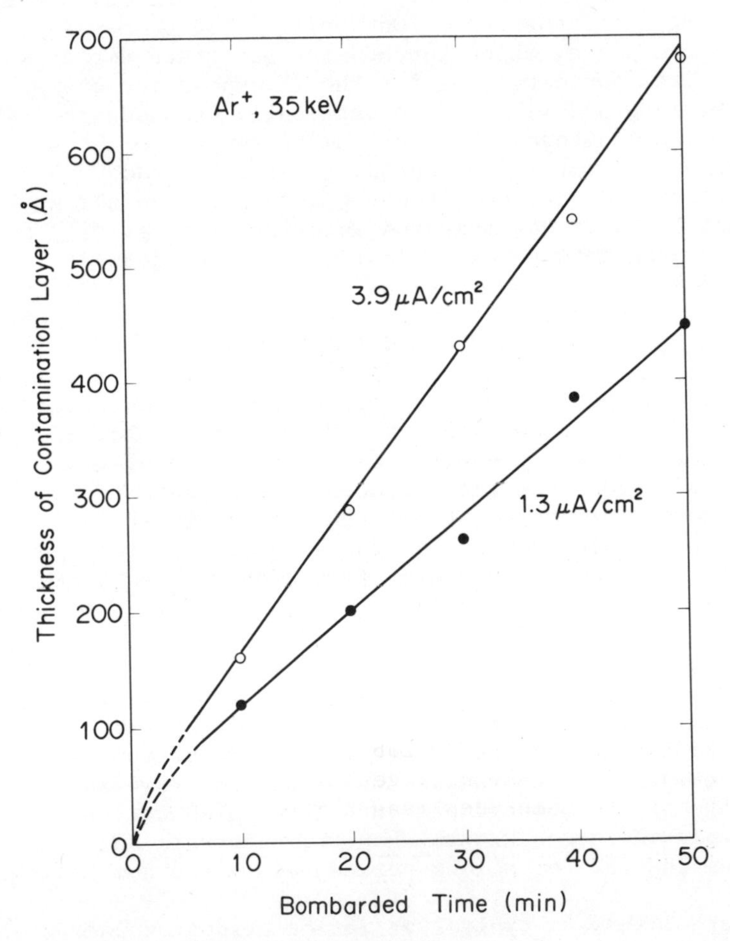

Figure 1. Contamination growth with constant ion current density at room temperature (ion energy: 35 keV).

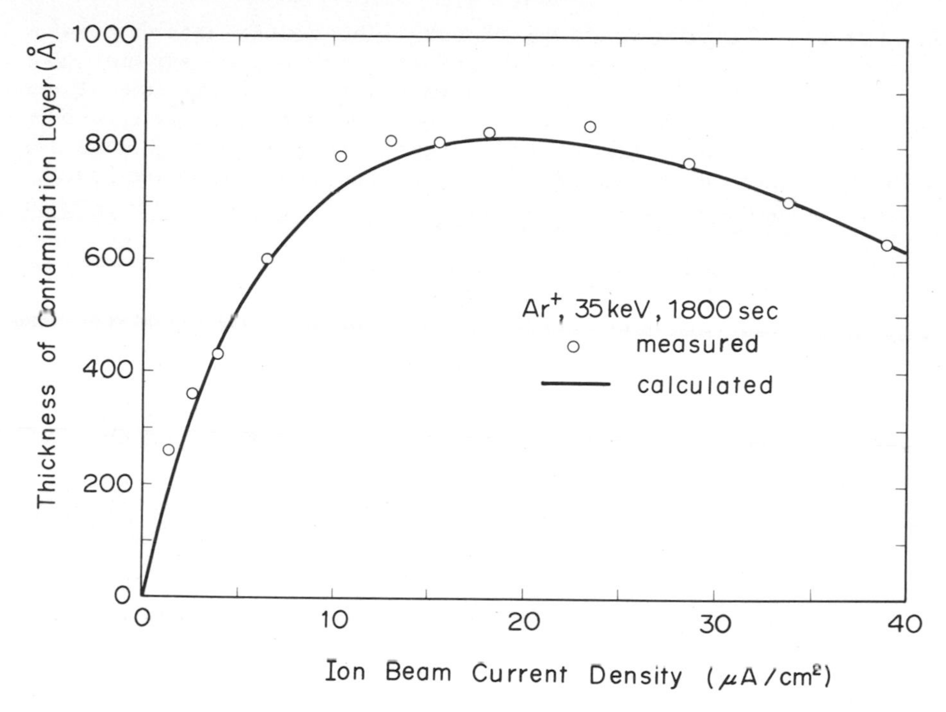

Figure 2. Variation of contamination thickness against ion current density (ion energy: 35 keV; bombarding time: 30 min.).

when the target appears to have been cleanly processed by the ion beam.

The contamination growth slightly depends on the ion energy, as shown in Figure 3. It is noteworthy that the contamination thickness is maximum at the ion energy of about 15 keV. In the case of an electron beam, it has been reported that the contamination in the high-energy region decreases with electron energy.[9]

Figure 4 shows a typical curve of contamination as a function of the substrate temperature. It can be seen that the use of a hot stage effectively reduces the contamination. The use of a liquid nitrogen trap surrounding the substrate also does this. In the experiment using the liquis nitrogen trap, the contamination layer was too thin for a thickness measurement and only the edge of the beam spot could be observed.

OTHER EXPERIMENTAL DATA ON CONTAMINATION BY ION BOMBARDMENT

Speidel[12] has reported temperature dependence of the contamination formed by lithium ion bombardment. The contamination growth decreases with the increase of substrate temperature, similar to Figure 4. The contamination growth rate with the beam current density of 0.14 μA cm^{-2} is 1 Å min^{-1} at room temperature, and it becomes 45 Å min^{-1} when the substrate temperature is - 80 °C and zero at 110 °C. Also the refractive index of the contamination layer, 2.35, has been reported by Speidel.

Möllenstedt and Speidel[15] have been reported that the contamination layer of 150 Å thickness was formed by the bombardment by ions of air with the beam current density of 1 μA cm^{-2} and the energy of 35 keV for 20 minutes.

There are a few other works secondarily concerned with the contamination by ion bombardment, but, as far as we know, no quantitative data have been published.

DISCUSSION

Typical data on contamination growth by electron and ion bom-

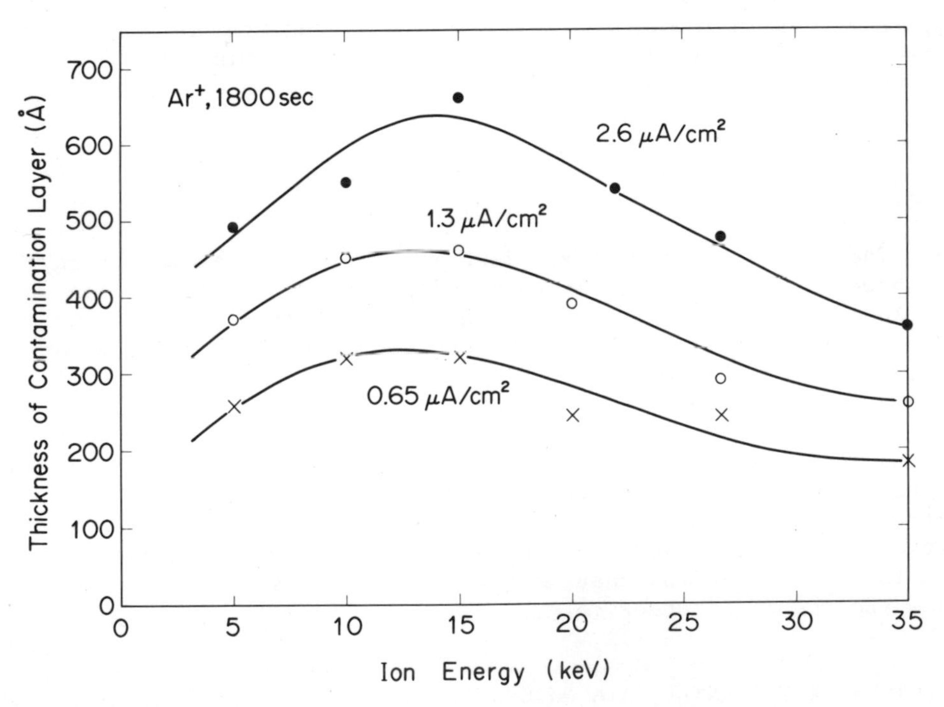

Figure 3. Contamination thickness versus ion energy (bombarding time: 30 min.).

bardment are compared in Table I. These data were obtained in the vacuum chambers evacuated by conventional diffusion pumps to the total pressure of about 10^{-5} Torr. The partial pressure of organic molecules on which the contamination growth directly depends is unknown. Thus the comparison is not rigorous. Nevertheless, it can be seen from the table that the contamination by ion bombardment as compared with electron bombardment occurs with smaller dose of the beam. (The contamination growth rate by ion bombardment decreases with the higher beam current density, as shown in Figure 2. On the other hand, the contamination growth rate by electron bombardment is almost proportional to the beam current density for this region.[1]) In other words, the contamination growth by ion bombardment as compared with electron bombardment is more "sensitive".

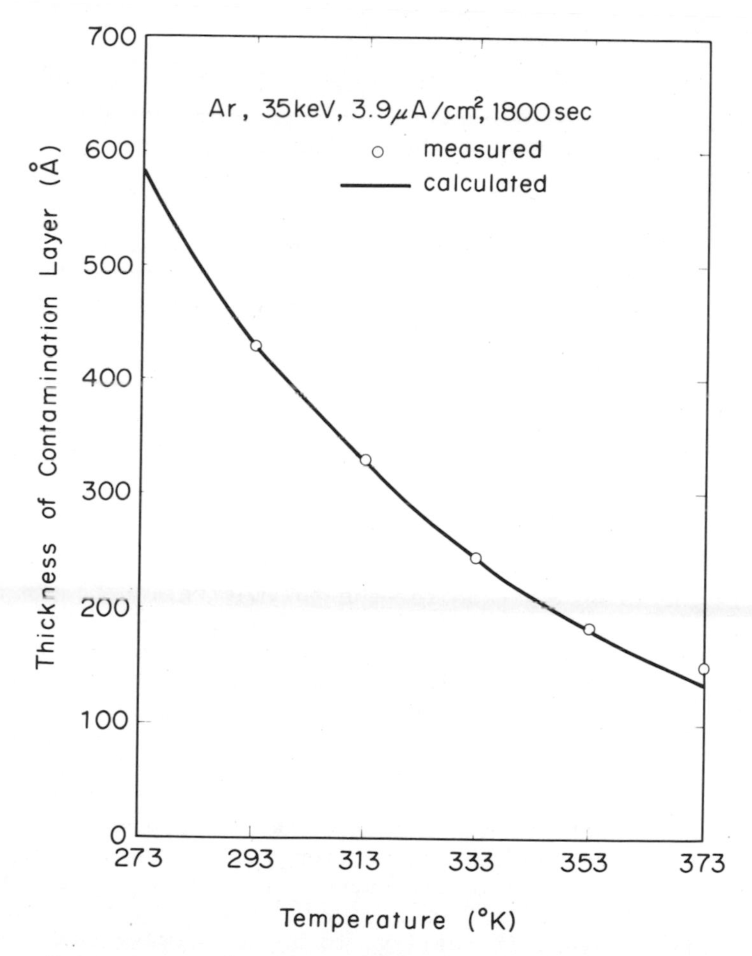

Figure 4. Temperature dependence of contamination thickness (ion energy: 35 keV; ion current density: 3.9 μA cm^{-2}; bombarding time: 30 min.).

Table I. Contamination Growth Data by Electron and Ion Bombardment

Spacies	Current Density ($A\ cm^{-2}$)	Growth Rate ($Å\ s^{-1}$)	Φ ($C\ cm^{-2}\ Å^{-1}$)	Author
elec.	5×10^{-2}	1.5	3.5×10^{-2}	Ennos[1]
elec.	1×10^{-4}	0.22	4.5×10^{-4}	Christy[6]
Li^+	1.4×10^{-7}	0.017	8.2×10^{-6}	Speidel[12]
Ar^+	1.3×10^{-6}	0.14	9.3×10^{-6}	Shimizu[14] et al.

(Φ: dose for 1 Å growth of the contamination layer)

The contamination growth by ion bombardment is calculated considering the sputtering effect by the use of a simple theory due to Christy.[6,13] We assume there are N organic molecules per unit area adsorbed on the surface, with a mean time of stay τ. When the surface is bombarded by an ion beam, some of the molecules remain on the surface as a result of polymerization and some of the molecules are sputtered away from the surface. The sputtering also affects the polymerized molecules. The rate of contamination growth dP/dt is given as a first-order approximation by

$$dP/dt=\sigma fN-\eta_1 f \tag{1}$$

where σ is the cross section for the polymerization, f the number of ions per unit area per unit time, and η_1 is the sputtering yield of the molecules tied due to polymerization on the surface. When the sputtering is superior, i.e. $\eta_1>\sigma N$, no contamination occurs. At the initial stage of ion bombardment, the rate of removal by sputtering effect depends on the number of polymerized molecules per unit area on the substrate.[25] However, once the substrate is covered with the contamination layer, the sputtering yield would be almost independent of the amount of polymerized molecules because only the surface region can escape from the polymerized layer by sputtering effect. Thus we can expect η_1 to be approximately independent of P. Furthermore, we must note the sputtered material from the polymerized layer, in reality, to be not always same the adsorbed molecule. Thus, η_1 is the apparent sputtering yield converted from the amount of sputtered material by one ion bombardment into number of the adsorbed organic molecules.

The number of molecules not yet polymerized on the surface decreases with the evaporation, the polymerization and the sputtering. If the surface is assumed to be incompletely covered by the molecules, the change in the number of molecules on the surface is expressed as follows;

$$dN/dt=F-N/\tau-\eta_2 f-(dP/dt+\eta_1 f), \tag{2}$$

$$\eta_2=\sigma_2 N \tag{3}$$

where F is the number of organic molecules which impinge on the surface from the surrounding vapor, per unit area per unit time, η_2 the sputtering yield of the adsorbed molecules, and σ_2 is the cross section of sputtering of the adsorbed molecules. The Equations (2) and (3) are inadequate under the condision of the low substrate temperature and higher partial pressures of organic molecules. Ling[8] has modified the Christy's theory to include such a condition by assuming unimoleculer adsorption which is expressed by the Langmuir equation. However, we shall use the Equations (2) and (3) in the following discussion, because the expression is simple and accurate for the contamination growth under normally used conditions.

On substituting Equations (1) and (3) into Equation (2) and integrating, we obtain

$$N=\frac{F}{\sigma' f+1/\tau}[1+K_0\exp\{-(\sigma' f+1/\tau)t\}] \tag{4}$$

where $\sigma'=\sigma+\sigma_2$ and K_0 is a constant which depends on the number of organic molecules present intially on the surface.

On inserting Equation (4) into Equation (1) and integrating, we obtain the thickness of the polymerized layer as follows;

$$d=\int_0^t v\frac{dP}{dt}dt$$

$$=\left(\frac{\sigma F}{\sigma' f+1/\tau}-\eta_1\right)vft+\frac{\sigma vfFK_0}{(\sigma' f+1/\tau)^2}[1-\exp\{-(\sigma' f+1/\tau)t\}] \tag{5}$$

where v is the volume of one molecule.

The above equation qualitatively agrees with the experimental result shown in Figure 1. In the early stage of ion bombardment, the growth is fast because the second term of the right-hand side of the above equation contributes to the growth; but the second term approaches a constant value with increasing the bombarding time, so the rate becomes constant as expressed in the first term. The contribution of the second term is about 50 Å (Figure 1), which is comparable with or less than the experimental error. Furthermore, the second term decreases with increasing f unless $\sigma' f \ll 1/\tau$, in which case the contamination growth is negligible compared with the adsorption. Therefore, we can neglect the second term when the total thickness is larger than several 100 Å. The mean mass of the adsorbed organic molecules is large compared with the mass of an incident argon ion. For example, the mass number of the diffusion oil which is a main source of the organic molecules is several hundreds. The sputtering of a fraction of the molecules, such as hydrogen atoms and methyl radicals, must also considered, but such an effect results not in a change of the number of adsorbed molecules but in a change of σ because the remainder of the molecules become chemically active. Thus, we shall negrect σ_2 in the following calculations.

From the above approximations, we obtain

$$d=\{\frac{vF}{1+(1/\sigma\tau f)}-\eta_1 vf\}t. \quad (6)$$

From Equation (6), we can estimate the value of ion current density for the continuous contamination growth (i.e. d>0) as follows;

$$f<F/\eta_1-1/\sigma\tau. \quad (7)$$

The curve shown in Figure 2 is calculated from Equation (6) by assuming

$$vF=1.1 \text{ Å s}^{-1}, \quad \sigma\tau=150 \text{ Å}^2 \text{ s}, \quad \eta_1 v=21 \text{ Å}^3.$$

The value of vF is a few times larger than in Christy's example. This may be due to the higher partial pressure of the organic molecules. The partial pressure of the organic molecules can be given by

$$p \simeq F(2\pi mkT_g)^{1/2} \quad (8)$$

where m is the average molecular weight, k is Boltzmann's constant, and T_g is the temperature of the surrounding vapor. The molecular volume v can be approximated to be about 600 $Å^3$ by assuming that the molecular weight is 400 and the density is one. Thus, the partial pressure of the organic molecules is estimated to be about 1×10^{-7} Torr. This value is close to the pressure of the diffusion pump oil at room temperature. The value of $\sigma\tau$ is several times larger than

Christy's, whose electron energy was 225 eV. On the other hand, the ion energy in our experiment was 35 keV (Figure 2). The cross section of polymerization for electron beam is estimated to decrease with an increase of the electron energy except for the very low-energy region. The contamination growth by electron bombardment in the high-energy region is almost inversely proportional to the square of the electron velocity.[9] The values of both τ in the Christy's and our experiments are not equal, but the difference may be not significant because the organic vapor sources in the experiments are similar. Therefore, we can conclude that the value of σ for ion beam is much larger than the value for electron beam of same energy. The value of $\eta_1 v$ is nearly equal to the value by assuming that the contamination is amorphous carbon (density$\simeq$1) and the sputtering yield is unity. The sputtering yield of carbon by argon ions is estimated to be 0.9-1 over our energy range.[22]

The energy dependence of the contamination growth by electron bombardment may suggest that the contamination growth depends on the energy loss at the surface. On the other hand, the phenomenon is more complicated with ion beams. The energy loss per unit path length of ions is nearly constant in the energy region of the experiments when both nuclear and electronic stopping are considered, and is proportional to the ion velocity when only the electronic stopping is considered.[21] The sputtering yield of the contamination by ion bombardment is estimated to increase with the ion energy over our energy range.[22] However, the sputtering effect not only causes a decrese of contamination growth but also creates chemical radicals. Therefore, the energy dependence shown in Figure 3 can be explained as follows. In the low-energy region, the increase of the cross section of polymerization with the increase of ion energy is dominant, and the increase of apparent sputtering yield is predominant in the high-energy region.

The value of τ depends on the substrate temperature, and the curve shown in Figure 4 is obtained by assuming

$$\sigma\tau = 1.2\exp(2800/RT) \quad \text{Å}^2\ \text{s} \qquad (9)$$

where R is the gas constant in cal K^{-1} mol^{-1} ($\simeq$1.98).

All the quantities assumed here are reasonable ones, but these are merely rough estimates because of experimental errors. Moreover, the following effects have not been considered in the calculations, and some of these can not be negligible under some conditions:

(a) The contribution of σ_2. In some cases, the sputtering of large organic molecules adsorbed on a substrate may be not negligible. In this case, Equation (6) is modified as follows;

$$d=\left\{\frac{vF}{1+\sigma_2/\sigma+(1/\sigma\tau f)}-\eta_1 vf\right\}t. \quad (10)$$

Also, Equation (7) is modified as follows;

$$f<\sigma/\sigma'(F/\eta_1-1/\sigma\tau). \quad (11)$$

Similar result has been given by Hirsch.[25]

(b) The effect of sputtering of the substrate. The true thickness of the contamination layer is thicker than the apparent value because only the height difference of the bombarded area from the masked area was measured.

(c) The rise of substrate temperature by ion bombardment. The temperature rise estimated by a simple calculation is only a few degrees and the effect on the contamination is less than the experimental error.

(d) The surface migration of organic molecules. This is not considered because the bombarded area is relatively large in our experiments. The surface migration is an important factor when the contamination is restricted in a small area of μm^2 or less.[11]

(e) The contribution of ions of organic molecules included in the incidnet beam. The pressure of argon in the ion source is estimated to be about 10^{-3} Torr, and the partial pressure of the organic molecules may be about 10^{-7} Torr. Thus, the organic component included in the beam is very small. Of course the cross section of ionization of organic molecules differs from the cross section of ionization of argon, and also the ions produced in the accelerating region must be considered, but the organic component may be less than 0.1 % of the total ions. Therefore, the number of organic ions incident on the surface is less than 10^{-14} mol cm^{-2} s^{-1} when the total ion current density is 1 μA cm^{-2}. On the assumption that the average molecular weight of the organic molecules and the density of the contamination layer are 400 and 1, respectively, this quantity corresponds to the contamination rate of 0.0004 Å s^{-1} or less, which is very small compared with the measured value.

(f) Recoil implantation of contaminating atoms. Some atoms such as carbon and oxygen are introduced into the substrate from the contamination layer by ion bombardment due to recoil or knock-ons of the atoms.[22] The phenomenon scarecely affects the calculations described above, but it makes more difficult to clean up the once contaminated surface by ion bombardment.

(g) The contribution of secondary electrons. The secondary

electrons emitted from the ion bombarded surface may contribute to the contamination growth. However, the contribution can not be appreciate because, as described above, the contamination by ion bombardment as compared with electron bombardment occurs with much smaller dose of the beam. The average energy of secondary electrons is estimated to be only a few eV,[24] and Mayer's experimental work[7] suggests that the contamination growth rapidly decreases with the decrease of electron energy in the low-energy region. Furthermore, the secondary electron emission is originally due to the ion bombardment, thus the whole process can be treated phenomenologically as one process of polymerization of the adsorbed molecules by ion bombardment.

(h) Contamination by Sputtered material. An ion bombarded sample can also be contaminated by the sputtered material. The surface contamination with polymerized layer of remaining organic molecules is quite serious in conventional vacuum, but the contamination by sputtered material becomes dominant in higher or "clean" (no presence of organic vapor) vacuum.

Although there remains some quantitative problems, however, the contamination growth model employed here is supported by the experimental results. The model is similar for that of electron bombardment case except for the sputtering effect, and the terms of sputtering effect appear in Equations (5),(6) and (10), i.e. the terms of η_1 and σ_2, decrease the contamination growth. Thus, the high contamination growth rate by ion bombardment may be explained by the large cross section of the polymerization of ions. Of course this is a phenomenological explanation because the sputtering effect, in reality, also contributes to the polymerization through creating the chemically radicals. Our model is supported by the fact that the contamination growth can be eliminated by heating the substrate or by the use of the cold trap surrounding the substrate; which is similar to the case of electron bombardment. However, it must be noted that heating the substrate may be less efective when the ion current density is high enough, although it is only estimated from Equation (6) and has not been ascertained experimentally. On the other hand, the use of a cold trap surrounding the substrate, which would lead to more practical difficulties in designing an ion beam apparatus, will always effectively reduce the contamination caused by the organic molecules.

CONCLUSIONS

From the experimental results and the consideration described above, we conclude that the high growth rate of contamination layer by ion bombardment is mainly due to the large cross section of polymerization by ions. When the ion dose is relatively small, the contamination grows rapidly with ion dose and depends strongly on the number of organic molecules initially adsorbed on the surface. With an increase in ion dose, the contamination growth becomes almost linear with the ion bombarding time and the contamination rate depends on the rate of organic molecules impinging on the surface from the surrounding vapor. In contrast, at high current densities, the contamination rather decreases with increasing ion current density because of the sputtering effect. Also the contamination rate is affected by the "stay time" which depends on the substrate temperature. Therefore, heating the substrate or using a cold trap surrounding the substrate effectively reduces the contamination.

REFERENCES

1. A. E. Ennos, Brit. J. Appl. Phys. 4, 101 (1953).
2. J. H. L. Watson, J. Appl. Phys. 18, 153 (1947).
3. V. E. Cosslett, J. Appl. Phys. 18, 844 (1947).
4. J. Hillier, J. Appl. Phys. 19, 226 (1948).
5. H. König and G. Helwig, Z. Phys. 129, 491 (1951).
6. R. W. Christy, J. Appl. Phys. 31, 1680 (1960).
7. L. Mayer, J. Appl. Phys. 34, 2088 (1963).
8. J. Ling, Brit. J. Appl. Phys. 17, 565 (1966).
9. H. Hashimoto, A. Kumao and K. Hosoi, in "Electron Microscopy 1968 - Proc. 4th Europ. Reg. Conf. Electron Microscopy, Rome" D. S. Bocciarelli, Editor, Vol.1, pp. 39-40, Tipografia Poliglotta Vaticana, Rome, 1968.
10. R. K. Hart, T. F. Kassner and J. K. Maurin, Phil. Mag. 21, 453 (1970).
11. K. H. Müller, Optik 33, 296 (1971).
12. R. Speidel, Z. Phys. 154, 238 (1959).
13. K. Shimizu, H. Kawakatsu and K. Kanaya, J. Phys. D 8, 1453 (1975).
14. K. Kanaya, H. Kawakatsu, S. Matsui, H. Yamazaki, I. Okazaki and K. Tanaka, Optik 21, 399 (1964).
15. G. Möllenstedt and R. Speidel, Z. Angew. Phys. 13, 231 (1961).
16. L. Holland, J. Phys. D 2, 767 (1969).
17. K. Kanaya, H. Kawakatsu, S. Matsui, H. Yamazaki, I. Okazaki and K. Tanaka, in "Proc. Electron and Laser Beam Symp., March 31 - April 2, 1965", A. B. El-Kareh, Editor, pp. 489-507, Alloyd General Corporation, Medford, Mass., 1965.
18. K. Kanaya, K. Shimizu, S. Matsui, H. Yamazaki, I. Okazaki and K. Tanaka, Bull. Electrotechnical Lab. 31, 316 (1967).
19. R. McKeever and A. Yokosawa, Rev. Sci. Instrum. 33, 746 (1962).
20. S. Tolansky, "Multiple-Beam Interferometry of Surfaces and Films", Oxford Univ. Press, London, 1948.
21. J. Lindhard, M. Scharff and H. E. Schiott, Mat. Fys. Medd. Dan Vid. Selsk. 33, no. 14, pp. 4-10 (1963).

22. K. Kanaya, K. Hojou, K. Koga and K. Toki, Japan. J. Appl. Phys. 12, 1297 (1973).
23. R. S. Nelson, Radiat. Eff. 2, 47 (1969).
24. M. Kaminsky, "Atomic and Ionic Impact Phenomena on Metal Surfaces", pp. 329-331, Springer-Verlag, Berlin, 1965.
25. E. H. Hirsch, J. Phys. D 10, 2069 (1977).

SURFACE CONTAMINATION AND CORROSION IN PYROTECHNIC ACTUATORS*

R. G. Jungst, R. K. Quinn, T. M. Massis
R. N. Roberts, and R. E. Whan

Sandia Laboratories
Albuquerque, New Mexico 87185

Accelerated aging studies of pyrotechnic actuators containing a mixture of titanium metal and potassium perchlorate powders have revealed that extensive corrosion of the metal pins and bridgewires used to ignite the device occurs in relatively short times. Analyses suggest that the corrosive process is associated with chlorine and organic contaminants. These can trigger corrosion by attacking the passivating oxide layer on the surface of metal alloys, thereby promoting reaction between the pyrotechnic and the metal. The contaminants may also react directly with the pyrotechnic to cause decomposition to corrosive species. Residual moisture in the pyrotechnic powder tends to enhance corrosion rates when contaminants are present. Tests run on several metal alloys have verified the corrosive action of certain organic solvent mixtures. Specifically, Tophet A®, Tophet C®, and Alloy 52 were corroded when exposed to combinations of alcohols and chlorinated organic liquids. Individual solvents did not cause corrosion. In each case, Tophet C® which contains about 25% iron was attacked faster than Tophet A® which is iron free. The gas phase in equilibrium with a boiling solvent mixture was more corrosive than the hot liquids or hot gases alone. In order to eliminate this corrosion problem, a new

*This work was supported by the U. S. Department of Energy.

cleaning procedure based on non-chlorinated solvents and boiling hydrogen peroxide has been incorporated into the production process and procedures for drying the pyrotechnic powder have been improved. This treatment is effective as demonstrated both by analysis of component surfaces and by additional aging studies on cleaned units.

INTRODUCTION

Stringent requirements on the reliability of pyrotechnic actuators over long lifetimes means that even low levels of corrosion after short periods of time are unacceptable. Materials of construction and methods of manufacture are chosen in order to avoid interfaces which are known to be incompatible. High temperature accelerated aging is used to verify that more subtle compatibility problems are not present and that corrosion-inducing contaminants have not been introduced during component assembly.

The configuration of the actuator considered in the present work is shown in Figure 1 and Table I gives the composition of some of the materials of construction. A pyrotechnic composition

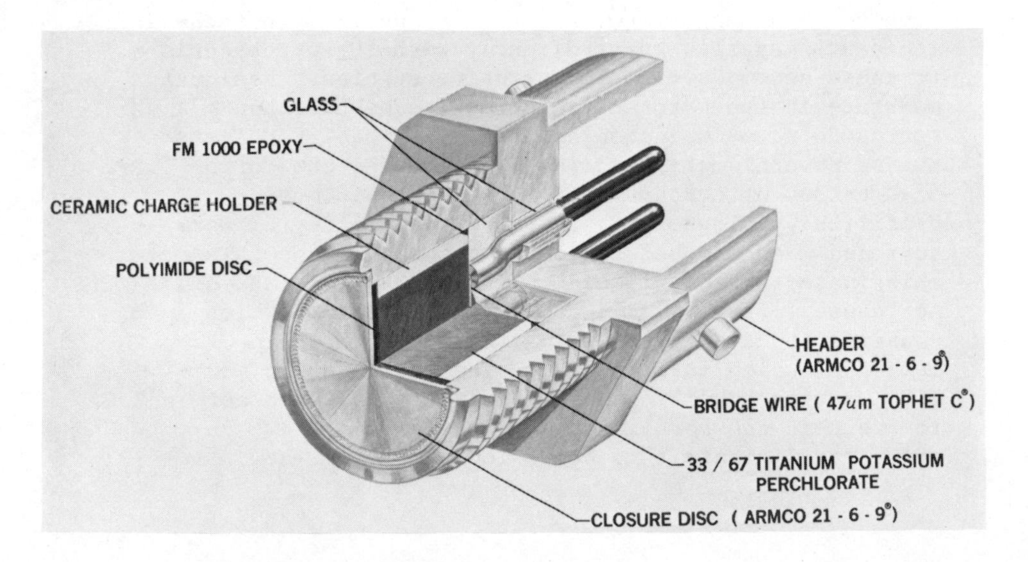

Figure 1. Cutaway view showing design of pyrotechnic actuator.

Table I. Nominal Composition of Materials Used in Actuator Construction.

Header	Glass	
Bridgewire	Tophet C®	60 wt % Ni, 16 wt % Cr, balance Fe
Pin	Alloy 52	50.5 wt % Ni, balance Fe

of 33 wt % titanium powder and 67 wt % potassium perchlorate is contained in the ceramic charge holder. Ignition of the pyrotechnic is achieved by electrically heating a Tophet C® bridgewire. The bridgewire is spot welded to two Alloy 52 pins embedded in a glass support. Once the unit is loaded and hermetically sealed with the closure disk, the critical area for corrosion is the bridgewire/powder interface where ignition of the pyrotechnic occurs.

During assembly of the component the following cleaning procedures were utilized:

1. Clean ultrasonically for 2-3 minutes in perchloroethylene (PCE)
2. Blow dry with nitrogen
3. Clean ultrasonically in water/detergent solution
4. Clean ultrasonically in deionized water
5. Blow dry with nitrogen
6. Clean ultrasonically in isopropyl alcohol (IPA)
7. Blow dry with nitrogen
8. Weld bridgewire
9. Bond charge holder
10. Spray with ethanol
11. Vacuum bake.

Development lots of this actuator underwent accelerated aging at temperatures up to 120°C for 12 months with no evidence of corrosive attack. Subsequent production units corroded over an average of 50% of the pin surface within 30 days at room temperature and so the processing and cleaning procedures were reevaluated in order to identify possible sources of contamination. A number of sophisticated surface analytical techniques were implemented to characterize corrosion products and to monitor contamination levels in actuators. The results of these studies and a revised cleaning process designed to minimize contamination will be described.

CORROSION CHARACTERIZATION

Figures 2 and 3 show scanning electron microscope (SEM) photomicrographs of the pin and bridgewire in uncorroded and corroded units respectively. Energy dispersive X-ray analyses of corroded areas (Figure 4) reveal high levels of chlorine. This is not accompanied by equally high amounts of potassium, indicating that it does not result from particulates of $KClO_4$ since authentic samples of $KClO_4$ or KCl show equal K and Cl peak intensities. Atomic ratios of chlorine to potassium are typically 5/1 or greater on both the corroded bridgewires and pins. A glass shard from the pin support is responsible for the high silicon peak in the EDAX spectrum of Alloy 52 in the lower right of Figure 4. The bridgewires also show enhanced surface concentrations of chromium in corroded areas. This increased level of chromium has been found in other cases for Tophet C® and has also been observed for corroded Tophet A®. Oxygen is the only element other than the alloy metals or chlorine which has been observed at high concentration in regions of corrosion. Examination of solid deposits by X-ray diffraction identified small amounts of TiO_2 and KCl, but the major portion could not be matched to a known diffraction pattern. A reasonable inference from this information is that the corrosion deposit consists of a mixture of metal chlorides, metal oxides or hydroxides, and metal oxychlorides.

SURFACE CONTAMINATION

Sources and Effects on Corrosion

Since examination of corroded areas consistently showed high levels of chlorine, contamination by any material containing this element was of interest. The pyrotechnic powder itself is formulated so as to initially be low in free chloride. Titanium metal is made by reduction of TiO_2 with hydrogen and so is essentially chloride free, while the $KClO_4$ contains about 100 ppm free chloride. Aging the blended powder at elevated temperature does increase the chloride level to a few hundred ppm initially, but the rate of increase then slows nearly to zero. No enhanced rate of chloride formation was found when aging the pyrotechnic with various component materials and no corrosion was observed in those tests. Additionally, the entire amount of chloride in the mass of pyrotechnic contained in one component is only about 6.7 microgram. This would all have to migrate to the pin and bridgewire to give the observed amount of corrosion, which is unlikely. Removal of all moisture from the finely divided titanium powder is very difficult and it is likely that this contributes to the corrosion by increasing the mobility of chloride ion in the actuator. Study

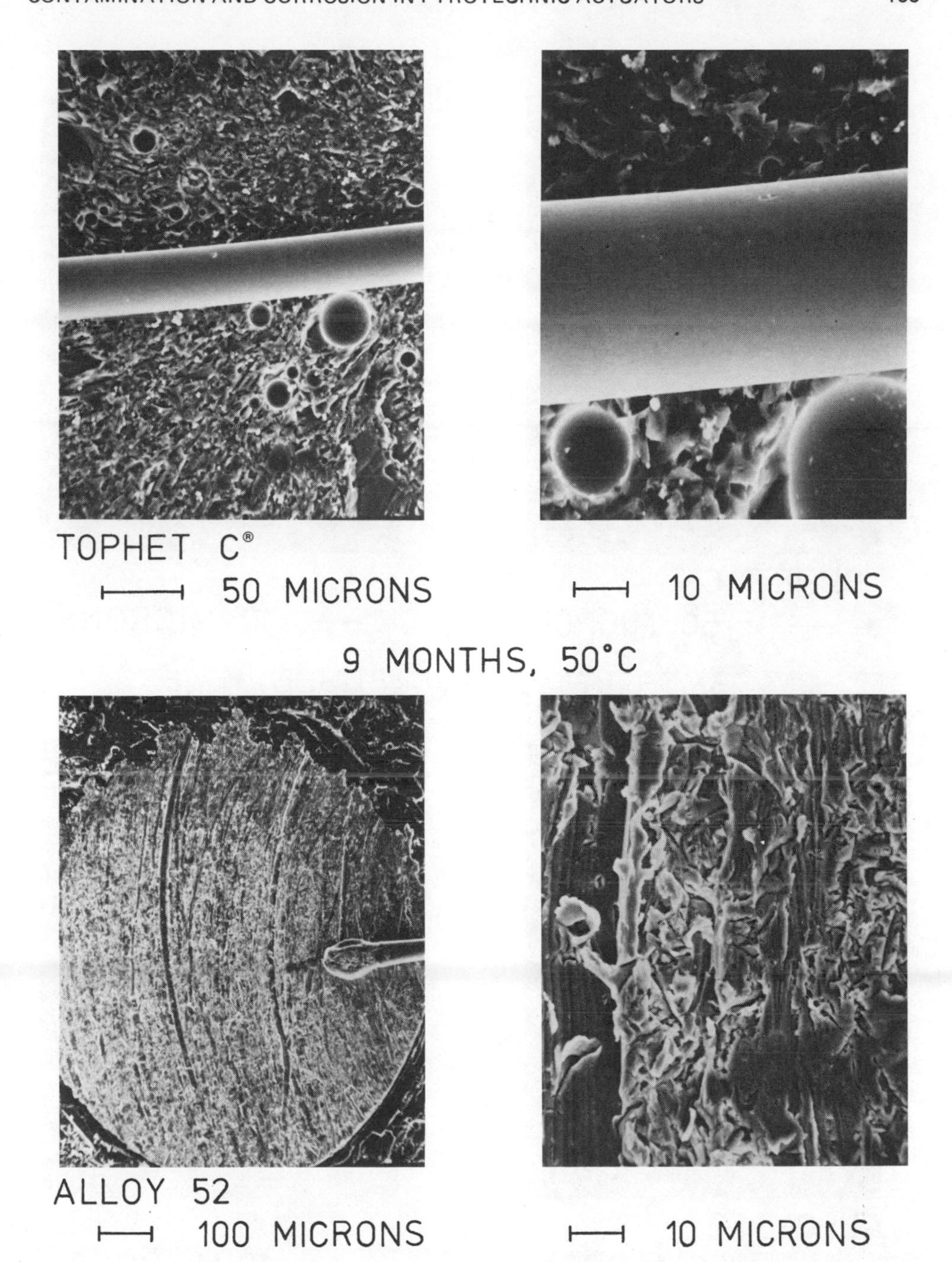

Figure 2. Scanning electron microscope photomicrographs of the pin and bridgewire in actuators showing no corrosion after aging.

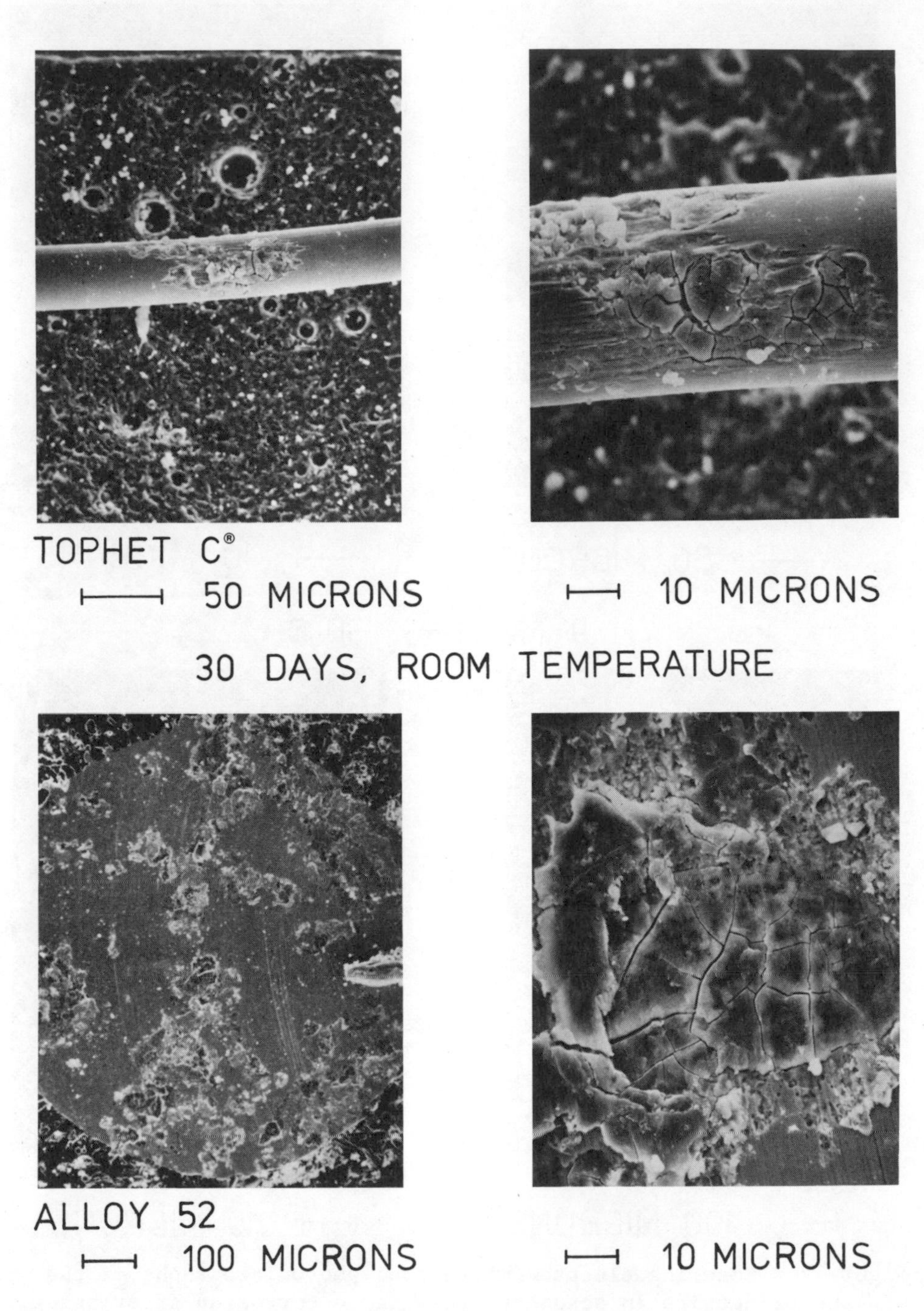

Figure 3. Scanning electron microscope photomicrographs of actuator pins and bridgewires which corroded during aging.

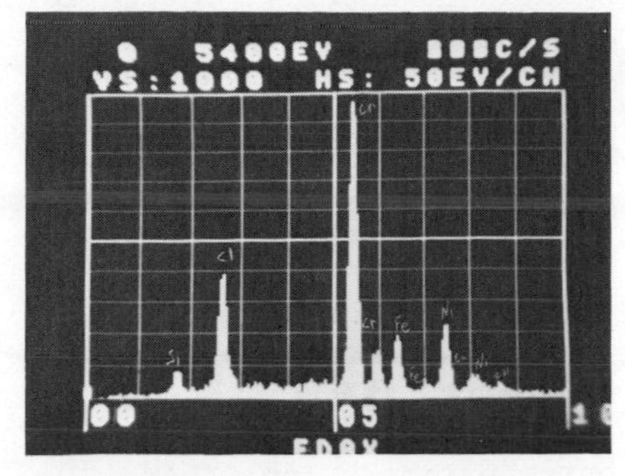

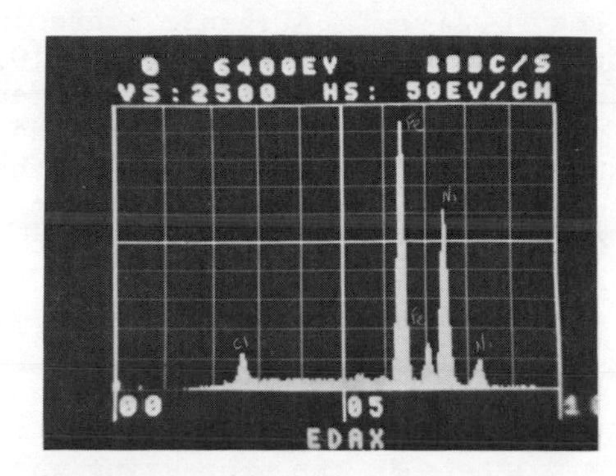

TOPHET C® ALLOY 52

30 DAYS, ROOM TEMPERATURE

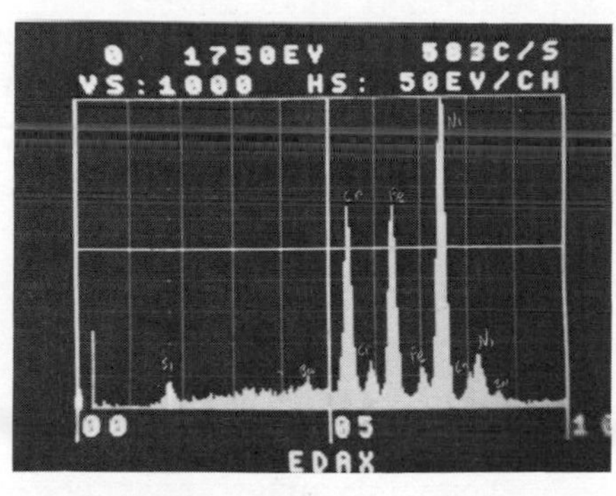

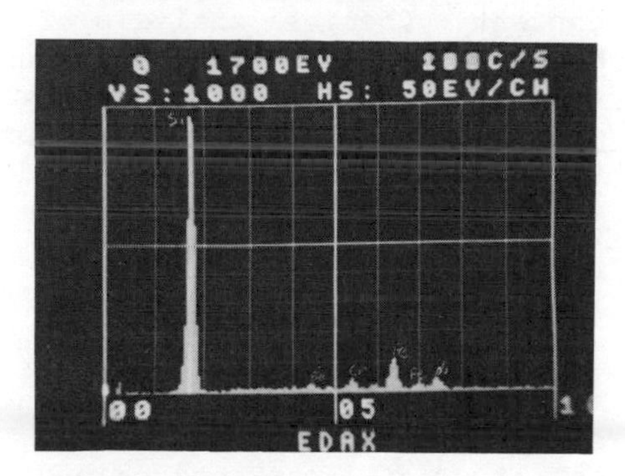

TOPHET C® ALLOY 52

9 MONTHS, 50°C

Figure 4. Elemental analysis of bridgewire and pin surfaces. The top two EDAX spectra are from corroded units as shown in Figure 3, and the lower two are from uncorroded units as shown in Figure 2.

groups of actuators that did not use a well dried pyrotechnic powder always corroded whereas units containing dried powder tended to be free of corrosion. In order to reduce moisture levels, the loaded units are vacuum dried just before welding the closure disk in place. This lowers the moisture content in the powder to less than 0.24 wt %. Organic contaminants could be reactive enough to generate free chloride from $KClO_4$ and slight amounts of corrosion have been seen on a test specimen of Alloy 52 which was purposely coated with products of epoxy adhesive cure and then aged in contact with a pressed pellet of the pyrotechnic powder. Acidic contaminants can trigger corrosion by attacking the passivating oxide layer on the surface of the metals. The system is not a simple one and several factors are undoubtedly operating simultaneously when actuators corrode. Contamination plus the presence of moist pyrotechnic powder is the environment which consistently causes the most rapid and extensive corrosive attack.

Steps in the processing where contamination by chlorine-containing compounds could occur were examined and modified if necessary to reduce that possibility. These included ensuring that electroplating solutions used to gold plate the exterior ends of the pins could not contact the interior of the unit, guarding against human contamination during handling and loading, and changing to a machining oil and epoxy adhesive with lower chlorine content. Chemical analysis of the glass material used in the header showed a negligible amount of chlorine so it was not changed. An additional area of concern was that chlorinated cleaning solvents might be contributing to the corrosion problem. Experiments to test this possibility will be described below.

Cleaning Solvent Reactivity

Exposure of test strips of the metal alloys in the actuator to cleaning solvents used in the processing was carried out in several ways. Simply immersing the metal in the solvent did not lead to significant corrosion. Figure 5 shows two other test configurations which did induce corrosion. In the first, the specimen and solvent were heated in an oven at $150^{\circ}C$ until all liquid just disappeared from the inner container. The outer flask was then capped and the sample left in the oven for a period of time. In the second, a metal strip was placed above a container of refluxing solvent so that it was in an atmosphere of condensing and evaporating liquid. The latter arrangement was the most effective at inducing corrosion.

In these experiments, PCE alone did not corrode any of the metals, but when mixed with alcohols corrosion occurred rapidly. This was true for Alloy 52, Tophet C®, Kovar® (29% Ni, 17% Co, balance Fe), and to a lesser extent Tophet A® (80% Ni, 20% Cr).

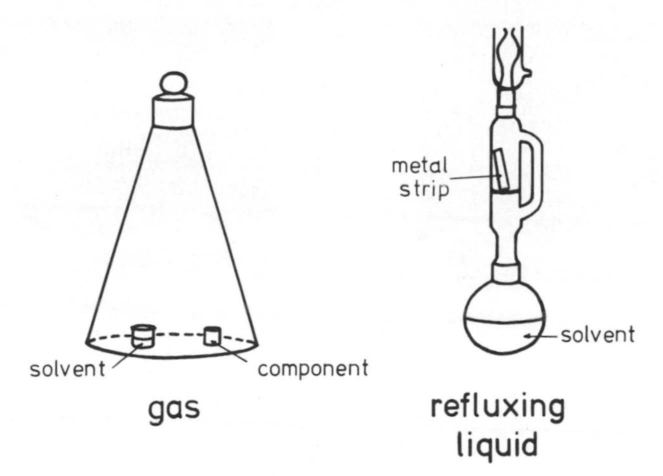

Figure 5. Apparatus for testing the ability of organic solvents to corrode metals.

Typical weight losses after two weeks exposure of a 0.0635 x 0.635 x 5.08 cm test strip to vapors of PCE/IPA (volume ratio 1/9) were 0.68% (σ = 0.26) for Alloy 52 and 1.78% (σ = 0.64) for Kovar®. This is equivalent to loss of a layer 0.0002 cm thick over the surface of the specimen of Alloy 52 and 0.0005 cm from Kovar®. Figure 6 shows SEM photomicrographs of Tophet C® wires after exposure to several solvent mixtures. A number of different alcohols including ethyl, isopropyl, cyclohexyl, and n-, s-, and t-butyl were found to form corrosive mixtures with PCE. For reasons which are not understood, methanol was the only alcohol tested which would not cause corrosion in these mixtures. Tests where the alcohol was replaced with a few other solvents such as N,N-dimethylformamide, cyclohexanone, and aqueous acetonitrile were also positive. Other chlorinated solvents such as trichloroethylene, 1,2-dichloroethane, trifluoro-1,1,2-trichloroethane, and 1,1,1-trichloroethane could be substituted for PCE and one still obtained a corrosive mixture.

Examination of the surface of Tophet C® wires with the electron microprobe after they had been corroded by PCE/IPA revealed some surface contamination by chlorine. All of the metals in the alloy are also present in amounts similar to those seen in uncorroded Tophet C®. The chlorine could not be preferentially associated with one alloy component from this experiment.

Precedent exists in the chemical literature for the reaction of PCE with alcohols, but the temperatures used were 500-600°C rather than the 80°C reflux temperature.[1] HCl was identified as a reaction product in the high temperature experiments by infrared spectroscopy and the main organic products were compounds formed by condensing the alcohol and alkene. Other authors have reported

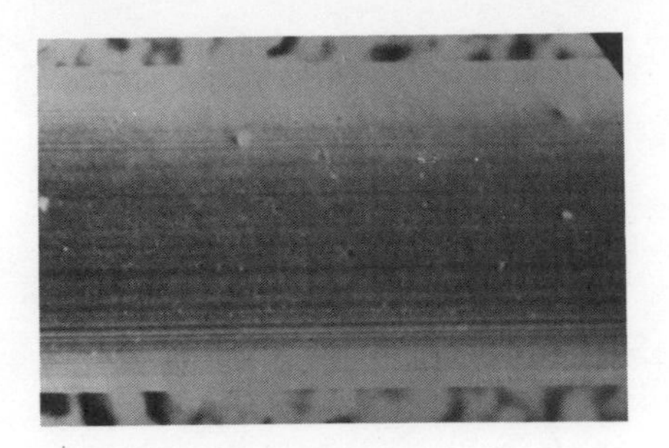

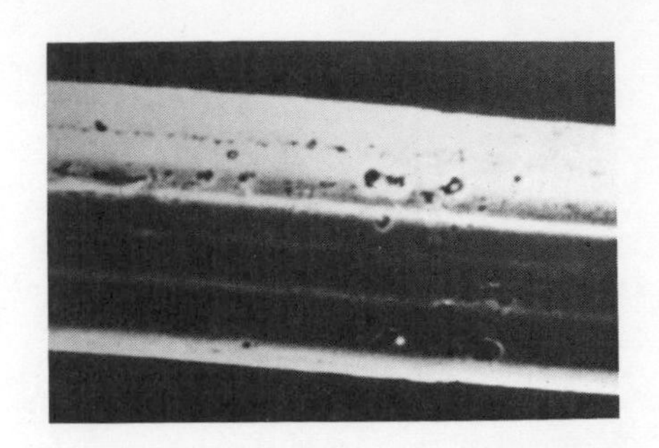

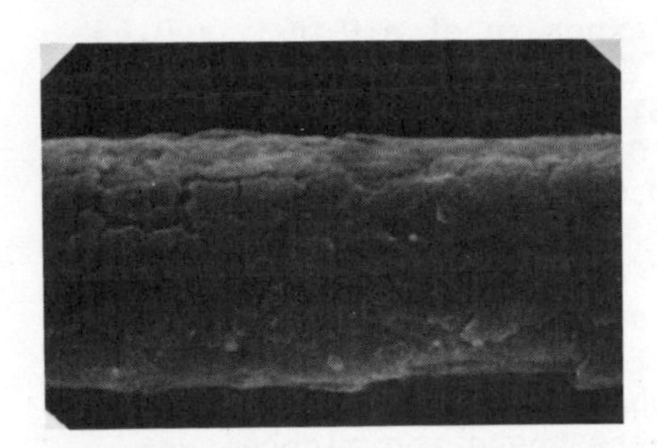

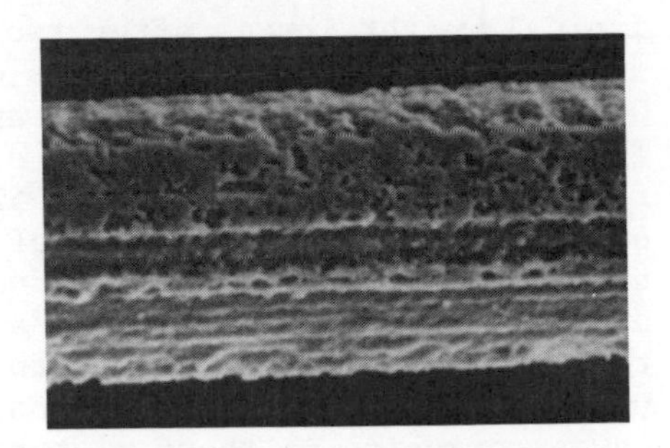

Figure 6. SEM photomicrographs of Tophet C® wires after exposure to several different solvent mixtures. IBA is isobutyl alcohol.

the corrosion of type 316 stainless steel by a CCl_4/isopropyl ether/acetone solvent mixture.[2] We have analyzed PCE/IPA mixtures by gas chromatography/mass spectroscopy after they have been used in corrosion tests of Tophet C® and Alloy 52. Some of the organic compounds which were present include pentachloroethane, trichloroisoprene, propylene, 1,1,2,2-tetrachloroethane, and 2-chloropropane.

Several other compounds (which could not be identified) were also found and these yielded fragments in the mass spectrometer which would be expected from chlorinated alcohols and ethers. These results imply that HCl is quite likely being generated in our reactions, but it has not been detected directly. Efforts to measure free HCl in the vapors of PCE/IPA while corroding an Alloy 52 strip failed. A Mine Safety Appliances Company detector (range = 0-70 ppm) was employed for these measurements.

Such experiments indicate that cleaning solvents would be reactive enough under certain conditions to cause corrosion in the actuators. If solvents were adsorbed strongly enough on metal surfaces or in the glass pin support, this would be a serious problem and alternate cleaning methods would be required.

Measurement of Surface Cleanliness

Consideration of the various possible sources of contamination pointed out the fact that a method was needed to monitor cleaning procedures. Auger electron spectroscopy (AES) was found to provide the most useful combination of convenience and high sensitivity to surface species. It was necessary to machine the chargeholder off the actuator flush with the pins in order to accommodate the AES electron analyzer geometry. Most spectra were run on the bridgewire or pin surfaces since beam charging effects made AES spectra difficult to obtain from the glass pin support. When the glass was studied, however, the trends appeared to be the same as observed on the metals.

Figure 7 shows an AES spectrum from the pin in an uncleaned actuator. High amounts of carbon are found as well as significant levels of chlorine and sulfur. In order to test how strongly organic cleaning solvents were retained on the surfaces, a unit was purposely contaminated with PCE by exposure in a Soxhlet extraction apparatus for 6-8 hours followed by vacuum drying at 110°C and <133 Pa pressure. Figure 8 shows the appearance of the AES spectrum taken from the pin region of this unit. The chlorine peak is very strong while carbon has been reduced somewhat from the uncleaned actuator. No evolution of adsorbed species from this actuator could be detected by mass spectroscopy during a heating ramp to 300°C and argon ion sputtering for 30 sec (∿50Å) still left detectable contamination. The PCE also penetrated into voids in the glass header since carbon and chlorine were still visible by AES in the bottom of a scratch purposely made in the surface. From a microscopic examination of the glass, it is estimated that about 10% of the header space consists of these bubbles.

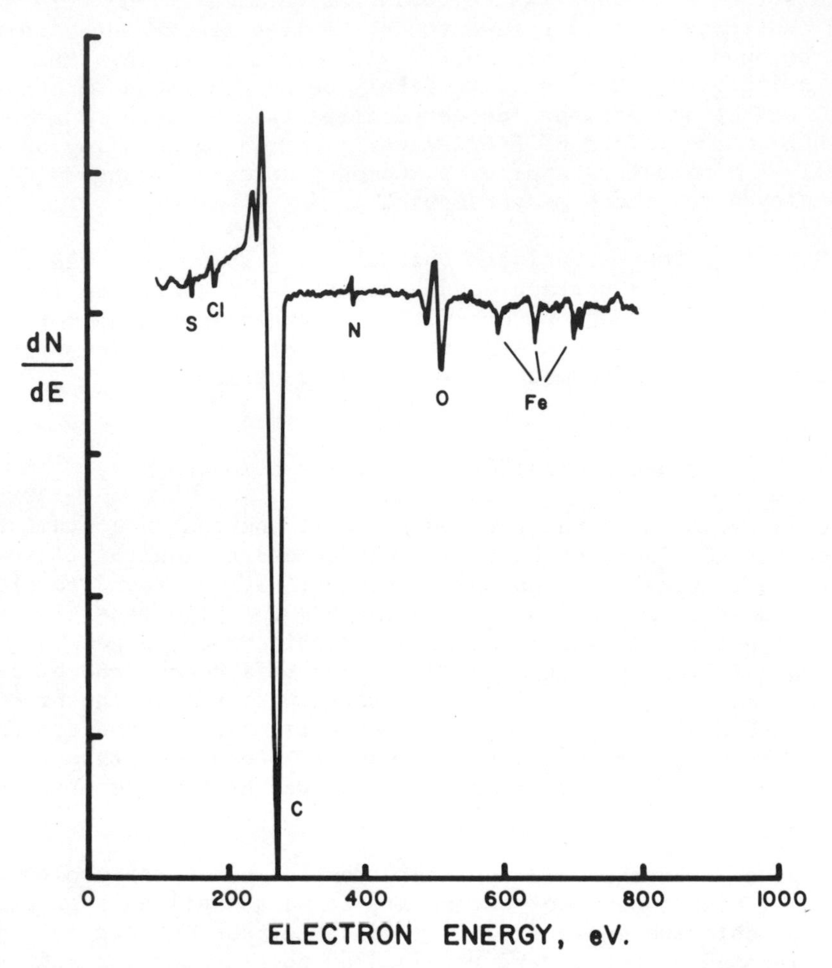

Figure 7. AES spectrum of the pin in an uncleaned actuator.

AES data show that the units as received are contaminated and that if PCE is used for cleaning, this chlorinated solvent can remain on the surface after moderate drying efforts. In order to clean the system, carbon- and chlorine-containing residue must be removed without introducing other contaminants.

CLEANING PROCEDURE

Several methods for more rigorously cleaning actuators were tested and the surfaces examined by AES. Since headers were routinely cleaned with PCE by the manufacturer, the new procedures were all tried on units which had been exposed to this solvent.

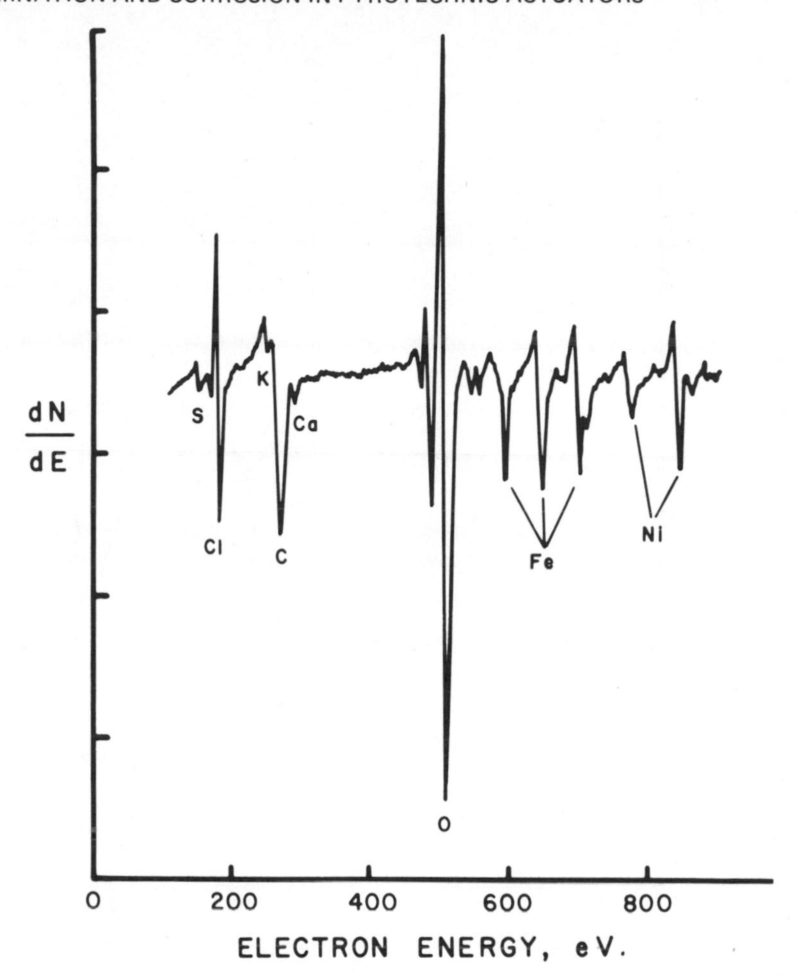

Figure 8. AES spectrum of the pin in an actuator intentionally contaminated with perchloroethylene.

More extreme vacuum bake out at 1.33×10^{-2} Pa and 450°C or extraction with alcohols in a Soxhlet apparatus for 6-8 hours followed by vacuum drying at 110°C for 16 hours were not successful in completely removing the chlorine-containing contamination. Alcohol extraction enhanced the AES carbon peak compared to vacuum bake out alone. A more effective procedure was treatment of the actuators with boiling hydrogen peroxide followed by a rinse with distilled water and drying at 65°C. This reduced carbon and chlorine to low levels in the AES spectrum (Figure 9). Another good method of cleaning was exposure to ozone generated by means of a UV lamp while heating to 80°C. This treatment reduced carbon to background levels but was not as good as peroxide for removing chlorine contamination. Figure 10 shows an AES spectrum from a unit cleaned in this way.

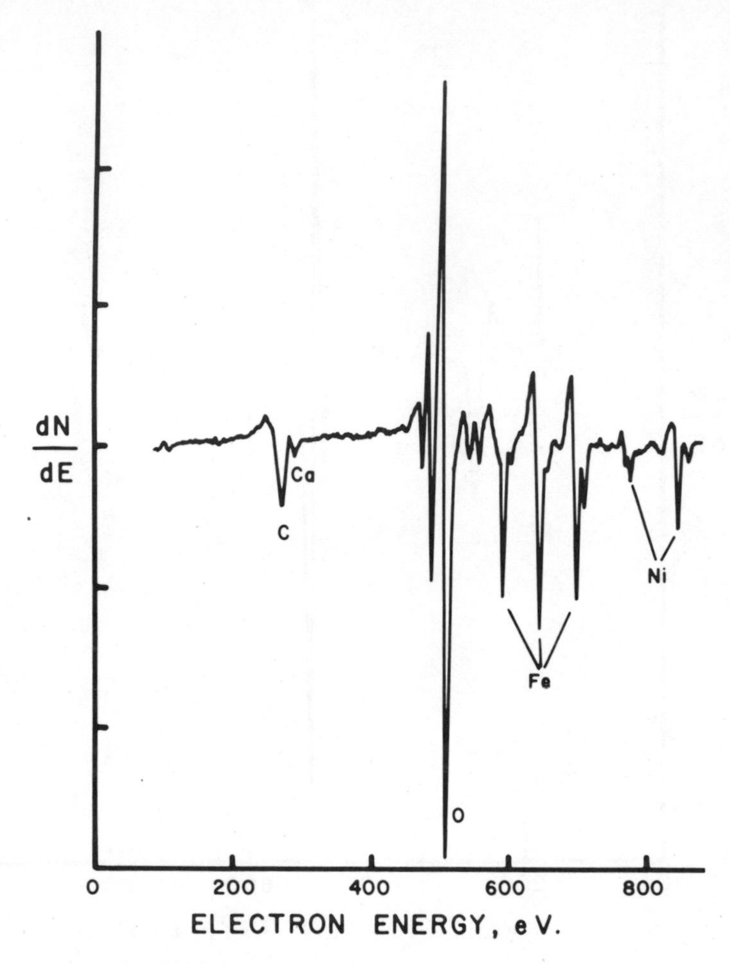

Figure 9. AES spectrum of the pin in an actuator cleaned with boiling hydrogen peroxide.

Since hydrogen peroxide more completely removed chlorine-containing residues and should also be more effective in reaching into pores in the glass material than ozone, it was incorporated into a new cleaning process. The concentration of H_2O_2 was set at 15% and the time at 10-12 minutes. Higher concentrations of peroxide or longer times caused some pitting of the metal surfaces while lower temperatures or shorter times of exposure left residual contamination which was visible in the AES spectrum. Hydrogen peroxide can be catalytically decomposed by metal concentrations of 10 ppm during this time interval so it is important to monitor peroxide concentration during cleaning.[3] Two H_2O_2 treatments may be necessary if the units being cleaned are excessively contaminated. One must also be aware of the fact that stabilizing chemicals added

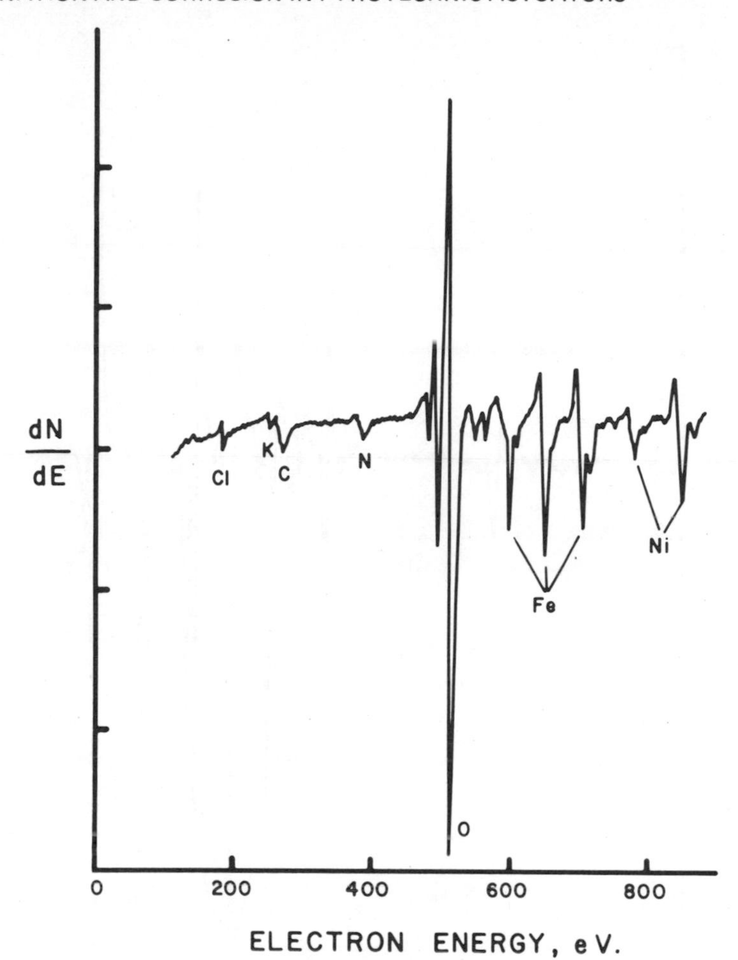

Figure 10. AES spectrum of the pin in an actuator cleaned with ozone.

to the H_2O_2 in low concentrations may themselves contaminate the component during cleaning. Tin has been observed on actuator surfaces treated with stannate-stabilized H_2O_2.

In general, for units thoroughly cleaned with peroxide, peak to peak ratios in the AES spectra are $O/Fe_{600} \geq 5$, $C/O \leq 0.3$, and $Cl/Fe_{600} \leq 0.3$. Fe_{600} refers to the iron peak at 600 eV in the AES spectrum. The thickness of the oxide layer is enhanced by peroxide cleaning compared to the initial oxide thickness and this serves to passivate the metal surfaces. Exact thicknesses of oxide are difficult to calculate, but are estimated from sputtering experiments to be at least 50Å. These experiments involved etching thin layers away with high energy argon ions followed by recording

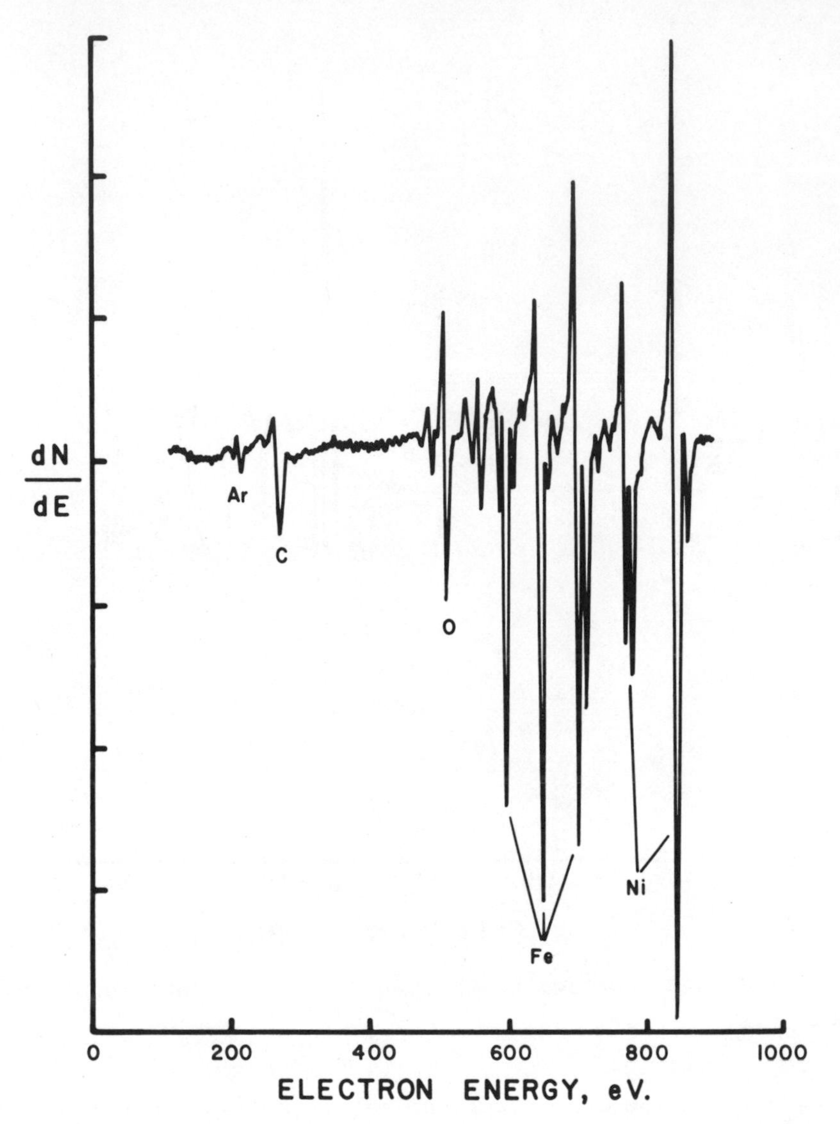

Figure 11. AES spectrum of the pin in an actuator cleaned with H_2O_2 and argon ion sputtered to remove about 50Å from the surface.

of AES spectra.[4] Sputtering also verifies that surface contamination has not simply been masked by the thicker oxide layer in the peroxide-treated units. Figure 11 shows the AES spectrum of a peroxide cleaned actuator pin after argon ion sputtering. Removal of the oxide exposes bare metal only.

The complete revised cleaning process for the actuators is

32 DAYS, 60°C
10 MICRONS

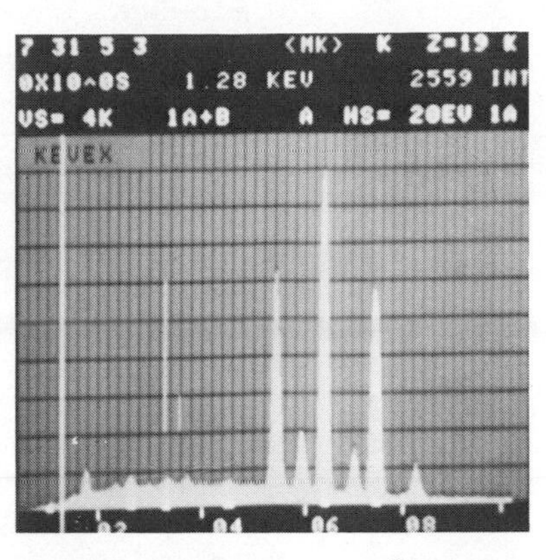

32 DAYS, 60°C

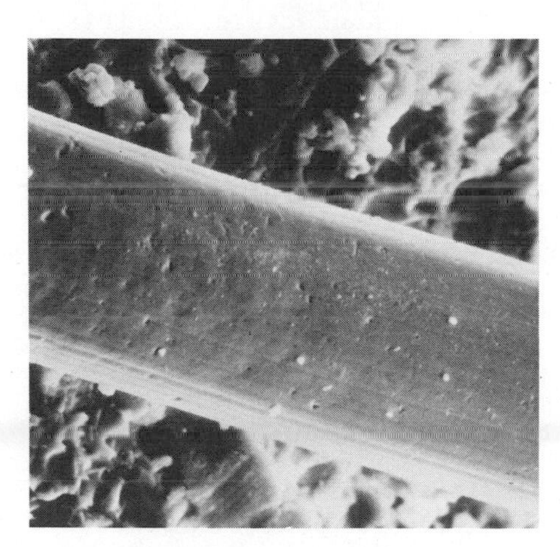

9 DAYS, 60°C
10 MICRONS

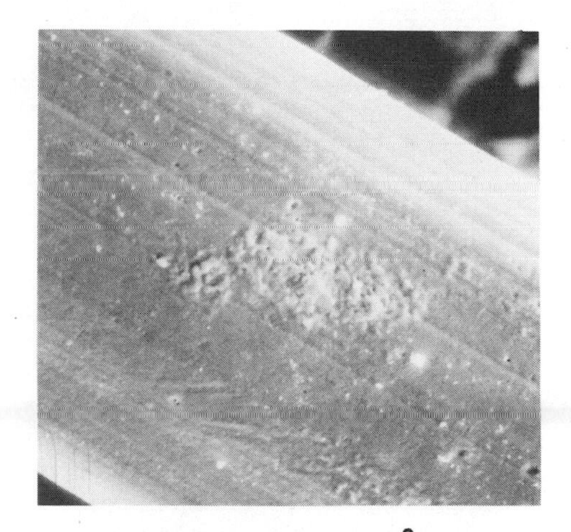

32 DAYS, 60°C
5 MICRONS

Figure 12. SEM photomicrograph of the bridgewire in an actuator cleaned with the H_2O_2 process and aged at elevated temperature.

shown below:

1. Clean ultrasonically in diacetone alcohol for 6 minutes
2. Clean ultrasonically in methanol for 6 minutes
3. Vacuum dry 1 hour at $100^{\circ}C$
4. Weld bridgewire
5. Immerse upright in boiling 15% H_2O_2 for 10-12 minutes
6. Rinse in distilled or deionized water
7. Vacuum dry 1 hour at $100^{\circ}C$
8. Bond charge holder
9. Clean ultrasonically in methanol for 6 minutes
10. Vacuum dry 1 hour at $100^{\circ}C$

Diacetone alcohol was chosen because it is the most effective non-chlorinated solvent of those tested for dissolving machining oil residues and volatiles from epoxy adhesive cure. Chlorinated solvents would do a better job of degreasing, but avoidance of contamination by chlorine-bearing compounds was considered to be of primary importance. Figure 12 shows an SEM photomicrograph of Tophet C® bridgewires in an actuator cleaned by the peroxide process and aged at elevated temperature. Although a few areas of mechanical damage can be seen, there is no evidence of corrosion. Similar observations were made in photomicrographs of the pin surfaces. Five production lots of actuators have been manufactured by this procedure. Accelerated aging studies at $110^{\circ}F$ have been carried out for up to 3 months with no corrosion being found. A limitation on using peroxide cleaning for other hardware is the possibility of attack by the H_2O_2 on certain materials unless the peroxide solution is so dilute that cleaning effectiveness is lost. Tests have already shown that silver braze used in sealing metals to ceramics is attacked quite rapidly by H_2O_2 solutions and could not be cleaned by the above procedure.

CONCLUSIONS

Corrosion in a pyrotechnic actuator has been studied and surface contamination by carbon- and chlorine-containing materials observed with AES spectroscopy. Removal of the possible sources of contamination and rigorous cleaning with hydrogen peroxide has lowered contamination levels substantially. In actuators where cleaning was thorough and excessive moisture has been removed from the pyrotechnic powder, no corrosion has occurred during accelerated aging.

ACKNOWLEDGEMENTS

The authors are grateful to Paul Holloway and Dennis Kramer who obtained the AES spectra at Sandia Laboratories. Gas chromatography/mass spectroscopy experiments were done by William Andrzejewski and studies on hydrogen peroxide stability by Kirk Shanahan.

REFERENCES

1. H. L. Schlichting and E. D. Weil, Belg. Pat. 622 421, Dec. 28, 1962. (Chem. Abstracts, 59, 11255f (1963).
2. A. Y. Ku and D. H. Freeman, Anal. Chem., 49, 1637 (1977).
3. K. L. Shanahan, "The Effect of Fe^{3+}, Cr^{3+}, Ni^{2+}, and Mn^{2+} on the Decomposition of Hydrogen Peroxide Solutions", Sandia Laboratories Report, SAND78-1778, March 1979. (Available DOE/TIC, Sandia Laboratories, Box 5800, Albuquerque, NM 87185, attn: R. P. Campbell).
4. N. R. Armstrong and R. K. Quinn, Surface Sci., 67, 451 (1977).

DECOMPOSITION OF HYDROGEN PEROXIDE CLEANING SOLUTIONS BY SELECTIVELY PLATED LEADFRAMES AND ITS IMPLICATIONS WITH RESPECT TO CIRCUIT YIELD

R. G. Fekula, W. J. Flood and D. L. Rehrig

Bell Telephone Laboratories, Inc. and
Western Electric Company
555 Union Boulevard, Allentown, Pa. 18103

Hydrogen peroxide is used extensively for cleaning electrical circuits because of its ability to oxidize a large number of inorganic and organic materials. During the pre-encapsulation cleaning of thin film integrated circuits assembled with selectively plated leadframes, sporadic violent hydrogen peroxide decomposition has been experienced. In addition, particles of metal have been noted in the peroxide bath after cleaning. The objectives of this work are: 1) to identify factors which influence the catalytic decomposition of the peroxide; 2) to establish to what extent the decomposition can be associated with selectively plated leadframes; 3) to determine if circuit yields are affected by the particles.

Many factors which influence the decomposition mechanism have been studied. The basic method of study involved measuring the hydrogen peroxide concentration as a function of time at the conditions set by the cleaning sequence. In addition, several other analytical techniques such as the laser microprobe and Auger electron spectroscopy were used to provide information about the substrates, solutions and precipitates.

The presence of thin Au on Ni in masked areas of a leadframe (extraneous Au) is observed on selectively plated leadframes. Spalling of this thin extraneous Au is a direct consequence of hydrogen peroxide cleaning and the deposition of spalled particles on active device surfaces leads to reduced circuit yields. Plating mask improvements on selectively plated leadframes reduced but did not eliminate the amount of extraneous Au on the frames.

INTRODUCTION

Hydrogen peroxide (H_2O_2) is extensively used for cleaning because of its ability to oxidize a large number of inorganic and organic materials.[1A] Silicon integrated circuits (SICs), film integrated (FICs) and hybrid integrated circuits (HICs) are all given a peroxide clean during processing.

On occasion, the H_2O_2 cleaning of IC assemblies prior to encapsulation has resulted in a violent decomposition of the solution. The decomposition problem has been particularly evident with the plastic dual inline product assembled with selectively plated lead frames.* This rapid decomposition increases peroxide consumption and reduces its concentration, thus reducing the effectiveness of the cleaning step. In addition, particulate matter has been noted in the peroxide bath after cleaning.

The objectives of this work are 1) to identify the factors which influence the catalytic decomposition of the peroxide, 2) to establish to what extent the decomposition can be associated with selectively plated leadframes, and 3) to determine if circuit yields are affected by the particles.

The scope of this investigation is extremely wide. Numerous effects were observed, but only the more significant facts are presented.

MECHANISMS OF DECOMPOSITION

The decomposition of hydrogen peroxide occurs via the reaction $2H_2O_2 \rightarrow 2H_2O + O_2\uparrow$.[1] The mechanism of decomposition depends on many factors including temperature, pH, and the type of catalyst present.[1] The presence of certain catalysts can accelerate this decomposition, especially at the higher temperatures ($\sim$90°C) normally employed in peroxide cleaning.

The basic mechanisms of catalytic decomposition can be classified into two types, (1) homogeneous and (2) heterogeneous reactions.[1] In a homogeneously catalyzed reaction there is an enhancement in the rate of the decomposition which occurs when the reactants and the catalysts are physically of the same state and phase, e.g., when a leadframe metal (solid) goes into solution and these metal ions then catalyze the peroxide decomposition. A heterogeneous reaction is a surface phenomenon. In this case there is an enhancement in the rate of decomposition brought about by the interface between the two phases. The decomposition of peroxide is catalyzed at the interface of the solution and the surfaces of the

---

*Selectively plated lead frames have Au plated only on areas where bonding and soldering occurs.

solid introduced into the solution. It is, however, very difficult to determine if one or both forms of decomposition are occurring either sequentially or simultaneously. Ions of the catalyst in the peroxide solution indicate the decomposition is at least in part homogeneous, whereas the absence of ions implies heterogeneous decomposition.

The rate of a homogeneous chemical reaction depends on the concentration of the reactants.[2] Factors which affect these concentrations are: 1) the dissolution rate of metal ions, 2) solution variables such as temperature, pH and stabilizers, 3) catalyst variables such as area, surface cleanliness, purity and oxidation state. Factors which affect the heterogeneous decomposition are: 1) surface area, both macroscopic and microscopic, 2) lattice imperfections such as screw and edge dislocations or point defects which intersect the surface, 3) geometric and electronic factors such as lattice spacing and atomic relationships, and 4) surface reactants including contamination or metal impurities at the surface.[3] Since surface contamination or reactants are a factor for both types of decomposition, a certain degree of uncertainty exists in identifying the reaction type.

EXPERIMENTAL METHOD

Cleaning specifications for FICs require that peroxide cleaning be performed in a 10% to 15% solution of H_2O_2 in water maintained at 90°C minimum for 15 to 20 minutes. For this work, a one-liter peroxide solution was heated on a hot plate in a 3 liter battery jar. When the solution temperature reached about 85°C, the concentration was measured (see below) and adjusted by addition of peroxide or water to obtain a 15.0 ± 0.3 wt % mix. When the solution temperature reached at least 90°C the concentration and pH were recorded, the test substrate introduced, and the timer started. The duration of the test was determined by the rate of decomposition. If decomposition was violent and fast, the test ended when the concentration fell to about two percent. If the concentration increased or decreased slowly the test was usually terminated after about 40 minutes.

The peroxide concentration was determined by measuring the solution refractive index with an American Optical Corporation Model 10402 Concentrimeter; the refractive index is then converted to weight percent peroxide from published tables.[1] A ceramic cup filter was used when necessary in conjunction with an aspirator to filter the decomposed peroxide for particulate matter removed from the test substrate. Calibration of the pH probes was performed at room temperature with buffered solutions. pH measurements of peroxide are subject to error when glass probes are used.[4] At room temperature the pH measured with glass probes is low and can be adjusted by addition of a correction factor. pH also requires correction because of the 100°C temperature. No correction factors were applied for this work and, therefore, the absolute pH may be in slight error.

Three experimental shortcomings were observed: 1) loss of solution due to evaporation; however, this effect was reproducible and significant only at the slower decomposition rates. 2) Constantly changing concentration; i.e., the decomposition occurs at surfaces for a heterogeneous reaction and therefore the solution concentration is not uniform throughout the beaker. This is a problem mainly when the decomposition is fast. 3) The time required to make the concentration measurement, approximately one minute; thus, at fast decomposition rates the number of data points is limited.

ANALYTICAL MEASUREMENTS

Several analytical techniques were used in this work. The laser microprobe was used to identify particles appearing in the decomposed solutions and to determine the bulk purity of the samples introduced into peroxide. X-ray diffraction was used to identify precipitates in the peroxide, X-ray fluorescence yielded thickness measurements of plated metal (e.g. Ni and Au) and Auger electron spectroscopy provided information on surface constituents. The surface area and morphology were examined optically and with a scanning electron microscope (SEM). Surface roughness was measured using a Sloan Instruments Corporation Dektak. The ionic concentration of the metallic ions in the peroxide solution was determined either by atomic absorption, or emission spectroscopy. The degree of accuracy required dictated the technique used. In two instances specific amounts of ionic contamination (metallic ion) were introduced intentionally to produce decomposition under controlled conditions. These concentrations were verified by atomic absorption analysis.

RESULTS

The change in concentration of peroxide as a function of time was investigated for 128 test pieces, hereafter called samples, inserted into the solution. The samples have been divided into two types for discussion purposes: (1) samples prepared in the laboratory and (2) actual production samples.

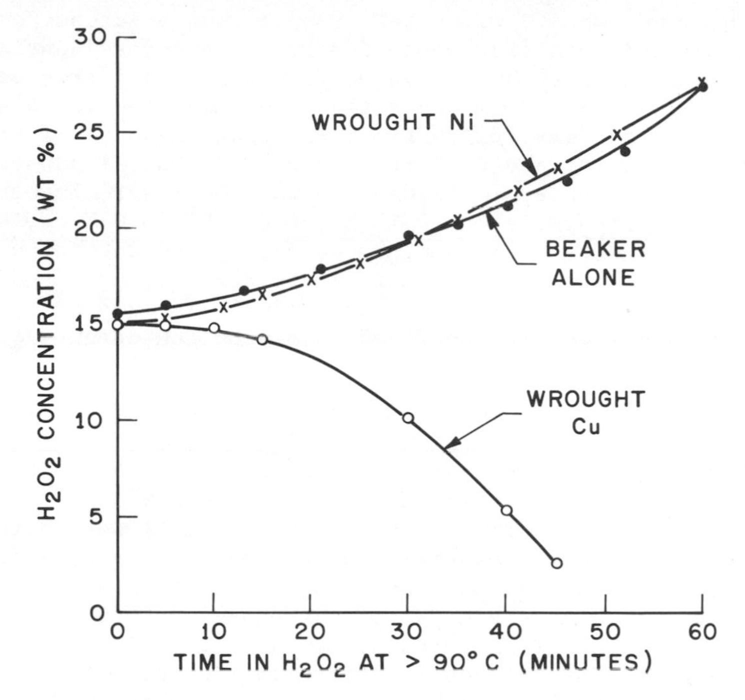

Figure 1. Change in concentration of peroxide showing effects of the beaker and specimens of wrought Cu and Ni.

LABORATORY SAMPLES

Decomposition without Samples

The first measurement of peroxide concentration as a function of time was performed without inserting a sample into the solution to determine the effect of the battery jar, thermometer, and pH probes on the peroxide at the test conditions. The data for this experiment is shown in Figure 1 as the curve labeled "beaker alone". In this case, the peroxide concentration is increasing with time. After 60 minutes the solution volume is reduced to 500 ml. Addition of DI water to obtain the original 1 liter volume produces a peroxide concentration within 1% of the original concentration. The concentration increase is therefore a result of reduced solution volume by evaporation of the H_2O. Any decomposition that may be occurring is insignificant compared to the evaporation rate. The pH at the beginning of the test was 4.4 and decreased to 3.6 after 45 minutes.

Decomposition with Ni

A second curve almost superimposed on the "beaker alone"

curve is obtained when a wrought Ni sample with a surface area of 185 cm^2 is placed in a fresh peroxide solution under identical conditions. In this test the pH ranged from 4.5 to 4.4 from beginning to end, respectively. From these two experiments we can see that: 1) the pH did not truly follow the concentration and 2) little if any decomposition is caused by the presence of the Ni substrate in the peroxide. Analysis of the Ni substrate by laser microprobe identified bulk impurities as 0.1% Mn, 0.03% Si, 0.01% Mg, 0.01% Co and 0.002% Cr with trace amounts of Cu, Al and Co on the surface. The purity for Ni was calculated to be 99.85%. Since the decomposition of the peroxide by the Ni was minimal, in this pH range, the rest of this study concentrated mainly on the effects of Cu and Au.

Cu Decomposition of Peroxide

The change in concentration of peroxide for 185 cm^2 of wrought Cu is also shown in Figure 1. The Cu purity was 99.998% with only 0.002% Ag in the bulk. Trace amounts of Ca and Al were detected on the surface. The decomposition of the peroxide occurs faster than the evaporation and therefore a net reduction in peroxide concentration occurs. At the start of the experiment (first 15 minutes), little change in peroxide concentration is noted. The decomposition is faster after the first 15 minutes but still requires an additional 30 minutes to decompose the peroxide to less than 3%. The solution pH changed from an initial 4.5 to final 6.6 after 45 minutes. A precipitate was noted in the solution at or near the end of the experiment. Filtering was performed and laser microprobe and x-ray analyses were made on the precipitate. The precipitate was identified as Cu and x-ray diffraction indicated that the copper was in the form of CuO. Approximately 10 ppm of Cu ions were detected in the solution by atomic absorption spectroscopy suggesting that the mechanism of decomposition was at least in part homogeneous.

Figure 2 shows the change in peroxide concentration for one evaporated Cu specimen having a surface area of 96 cm^2. Solution samples were removed from the beaker during the test and analyzed for Cu by emission spectroscopy. The approximate Cu concentration is shown for the times at which the sample was removed. Before the test started, a background Cu concentration of about 0.03 ppm was found in the peroxide. The Cu concentration does not change appreciably during the first 20 minutes and the peroxide concentration changes only slightly. The dissolution of Cu into solution is slow at first just like the decomposition rate. The final Cu concentration is too low since a precipitate is present at this point and therefore reduces the Cu concentration in solution. D. D. Eley and D. M. MacMahon[5] observed an initial induction period and subsequently a reaction which appeared to be coming from the solution as well as the Cu surface. This is in good agreement with present observations.

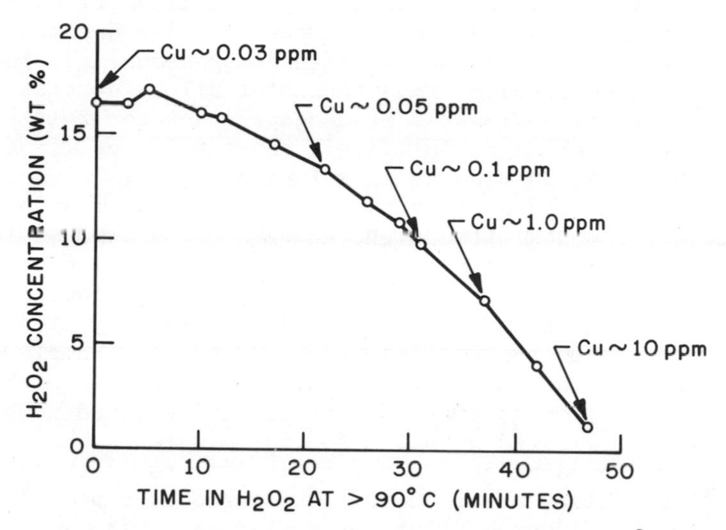

Figure 2. Change in concentration of H_2O_2 for 96 cm^2 of Cu. Cu ion concentration determined by spectrographic analysis.

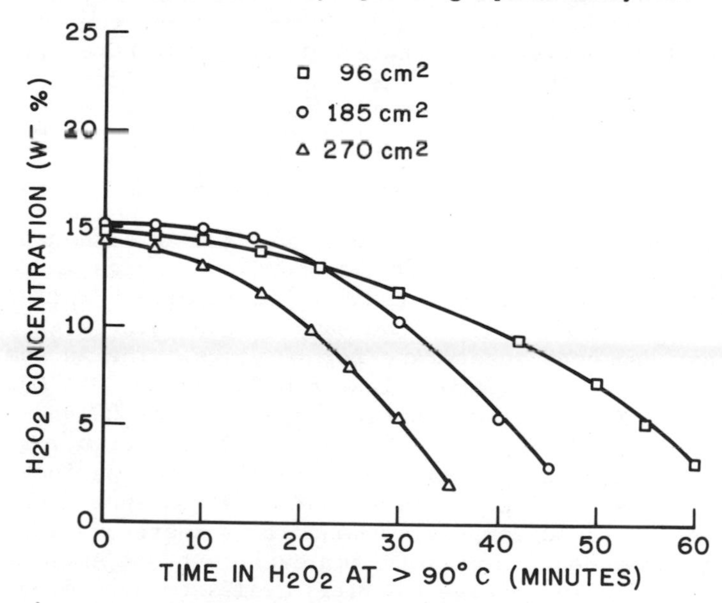

Figure 3. Change in concentration of H_2O_2 for three different areas of wrought Cu.

Although the presence of Cu ions in the peroxide solution suggests a homogeneous decomposition mechanism, heterogeneous decomposition may also be occurring. Figure 3 illustrates the effect of an increase in Cu surface area on the peroxide decomposition. This area effect may be caused by both homogeneous and heterogenous factors. The item of major importance for all Cu samples tested is, however, the relative slowness of decomposition for fairly high Cu area to solution volume ratios, i.e., about 0.5% per minute for 270 cm^2 of Cu after 20 minutes in solution.

The oxidation state of the metal catalyst was mentioned earlier as a factor which affects the homogeneous decomposition rate. The surface of Cu is easily oxidized and therefore a series of experiments was performed to show the effect of surface oxidation. Three samples with 96 cm^2 of evaporated Cu were subjected to peroxide cleaning after the following pre-treatment. Sample number one was pickled in a 10% HCl solution, number two was pickled followed by oxidation at 200°C for three minutes, and number three was pickled and oxidized at 300°C for ten minutes. The surfaces changed color from a bright Cu after pickling to rusty colored after 200°C and black after 300°C. The three decomposition curves were discernibly different, but the differences were small, less than 2% after 30 minutes. The sample subjected to peroxide immediately after cleaning (number one) showed the fastest decomposition, the 300°C treated sample (number three) the second fastest and the slowest was treated at 200°C for three minutes. The oxidation process tends to decrease the peroxide decomposition rate.

Au Decomposition of Peroxide

The decomposition of peroxide by wrought Au is dependent on macroscopic surface area. Figure 4 shows the change in concentration of peroxide for various areas of "as received" rolled wrought Au (not previously cleaned). The samples represent a wrought Au surface that had been in storage for 10 years. The Au samples were 99.96% pure. Bulk impurities of 0.02% Ag, 0.005% Cu, 0.01% Pd, and 0.002% Rh and trace surface impurities of Ca and Be were detected by laser microprobe. Decomposition of the peroxide occurs in approximately 15 minutes for 185 cm^2 of surface area.

As mentioned earlier surface contamination itself could be responsible for the decomposition and it is therefore important that sample treatment be uniform for each test. A Au sample was subjected to five successive cleaning cycles to determine the effect of the surface contamination on the peroxide. The results in Figure 5 show that the "as received" decomposition is rapid for 185 cm^2 of wrought Au, but decreases with successive exposure to new solutions. After the third exposure, no further change in decomposition rate is observed. A fifth exposure was made after allowing the substrate to remain in a petri dish at room condi-

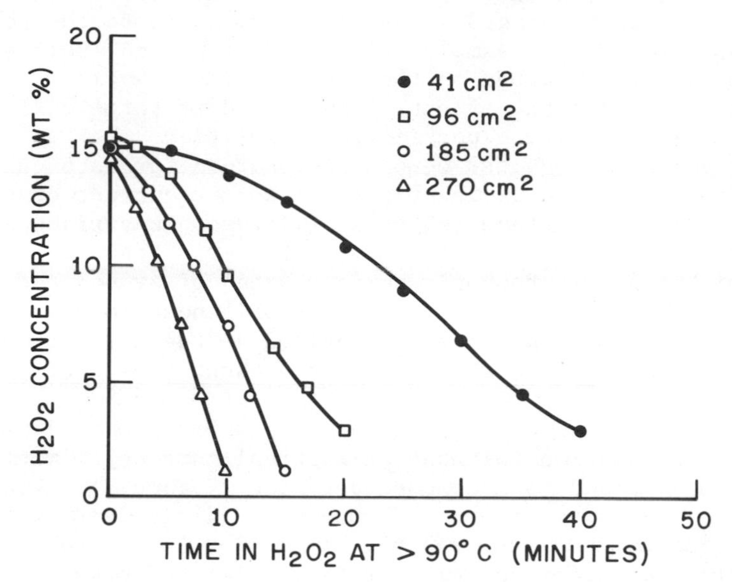

Figure 4. Change in concentration of H_2O_2 for four different areas of wrought Au.

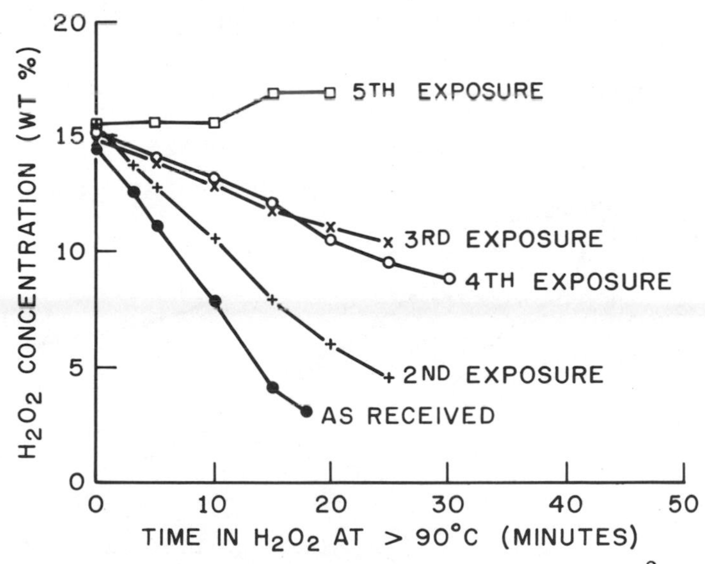

Figure 5. Change in concentration of H_2O_2 for 185 cm^2 of wrought Au with five successive H_2O_2 exposures.

tions for a period of four weeks. Little decomposition was evident thus suggesting that the factors responsible for the fast decomposition were changed by previous exposures to the peroxide. Auger analysis of this sample and the "as received" sample found significantly more carbon on the "as received" sample.

Figure 6 shows the effect of Au plated at three different current densities on the peroxide decomposition. Dektak measurements of surface roughness were also recorded for each current density and are shown on the figure. As the current density is increased, the surface roughness (microscopic surface area) increases and the peroxide decomposes faster. The samples were made from 96 cm^2 (macroscopic surface area) of evaporated Ti and Pd followed by plated Au of sufficient thickness to completely mask the Pd. A second Au plate was deposited at the current densities shown in the figure. Carbon could be present from entrapped plating salt residue.

Au plating that occurs when a metal substrate is immersed in a Au plating solution (without an external current) "immersion Au" has also been found to decompose peroxide (Figure 7). Wrought Ni, alone, was previously shown to cause very little decomposition of the peroxide for the as-mixed pH. Immersion Au plated in a cyanide bath on an identical wrought Ni sample, and on a plated Ni sample caused violent and rapid decomposition of the peroxide. The physical properties of this immersion Au have not been determined and therefore a complete understanding is not possible at present. However, it is believed that this immersion deposit, present on

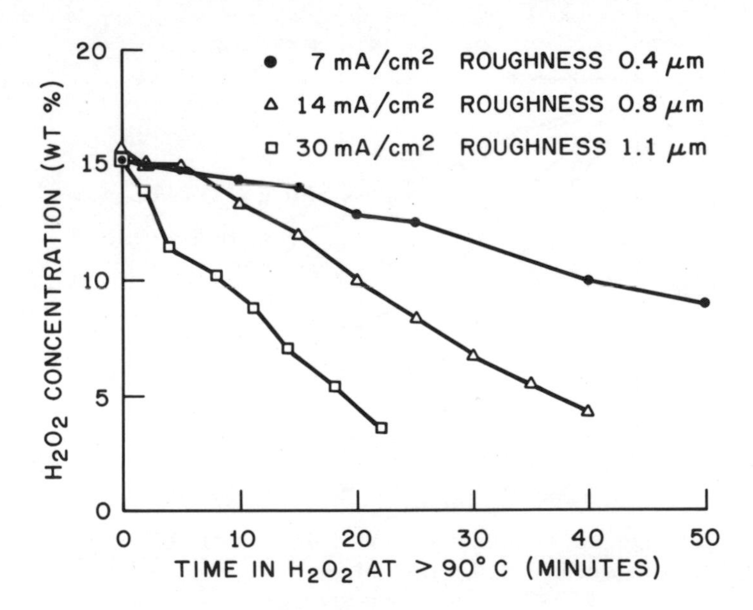

Figure 6. Change in concentration of H_2O_2 for 96 cm^2 of plated Au on TiPd at various current densities.

lead frames, is a contributing factor to H_2O_2 decomposition and solution particulates in the production environments as described below.

H_2O_2 Decomposition on Production Leadframes

The cost of gold has initiated the development of the selective plating process[6,7] to reduce the amount of Au applied to the lead frame; i.e., the tip of the lead for bonding and the shank of the lead for solderability. The remaining area consists of plated Ni and a small amount of exposed Cu. The selectively plated leads connect with the substrate carrying the I.C.

As the selective or spot plated frame use increased, peroxide consumption increased and the frequency of violent peroxide reactions also increased. Other data[8,9] indicated that along with the violent decomposition, particulate matter was found both in the solution and on the active device surface. A yield loss was attributed to these particles (believed to be Au) and they were believed to originate from the selectively plated lead frame.

Spot plated lead frames sometimes exhibit a yellowish tint

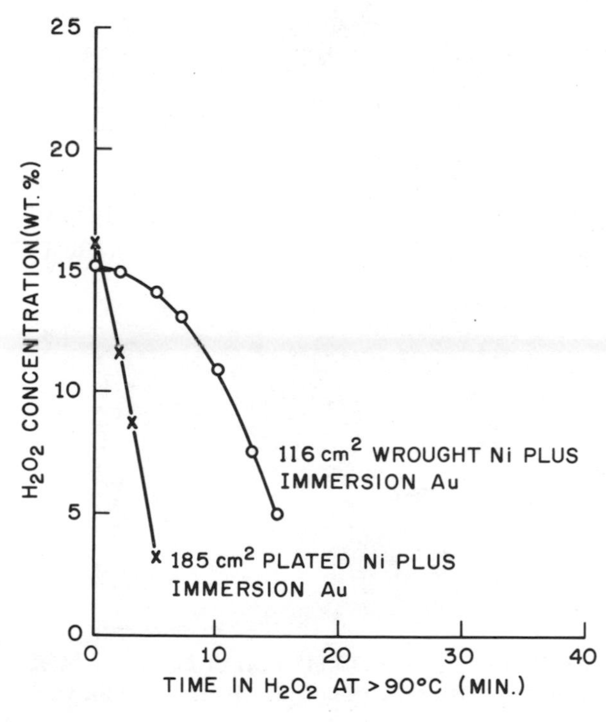

Figure 7. Change in concentration of H_2O_2 for plated and wrought Ni with immersion Au.

in the Ni area of the frame. Auger electron spectroscopy analysis showed the presence of Au over the entire surface area where there should be only Ni. It is probable that this deposit originates from immersion plating, which occurs when the lead frame is exposed to plating solution, without current, during the index cycle of the plating facility. No matter what type plating mask is used the deposit is always present on selectively plated frames as a very low density, sometimes discontinuous film, up to 750Å thick. This information was obtained through X-ray fluorescence examination.

Figure 8, shows two photographs of the same area of a lead frame and illustrates the poor adherence of extraneous Au. The top photo shows a cross scratched into the frame at an area containing the extraneous Au. The lower photo is the same area of the

Figure 8. Spalling of extraneous Au by H_2O_2.

frame after exposure to H_2O_2. Spalling is apparent and particles were observed in the solution. The actual decomposition rates for 370 cm^2 of these same frames and for 185 cm^2 in 1 liter of H_2O_2 solution are shown in Figure 9. The decomposition was violent and the particles filtered from the solution were determined to be 95 to 99% Au by the laser emission microprobe.

To determine the effect on product yields, these frames were used to fabricate actual devices. Two groups of medium scale integrated circuits were assembled with half receiving the peroxide cleaning and the other half not cleaned. Each half contained over 2000 circuits. In both groups, functional tests showed a 2:1 difference in the total test failure rates with the cleaned groups having the higher failure rate. The difference is attributed to the particles of extraneous Au being dislodged from the frame and subsequently being deposited on the active device area. However, no particles were verified on the decapsulated devices due to the interference of the residual encapsulant.

The factor causing the actual decomposition of the peroxide is not nearly so evident. An attempt is made to answer the question through Figure 10. The curves shown are taken from the previous data presented and, as indicated, Ni is not a factor at these conditions. Wrought Au and wrought Cu do cause decomposition but at a relatively slow rate compared to the production frames in Group 2. The fast decomposition rates occur with high microscopic surface area materials namely the immersion Au on plated

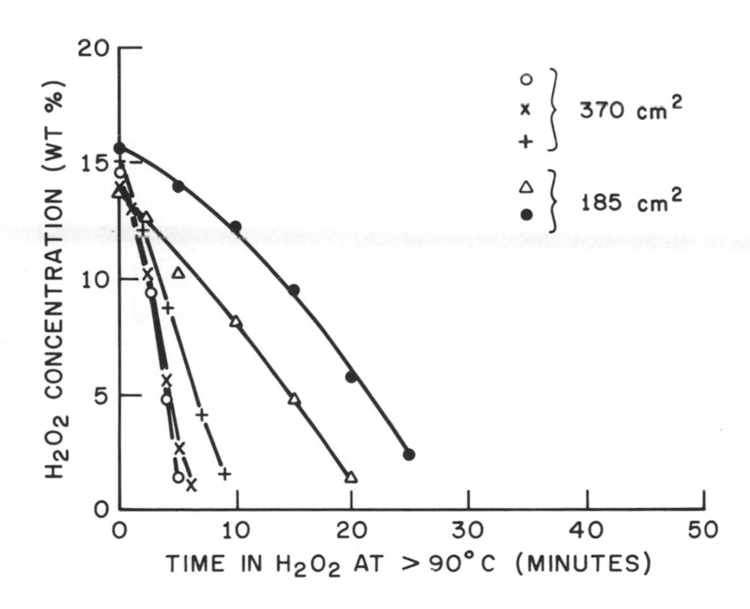

Figure 9. Change in concentration of H_2O_2 for production leadframes.

Ni. Since the immersion Au is removed in the peroxide the effective areas are also dramatically increased. There also remains the possibility that the rates for Au are controlled by entrapped plating salts (organic). It must be concluded that the decomposition is due to extraneous Au and/or C. The magnitude of the problem can be illustrated by examining the ratio of the production lead frame macroscopic surface area to peroxide solution volume, which on the product line is approximately 2000 cm^2/liter for the most common frame type. This large area is made up of less than 10 cm^2/liter of exposed Cu, 450 cm^2/liter of electroplated Au and 1540 cm^2/liter of electroplated Ni covered with varying amounts of extraneous Au. The magnitude of the decomposition problem on the product line can be seen when considering that the experimental data presented was obtained with a macroscopic surface area to solution ratio that is 1/10 that of production.

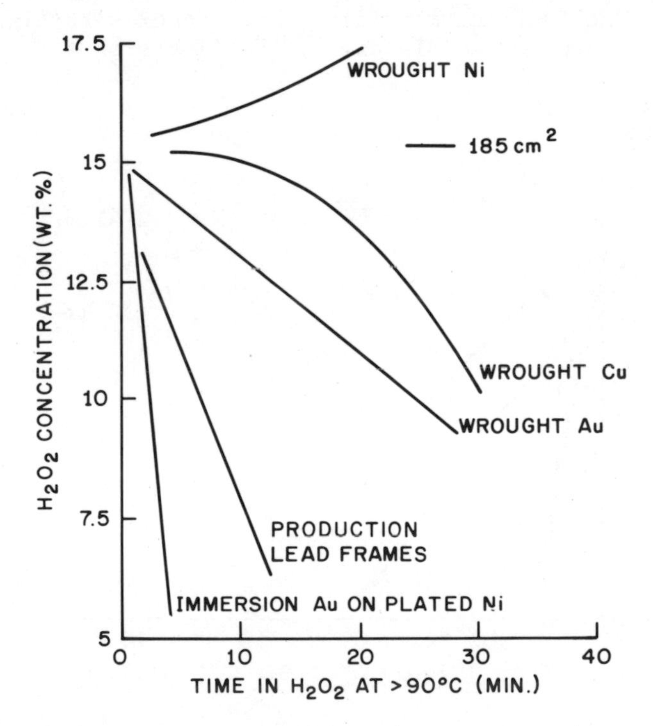

Figure 10. Change in concentration of H_2O_2 comparison for various materials.

SUMMARY

The factors which adversely affect H_2O_2 cleaning and product yield are:

1. Extraneous Au and/or C are the major contributors to H_2O_2 decomposition.
2. Surface area, both macro-and microscopic, increases the peroxide decomposition rate.
3. The spalling of extraneous gold from selective plated lead frames and the subsequent deposition of these particles on device surfaces is a direct consequence of peroxide cleaning.

Those factors which do not affect H_2O_2 decomposition are:

1. Wrought Ni or plated Ni.
2. At the present time, copper is not a factor in the decomposition of peroxide. It may become a factor as the amount of exposed bare copper, especially plated copper, increases in use.

Generally, the basic technical problem of H_2O_2 decomposition has not been solved; however, knowing its major cause allowed us to alter the processing sequence and thus eliminate the yield problem.

ACKNOWLEDGMENT

The authors would like to thank F. Minardi, J. M. Morabito and C. A. Steidel for their support and guidance on this work.

REFERENCES

1. W. C. Schumb, C. N. Satterfield, and R. L. Wentworth, "Hydrogen Peroxide," Reinhold Publishing Corp., New York, 1955.

1A. W. Kern and D. Puotinca, RCA Rev. 31 (2), 187 (1970).

2. M. J. Sienko and R. A. Plane, "Chemistry," 4th Edition, McGraw-Hill, New York 1971.
3. J. M. Thomas and W. J. Thomas, "Introduction to the Principles of Heterogeneous Catalysis," Academic Press, New York 1967.
4. Kirk-Othmer, Editor, "Encyclopedia of Chemical Technology," Second Edition, Volume 11, John Wiley and Sons, Inc., 1966.
5. D. D. Eley and D. M. MacMahon, Catalysis 14, 193-200 (1969).
6. D. R. Bush, J. N. Lesyk and H. J. Litsch, "Characterization of the Gold Spot Plating Process for Lead Frame Bondability," unpublished Bell System Data, Allentown, October 13, 1975.
7. D. L. Rehrig, The Western Electric Engineer, Allentown, 22 (2), 48 (April 1978).
8. D. L. Rehrig, "Contamination Resulting from Hydrogen Peroxide," unpublished Bell System Data, Allentown, May 27, 1976.
9. M. L. White, (1977), personal communication.

ELLIPSOMETRIC OBSERVATIONS OF ALUMINIUM HYDROXIDE FILMS GROWN IN WATER VAPOR

W.E.J. Neal and A.S. Rehal

The University of Aston in Birmingham
Gosta Green
Birmingham B4 7ET, United Kingdom

The paper describes ellipsometric investigations of aluminium hydroxide films grown on aluminium surfaces which have been exposed to water vapor. The types of samples used for the investigations have been:-

1. Commercial aluminium sheet which had been etched by Argon ions and exposed to water vapor in a vacuum system at 25 °C.

2. As supplied commercial aluminium sheet which had been chemically or mechanically cleaned and exposed to saturated water vapor at 70 °C.

3. Freshly deposited films exposed to water vapor in a vacuum system at 25 °C.

The optical constants for aluminium hydroxide at a wavelength of 549 nm for films up to 65 nm thick were found to be n = 1.55 to 1.60, k = 0.

For hydroxide films which were grown on gold surfaces (by depositing aluminium films on gold and then exposing to saturated water vapor at 70 °C) the best values of optical constants were found to be:-

n_1 = 1.58±0.03, k_1 = 0 for films< 80 nm thick
n_1 = 1.58±0.03, k_1 = 0.0009 for films> 80 nm thick

INTRODUCTION

The susceptibility of Aluminium-Zinc-Magnesium alloys to stress corrosion often restricts their more widespread use with the result that much research effort has been put into determining the mechanism of stress-corrosion cracking. The properties of corrosion layers of hydroxides or oxides on metals are therefore of interest to workers in many branches of science and engineering. Tronstad[1] was one of the earliest workers to study the growth of corrosion layers on materials by ellipsometry. Following his pioneering work the technique has been employed by many authors [2,3,4,5] in conjunction with oxidation and corrosion studies.

The work reported here forms part of a programme of investigation using the combined techniques of ESCA (XPS) electron microscopy and ellipsometry to study the composition, optical and structural properties of films grown on aluminium and aluminium alloys exposed to various environmental conditions. The optical properties of aluminium hydroxide films grown on pure aluminium when exposed to saturated water vapor will be considered here, particularly in the early stages of growth. Alwitt[6] and Vedder and Vermilyea[7] have made extensive studies on the reaction of aluminium with water (in the liquid phase). The reaction of aluminium with water at 70 ^{o}C produces a duplex film consisting of pseudo-boehmite on aluminium oxyhydroxide, similar to boehmite (AlOOH) but containing more water, and bayerite ($Al(OH)_3$). Bayerite crystals take the form of pillars or cones and pseudo-boehmite appears as needles when viewed by transmission electron microscopy or platelets when viewed by scanning electron microscopy[6].

There has been much less reported work on the reaction of aluminium with water vapor or the formation and structure of aluminium hydroxide films on aluminium formed at low exposures. Such studies require ultra high vacuum facilities and surface sensitive techniques and we have used the complementary techniques of ESCA with ellipsometry. In some instances it has also been possible to combine transmission electron microscopy (TEM), scanning transmission electron microscopy (STEM) and scanning electron microscopy using the JEOL 100C Temscan Instrument. It is possible with the instrument to compare surface features with underlying microstructure for precisely located areas.

OPTICAL MEASUREMENTS

An outline of the technique of ellipsometry is included in a review on the "Application of ellipsometry to surface films and film growth" by Neal[8]. Essentially, in measurements on surface film growth, the optical properties of a surface are determined before, during and after the growth of the layer or film and the instrument is particularly appropriate in studies of highly reflecting materials such as metals. In most circumstances when an incident beam of plane polarized light is reflected from a metal surface the reflected beam is elliptically polarized. The optical constants of a surface can be determined from the orientation of the ellipse with respect to the plane of incidence and the ratio of the major to minor axis. The presence of a thin layer on the surface modifies the optical constants and in turn the characteristics of the elliptically polarized light in the reflected beam.

In the present work a chopped light source was used in conjunction with a phase sensitive detector. For most of the investigation the radiation wavelength was kept at 549 nm with a compensator in the reflected beam. On occasions when a range of wavelengths was used the compensator was removed and the method of analysis due to Beattie and Conn[9] was employed.

The basic equation of the instrument is given by:

$$r_p / r_s = \tan \psi \exp i \Delta \quad (1)$$

where r_p and r_s are reflection coefficients for light with electric vectors parallel and perpendicular to the plane of incidence respectively. The parameters ψ and Δ are obtained from instrument settings of polarizer and analyzer at extinction. The refractive index is written as $n - ik$ and the optical constants n and k for a film free surface can be computed from ψ and Δ and the angle of incidence ϕ_o. The angles ψ and Δ can be used to characterise a surface. The parameters change with the growth of a surface layer and are indicative of changes in the characteristics of the elliptically polarized light. In the early stages of growth the changes in ψ and Δ are proportional to film thickness (t) and can be written:

$$\psi = \bar{\psi} + At \quad (2)$$

$$\Delta = \bar{\Delta} + Bt \quad (3)$$

where $\bar{\psi}$ and $\bar{\Delta}$ relate to a clean surface and ψ and Δ to a film covered surface. A and B are constants and depend on the optical constants of the film and base metal and on the angle of incidence. For films thicker than about 10 nm, more exact equations, derived by Drude[10] almost 100 years ago, and given by Ditchburn[11] in a suitable form for computation, must be used. Figure 1 illustrates reflections from an ideal film covered surface in which multiple reflections take place in the film. The phase of the reflected

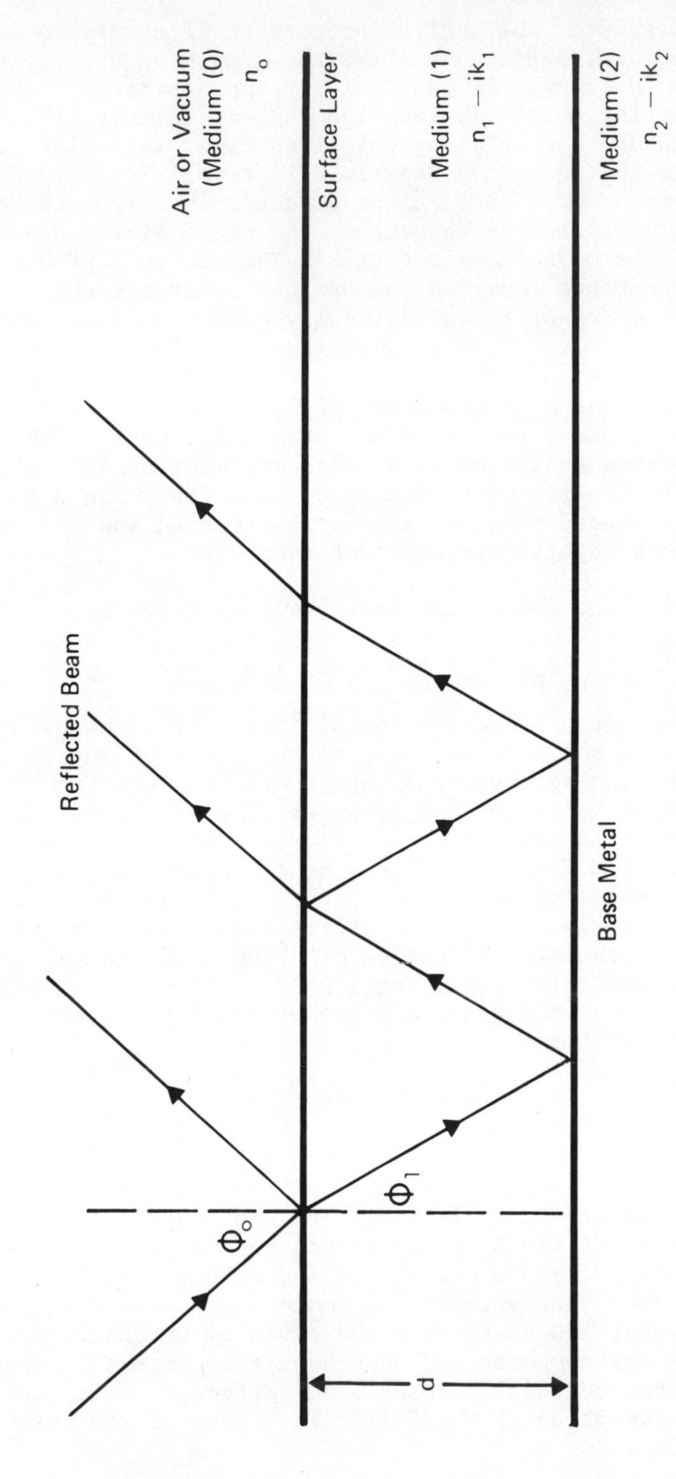

Figure 1. Reflection and transmission of light by a surface layer.

beam is thus dependent on film thickness. In the case of a non absorbing film in which $k_1 = 0$ on a substrate of known optical constants (n_2 and k_2) two unknown quantities n_1 and t for a surface film can be determined. In the case of films which absorb the electromagnetic radiation there is a basic difficulty in that three unknown quantities n_1, k_1 and t need to be determined and only two instrument parameters (ψ and Δ) are measured. Several methods have been suggested to increase the number of instrument readings and these are discussed by McCrackin and Colson[12].

HYDROXIDE GROWTH ON ALUMINIUM

Figure 2 shows a plot of computed variations in ψ and Δ as a layer of aluminium hydroxide ($n_1 = 1.58, k_1 = 0$) forms on aluminium for light of wavelength 549 nm (the angle of incidence = 64.20).

Figure 3 shows a plot of computed variations in ψ and Δ for layers of aluminium hydroxide on gold. For a non absorbing corrosion layer a closed loop is formed as the layer thickness increases in that the ψ and Δ values repeat themselves. For the growth of an absorbing layer a spiral would be formed. Even for non-absorbing films which is the case for most oxides and hydroxides, in order to obtain both n_1 and t it is necessary to know the optical constants n_2 and k_2 of the underlying base metal, i.e. the film free surface instrument readings ψ and Δ (equations (2) and (3)) need to be known. In studies of corrosion layers particularly on reactive metals such as aluminium, problems are experienced in obtaining the characteristics of a film free surface. Even in a vacuum system held at a pressure of 10^{-6} Pa layers form on a freshly evaporated film in a matter of minutes. This poses a problem when values of corrosion thicknesses are required in the early stages of growth. Various approaches have been made to produce film free surfaces in order to measure subsequent growth of corrosion layers when exposed to different environments. In some cases fresh surfaces have been prepared by evaporating a metal in ultra high vacuum[13,14]. In other instances clean surface areas have been prepared by bombarding a corroded surface with positive argon ions to remove existing corrosion layers, prior to subsequent growth in a chosen environment. We have carried out this procedure for sheets of commercial aluminium. In addition we have used freshly evaporated aluminium films, on aluminium sheet. The films were then exposed to an oxygen or water vapor environment. Ellipsometric observation on fresh films or bombarded surfaces provides information from which n_2 and k_2 can be directly calculated. We have also employed a procedure of using a non reactive metal such as gold or platinum on which to form a corrosion layer. This enables the optical constants and thickness of corrosion layer growth to be determined without the need for optical measurements in an ultra high vacuum system.

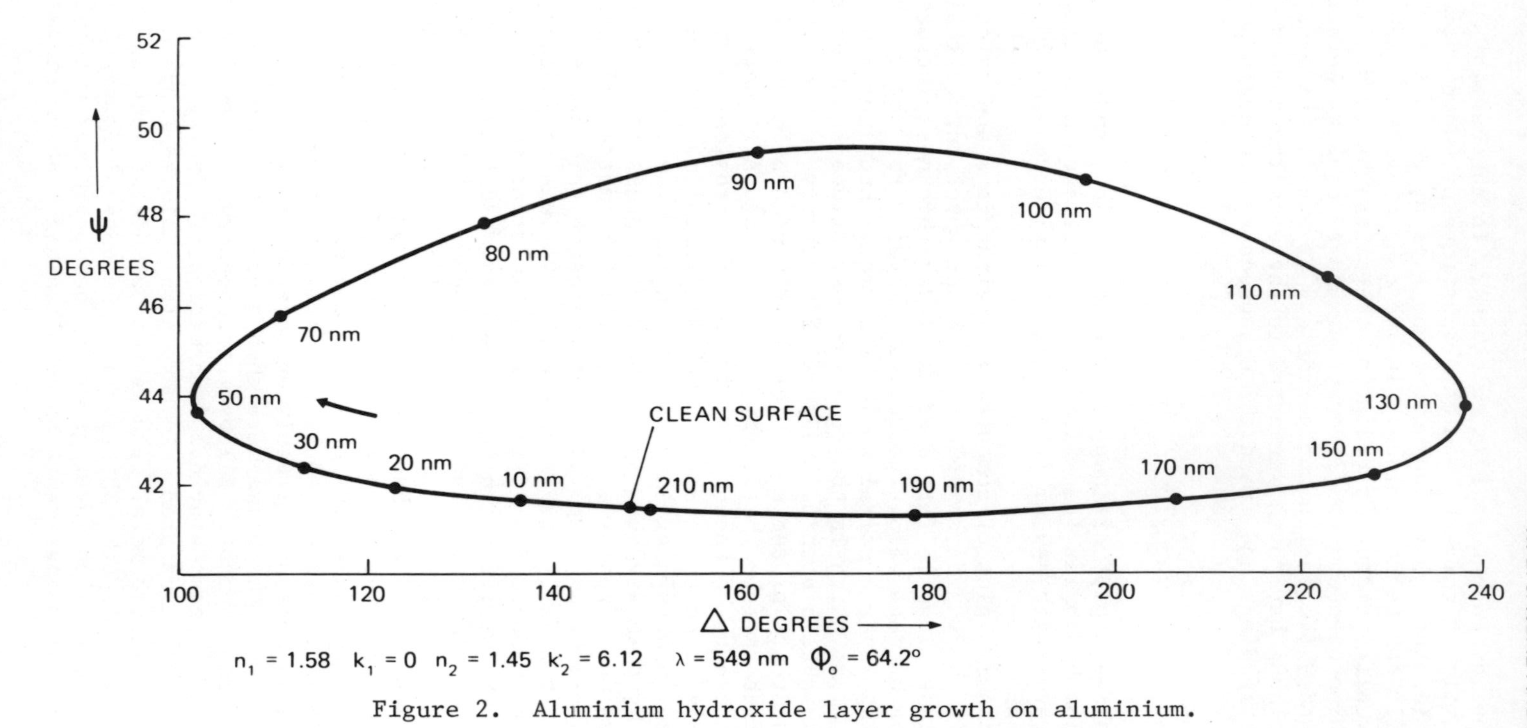

$n_1 = 1.58$ $k_1 = 0$ $n_2 = 1.45$ $k_2 = 6.12$ $\lambda = 549$ nm $\Phi_o = 64.2°$

Figure 2. Aluminium hydroxide layer growth on aluminium.

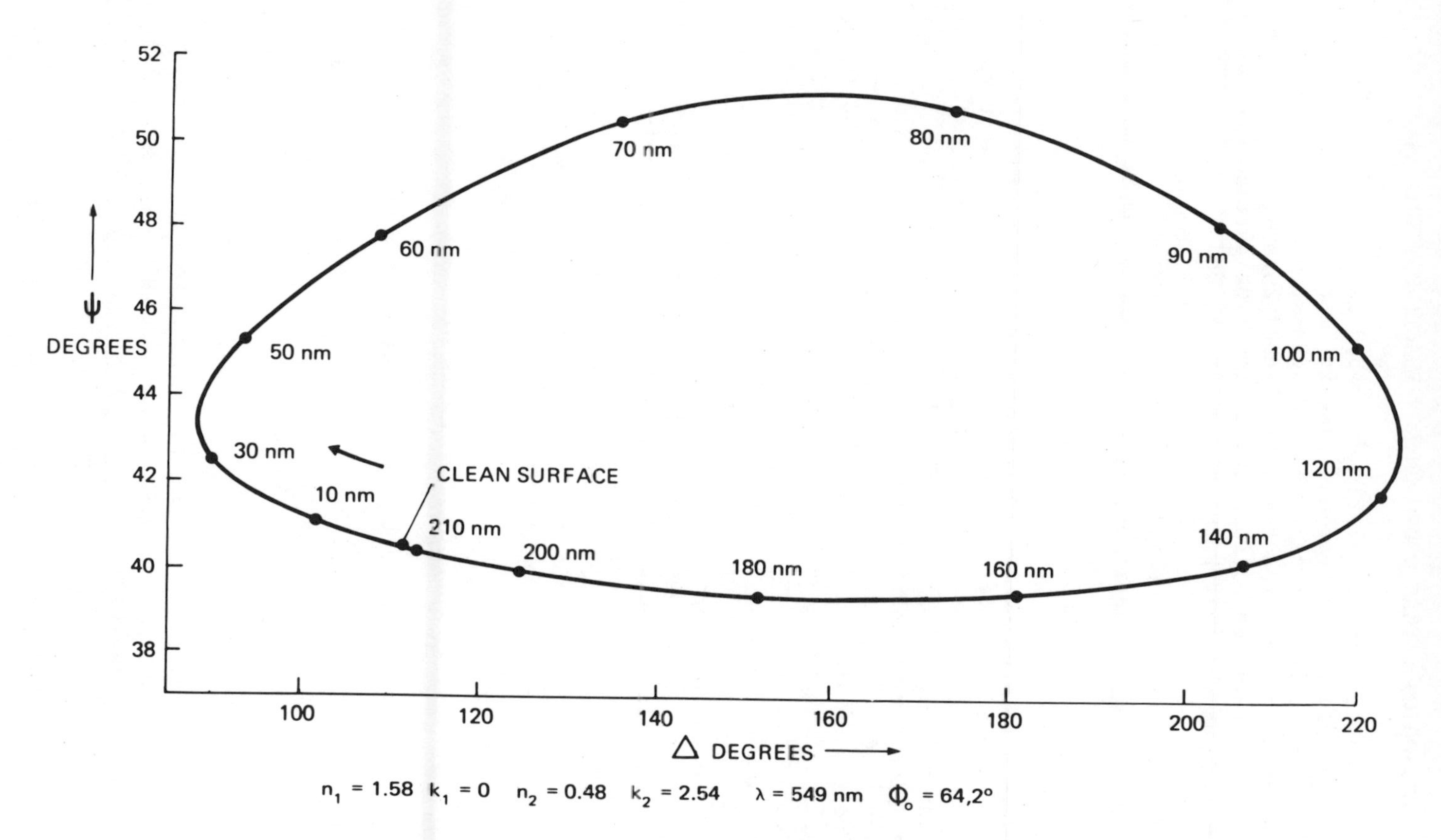

Figure 3. Aluminium hydroxide layer growth on gold.

CORROSION LAYERS GROWN ON NON REACTIVE METALS

The procedure can be briefly described as follows: Thick films of a highly reflective, non reactive metal such as gold or platinum are deposited on substrates of glass, quartz or sapphire. The film forms the base and the optical constants n_2 and k_2 for the base are determined ellipsometrically. The surface will be characterised by the instrument parameters ψ_s and Δ_s (corresponding to ψ and Δ of equations (2) and (3)). A film of the reactive metal whose corrosive properties are to be examined is then superimposed on the non reactive base. Its thickness can be determined during deposition by a quartz crystal monitor or after deposition by an interference method suggested by Tolansky[15] using fringes of equal chromatic order. We used the latter method. The reactive metal is then exposed to the appropriate environment and the corrosive reaction permitted to proceed until all the reactive metal has been converted to films of the corrosion material. The instrument parameters ψ_F and Δ_F for the composite surface of base material plus corrosion films are determined. Since n_2 and k_2 are known from values of ψ_s, Δ_s and ϕ_o (the angle of incidence) if the corrosion layer is non-absorbing both its refractive index n_1 and its thickness t can be determined. A check on the thickness of the corrosion layer can also be made by the Tolansky[15] method. In the case of absorbing layers it is necessary to determine the thickness independently of the ellipsometer if the refractive index of the layer is not known, as mentioned earlier. Measurement both of the reactive metal thickness and the corrosion thickness enabled volume changes from the metal to corrosion layer to be estimated. In the present investigations the base metal was gold and aluminium hydroxide films of various thicknesses were grown by exposing aluminium films of varying thicknesses (deposited on the gold) to water vapor at 70 ^{o}C.

RESULTS AND DISCUSSION

Aluminium Hydroxide Grown on Aluminium

Table I shows the variation of ψ and Δ for a sample of commercial aluminium sheet which has been etched with Ar^+ ions. The radiation wavelength was 549 nm.

Table I. Etching of 'As Received' Aluminium Sheet.

Δ	123.4	138.95	138.80	143.37	143.56
ψ	39.9	42.29	42.85	42.89	43.0
Etch time (mins)	0	30	60	150	210

The total oxide thickness on 'as received' sheet was calculated to be 20 nm.

Table II shows the variation of ψ and Δ for a sample of commercial aluminium sheet which had first been heated at 350 °C for 15 mins. in air and then etched with Ar^+ ions. The radiation wavelength was 549 nm.

Table II. Etching of Pre-heated Aluminium Sheet.

Δ	115.13	138.90	139.45	140.09	141.43	143.47
ψ	37.2	42.85	42.90	42.86	43.0	43.0
Etch time (mins)	0	90	120	150	210	300

The total oxide thickness prior to etching was calculated to be 28 nm.

In both cases the accelerating voltage was 4kV and the beam current 20μA.

Table III shows the variations in ψ and Δ as the etched aluminium sheet (from Table II) was exposed to water vapor at 25 °C. Values of hydroxide layer thickness calculated from ellipsometry and ESCA measurements are also given.

Table III. Hydroxide Growth on Aluminium Sheet.

ψ	Δ	Exposure (Langmuirs)	Hydroxide Thickness, nm Ellipsometry	ESCA*
43.0	143.47	0	0	0
43.1	142.92	10	0.52	0.51
43.15	142.78	30	0.66	0.56
43.18	142.65	90	0.8	0.65
43.18	142.40	200	1.06	0.69
43.20	142.29	900	1.17	
43.20	141.94	$2x10^3$	1.50	0.86
43.25	141.77	$7x10^3$	1.70	1.01

*These thicknesses were obtained assuming electron escape depths of 2.0 nm and 1.65 nm from pure aluminium and aluminium hydroxide respectively.

Table IV shows the variations in ψ and Δ for hydroxide growth at 25 °C on aluminium film which had been deposited on etched aluminium sheets with a system base pressure of 10^{-8} Pa.

Table IV. Hydroxide Growth on Aluminium Film.

ψ	Δ	Exposure (Langmuirs)	Thickness nm Ellipsometry
44.18	144.1	0	0
44.18	143.56	10	0.52
44.12	143.33	70	0.85
44.25	142.96	300	1.12
44.26	142.77	10^3	1.33
44.4	142.45	$6x10^3$	1.62

From tables I and II it can be seen that it is possible to obtain values of ψ and Δ, after etching which can be used to characterise a surface prior to subsequent tests. In some samples of sheet which were not previously heat treated differing values of ψ could be obtained and was attributed to differences in surface structure. The rates of hydroxide growth on films and etched surfaces in the early stages of growth were similar.

A series of tests was also made on aluminium films and on commercial aluminium sheets which were either chemically or mechanically cleaned and then exposed to saturated water vapor at 70 ^{o}C for between 150 and 200 minutes. The average values of refractive index for aluminium hydroxide films up to 65 nm thick at a wavelength of 549 nm with k assumed to be zero were:

Etched Sheet and Films $n = 1.55\pm0.05$
Sheet (Chemically Cleaned) $n = 1.55\pm0.05$
Sheet (Mechanically Cleaned) $n = 1.60\pm0.05$

ALUMINIUM HYDROXIDE GROWN ON GOLD

Figures 4 and 5 show computed variations in the values of ψ and Δ for films of differing refractive indices on a gold surface. The experimental values found for aluminium hydroxide films of 8 different thicknesses are also given. The instrument settings for a film free gold base were $\psi_S = 40.7^{o}$ and $\Delta_S = 119.6^{o}$ at a radiation wavelength of 549 nm and an angle of incidence of 64.2^{o}. Values of the optical constants of the gold base were computed using the measured values of ψ_S and Δ_S and found to be $n_2 = 0.48$ and $k_2 = 2.54$. The experimental points were obtained by exposing the aluminium films to saturated water vapor at 70 ^{o}C until all the metal had been converted to the hydroxide in each case.

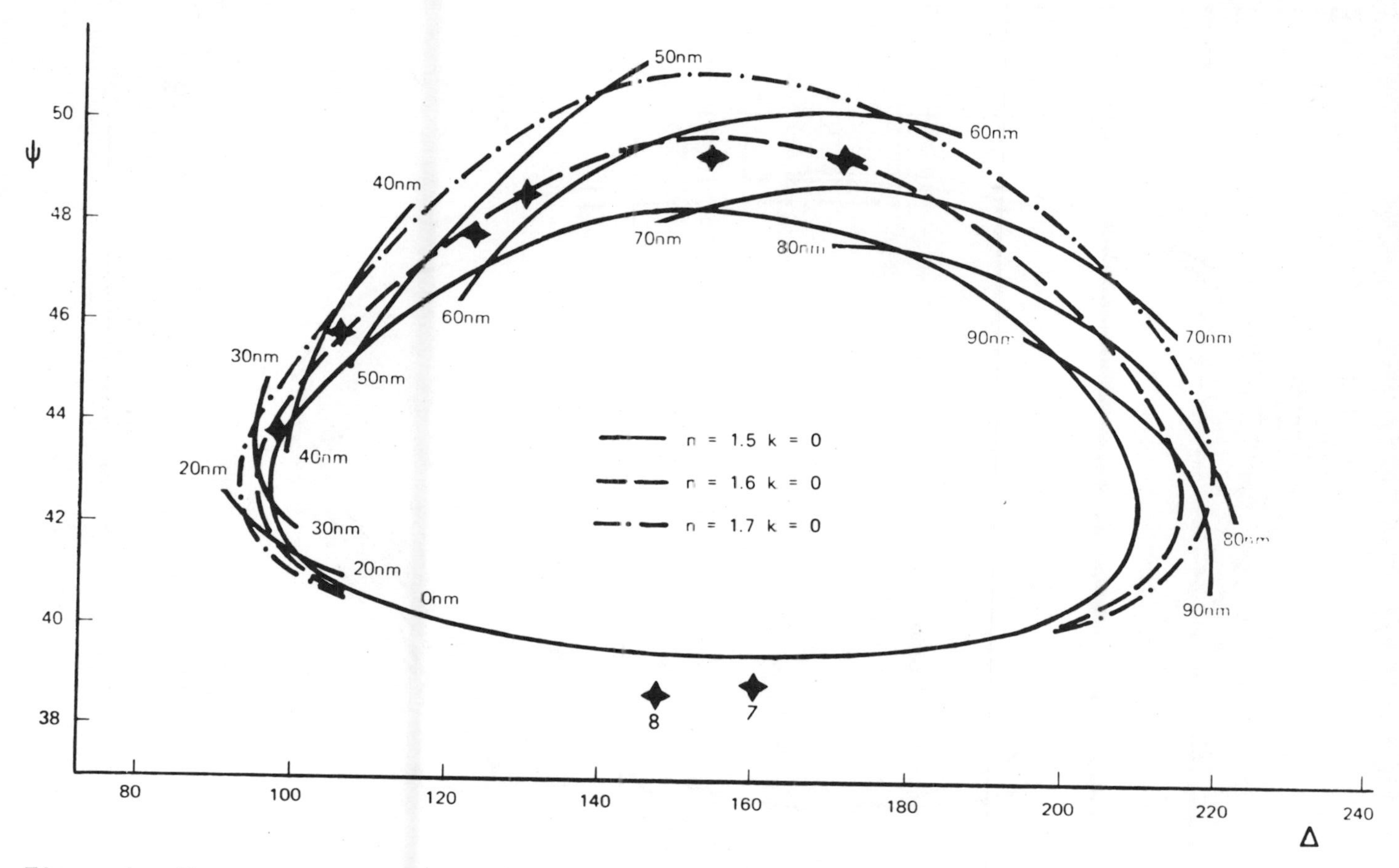

Figure 4. The graph of ψ against Δ for films of different refractive indices on a gold substrate.

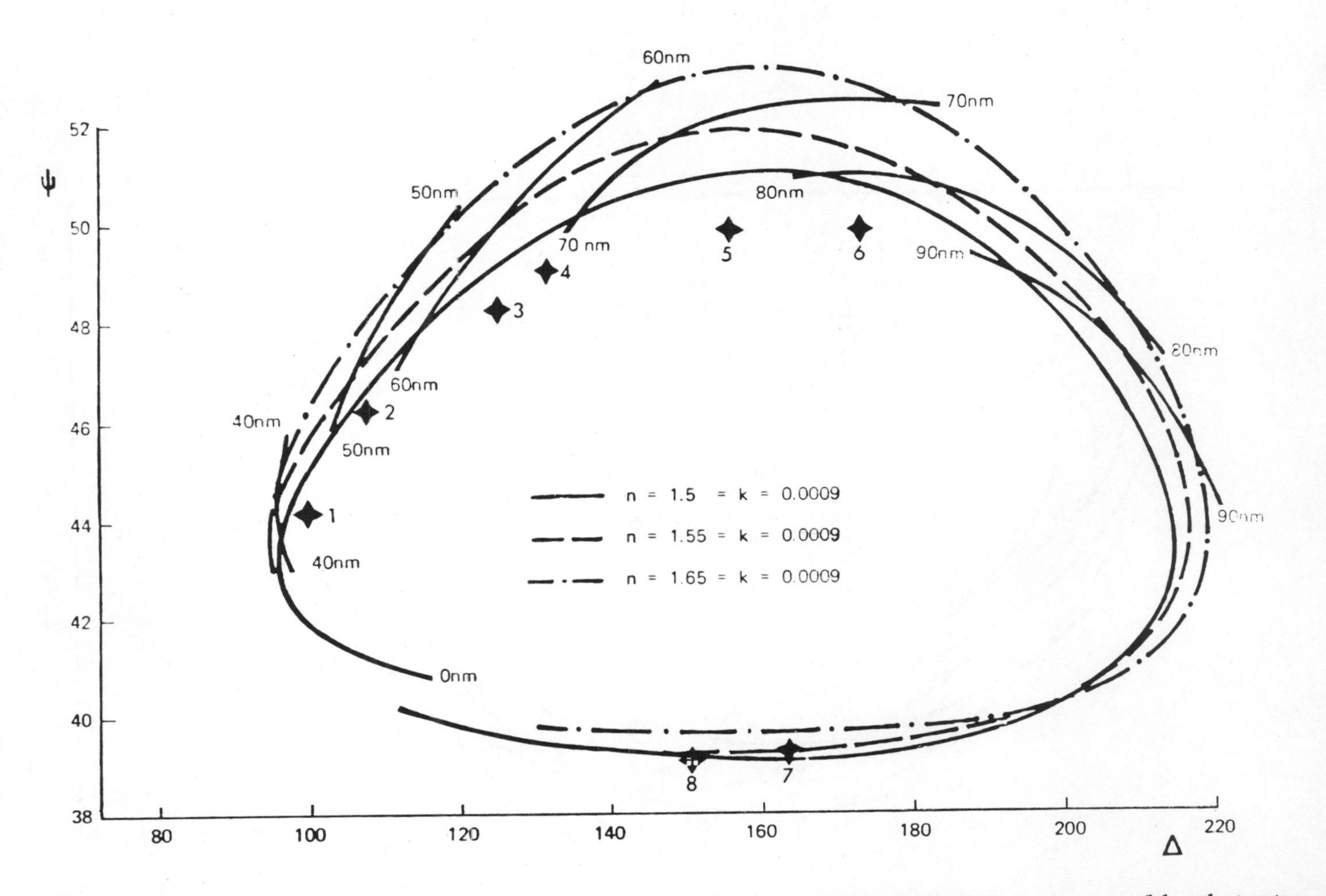

Figure 5. The graph of ψ against Δ for films of different refractive indices on a gold substrate.

The values of ψ_F and Δ_F, given in figures 4 and 5, were then computed using a two layer programme and assumed values for the hydroxide optical constants. In all cases in figure (4) $k_1 = 0$ with n varying between 1.5 and 1.7. In each case a complete loop is formed and values of ψ_F and Δ_F repeat themselves for hydroxide films greater than 212 nm.

The experimental points for two films (7 and 8) do not lie on any of the computed closed loops. Figure 5 shows similar plots but with an assumed slight absorption for the hydroxide film with $k_1 = 0.0009$ in each case.

The optical constants for aluminium hydroxide which best fit the experimental results for a wavelength of 549 nm are:

For films < 80nm $n_1 = 1.58 \pm 0.03$, $k_1 = 0$
For films > 80nm $n_1 = 1.58 \pm 0.03$, $k_1 = 0.0009$

These results compare well with those obtained for growth on etched sheet and on fresh aluminium films deposited on etched sheet. In ellipsometric examination of surfaces there is no resolution in the plane of the surfaces so that thicknesses relate to average values over the area examined (in this case $2mm^2$).

In this work it was also found that the ratio of hydroxide layer thickness to metal thickness decreased from 1.6 for films 3 nm thick to 1.4 for films 18 nm thick.

Values of optical constants for aluminium hydroxide which we have obtained from growth on aluminium films and sheet and on gold can be compared with films prepared by reacting electropolished aluminium with hot water and given by Barrett[16] as n = 1.60, $k_1 = 0$. As already mentioned the latter technique could be used with any non reactive base metal but gold and platinum would be preferable since with most other metals thin oxide layers would always be present. Computer calculations show that by using platinum (as the base) for the examination of aluminium hydroxide corrosion layers the ψ sensitivity would be improved in the early stages of growth whilst the Δ sensitivity would not be greatly reduced. We have also examined aluminium hydroxide films prepared directly on glass substrates. In such cases $n_2 = 1.5$, $k_2 = 0$ and the ψ and Δ sensitivities are greatly reduced. Figures 6 and 7 are typical scanning electron microscopy (SEM) micrographs of aluminium hydroxide grown on thick aluminium films which were deposited on gold and glass. The aluminium was exposed to saturated water vapor at 70 °C. The spongy nature and island growth of the hydroxide films can be seen.

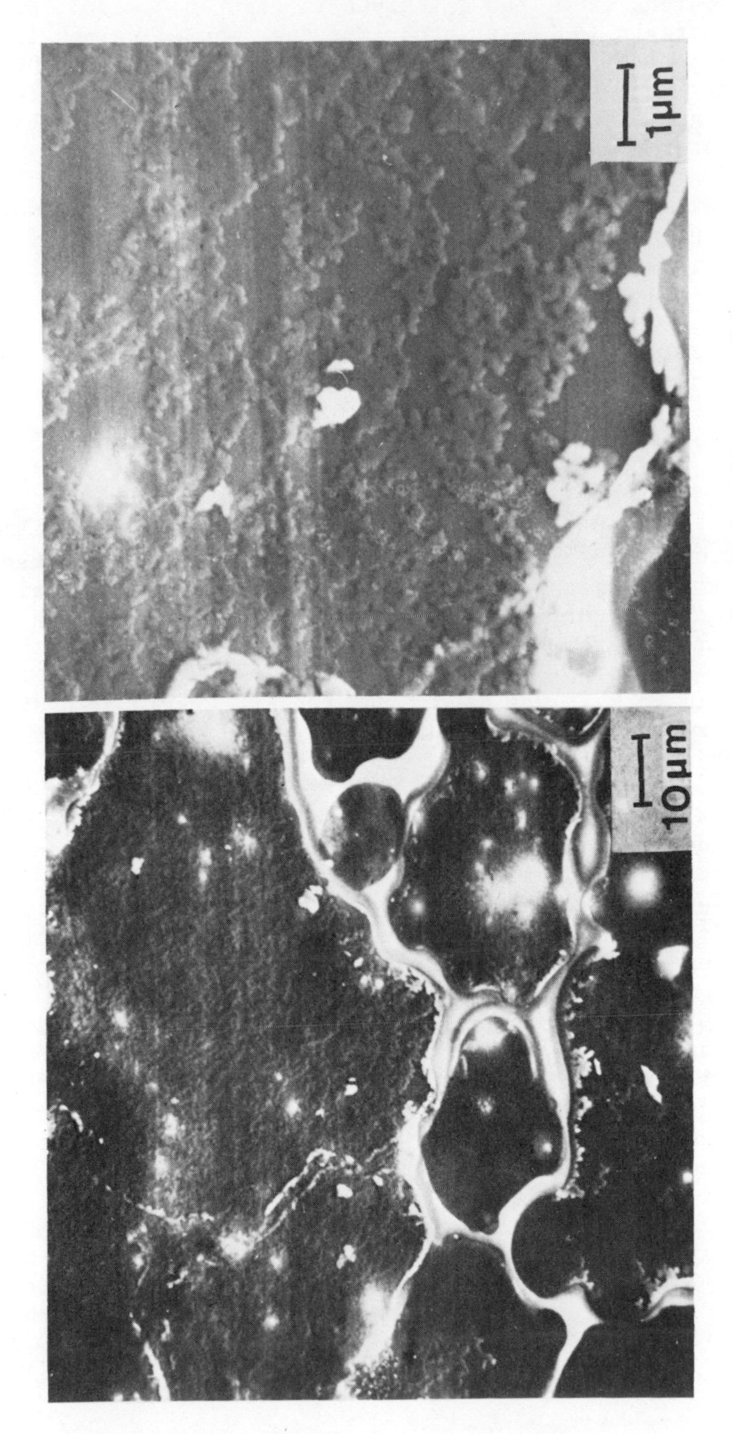

Figure 6. SEM micrograph of aluminium hydroxide on gold.

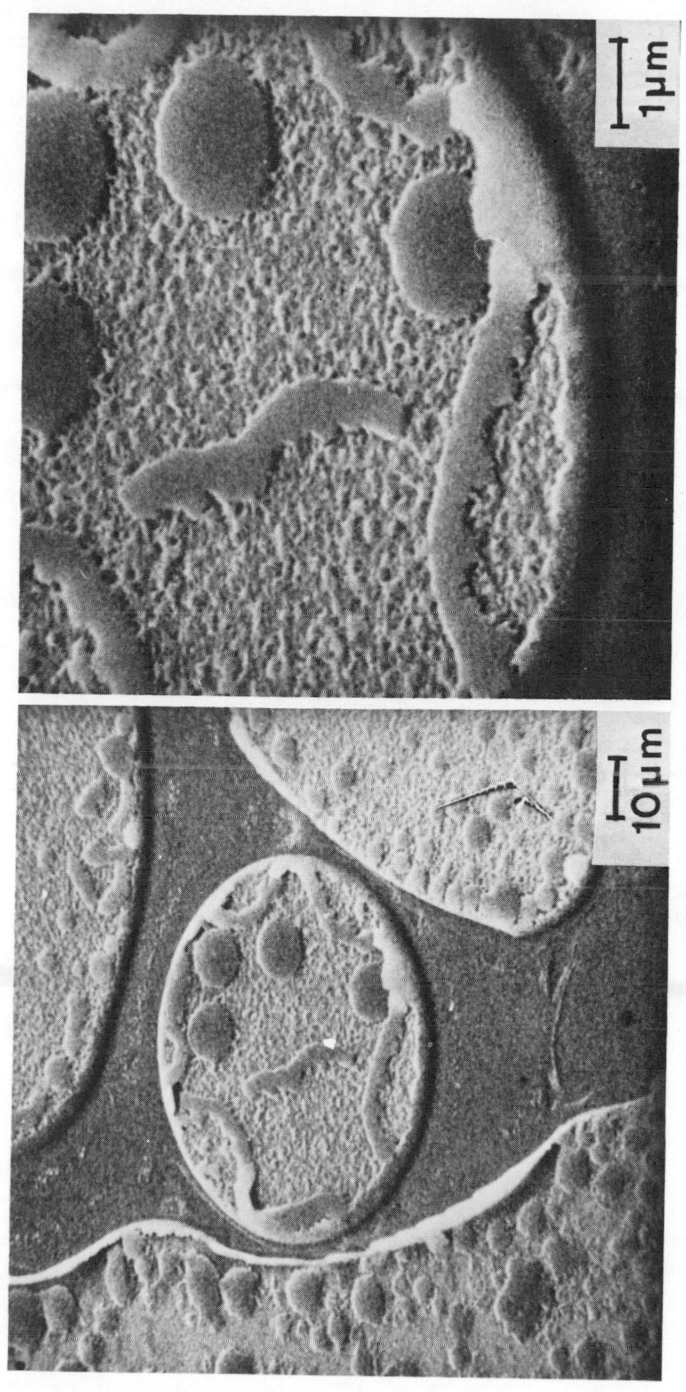

Figure 7. SEM micrograph of aluminium hydroxide on glass.

Figure 8 shows SEM micrographs of aluminium hydroxide growth on aluminium sheet formed when the sheet was exposed to saturated water vapor at 70 °C. Figure 8a was taken after 20 minutes exposure and shows islands of aluminium hydroxide with pillars of bayerite. Figure 8b was taken after a 1 hour exposure and shows the joining together of islands to form a continuous hydroxide film. Figures 8c and 8d are higher magnifications of an island. Figure 8c shows boehmite (island) together with bayerite pillars. Figure 8d is a micrograph taken in the reflective mode and shows the bayerite growing into the aluminium substrate.

SUMMARY

Two methods have been used for the optical characterisation of aluminium surfaces prior to the growth of aluminium hydroxide layers by exposure to water vapor:-

1. Freshly deposited aluminium films.
2. Ion etched aluminium sheet.

For films less than 65 nm thick and assuming no absorption the refractive index for aluminium hydroxide was found to lie between 1.55 and 1.60 for a wavelength of 549 nm.

Another method for determining the optical constants of hydroxide films was employed by depositing films of aluminium on gold and subsequently exposing them to water vapor. The optical constants of aluminium hydroxide for a wavelength of 549 nm were found to be:

(a) For films < 80 nm, $n_1 = 1.58 \pm 0.03$, $k_1 = 0$
(b) For films > 80 nm, $n_1 = 1.58 \pm 0.03$, $k_1 = 0.0009$

The latter technique has been shown to be a viable alternative in the determination of optical constants of corrosion layers of reactive metals when problems of surface characterisation exist and when optical measurements of a reactive metal in ultra high vacuum systems are not possible.

In addition to the work described, another part of the experimental programme has been to make measurements of oxide growth on aluminium in vacuo, and the investigations will be extended in the next phase to include oxide and hydroxide growth on aluminium alloys, again with the emphasis being on the early stages of growth. The investigations have been undertaken as part of a collaborative programme with the Alcan Research Laboratories, Banbury, U.K.

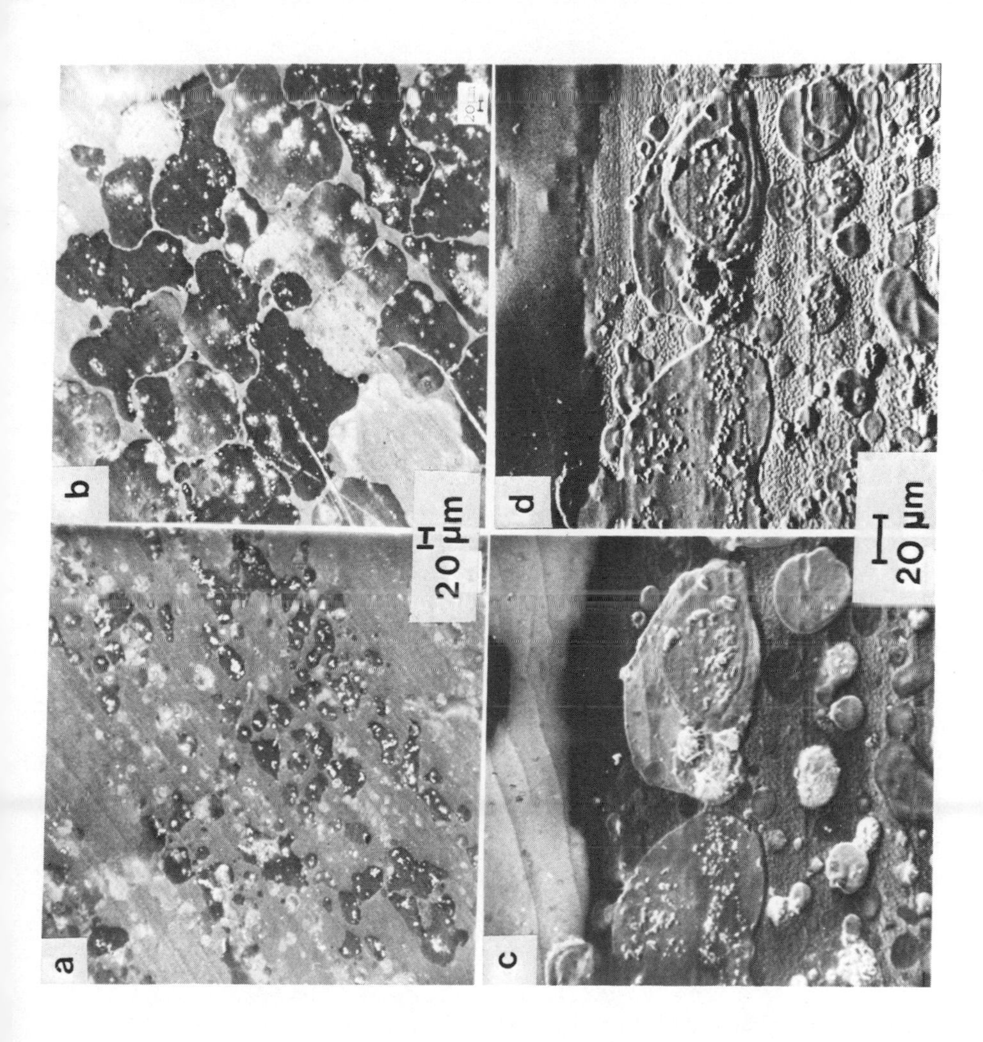

Figure 8. Hydroxide growth on aluminium sheet.

ACKNOWLEDGEMENTS

The authors wish to thank Dr. G. M. Scamans for his assistance in preparing the electron micrographs and to Alcan Research Laboratories for permission to reproduce Figures 6, 7 and 8. One of us, A. Rehal, wishes to acknowledge a studentship from the Science Research Council.

REFERENCES

1. L.Tronstad, Trans. Faraday Soc. 29, 502 (1933).
2. M. A. Barrett and A. B. Winterbottom, in "First International Congress on Metallic Corrosion," p. 657, London, Butterworths 1962.
3. J. Kruger and P. C. S. Hayfield in "Handbook on Corrosion Testing," W. H. Ailor, Editor, p. 783, Wiley, New York (1971).
4. P. C. S. Hayfield, Surface Science, 56, 488 (1976).
5. R. Aguado-Bombin and W. E. J. Neal, Thin Solid Films, 42, 91 (1977).
6. R. S. Alwitt in "Oxides and Oxide Films," J. W. Diggle and A. K. Vijh, Editors, Vol. 4, p. 169, Marcel Dekker, New York, 1976.
7. W. Vedder and D. A. Vermilyea, J. Electrochem. Soc. 115, 561 (1968).
8. W. E. J. Neal, Surface Technol., 6, 81 (1977).
9. J. R. Beattie and G. K. T. Conn, Phil. Mag. 46, 222 (1955).
10. P. Drude, Ann. Phys. Leipzig, 32, 584, (1887); 34, 489 (1888); 39, 481 (1890).
11. R. W. Ditchburn, J. Opt. Soc. Am., 45, 743 (1955).
12. F. L. McCrackin and J. P. Colson in "Ellipsometry in the Measurement of Surfaces and Thin Films," E. Passaglia, Editor, p. 61, National Bureau of Standards, Washington, D.C., (1964).
13. R. W. Fane and W. E. J. Neal, J. Opt. Soc. Am., 60, 790 (1970).
14. J. H. Halford, F. K. Chin and J. E. Norman, J. Opt. Soc. Am. 63, 786, (1973).
15. S. Tolansky, "Multiple Beam Interferometry," Clarendon Press, Oxford, (1969).
16. M. A. Barrett in "Ellipsometry in the Measurement of Surfaces and Thin Films," E. Passaglia, Editor, p. 213, National Bureau of Standards, Washington, D.C. (1964).

CONTAMINATION ON ZINC AND ALUMINUM SURFACES AFTER EXTENDED URBAN EXPOSURES

G. B. Munier, L. A. Psota, B. T. Reagor,
B. Russiello, and J. D. Sinclair

Bell Telephone Laboratories
Holmdel, New Jersey 07733

Water soluble contaminants on electromechanical telephone switching equipment exposed for up to 40 years to the New York City environment have been analyzed by a comination of methods. Samples were collected from surfaces by three successive extractions with filter paper squares moistened with distilled water. Elements having atomic number greater than 11 were identified by energy dispersive X-ray analysis. Selected anions were identified with microchemical tests and infrared spectroscopy and determined quantitatively with ion selective electrodes. Nitrate and sulfate concentrations of samples from one location were measured with nitrate and lead ion selective electrodes (the latter for titration of sulfate). Samples from several locations were analyzed with a chloride selective electrode. On zinc surfaces, chloride concentrations averaged 27 $\mu g/cm^2$ for typical locations, and sulfate and nitrate concentrations averaged 48 and 6 $\mu g/cm^2$ respectively. On aluminum surfaces, chloride concentrations averaged <2 $\mu g/cm^2$, and sulfate and nitrate concentrations averaged 25 and 3 $\mu g/cm^2$, respectively.

The minimum relative humidity at which significant moisture acquisition occurred and the overall moisture pickup characteristics of the contaminants at 100 percent RH were determined gravimetrically. Moisture pickup of contaminants on zinc and aluminum surfaces occurs above 26 and 46 percent RH, respectively. Using the analytical data and hygroscopicity information, environmental parameters that ensure continued satisfactory operation of equipment can be established.

INTRODUCTION

The cumulative contamination of electronic devices and machines during long term exposure to urban atmospheres is of particular concern to the electronics industry. The contamination may be accumulated dusts or a complex film of corrosion products possibly containing oxides, sulfides, sulfates, oxy anions of nitrogen, chlorides, and other substances, as well as dusts.[1-5] Contamination can lead to electrical leakage through conducting substances in the contaminant[6-7]. Electrical leakage is often enhanced at high relative humidities, particularly if the contaminant is moderately hygroscopic. In some cases, a specific minimum relative humidity can be identified at which moisture acquisition will occur.[8] Exposure of contaminated devices to an environment in which the "critical relative humidity" (CRH) of the contaminant(s) is exceeded may lead to moisture acquisition and device failure.

Contaminants that produce electrical leakage failures in the field are usually water soluble, and samples can be quantitatively collected by water extraction. Of course, dusts and some corrosion products can be collected by brushing or gentle scraping, but in many cases the important contaminant is a tenacious film. Often the equipment is operational and must be sampled non-destructively in the field.

Convenient procedures for analyzing contaminants, mapping their distribution at an equipment site, and measuring their moisture pickup characteristics are valuable in assessing the causes of failure of equipment and in determining procedures for dealing with the problem. In some cases, device modification may be appropriate, while in other situations, particularly in hazardous environments, improved humidity control or air filtration may be suitable.

This paper presents the methods that were used to analyze problems occurring in electromechanical switching equipment that has been exposed for up to 40 years to an urban environment. Contamination data for equipment in one building (hereafter designated as Building I), which developed switching problems during and after a high humidity episode, are discussed in detail.

EXPERIMENTAL

Water extraction of equipment surfaces can be conveniently accomplished with small sections of filter paper that have been moistened with distilled water. In this study, square sections (one-half inch on a side) of Whatman 3MM filter paper were used. Extraction of zinc and aluminum surfaces was convenient and appropriate to the equipment being examined. The papers were removed from the surfaces when nearly dry. The procedure was carried out three times on exactly the same sample area. Experience has shown that, for most contaminants, an essentially complete extraction of water soluble substances is accomplished with three extractions.

Elements having atomic numbers greater than 11 were identified in a scanning electron microscope by energy dispersive X-ray techniques (SEM/X-ray). These analyses were carried out on the residue from a distilled water extract of several papers. Low atomic number elements and multielement anions were identified by standard microchemical tests or infrared spectroscopy. In the urban environment involved in this study, the anions of particular concern are chloride, sulfate, and nitrate. Quantitative analyses of these anions were carried out with Orion specific ion electrodes.[9-10]

Forty-six samples (each consisting of three papers) from a single floor in Building I were randomly selected from more than 75 samples from the floor for nitrate, sulfate, and chloride determinations. Several hundred samples from a number of nearby buildings were also analyzed for chloride. The analysis for each anion was accomplished after redissolving the soluble substances collected in the sampling papers in 10 ml of distilled water.

Chloride ion concentrations in the 10 ml solutions were monitored with a chloride selective electrode using standard procedures.[9,10] The solutions were then divided in half for nitrate and sulfate analysis. Nitrate was monitored with a nitrate electrode,[9-10] and again standard procedures were followed using potassium fluoride as the ionic strength adjustor. The low levels of nitrate encountered approached the sensitivity limit of the method. Sulfate concentrations were determined by titration with standard lead perchlorate using a lead electrode to monitor the end point.[11-12] The 5 ml unknown solution was diluted in each case with 5 ml methyl alcohol. The electrode manufacturer recommended plotting electrode potential versus quantity of lead perchlorate added to determine the endpoint, this being taken as the point of greatest inflection. At the low concentrations of sulfate found, relative to the detection limits of the procedure, the point of greatest inflection in these particular solutions was not always readily apparent and in some cases was ambiguous. Expermentation with standard solutions indicated that if the endpoint was selected to be -200 mV, adequately satisfactory agreement between known and measured concentrations was obtained. Significant concentrations of chloride, sulfate, and nitrate were measured for the blank paper squares, and appropriate corrections were made for each sample .

Moisture pickup characteristics of the residues of some of the paper extracts from alumium and zinc plated surfaces and of a green corrosion product contaminating some surfaces (collected by brushing) were measured by gravimetric methods using a duPont 951 Thermogravimetric Analyzer coupled to a 990 Thermal Analyzer. Details of the method are given elsewhere.[8] Continuous moisture pickup rates for the various materials at nearly 100 percent RH at 23°C were determined by passing a stream of air, which had been moisturized by bubbling it through distilled water, through the microbalance sample tube (2.5 cm diameter by 12 cm long). Weight was recorded graphically as a function of time, and each run was allowed to proceed until the rate of weight gain approached zero or reached an extended steady state.

CRH measurements were made using the same balance with a modified air stream apparatus. A minimum RH in the balance housing (maintained at 23°C) of 22 percent was achieved by passing a stream of dry air through a dispersion frit submerged in distilled water cooled to 0°C. The water was warmed from 0°C to nearly 23°C at a rate of 0.25°C/min, thereby increasing the RH from 22 to nearly 100 percent. A recording of weight versus time was translated to weight versus RH. The relative humidity at which an onset in weight gain occurred was taken to be the CRH. Samples were equilibrated for several days or weeks with the laboratory atmosphere (RH approximately 40 to 50 percent) before they were analyzed in the microbalance.

RESULTS AND DISCUSSION

Table I summarizes the types of surfaces extracted, the year of their manufacture, and the measured chloride, sulfate, and nitrate concentration levels for the 46 samples from Building I. For samples 1 to 5, the three papers used to extract the surface were stored and measured separately to verify that nearly complete extraction of the surface was achieved with three papers. For all other samples, the three papers were combined for storage and measurements. In the cases of nitrate and chloride, this procedure established that three papers provide essentially complete surface extraction, but in the case of sulfate, another problem was indicated. The total sulfate levels from the three papers for samples 1 to 5 were very high relative to other 1947 vintage surfaces. In the case of chloride, the concentrations of samples 1 to 5 were not extraordinary. This result suggests that an interfering effect, caused by a species other than sulfate, increased the apparent concentrations to unrealistically high levels. The effect appeared to have a similar magnitude for each titration whether the three papers were run simultaneously or separately. The sulfate concentrations indicated for samples 6 to 46 thus include this effect only one time, while samples 1 to 5 include it three times, once for each paper. An accurate correction factor for the effect was not readily derivable, and thus it was decided that the sulfate concentrations for samples 1 to 5 should be based on the extract from the first paper only. These values are given in parentheses. Concentrations for samples 6 to 46 are high by perhaps 7 to 15 μg/cm^2. Concentrations for samples 1 to 5 may or may not be high, but they are certainly lower than they would have been had all three papers been analyzed simultaneously. Nitrate concentrations were found to be very low. Samples 1 to 19 were analyzed and found to have concentrations below the region of linearity in the electrode calibration plots from which concentrations were determined. We originally anticipated that at least a few samples would have concentrations in the linear regime but found after completing 19 analyses that this would probably not be the case. For the remaining samples, concentrations were determined by extrapolation into the nonlinear regime. The data were treated very conservatively, based on results for standard solutions, so the concentrations listed are lower limits for nitrate. Unfortunately, only an upper limit can be set for samples 1 to 19 because the solutions were discarded before the extrapolation procedure was initiated.

Table I. Contaminant Surface Concentrations

Sample No.	Surface Metal	Year of Manufacture of Sampled Surface	Chloride Concentration (μg/cm^2)	Sulfate Concentration (μg/cm^2)	Nitrate Concentration (μg/cm^2)
1.	Zn	47	39	127(65)*	<31
2.	Zn	47	28	136(56)	<31
3.	Zn	47	27	102(38)	<31
4.	Zn	47	22	94(47)	<31
5.	Zn	47	49	131(64)	<31
6.	Zn	47	34		<31
7.	Zn	47	29		<31
8.	Zn	47	41	41	<31
9.	Zn	47	22		<31
10.	Zn	64	7	33	<31
11.	Zn	64	6	40	<31
12.	Zn	64	8	39	<31
13.	Zn	64	6	34	<31
14.	Zn	47	42	60	<31
15.	Zn	47	30	58	<31
16.	Zn	47	27	28	<31
17.	Zn	47	21	46	<31
18.	Zn	47	35	58	<31
19.	Zn	47	21	57	<31
20.	Zn	47	52	50	15
21.	Zn	47	37	63	18

Table I. Contaminant Surface Concentrations - Cont'd.

Sample No.	Surface Metal	Year of Manufacture of Sampled Surface	Chloride Concentration ($\mu g/cm^2$)	Sulfate Concentration ($\mu g/cm^2$)	Nitrate Concentration ($\mu g/cm^2$)
22.	Al	47	4	27	5
23.	Al	56	<2	26	2
24.	Al	56	7	39	7
25.	Al	57	<2	28	3
26.	Zn	47	45	64	8
27.	Zn	47	42	50	8
28.	Al	47	<2	29	2
29.	Al	47	<2	29	2
30.	Al	56	<2	26	0
31.	Al	56	<2	15	8
32.	Zn	47	41	51	6
33.	Zn	47	58	99	15
34.	Al	56	<2	16	0
35.	Al	56	<2	14	0
36.	Zn	57	5	27	2
37.	Zn	57	11	40	4
38.	Zn	57	10	39	0
39.	Zn	57	14	27	4
40.	Zn	56	5	39	0
41.	Zn	56	10	52	1
42.	Zn	56	12	27	0
43.	Zn	56	10	51	4
44.	Zn	47	48	51	7
45.	Zn	47	27	39	4
46.	Zn	47	45	51	8

*Values in parentheses are the concentrations extracted by the first paper.

An additional 38 paper extract samples that had been collected at Building I were separated into two groups. All the samples from zinc surfaces were combined and extracted in one solution, as were all the samples from aluminum surfaces. The distilled water extracts were filtered and the filtrates evaporated to dryness and weighed. By this procedure, we determined that the average weight of corrosion salts extracted from zinc surfaces (after subtracting an experimentally determined value for the solids extracted from three paper blanks) was 117 $\mu g/cm^2$, while the average for aluminum surfaces was 61 $\mu g/cm^2$.

SEM/X-ray analyses of these extracts were carried out to make possible an estimate of the predominant cations associated with the chloride, sulfate, and nitrate anions. From these results, the weight of extractable material could be estimated from the determined salt concentrations and compared with that from the weighings.

Zinc is an excellent scavenger for chloride and was presumed to be the only significant cation associated with chloride on zinc surfaces. Sulfate is a significant component of the dust at the urban location of this building, and, on the basis of SEM/X-ray analysis and air quality data available from the Environmental Protection Agency,[13] the associated cations were approximated as 1/4 $(NH_4^+)_2$, 1/4 $(Na^+)_2$, 1/4 Ca^{+2}, and 1/4 Zn^{+2}. The cations associated with nitrate on zinc surfaces were, in similar fashion, approximated as 1/4 (NH_4^+), 1/4 (Na^+), 1/4 $(Ca^{+2})_{1/2}$ and 1/4 $(Zn^{+2})_{1/2}$. Based on these assumptions and the average chloride, sulfate, and nitrate concentrations from Table II, the average extract was calculated to weigh 128 $\mu g/cm^2$, which is judged not to be significantly different from the 117 $\mu g/cm^2$ value measured experimentally for a different set of samples collected from the same location. The reasonably close agreement is evidence that the water extractable matter can be accounted for as chloride, sulfate, and nitrate salts.

Table II. Average Contaminant Surface Concentrations ($\mu g/cm^2$)

	Chloride	Sulfate	Nitrate
Surface:			
Zinc	27	48	6
Aluminum	<2*	25	3

*This number is based on all values <2 $\mu g/cm^2$ being 0; if all values less than 2 are assumed to be 1, the average is 2.

SEM/X-ray analysis of extracts from the aluminum surfaces and the air quality data[13] indicate that chloride, sulfate, and nitrate are present primarily as ammonium, sodium, and calcium salts. If the ammonium, sodium, and calcium ions associated with chloride and nitrate are presumed to be present as 1/3 (NH_4^+), 1/3 Na^+, and 1/3 $(Ca^{+2})_{1/2}$ and those associated with sulfate are presumed to be present as 1/3 $(NH_4^+)_2$, 1/3 $(Na^+)_2$, and 1/3 (Ca^{+2}), then the weight of extract contributed by all the salts is calculated to be 42 $\mu g/cm^2$. The total extract weight of 61 $\mu g/cm^2$ measured experimentally indicates that approximately 70 percent of the extractable material can be accounted for as chloride, sulfate, and nitrate salts. It is not certain whether the remaining 30 percent

Table III. Moisture Pickup Characteristics

Sample	Critical RH, percent	Initial Rate of Moisture Pickup at 100 Percent RH (percent/min)	Rate of Moisture Pickup at Termination at 100 Percent RH (percent/min)	Average Weight Gain Rate at 100 Percent RH (percent/min)	Total Weight Gain (percent)	Sample Weight (mg)
Green corrosion product	80					
Nickel sulfate	91			0.001	0.144	
Extract from zinc surfaces	26	0.49	0.048	0.060	153	5.1
Zinc chloride	<10	0.28	0.19	0.256	8.7	34
Extract from aluminum surfaces	46 (35)*	0.79	0.02	0.075	84	1.9
Calcium chloride	36	0.34	0.34	0.338	8.1	53

*A minor component was responsible for a slight moisture pickup at this RH.

represents unaccounted for species, inappropriate assumptions in the calculations, experimental error, or a combination of these, although, in view of some of the experimental difficulties and the fairly crude nature of the approximations, the first seems unlikely. Other inorganic anions would very likely have been detected if they were present in sufficient quantity to account for 30 percent of the total weight.

Previously, several other buildings in the same urban environment as Building I having equipment with similar vintage distributions were sampled and analyzed for chloride ion concentrations on surfaces. The results for zinc surfaces are as follows: Building II - 34 $\mu g/cm^2$; Building III - 23 $\mu g/cm^2$; Building IV - 27 $\mu g/cm^2$; and Building V - 24 $\mu g/cm^2$. The value for Building I (27 $\mu g/cm^2$) is identical to the average from the other locations (27 $\mu g/cm^2$), indicating that the chloride contamination in Building I is typical of this urban environment.

The moisture pickup characteristics of the green corrosion product that contaminated switching equipment springs and contacts and the residues from the moistened paper extracts from zinc surfaces and aluminum surfaces are summarized in Table III, along with similar data for some pertinent salts. SEM/X-ray analysis and microchemical tests of the green corrosion product indicated that it was rich in nickel sulfate. The urban atmosphere to which the equipment was exposed is known to contain high levels of sulfur dioxide and particulate matter that is rich in sulfate salts. The effective CRH of the green corrosion product was approximately 82 percent. The critical RH of a pure sample of nickel sulfate was 91 percent, which is reasonably consistent with the chemical analysis. This rather high CRH indicates that the nickel sulfate corrosion product will not pose a moisture hazard in normal building environments. Consideration of the temperature-humidity conditions that existed during a series of equipment malfunctions in a high humidity period indicates that the 91 percent RH level was exceeded at equipment surfaces. The acquired moisture provided low resistance leakage paths which led to equipment problems. Unfortunately, an insufficient amount of the green corrosion product was available to measure the overall moisture pickup characteristics.

The water soluble contaminants extracted from zinc surfaces exhibited a low critical RH and a moderate rate of moisture pickup at 100 percent RH. The moisture pickup of zinc chloride, the major contaminant on these surfaces, readily accounts for the moderate tendency of the extract to acquire moisture.

The water extract from aluminum surfaces, which was found to be a mixture of several salts as discussed above, is moderately hygroscopic. The measured critical RH is lower than that for several of the salts that could be present and for which data is available,[8,14] including ammonium sulfate (80 percent), sodium sulfate (approximately 90 percent), sodium chloride (75 percent), sodium nitrate (74 percent), and ammonium chloride (77 percent). Calcium chloride, which is also likely to be present, has a critical RH of approximately 35 percent and may be the major contributing factor to the low critical RH of the extract.

CONCLUSIONS

This study has shown that a thorough analysis of contaminant concentrations on operating electrical equipment can be accomplished using paper sampling methods and routine analytical procedures. By combining analytical results with knowledge of the CRH, which can be readily determined by instrumental gravimetric methods or found in one of several compilations, and the moisture pickup rates, which can also be experimentally determined, useful conclusions about the environmentally induced causes of equipment malfunction and recommendations about environmental parameters that will reduce equipment degradation can be made.

ACKNOWLEDGMENTS

The authors wish to express their appreciation to C. J. Weschler, C. A. Russell, and P. C. Milner for very helpful discussions during the course of this work.

REFERENCES

1. E. J. Bauer, B. T. Reagor, and C. A. Russell, ASHRAE J., p. 53 (October, 1973).
2. (a) C. J. Weschler, paper number 225 presented at the annual meeting of the American Chemical Society, Miami, September, 1978; (b) C. J. Weschler and M. V. Walker, to be published.
3. R. K. Patterson and J. Wagman, J. Aerosol Sci., 8, 269 (1977).
4. B. P. Leaderer, J. Air Pollut. Contr. Assoc., 28(4), 321 (1978).
5. C. J. Weschler, Environ. Sci. Technol., 12(8), 923 (1978).
6. R. P. Frankenthal and W. H. Becker, paper number 246 presented at the Electrochemical Society Meeting, Pittsburgh, October 15-20, 1978.
7. D. W. Rice, P. B. P. Phipps, and R. F. Tremoureuz, paper number 9 presented at the Electrochemical Society Meeting, Philadelphia, May 8-13, 1977.
8. J. D. Sinclair, J. Electrochem. Soc., 5, 734 (1978).
9. R. P. Buck, Anal. Chem., 50(5), 17R (1978).
10. W. Gallay, H. Egan, J. L. Monkman, R. Truhaut, P. W. West, and G. Widmark, "Environmental Pollutants Selected Analytical Methods (Scope 6)," pp. 236, 248, Ann Arbor Science Publishers Inc., Ann Arbor, Michigan.

11. H. Clyster and F. Adams, Anal. Chim. Acta, 92, 251 (1977).
12. J. W. Ross and M. S. Frant, Anal. Chem., 41, 967 (1969).
13. "Air Quality Data for Nonmetallic Inorganic Ions 1971 Through 1974" from the National Air Surveillance Networks, EPA-600/4-77-003, January, 1977.
14. Kh. Balarev, Kh. Stoeva, Kh. Stoev, God. Vissh. Khim. - Tekhnol. Inst., Burgas. Bulg., 10(10), 623 (1973).

$^{238}PuO_2$ SURFACE CONTAMINATION OF RADIOISOTOPIC HEAT SOURCES

D. R. Schaeffer and C. O. Brewer

Mound Facility*
Miamisburg, Ohio

Radioisotopic thermoelectric generators (RTG's) employ the self-heating characteristic of decaying radioactive material (in this case, plutonium-238) to generate electricity. The highly radioactive fuel presents the disadvantage of requiring isolation to prevent environmental contamination. Isolation is achieved by sealed metallic containers which are inert to their environment; however, the exterior surfaces of the containers have been exposed to plutonium and must be cleaned to meet strict health physics requirements. Background information concerning the development of cleaning techniques will be presented. Some of the factors which must be considered are isolation of the parts from personal contact, self-heating of the items to be cleaned, and in most instances removal from an air environment while at elevated temperatures.

Surface contamination and cleaning characteristics of two radioisotopic heat sources will be discussed. The Milliwatt Generator is a small (4½ watts) heat source which is successfully cleaned by hand in a series of hot acid baths.

The Multi-Hundred Watt Isotopic Heat Source presents additional problems in removing the surface contamination because of its large size (100 watts) and its grit-blasted surface. A study has characterized the behavior

*Mound Facility is operated by Monsanto Research Corporation for the Department of Energy under Contract No. EY-76-C-04-0053.

of the plutonium during aging of the surface at the heat source service temperature of 1350°C. Results from this study show that normal decontamination effectively removes the superficial plutonium but does not extract the plutonium which is deep within the grit-blasted structure. Subsequent heating results in migration of microcurie amounts of plutonium out of the grit-blasted surface.

INTRODUCTION

Radioisotopic heat sources are capsules of radioactive material, in this case plutonium-238, which inherently emits heat during the decay of the radioactive element. The heat sources are used in remote areas to provide heat for thermoelectric generators or to keep instrumentation at operating temperatures. The highly radioactive fuel ($^{238}PuO_2$) possesses the disadvantage of requiring isolation to prevent environmental contamination. This isolation is provided by multiple metallic containers; however, the exterior surfaces of these containers are exposed to the plutonium fuel during assembly and thus become contaminated with sufficient plutonium to be a health hazard. Removal of this plutonium is required to meet strict health physics requirements.

Radioisotopic heat sources present unique characteristics which make removal of residual plutonium difficult. Human contact, such as through inhalation, ingestion, or puncture absorption, must be excluded because of health considerations. Shielding of personnel from the neutron and gamma radiation emitted from larger quantities of plutonium is essential. The inherent heat produced restricts hand contact and could cause oxidation of the refractory encapsulants. Some heat sources have thin (0.002 to 0.005 inch) disks which must retain their integrity and thus hamper the quicker, more vigorous methods of decontamination.

During the fifteen years Mound Facility has been producing radioisotopic heat sources, many different cleaning or decontamination methods have been studied and attempted. Some of these methods are summarized in Table 1. Evolution of decontamination procedures has resulted in a similar procedure for heat sources encapsulated in refractory metals or noble metals. This procedure combines the use of hot acid baths and scouring powder scrubs. The acid solution contains two parts HF, two parts HNO_3, and five parts H_2O and is used in either a fuming or a boiling condition. Although the procedure is the same, there are variations which are dependent on the size (wattage) of the heat source. The procedure and its variations are best depicted by detailing the decontamination effort and results for a small (4-watt) and a large (100-watt) heat source.

Table I. Summary of Decontamination Techniques for $^{238}PuO_2$ Heat Sources

Technique	Procedure	Capability to Remove $^{238}PuO_2$	Comments
Tumbling	Used alumina, silica sand, and scouring powder--and water slurries of each--for times of up to 24 hr.	poor[a]	Minimizes contact, but ineffective as decontaminate.
Grit blast		?	Potentially harmful to delicate components
Abrade surface with steel wool	Hand scrub with scouring powder and water.	good	Used only on test capsules without delicate parts.
Scrub with scouring powder	Hand scrub surface using moist rag or cheese cloth, or wet stiff brush	good	Will decontaminate; however, speed increased when used in conjunction with acid baths. Presently used as alternating step with acid bath or as final step.
Dissolve $^{238}PuO_2$	Acid bath of 5:1:1 HF: HNO_3: CH_3COOH at room temperature. Acid diluted with 50 vol % H_2O.	good	Some surface removal of tantalum and tantalum alloys.
Dissolve $^{238}PuO_2$	Acid bath of 5:1:1 HF:HNO_3:H_2SO_4 at room temperature. Acid diluted with 50 to 90 vol % H_2O.	excellent	Excellent decontamination capability offset by removal of the surface of tantalum alloy encapsulants.
Dissolve $^{238}PuO_2$	Acid bath of 2:2:5 HF:HNO_3:H_2O at 70 and 80°C (fuming).	very good	Minor etching of tantalum and tantalum alloys. Presently used for refractory metal and noble metal radioisotopic heat sources in conjunction with scouring powder scrub.

Table I. Continued

Technique	Procedure	Capability to Remove $^{238}PuO_2$	Comments
Form soluble complex	Baths of acetic or oxalic acid.	poor[a]	Will form soluble complexes with low-fired $^{238}PuO_2$ only. Ineffective with the high-fired $^{238}PuO_2$ used in heat sources. With oxalic acid solutions plutonium bearing compounds could precipitate out, causing recontamination.
Ultrasonic	Ultrasonic cleaning in above acid baths.	good	Little or no improvement over use of acid baths alone. Damage to delicate parts observed.
Ultrasonic	Ultrasonic cleaning in commercially available nuclear decontaminating solutions.	fair	Capability for decontaminating less than developed acid solutions.
Shielding	Exterior surface covered with nickel foil to reduce contact of surface with plutonium.		Contamination reduced.
Strippable Paint	Cover clean surface and use as shield or use to remove contaminate already on the surface.		Effectively used in $^{238}PuO_2$ applications other than heat sources. The strippable paints are unable to withstand the temperatures generated by heat sources.

[a]The techniques that are rated "poor" are incapable of lowering the contamination sufficiently. Other ratings relate to speed and convenience of contaminate removal.

DECONTAMINATION OF THE MILLIWATT GENERATOR HEAT SOURCE

The Milliwatt Generator heat source, shown in Figure 1, is a 4-watt (7.1 g of $^{238}PuO_2$) source encapsulated in T-111 (2 wt % hafnium, 8 wt % tungsten/tantalum). Decontamination is accomplished in a fumehood with a series of four baths of acid consisting of two parts HF, two parts HNO_3, and five parts H_2O. The acid is contained in Teflon* beakers and heated till fuming. The heat sources are individually submerged in each of the series of acid baths for approximately 2 min and are rinsed with distilled water between each bath. After the final rinse, 50% of the capsule surface is wiped with an asbestos filter wipe (called a swipe) and checked on an alpha counter for "wipable" contamination. Any residual $^{238}PuO_2$ is removed with scouring powder on moist cheese cloth, until the sources are decontaminated to levels of <10 cpm of alpha radiation. Subsequent heating of the source to above its service temperature does not release any significant amount of $^{238}PuO_2$ from the T-111 liner.

$^{238}PuO_2$ CONTAMINATION OF THE MULTIHUNDRED WATT HEAT SOURCE

Decontamination Procedure

The Multihundred Watt isotopic heat source contains 2400 watts (3810 g) of $^{238}PuO_2$ divided into 24 individual 100-watt, iridium-encapsulated sources, as exemplified in Figure 2; these are called Post Impact Shell Assemblies (PISA's). Decontamination of the PISA's employs the same acid solution (two parts HF, two parts HNO_3, and five parts H_2O) and scouring powder scrub as the Milliwatt heat source; however, the techniques using this procedure differ. While the larger heat source requires shielding of personnel from the neutron and gamma radiation which demands that the sources be decontaminated by manipulators in the cell, the PISA's are initially scrubbed with scouring powder on moist rags or a stiff brush to remove the gross contamination. Each PISA is then soaked in the boiling acid for 2-5 min and rinsed with distilled water. This scrub/acid cycle is repeated three times before a measure of the "wipable" contamination is made. This method reduces the contamination level to <100 cpm; however, it was noted that increases in the level of contamination occurred with time at the service temperature. A test, which is subsequently described, was designed to assess the magnitude and behavior of this remnant plutonium.

*Teflon is a trademark of du Pont Company.

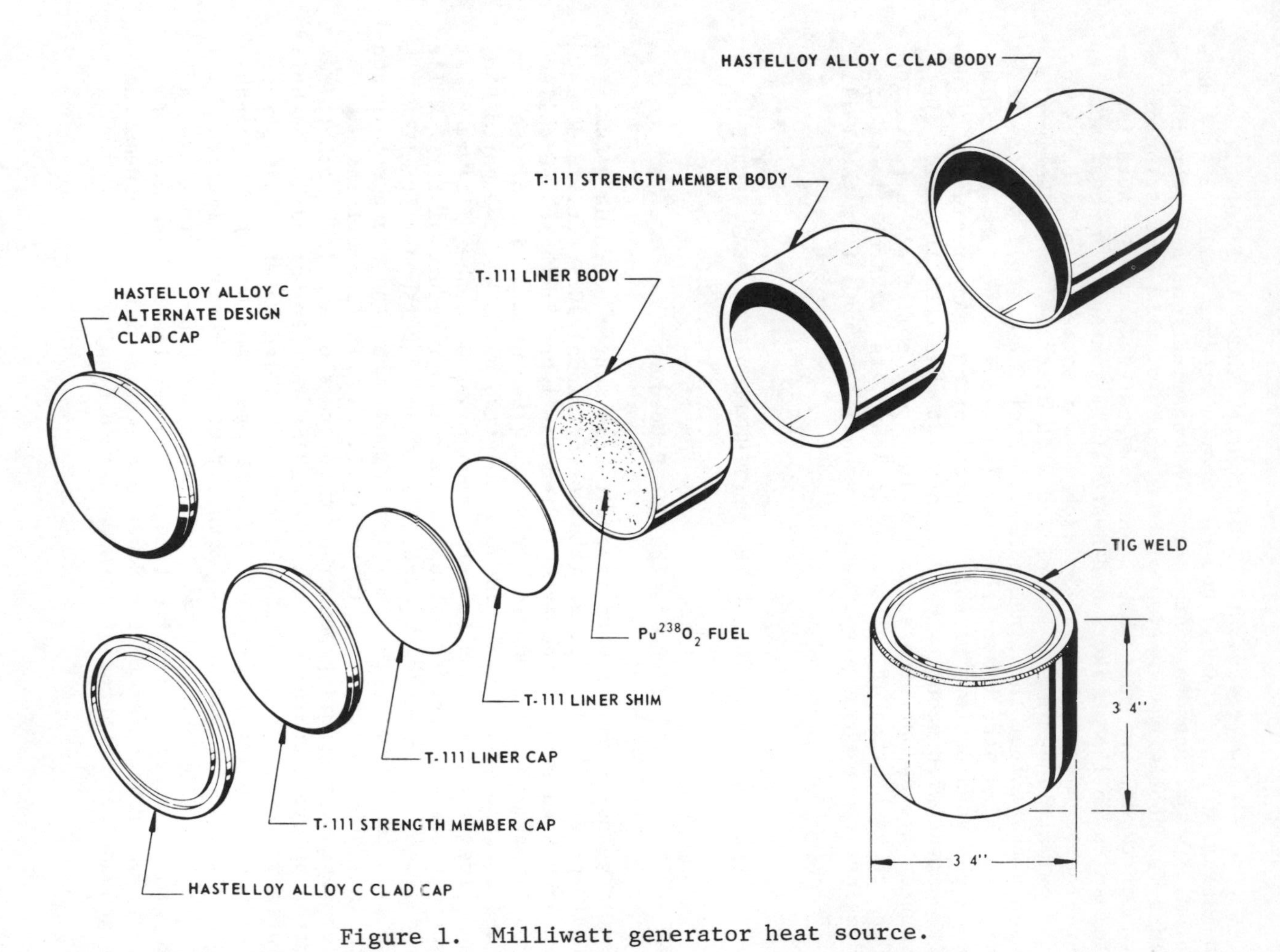

Figure 1. Milliwatt generator heat source.

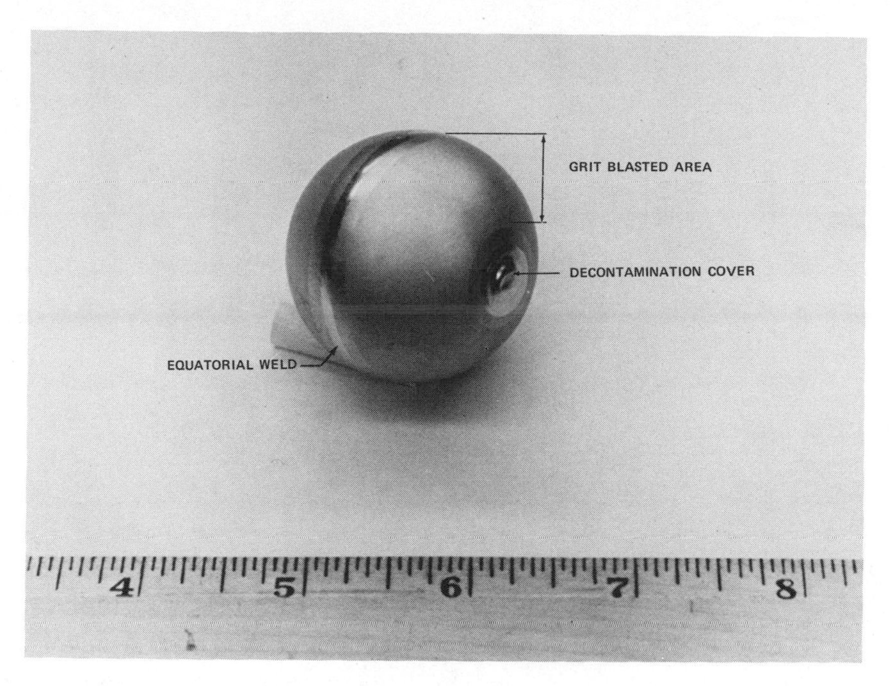

Figure 2. Photograph of the primary impact shell assembly (PISA). It consists of two iridium hemishells which, when welded together, form the container for a $^{238}PuO_2$ sphere.

Characterization of the Residual $^{238}PuO_2$

A PISA was heated for various periods of time at 1350°C (the service temperature) and the amount of distribution of the plutonium on the surface was characterized after each heat treatment. After a total of 1535 hr at temperature, the distribution of plutonium in the cross section of the iridium wall of the PISA was determined by removing layers of iridium from the surface by electropolishing and determining the plutonium content in each layer.

The PISA used in this test is shown in Figure 2 and was fabricated in the ^{238}Pu alpha contaminated "hot" facilities normally used in production. It was fabricated in the same manner as the production PISA's with the exception that just prior to welding, the $^{238}PuO_2$ fuel sphere was replaced by a ThO_2 simulant sphere. After fabrication, the PISA was decontaminated by the normal procedure and then helium leak checked and examined for surface alpha contamination. Measurements of the surface alpha contamination were made by first measuring the amount of $^{238}PuO_2$ that could be wiped from the surface. The second method of determining the alpha

Figure 3. Photograph of the graphite impact shell (GIS) which is used to contain the PISA. After the PISA is placed into the cavity of the body of the GIS, the cap is screwed into the body thereby providing impact protection.

contamination involved direct measurement of the alpha radiation emitted from various portions of the PISA. As in normal production, the PISA was then inserted into a graphite impact shell (GIS), which is shown in Figure 3. The PISA and surrounding GIS were then subjected to heat treatments at 1350°C in a vacuum furnace for various periods of time up to a total of 1535 hr.

After each heat treatment, measurements were made of the alpha contamination on the surfaces of the PISA, GIS, and furnace. Direct determinations of the alpha contamination were also made on the PISA and GIS after each heat treatment but unlike the wipable measurements, they were made after the PISA and GIS were again decontaminated. Decontamination of the PISA involved scrubbing with scouring powder and water* with the resultant mixture examined for its

*The decontamination of the PISA by scouring powder and water differs from the initial acid decontamination in that the scouring powder and water removes only the surface plutonium, whereas the boiling acid decontamination dissolves the $^{238}PuO_2$ and can thus remove it from deeper within the grit blasted structure.

plutonium content by liquid scintillation. The GIS was decontaminated by wiping it with a dry rag. After each heat treatment, the PISA was also helium leak checked, and after all but the first heat treatment (100 hr at 1350°C) the PISA was examined for surface defects by dye penetrant.

After a total of 1535 hr at temperature, the distribution of plutonium in the cross section of the iridium wall in the grit blasted area was measured. This was accomplished by removing "layers" of iridium from the outside surface of the PISA by electropolishing and determining the plutonium content in the removed layer by finding the amount of plutonium in the electrolyte used to remove the layer. The plutonium distribution was also characterized by measuring the direct alpha radiation emitted from the newly exposed surface.

Removal of iridium from the surface of the PISA was accomplished by clamping the PISA into the fixture described in Figure 4, which allowed a portion (0.11 in.^{2}) of the PISA's outer surface to be in contact with an electrolyte held in a reservoir at the top of the fixture. Ten volts were then applied between the PISA and an electrode submerged in an electrolyte consisting of 130 g of KCN in one liter of water. A depth micrometer was used to measure the thickness of the removed iridium layer by measuring the distance between the top of the fixture and the "top" of the PISA. The electropolishing procedure resulted in the removal of about 0.1 mil after 5 min of polishing.

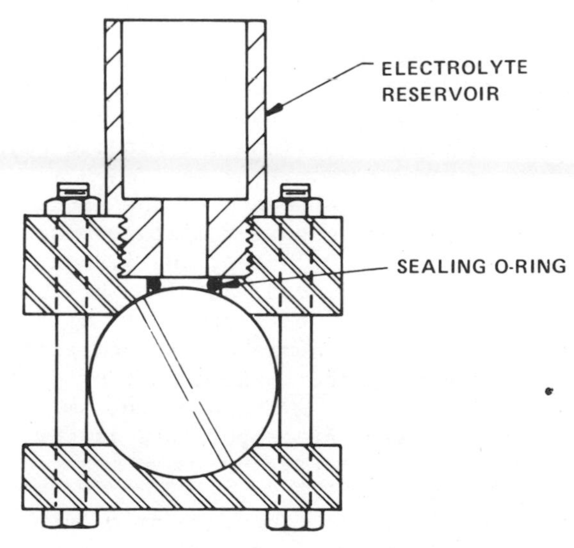

Figure 4. Sketch of test fixture used to remove a portion of the PISA surface.

After electropolishing the iridium surface, the electrolyte and subsequent reservoir washings were submitted for liquid scintillation assay to determine their plutonium content and, thus, the plutonium content of the removed surface layer. Measurements of the direct alpha radiation emitted from the newly exposed iridium surface were accomplished by removing the reservoir and sealing O-ring and replacing them with a new O-ring and mask. The O-ring and mask ensure that only the newly exposed surface was measured for alpha radiation. The direct radiation measurements of the iridium surface were relative rather than quantitative measurements of the amount of plutonium present on the surface.

Results of the PISA Residual Contamination Study

A relatively substantial increase in the alpha contamination on the external PISA surface occurs with time at constant temperature as proven by increases in both wipable and direct alpha radiation. The averaged data obtained from the swipes of the PISA are plotted versus time at constant temperature in Figure 5. The largest increase in wipable alpha contamination occurred in the two grit blasted areas (see designated regions in Figure 2) located between the decontamination covers and the equatorial weld. The averaged results from the measurements of the direct alpha radiation emitted from the various portions of the PISA are displayed in Figure 6. The greatest increase of direct alpha radiation was from the general area of the decontamination covers with the second largest increase from the grit blasted areas. Further direct measurements of both decontamination cover areas using masks to isolate the radiation emitted from the various regions in the area found the decontamination covers, the non-grit blasted areas encircling the decontamination covers, and the surrounding grit blasted areas to emit 3%, 15%, and 82% of the direct alpha radiation, respectively. Therefore, the grit blasted areas of the PISA experienced the largest increase in direct as well as the wipable alpha radiation.

Liquid scintillation measurements of the alpha radiation emitted from the water/scouring powder mixtures used to decontaminate the PISA were 2×10^6, 4.03×10^7, and 4.96×10^7 disintegrations per minute after the 300, 700, and 1535 hr treatments, respectively. The total of 9.2×10^7 corresponds to 41.4 μCi of plutonium which was removed from the surface. An additional 0.9 μCi was removed by the swipes making the total amount of plutonium which was removed roughly 42.3 μCi.* A graph of the amount of plutonium which was removed versus time at temperature is given in Figure 7.

*42.3 μCi represents only the minimum amount of plutonium that was removed, since it was impossible to measure the plutonium remaining on the rags used for decontamination.

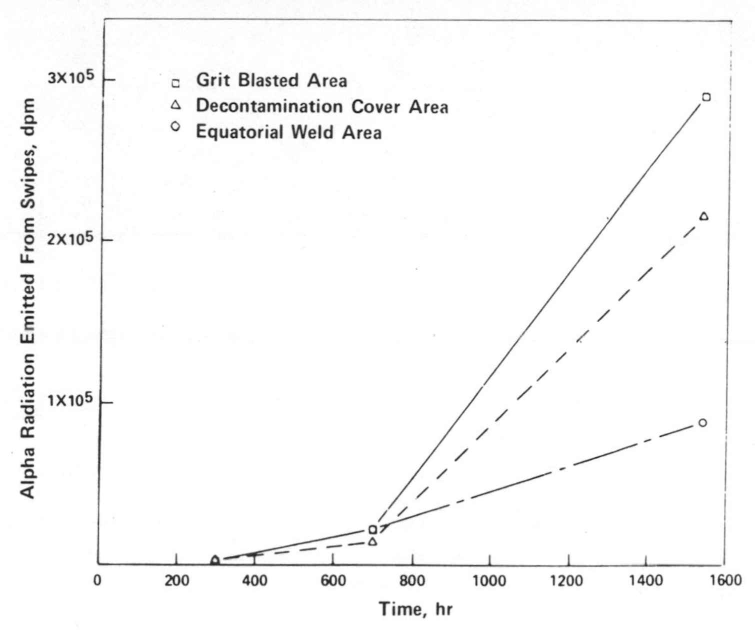

Figure 5. Graph of the wipable alpha contamination present on various regions of the PISA versus the time the PISA was exposed to 1350°C. The amount of wipable alpha contamination was determined by taking swipes.

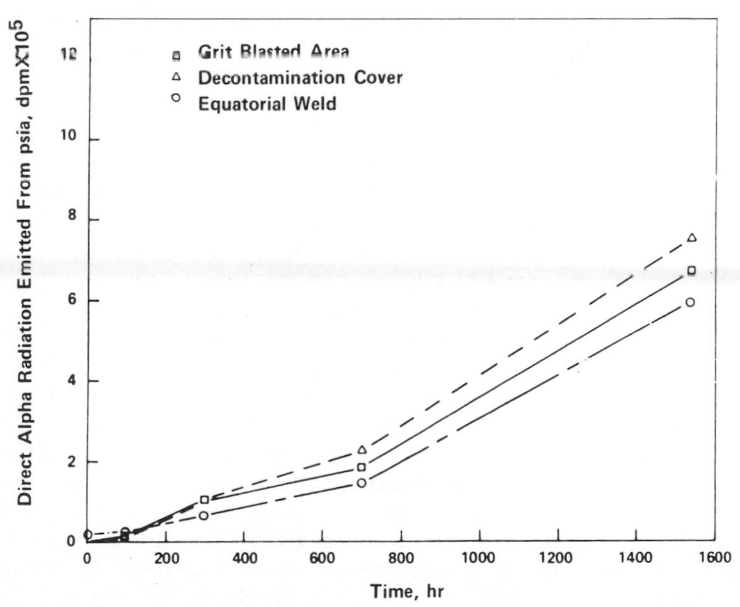

Figure 6. Graph of the direct alpha radiation emitted from the various areas of the PISA versus the time the PISA was exposed to 1350°C.

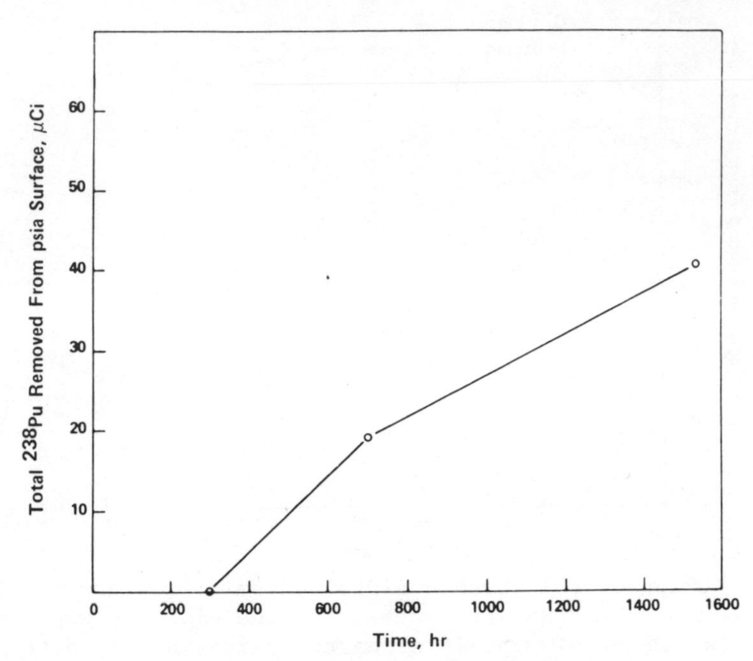

Figure 7. Graph of the total amount of ^{238}Pu removed from the surface of the PISA as a function of the time the PISA was exposed to 1350°C. The ^{238}Pu was removed from the surface during decontamination and swipe measurements.

The determination of the cross-sectional distribution of plutonium in the grit blasted area of the PISA wall showed a concentration of plutonium at the exterior edge. The direct alpha radiation measurements of the iridium surface at various depths into the wall is displayed in Figure 8 and shows that the plutonium is concentrated in the grit blasted structure located on the outside 0.1 to 0.2 mils of the PISA wall. Alpha radiation measurements of the electrolyte solutions used to remove the iridium surface layers are displayed in Figure 9. These data reveal a similar plutonium concentration in the grit blasted structure as found during the direct alpha radiation measurements at various depths into the wall. The amount of plutonium contained in the grit blasted structure (the outer 0.15 mils), which was delineated by the sealing O-ring was about 0.92 μCi, which corresponds to approximately 55 μCi in the grit blasted structure of the entire PISA. Since about 42 μCi of ^{238}Pu was removed from the grit blasted structure during the 1350°C heat treatments and subsequent decontaminations, a total of roughly 97 μCi ^{238}Pu was originally in the grit blasted structure.

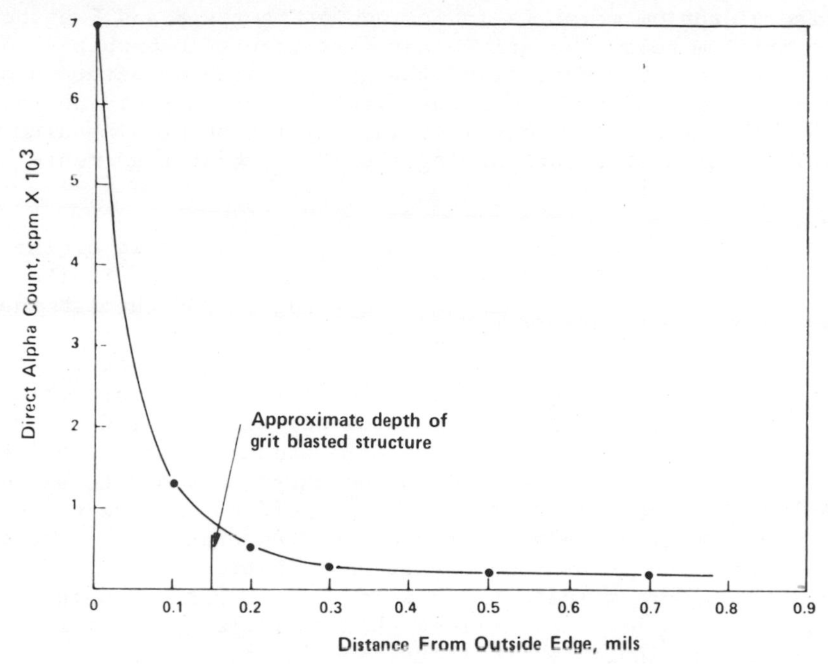

Figure 8. Graph of the direct alpha radiation coming from the PISA surface as a function of depth into the wall.

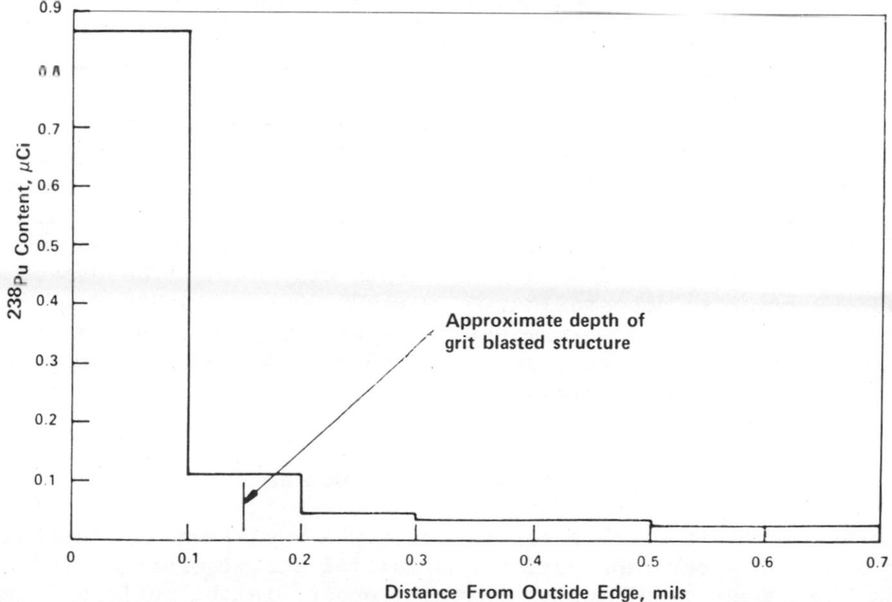

Figure 9. Distribution of plutonium in the outside portion of the PISA wall. The amount of plutonium corresponds to the alpha radiation emitted from the electrolyte used to remove the "layers" from 0.11 in.^{2} of the surface.

The plutonium which is indicated in Figures 8 and 9 as being in the iridium below the grit blasted structure is probably an anomaly, due to slow removal of the grit blasted structure underneath the O-ring used to seal the electrolyte reservoir to the PISA. However, the much less probable inward diffusion of plutonium from the grit blasted structure during the 1350°C heat treatments would cause a similar effect.

Dye penetrant examinations and helium leak checks of the PISA determined that the iridium sphere did develop faults in its exterior surface, but that the faults did not extend through the wall. The dye penetrant examinations revealed numerous pores on one end of the PISA in the area between the decontamination cover and the grit blasted area during the first examination which occurred after 300 hr at 1350°C, and in the same area on the other end of the sphere after 1535 hr. These pores were profuse, had a diameter of less than 0.001 in., and were deep enough to continue to bleed dye strongly. The dye penetrant examination did not reveal any other defects except after 1535 hr at constant temperature when several cracks were found in both of the decontamination covers and three 0.015-in. long crack indications were found about 0.050 in. from the equatorial weld. Helium leak checks of the PISA after all the heat treatments indicated that the capsule had retained its integrity; therefore, the defects found by dye penetrant did not extend through the capsule wall and cause the observed alpha contamination of the PISA surface.

Alpha contamination of the GIS was confined to the interior cavity which was in contact with the PISA and to the threads; there was no direct [the minimum limit of detection was 200 disintegrations per minute (dpm)] or wipable alpha contamination detected on the outside of the GIS or in the vacuum furnace. Table II lists the wipable contamination of the inside of the GIS. Direct alpha radiation emitted from the interior of the GIS was detected only after the total of 1535 hr at 1350°C was completed. The direct readings prior to decontamination were 30,000 to 160,000 dpm from the cavity and 4000 dpm from the threads; after decontamination (wiping with dry rag) they were 6000 to 80,000 dpm from the cavity and 400 dpm from the threads.

Discussion of PISA Residual Contamination

The level of wipable and direct alpha radiation on the PISA surface just after fabrication and initial decontamination indicates that the amount of plutonium on and embedded in the surface is very low; however, subsequent aging of the PISA at the service temperature (1350°C) results in the increase of both wipable and direct alpha contamination on the exterior surface. The source of this plutonium as determined by the area with the highest wipable and

Table II. Wipable Alpha Contamination of GIS as Measured by Swipes

Area	Before Heat Treatment (dpm)	After 100 hr at 1350°C (dpm)	After 300 hr at 1350°C (dpm)	After 700 hr at 1350°C (dpm)	After 1535 hr at 1350°C (dpm)
Internal Body	0	114	70	966	30,000
Internal Cap	0	4	48	168	10,000
Body Threads	0	8	10	80	6,000
Cap Threads	0	0	30	4	3,800
External Body*	0	30	20	4	3,800
External Cap*	0	12	22	6	0

*Alpha contamination on the exterior portion of the GIS after the shorter thermal exposures is probably due to the surrounding furnace which was measuring 500 dpm prior to the test.

direct alpha radiation was the grit blasted area. Proof that the grit blasted area was the source of the plutonium was obtained by determining the plutonium cross-sectional distribution in the PISA wall. This distribution was characterized by a concentration of plutonium in the grit blasted structure with little or no plutonium below it. The amount of plutonium in the grit blasted structure of the entire PISA was found to be approximately 55 μCi, which is slightly larger than the 42 μCi of plutonium which had been removed from the PISA surface during all of the 1350°C heat treatments; therefore, the grit blasted structure does contain enough plutonium to cause the observed surface contamination.

The concentration of plutonium in the grit blasted structure and the fact that the PISA did not develop any leaks dictate that the plutonium originally came from a source outside the PISA, rather than from within it. This original source is the plutonium environment in which the PISA is fabricated. Subsequent acid decontamination does not remove the plutonium which is deeply embedded within the grit blasted structure and this plutonium effuses during heating causing the observed surface contamination.

CONCLUSIONS

Removal of $^{238}PuO_2$ from noble and refractory metals is possible using a combination of a hot acid solution: two parts HF, two parts HNO_3, and five parts H_2O; and a scouring powder scrub. However, the $^{238}PuO_2$ is not entirely removed from a grit blasted surface but remains entrapped in the distorted surface. Subsequent heating of the capsule results in effusion of the $^{238}PuO_2$ from the grit blasted structure. In the case of the Multihundred Watt heat source, the amount of this residual $^{238}PuO_2$ was not enough to present a health hazard and was contained by the secondary capsule.

ACKNOWLEDGEMENTS

The authors are indebted to D. L. Fleming who has contributed extensively to heat source decontamination efforts and C. E. Burgan who adapted the decontamination procedure to the Milliwatt Heat Source.

MICROBIAL CONTAMINATION OF SURFACES

J. W. Costerton and G. G. Gessey

Department of Biology
University of Calgary
Calgary, Alberta, Canada T2n 1N4

Adherent bacteria predominate in natural water systems and therefore they quickly colonize submerged surfaces in industrial systems in which this water is used for heat exchange. This population is dependent on dissolved organic and inorganic molecules in the flowing water for its nutrititive requirements and on polysaccharide fibers for its continued adhesion. Our developing understanding of this adherent mode of growth, in both natural and industrial systems, offers the immediate prospect of its control by manipulation of nutrients or by interference with the polymerase-polysaccharide mechanism of adhesion used by these bacteria. This control is of paramount importance in the development of more efficient usage of our energy resources and of the development of a better interface between industrial and ecological concerns.

INTRODUCTION

There is a developing concensus in current microbiology that is of paramount importance in surface contamination. In a very wide variety of natural systems from soil[1], to water[2], to animal intestines[3] and lungs[4] it is becoming apparent that the natural state of many, if not most, of the bacteria in these ecosystems is in a complex fibrous mat adherent to a surface[5]. These adherent bacterial populations have been shown to be very numerous[6], and to be both taxonomically distinct and specialized in their physiological characteristics[7], and in their use of extracellular carbohydrate fibrils[8] to mediate their attachment to surfaces[5]. The study of these distinct adherent populations is proceeding at a steadily accelerating rate in both natural systems and disease states[9].

ADHERENT BACTERIA IN NATURAL AQUATIC SYSTEMS

We have used epifluorescence microscopy[10] to quantitate the bacteria in the slime film from submerged surfaces in a wide variety of streams and rivers, and from pristine mountain brooks to heavily polluted muddy canals, and we have always found that the bacteria on a cm^2 of surface are at least 200 times as numerous as those in a cm^3 of the flowing water[6]. A careful examination of this adherent population by electron microscopy has shown that it is embedded in a fibrous matrix of acid polysaccharide fibers[11] that tends to connect the bacteria to each other to form microcolonies of the same species (Figure 1), and to connect these microcolonies to the surface to form a coherent slime film. Very often these bacterial microcolonies are interspersed among algae to form a mixed adherent population (Figure 2). As the algae decompose, the heterotrophic bacteria in the slime film simply take up the available nutrients and thus convert algal biomass to bacterial biomass[12].

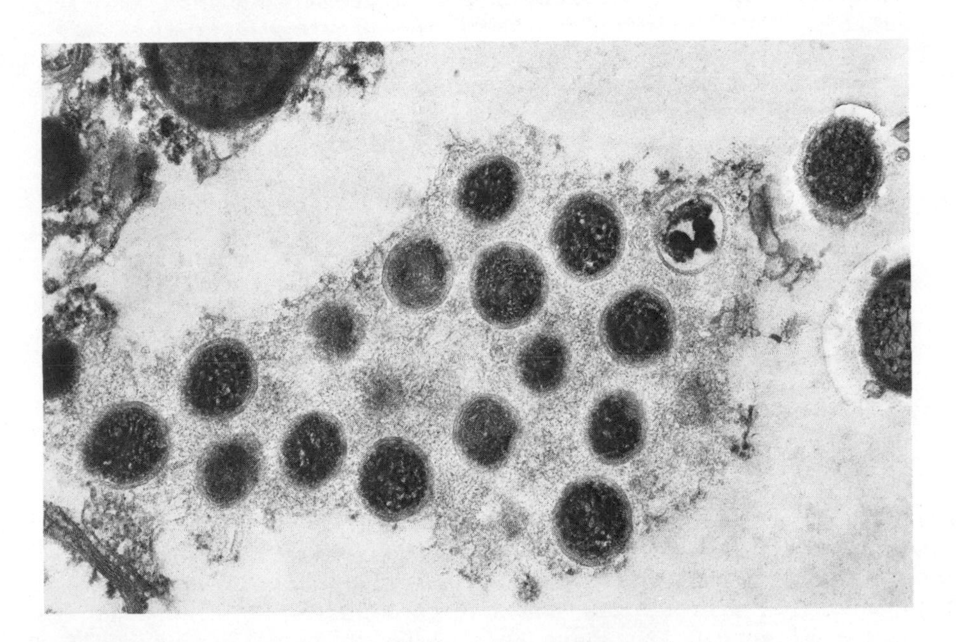

Figure 1. Electron micrograph of a slime-enclosed microcolony of aquatic bacteria of a single morphological type which has been stained with ruthenium red to reveal the polysaccharide glycocalyx fibers.

Sepers[13] has concluded that the only aquatic biota capable of taking up significant amounts of dissolved organic compounds, at natural environmental concentrations, are the bacteria. Certainly adherent bacteria are surpassingly well equipped for this uptake because their enveloping layer of carbohydrate fibers acts as an ionic exchange matrix to trap organic molecules[12] and the bacteria have very efficient permeases for subsequent transport into the cell[14]. Furthermore, the tendency of organic molecules to concentrate at submerged surfaces in aquatic environments[15] places the adherent heterotrophic bacterium in a very advantageous position vis-a-vis the availability of organic nutrients. Thus the adherent habit is very suitable for bacteria in aquatic systems because they simply attach to a submerged surface near a source of dissolved organic molecules (e.g., a dead worm) and maintain this ideal position by their adhesion.

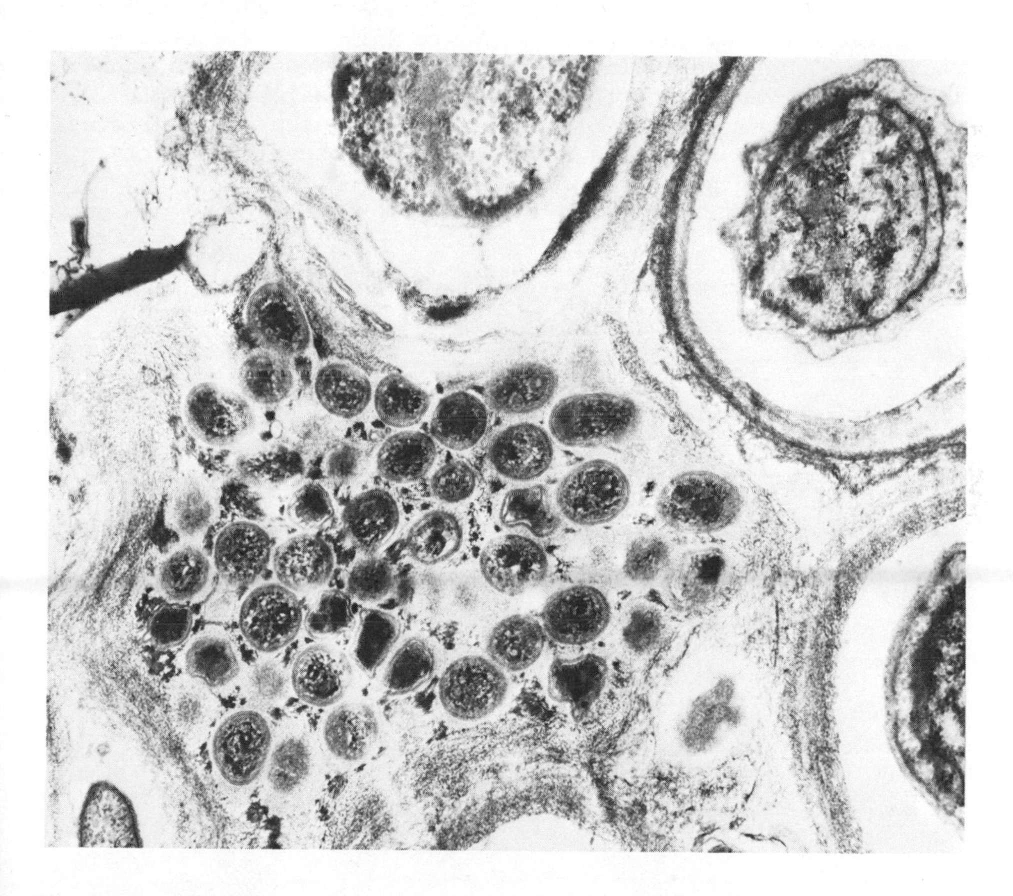

Figure 2. Electron micrograph of a slime-enclosed microcolony of aquatic bacteria which lies between the well-developed slime sheaths of several blue-green algae in an adherent matrix in an alpine stream.

At any given time some of the bacteria in an aquatic system are moving with the flowing water and many of these planktonic organisms favour the adherent growth habit. Consequently an inert surface introduced into this system is quickly colonized by bacteria which adhere to it by means of their anionic carbohydrate fibers (Figure 3). Soon the initial cell divides and forms a microcolony whose enveloping fibers trap both organic molecules and inert clay particles (Figure 4) and a slime film is built up that may eventually include algae and reach considerable proportions (Figure 5).

Thus, when we harness a natural stream for a "once-through" heat exchanger, we present a very large area of submerged surface to the mobile members of a bacterial population whose natural tendency is to adhere. Once adherent, the bacteria are easily sustained by their preferred access to dissolved organic molecules in the water and their burdgeoning growth first forms microcolonies and then a coherent slime mass that can assume massive proportions and reduce the Eh of the surface to an anaerobic level. Therefore bacterial fouling of heat-exchange systems is a product of the pronounced natural tendency of aquatic bacteria to attach themselves to submerged surfaces.

Figure 3. Electron micrograph of an aquatic bacterium forming an attachment to a plastic substrate in an alpine stream by means of its ruthenium red-positive glycocalyx fibers.

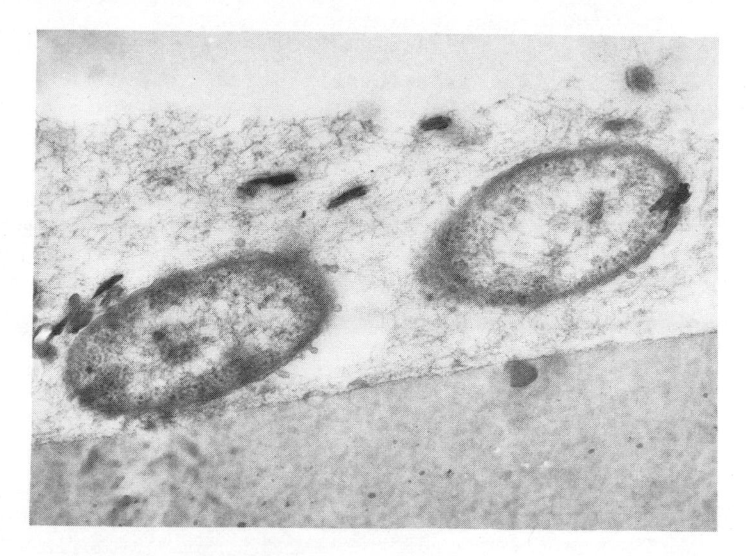

Figure 4. Electron micrograph showing the development of a bacterial microcolony on a plastic substrate in a mountain stream. Note that the fibrous matrix which attaches these bacteria to each other and to the surface also serves to trap dark flake-like clay particles.

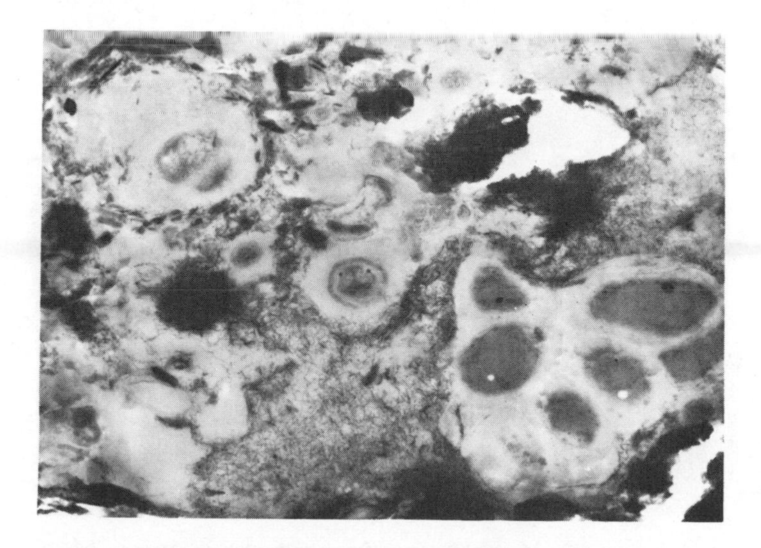

Figure 5. Electron micrograph showing a small part of a confluent adherent microbiological matrix recovered from a natural system. Note that the bacteria and algae live in lacunae in a coherent slime mass.

MECHANISMS OF BACTERIAL ADHESION

While fine hair-line protein structures called pili have been implicated in the specific adhesion of some pathogenic bacteria to host tissues[16] the enormous shear forces operative on adherent bacteria in aquatic systems dictate a more robust attachment mechanism dependent on large numbers of polysaccharide fibers[5]. Perhaps the best model for this mechanism is the _Streptococcus mutans_ systems where the attachment of this dental pathogen to the tooth surface has been shown to be dependent on its production of two distinct polysaccharides (a glucan and a fructan) whose monomers are derived from the enzymatic breakdown of sucrose[17]. The fibrous polysaccharide polymers that surround the cells of many bac-

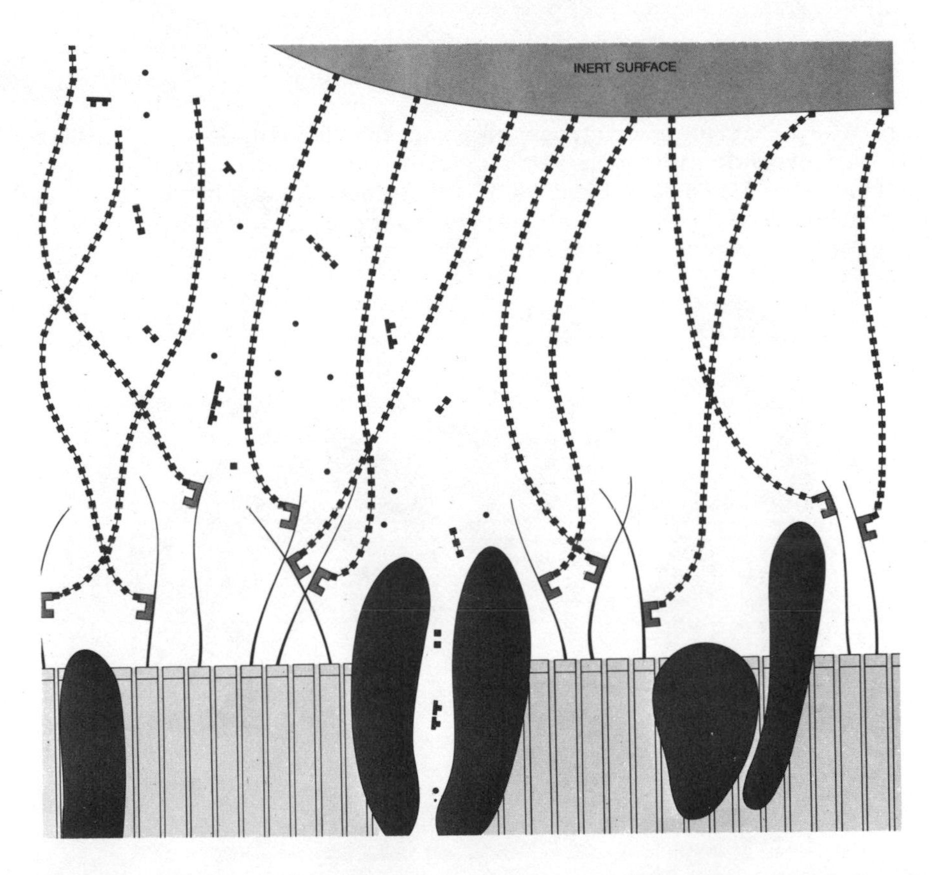

Figure 6. Schematic diagram showing the production of polysaccharide polymers that attach the bacterium to an inert surface. These polymers are produced by polymerases which are shown adherent to the hair-like distal portions of the lipopolysaccharide molecules of the outer cell wall.

teria have been characterized and largely found to be mannans, glucans and uronic acids[5] and the actual spatial arrangement of these fibers has been analyzed by X-ray crystallography[18] and found to be a highly-ordered radial pattern with very little space between the fibers. This radial glycocalyx of fibers has been stabilized by antibodies[19] and by lectins[20] and shown to enclose the cell completely and to comprise a capsule up to 2 μm in width.

The production of the polysaccharide glycocalyx fibers that mediate attachment to surfaces is under a measure of physiological control[21] and is the specific function of polymerases which are located at the outer surface of the bacterial cell (Figure 6). These polymerases produce fibrous polysaccharides from monomeric or dimeric subunits and thus build up the structure that mediates adhesion. Several of these polymerases have now been isolated[22] and the pure polymer can be produced "in vitro" for detailed chemical analysis. In at least one instance the polymerase can leave the surface of a bacterial cell and travel some distance in association with the nascent polymer[5] so that pristine surfaces may be colonized, at some distance from the closest bacterium, by bacterial glycocalyx fibers. In the dental system, the resulting complex of bacteria, polysaccharide fibers and polymerases can build

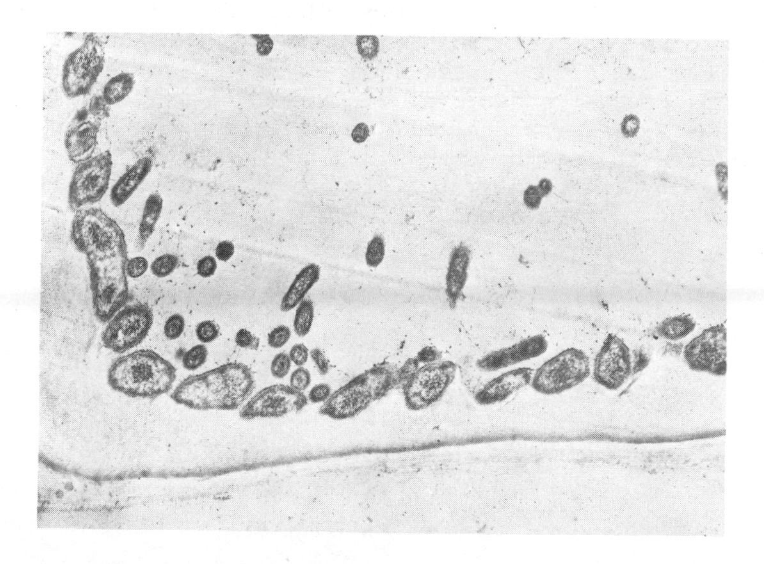

Figure 7. Electron micrograph of a decomposing leaf showing one morphological type of bacteria that colonizes and digests the cellulose cell wall. This primary population is in turn colonized by a population of much smaller cells and the system appears to function as a simple "consortium".

up so fast that a clean tooth surface can become totally covered with plaque and effectively anaerobic in just one day[23]. We have presented morphological evidence[5] that these thick and complex biological matrices may have a well-defined substructure in that one species may colonize a surface because of a specific association of its polysaccharide glycocalyx with the surface, and then a second species may attach to the first layer of cells by interaction of their glycocalices (Figure 7), and then subsequent layers of bacteria may attach. If the physiological activities of these layers of bacteria are correlated (as in methane production) the mixed population is called a "consortium".

The precise nature of the effective chemical and physical associations of the bacterial polysaccharide fibers with the inert surface to which they adhere is not well understood but the tenacity of their adhesion attests to the effectiveness of the mechanism.

BACTERIAL FOULING OF INDUSTRIAL HEAT-EXCHANGE SURFACES

The pronounced natural tendency of aquatic bacteria to adhere to surfaces has been discussed above and is perhaps typified by Zobell's old observation[24] that the bacteria in a sea water sample almost all adhere to the walls of the container if allowed to stand for one hour after collection. Hence we must expect that the population of bacteria that enters a heat-exchanger from a natural stream will display the same avidity for attachment to surfaces. By its very nature a heat-exchanger presents a very large surface area for bacterial colonization and this colonization interferes to a major extent with the physical process of heat exchange by retarding water flow in the few microns immediately adjacent to the colonized surface (Figure 8) and thus

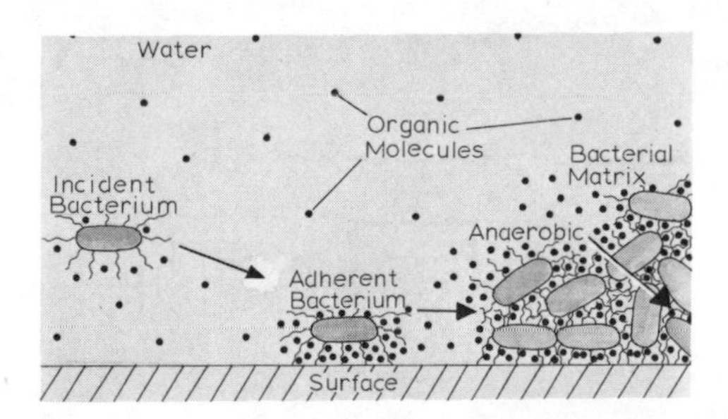

Figure 8. Schematic diagram showing the entry of a planktonic bacterium into an uncolonized system, its adherence to a submerged surface and effective trapping of organic molecules, and its eventual formation of a microbial matrix whose inner portion becomes anaerobic.

producing an insulating layer of hydrated polysaccharide slime. Because an adherent bacterial matrix only 12 μm thick can reduce the Eh at the colonized surface to very anaerobic levels[23] bacterial fouling produces ideal conditions for metal corrosion as well as interfering with heat exchange itself.

The determination of the extent of bacterial fouling of an industrial system is made very difficult by the cryptic design of these systems and the fact that the bacteria are enclosed in slime matrices so that colony counts are not useful. Thus, the only presently available and accurate method of enumerating bacteria in these systems is direct counting by epifluorescence microscopy of material recovered by aseptic scraping[6], but the important surfaces in these systems are very cryptic and sample recovery is difficult. A new and very sensitive assay can detect the lipopolysaccharide[25] that gram-negative bacteria shed into their milieu[26], and can therefore detect the presence of these organisms "up stream", and may therefore be very useful for monitoring bacterial fouling. This very sensitive assay for the presence of the gram-negative bacteria that predominate in aquatic systems[27] may also be very useful for monitoring the bacterial content of water used to wash delicate industrial products.

Now that an accurate perception of the origins of bacterial fouling and of the mechanism of bacterial adhesion have evolved, several exciting strategies for the control of this phenomenon are presented to us. First, the incident planktonic bacteria must find a surface to which they can adhere and studies of the comparative resistance to bacterial colonization of a variety of metals and plastics is currently underway in our laboratory. Second, the bacteria are dependent on their externally-located

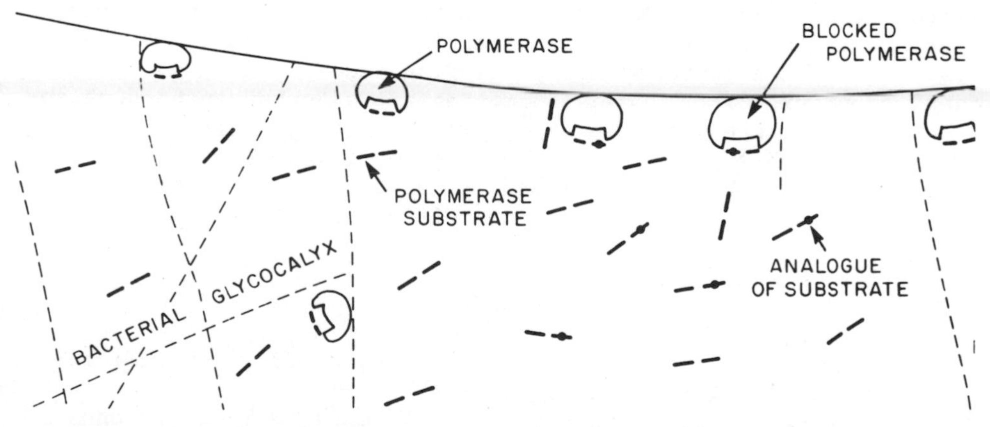

Figure 9. Schematic diagram showing the formation of polysaccharide glycocalyx fibers by polymerases and the effect of blocking the enzyme with analogues of the substrate.

polymerases for their adhesion to surfaces and, being enzymes, these polymerases are susceptible to inhibition by substrate analogues (Figure 9). Thus a battery of analogues of the substrates of the most common fouling bacteria could be used both to saturate potential binding sites on the surface and to inhibit the polymerases so that the fibrous polysaccharides that mediate adhesion in these systems could not be produced. Third, once the adherent bacteria have colonized the surfaces within the system they will develop slime film in direct proportion to the amount of dissolved organic nutrients available in the flowing water, with some limitation by oxygen availability. Therefore, the limitation of an essential nutrient (either organic or inorganic), by pretreatment of the water will have the effect of controlling the extent of bacterial fouling. In open systems (cooling towers) photosynthesis by algae introduces organic matter into the system and it may be useful to limit this process by light exclusion. These three strategies are all based on the prevention or control of colonization and all would be facilitated by basic research on the detailed chemistry of the bacterial glycocalyx polysaccharides and the mechanism of their adherence to inert surfaces.

Once bacterial colonization has been established in an industrial system we must direct our attention to its removal. In this respect natural systems offer some encouragement because the number of cells in the adherent bacterial populations of many streams fluctuate between wide extremes[6] and the bacterial matrices occasionally slough almost completely. Pulse chlorination has been relatively effective in both killing adherent bacteria and effecting their removal from the system by sloughing of the bacterial matrix but this procedure is presently subjected to increasingly critical ecological scrutiny. Therefore other methods should be monitored for their efficiency in both killing the bacteria and removing the adherent matrix. The present empirical approach to the removal of the bacterial matrix from industrial systems can be altered by our developing understanding of the chemical nature of the adhesion mechanism. Agents can be developed which interfere with the association of the polysaccharide fibers with the surface, or with the integrity of the polymer itself, so that the bacterial matrix can be removed with a consequent dramatic improvement in the efficiency of heat exchange.

REFERENCES

1. H.C. Bae, E.H. Cota-Robles and L.E. Casida Jr., Appl. Microbiol. 23, 637 (1975).
2. G.G. Geesey, W.J. Richardson, H.G. Yoemans, R.T. Irvin and J.W. Costerton, Can. J. Microbiol. 23, 1733 (1977).
3. D.C. Savage, Annu. Rev. Microbiol. 31, 107 (1977).

4. J.W. Costerton, M.R.W. Brown and J.M. Sturgess, in "Pseudomonas aeruginosa: Clinical Manifestations and Current Therapy", (R.G. Doggett, ed.) Academic Press, New York (in press).
5. J.W. Costerton, G.G. Geesey and K.-J. Cheng, Scientific American 238, 86 (1978).
6. G.G. Geesey, R. Mutch, J.W. Costerton and R. Green, Limnol. Oceanogr. 23, 1214 (1979).
7. K.-J. Cheng, R.P. McCowan and J.W. Costerton, J. Clin. Nutr. (in press).
8. W.A. Corpe, Develop. Ind. Microbiol. 11, 402 (1970).
9. K.C. Marshall,"Interfaces in Microbial Ecology,"Harvard University Press, Cambridge, MA, 1976.
10. R.J. Daley and J.E. Hobbie, Limnol. Oceanogr. 20, 875 (1975).
11. M. Fletcher and G.D. Floodgate, J. Gen. Microbiol. 74, 325 (1973).
12. T.I. Ladd, J.W. Costerton and G.G. Geesey, in "Native Aquatic Bacteria: Enumeration, Activity and Ecology", J.W. Costerton and R.R. Colwell, Editors, ASTM Press, Philadelphia (in press).
13. A.B.J. Sepers, Hydrobiologia 52, 39 (1977).
14. J.H. Weiner and L.A. Heppel, J. Biol. Chem. 246, 6933 (1971).
15. C.E. Zobell, J. Bacteriol. 46, 39 (1943).
16. N. Kumazawa and R. Yamagawa, Infec. Immun. 5, 27 (1972).
17. R.J. Gibbons and J. van Houte, Ann. Rev. Med. 26, 121 (1975).
18. R. Moorhouse, W.T. Winter, S. Arnott and M.E. Bayer, J. Mol. Biol. 109, 373 (1977).
19. M.E. Bayer and H. Thurow, J. Bacteriol. 130, 911 (1977).
20. D.C. Birdsell, R.J. Doyle and M. Morgenstern, J. Bacteriol. 121, 726 (1975).
21. K.-J. Cheng, D.E. Akin and J.W. Costerton, Fed. Proc. 36, 193 (1977).
22. R. Chan and J.W. Costerton (in preparation).
23. W.A. Coulter and C. Russell, J. Appl. Bacteriol. 40, 73 (1976).
24. C.E. Zobell and D.O. Anderson, Biol. Bull. 71, 324 (1936).
25. S.W. Watson, T.J. Novitsky, H.L. Quinby and F.W. Valois, Appl. Environ. Microbiol. 33, 940 (1977).
26. E. Work, Folia Microbiologica (Prague) 12, 220 (1967).
27. J.W. Costerton, J.M. Ingram and K.-J. Cheng, Bacteriol. Rev. 38, 87 (1974).

SURFACE CLEANING IN THIN FILM TECHNOLOGY*†

D. M. Mattox

Sandia Laboratories

Albuquerque, New Mexico 87185

INTRODUCTION

As a practical matter, a "clean surface" is one that contains no significant amounts of undesirable material. In some cases, such as the adhesion of deposited thin films, foreign material on the surface may be desirable or necessary. Adhesion is often equated to surface cleanliness,[+] so surface treatments or the deliberate addition of foreign material which improves adhesion may be considered cleaning techniques.

[+]Equating adhesion, which is a gross effect, to bonding or cleanliness may be very misleading. Failure of adhesion may be more related to fracture mechanisms than to bonding. In thin films the intrinsic stress may result in adhesive failure even though chemical bonding may be high; also the interfacial morphology may lead to easy fracture, even though the chemical bonding is strong. Refer to D. M. Mattox, "Thin Film Adhesion and Adhesive Failure - A Perspective," Adhesion Measurement of Thin Films, Thick Films, and Bulk Coatings, K. L. Mittal, Editor, pp. 54-62, American Society for Testing & Materials, Philadelphia, 1978.

*Work supported by the U. S. Department of Energy.

†This is an abbreviated version of the full paper published in Thin Solid Films 53, 81-96 (1978).

Surface contaminants may be classed as (1) reaction layers, (2) adsorbed layers, (3) variable composition contaminants, or (4) particulate contaminants. Reaction layers may be in the form of oxides, carbides, or other such layers. The material to form these layers may come from the external environment or from the bulk of the material. Adsorbed layers form by adsorption of material from the environments or diffusion over the surface from the surroundings. Variable composition contaminants may occur as layers or regions near the surface, such as the diffusion of a minor constituent from the bulk to the surface region giving a high surface concentration or may be in the form of second-phase material strongly bound on the surface. Particulate contaminants, such as dust, are loosely bound contaminants whose origin is usually from the external environment. In some cases, surface contamination may result from the "cleaning" techniques used.

To design a cleaning process one must realize where the origin of contaminants may lie. In a typical processing sequence, the following factors may be important in cleaning and contamination control: (1) as received condition, (2) time, (3) component design, (4) processing, (5) handling, and (6) storage.

In addition to designing a cleaning procedure to remove undesirable contaminants from the surface, one must be concerned with the effect of the cleaning procedure on the surface. The cleaning procedure may change the surface composition to something which may be undesirable. Cleaning may also affect the surface morphology by selective etching, thermal faceting, etc. The cleaning process may leave undesirable residuals on the surface.

Cleaning processes may be categorized as specific or general. Specific cleaning processes are designed to remove specific types of contaminants without changing the surface of interest; general cleaning techniques often involve removing some of the surface material, as well as contaminants. Etching procedures such as sputter cleaning, electropolishing, and chemical etches, fall into this category as does mechanical abrasion. Often a cleaning procedure will involve a combination of procedures used in sequence. Unless the cleaning procedure is designed to take advantage of the strong points and weak points of each step, the cleaning steps may defeat each other.

CLEANING - OXIDE SURFACES

In the deposition of thin films on oxide substrates, it is generally considered that chemical reaction between film and surface is a necessary criterion for adhesion. Thus, metals which are strong oxide formers should have good adhesion to oxide surfaces by forming interfacial oxide regions, but metals such as gold, which do not form stable oxides, should have poor adhesion to oxide surfaces. Generally, it is found that sputter deposition of films gives better adhesion than vacuum deposition. This may be attributed to the plasma cleaning and higher energy of the depositing atoms, as well as the heating and electron/neutral atom bombardment of the substrate which normally accompanies sputter deposition. The deposition gaseous environment can also affect the adhesion of metal films to oxide surfaces. Oxygen is sometimes used to generate an interfacial layer conducive to good adhesion. Substrate surface chemistry also may have an important effect on the adhesion of metal films to oxide surfaces. Selective leaching of materials from the surface may result in adhesion problems.

Other procedures may be used to modify the surface chemistry to improve adhesion. The most widely used technique is to use multilayer metallization techniques where a thin layer of reactive material such as titanium or chromium is deposited in contact with an oxide and then the desired metallization, usually gold, is deposited as the upper layer. The "adhesive layer" chemically reacts with the substrate and alloys with some of gold to give good adhesion to both materials.

In order to obtain good adhesion, it is necessary that the surface be free of contaminants which inhibit the interaction between the surface and the depositing film atoms. The most commonly encountered contaminants on oxide surfaces are hydrocarbons, hydroxyl ions, and alkali halides.

The substrates should be cleaned in a good detergent or scrubbed with an abrasive cleaner in order to remove water-soluble contaminants and some organic contaminants. Mechanical scrubbing is more effective than ultrasonic agitation. Solvent cleaning is widely used for cleaning insulator surfaces, but with the advent of new industrial safety standards, many of the organic solvents are difficult to use in production environments. Generally, solvent cleaning adds little to good wash followed by air firing, oxygen plasma cleaning, glow discharge, or sputter etching of an oxide surface.

A number of cleaning procedures may be used to remove oxidizable contaminants such as hydrocarbons. These include boiling in hydrogen peroxide (H_2O_2), hot oxidizing solutions such as nitric acid or potassium dichromate, exposure to an oxygen plasma, high-temperature air fire, or exposure to short wavelength ultraviolet radiation in the oxygen (UV/O_3 cleaning). If the oxidation product is volatile such as CO or CO_2, the contaminant volatilizes; if the product is soluble, it may be dissolved in the cleaning solution. The air fire, the oxygen plasma treatment, and the UV/O_3 treatment have the advantage that organic contaminants are reactively volatilized and residuals from cleaning solutions are avoided.

Exposure of oxide surfaces to an inert gas discharge (glow discharge cleaning) has long been used to clean the surface prior to film deposition. The cleaning mechanism is poorly defined since the surface is being bombarded with ions, electrons, and high-energy neutrals, as well as radiation from the plasma. Exposure of an insulating surface to a dc plasma shielded from the cathode allows a small negative potential to build up on the surface (wall potential) due to the higher mobility of electrons compared to the ions in the plasma. This allows a low energy "ion scrubbing" of the surface to take place. If the surface is not shielded from the cathode, it will also be bombarded by high energy electrons and high energy neutrals from the cathode.

Sputter cleaning or sputter etching has the advantage that it may be done in situ in the deposition system just prior to the deposition process. In sputter cleaning an insulator surface, an rf potential is impressed on the surface and the surface is then alternately bombarded with ions and electrons. The ion bombardment results in removal of material by a momentum transfer process called sputtering. Sputter cleaning is an effective cleaning technique, but problems may arise because of backscattering of sputtered material to the surface, particularly at high pressures. In some cases, gas incorporation in the sputtered surface may be released during subsequent processing, causing loss of adhesion of a film deposited on a sputter-cleaned surface. This can be avoided by heating the substrate during or after cleaning.

In the ion plating process, the sputter cleaning is an integral step in the deposition process. In this technique, the film deposition is begun while the surface is being sputter-cleaned. A film results only when the deposition rate is greater than the removal rate. For insulators, the substrate potential must be rf to prevent charge buildup which would halt dc sputter cleaning.

Various chemical etching techniques are also used to clean ceramic type material. Chemical etching has also been used to improve film adhesion by changing the surface morphology. If high alumina ceramics are etched in fused NaOH, it is found that the adhesion of copper to the surface is greatly improved. In this case, the improved adhesion is attributed to microroughening of the surface improving the mechanical interlocking.

Particulate contamination may best be removed from surfaces by blow-dusting. In some cases an ionized gas may be used to keep from charging the surface. Solvent sprays may also be used.

CLEANING - METAL SURFACES

The adhesion between deposited metal films and metal surfaces depends upon the interfacial morphology, the nucleation, chemical reaction, and diffusion in the interfacial region. It has often been assumed that to obtain good adhesion, there must be mutual solubility in the metal-metal system. Solubility is a desirable characteristic but is not necessary, as has been demonstrated in good adhesion between completely insoluble materials. A lack of solubility will affect the nucleation of the depositing atoms on the surface and thus affect the interfacial morphology. The extent of interactions in the interfacial region may also be important to adhesion. The growth of extensive brittle intermetallic layers may give poor strength. Kirkendahl porosity, which weakens the interface, may occur in cases of asymmetric diffusion rates.

The presence of surface contaminants may prevent diffusion and reaction between the depositing atoms and the surface. Very thin carbon or oxide (10-100 Å) layers may provide effective surface barrier layers which prevent good bonding or adhesion. Since surface oxides and adsorbed hydrocarbons are common contaminants on metal surfaces, the elimination of these materials are often a prime concern in thin film deposition technology.

The surface contaminants on metal may be removed by a variety of techniques which may depend upon the metals involved. As with the oxide surfaces, the metal surfaces should be detergent and solvent cleaned to rid the surface of soluble contaminants. Gross contamination, such as thick oxides, may be removed by abrasive cleaning, etching, or fluxing. Etching may be performed by chemical etching or electropolishing, depending on the metals involved. Various fluxes may be used to dissolve or float away thick oxides. These fluxes are often molten salts. After abrasive cleaning, etching, or fluxing, care must be taken to remove residuals from the surface by rinsing.

The most universally applicable cleaning technique for removing trace contamination is sputter cleaning by ultrahigh vacuum heating to desorb gases incorporated in the surface during cleaning. For conductive surfaces, the applied sputtering potential may be dc or rf, although rf is preferred. When using sputter cleaning, care must be taken to flush away the sputtered contaminants by using a high system throughput. The discharge parameters may be used as an indicator of cleaning and system condition. The power input to the system is sometimes used as a measure of cleaning efficiency. Often it is not realized that sputter cleaning can do more harm than good if the proper conditions are not used. For sputter cleaning to be effective on reactive materials, the sputtering environment must be very free of contaminating gases. An advantage of sputter cleaning is that it may be done in situ just prior to film deposition in many systems. When an object is being sputter-cleaned, care must be taken that the surface is not contaminated by material sputtered from holders and shields. As was discussed earlier, the ion plating process has sputter cleaning and surface heating as an integral part of the deposition process so that recontamination problems are minimized.

Vacuum or hydrogen firing may also be used to clean metal surfaces. A number of chemical and electrochemical cleaning techniques exist for particular metals and contaminants. Glow discharge cleaning in an inert gas or oxygen may be used to remove hydrocarbon contamination from metal surfaces. An oxygen discharge may cause excessive oxidation of some metals which do not form passive oxides. Recently, "reactive plasma cleaning" has been used to generate clean metal surfaces at a lower power input than is possible with sputter cleaning alone. Gold presents a unique case for cleaning a metal. Since gold does not readily oxidize, it may be cleaned by the oxidizing treatments discussed under the cleaning of oxides.

CLEANING - ORGANIC POLYMERS

Often with organic materials the surface is composed of low molecular weight fractions which fracture easily. To obtain good adhesion, the surface must be modified in such a way that it does not represent a zone of weakness at the interface; thus, particularly in the case of organics, cleaning means modifying the surface region in a desirable way.

Some undesirable organics may be removed from the surface by a detergent wash or solvent clean. Solvents may be absorbed into the polymer structure, causing it to expand and possibly craze on drying, and may leach some of the low molecular weight fractions from the surface.

Generally, surface treatments of organics conducive to good adhesion of deposited thin films may be classed as (1) roughening, (2) activation, or (3) cross-linking. Surface roughness and mechanical interlocking has been shown to be an important factor in adhesion. The roughness may be obtained mechanically such as with abrasion, grit blasting, etc., or may be obtained by preferentially etching the organic surface.

Activation of some polymer surfaces may be accomplished by chemical treatments where organometallic complexes are formed in the surface regions. Such activation treatments are the iodine treatment of nylon and the sodium treatment of teflon. Activation may also be accomplished by the addition of a small amount of material which reacts both with the substrate surface and the deposited film (primer). These materials may be highly polar, low molecular weight organics.

Cross-linking may be used to increase the strength of the surface material and to reduce the amount of undesirable low molecular weight components. Cross-linking can be encouraged by (1) corona discharge, (2) exposure to a plasma, and (3) exposure to ultraviolet radiation. These treatments are probably not fundamentally different since the ultraviolet radiation from a corona, gas discharge, or ultraviolet lamp causes bond scission and adds carbonyl groups to the polymer surface. In thin film technology, exposure of the polymer surface to a gas discharge can often be done in situ before film deposition.

Surface modification may also occur during film deposition. If the film is being deposited by sputter deposition, the depositing atoms have much higher than thermal energies and may be implanted into the surface to an appreciable depth. These physically incorporated atoms may then act as nucleating and bonding sites for further atom deposition. A problem with sputter deposition of films on plastics is the heating by secondary electrons which may be present. This electron bombardment can be reduced by the use of magnetron type sputter deposition apparatus.

MONITORING SURFACE CLEANING

The best monitor for cleaning is subsequent processing or usage. This criteria is often not acceptable both from a practical and an aesthetic standpoint, so some type of in-process testing and acceptance specification is desired. Unfortunately, most testing affects the surface, so testing must usually be performed on representative surfaces which then qualify the untested surfaces. This means that the surface to be tested must be shown to be representative of the surface to be utilized and that rigid cleaning specifications must be provided so that all surfaces are cleaned alike.

Monitoring techniques may be classed as direct or indirect. Direct monitoring involves identifying the surfaces species directly. This may be done by various surface analytical techniques such as Auger electron spectroscopy (AES), appearance potential spectroscopy (APS), ion scattering spectrometry (ISS), multireflectance infrared analysis, and others. Of prime concern are the level of detection of the technique and the value judgment as to whether the amount and type of detected species are important. Indirect monitoring utilizes the behavior of the surface to imply the presence of a contaminant. Again, the technique may not indicate whether the type and amount of contaminant is harmful or not. The most generally used indirect monitoring techniques for detecting low surface energy contaminants such as hydrocarbons, may be classed as (1) wetting, (2) contact angle, (3) coefficient of adhesion/friction, (4) adsorption, (5) nucleation, and (6) adhesion.

Ideally one would like to generate an atomically clean surface, then contaminate the surface in a known way while monitoring the effect on subsequent processing and usage. This would allow the delineation of the values for significant amounts of undesirable materials. Unfortunately, in most cases this is impossible or impractical.

CLEAN SURFACE STORAGE

An often neglected aspect of cleaning is that of storage. After an elaborate cleaning procedure, it is not too unusual to find that the surface is stored in a plastic bag or tray in-process or in a contaminated atmosphere in use; therefore, storage and packaging must be considered as part of the cleaning process. In-process storage normally means keeping contaminants away. Storage in an ultraclean, controlled environment is usually unnecessary. The problem is to identify the undesirable contaminants and eliminate them from the storage environment.

Dust may be minimized by storing the cleaned surface in a closed container or in a clean bench. One of the most common contaminants is hydrocarbon vapors adsorbed from the air. Adsorptance of these contaminants may be minimized by storing in freshly oxidized aluminum containers which preferentially adsorb the hydrocarbons. A disadvantage of this technique is the extra steps involved in periodically stripping and oxidizing the metal surfaces. The UV/O_3 cleaning technique may be used to keep oxide and gold surfaces clean indefinitely in an ambient environment. Cleaned surfaces should be stored in clean glass containers at least.

Often clean surfaces are sealed in a hermetic device. In some cases, these surfaces are contaminated by the sealing operation which frequently involves heating of other nonclean surfaces. Before sealing, all interior surfaces should be as clean as possible. The design of hermetically-sealed components should exclude sources of contamination, such as adhesives and lubricants, which may be sources of contamination.

Since thin film systems normally have a high surface-to-volume ratio, they may be very sensitive to surface contaminants such as corrosive agents. For instance, the Ti-Au metallization system is very sensitive to chlorine contamination; thus, in-process and in-use care must be taken to eliminate chlorine or halogenated organics from the cleaning process and the storage environment.

CONCLUSION

Cleaning a surface involves not only the generally accepted "art" but also science in the form of understanding a surface and its effects on the desired application. Often an "atomically clean" surface is not necessary nor even desirable. One of the key problems in cleaning is to define what are significant amounts of undesirable material. Here experience and "art" are often the most reliable indicator, but work should be done to quantify the answer.

NOTE: All references and figures may be found in the full published text.

Part II
Cleaning of Surfaces

UV/OZONE CLEANING OF SURFACES: A REVIEW

John R. Vig

US Army Electronics Technology and Devices
Laboratory, ERADCOM
Fort Monmouth, New Jersey 07703

The ultraviolet (UV)/ozone cleaning procedure is an effective method of removing a variety of contaminants from surfaces. It is a simple-to-use dry process which is inexpensive to set up and operate. It can rapidly produce clean surfaces, in air, at ambient temperatures. By placing properly precleaned surfaces within a few millimeters of an ozone producing UV source, the process can produce clean surfaces in less than one minute.

This paper reviews the UV/ozone cleaning method, including the variables of the process, the types of surfaces which have been successfully cleaned, the contaminants which can be removed, the construction of a UV/ozone cleaning facility, the mechanism of the process, and successful applications to date.

INTRODUCTION

The ability of ultraviolet (UV) light to decompose organic molecules has been known for a long time, but it is only in the last few years that UV cleaning of surfaces has been explored.

In 1972, Bolon and Kunz[1] reported the ability of UV light to depolymerize a variety of photoresist polymers. The polymer films were enclosed in a quartz tube, the tube was evacuated, then back-filled with oxygen. The samples were irradiated with UV light from a medium pressure mercury lamp which generated ozone. The polymer films, which had been several thousand angstroms thick, were successfully depolymerized in less than an hour. The major products of depolymerization were found to be water and carbon dioxide. Subsequent to depolymerization, the substrates were examined by Auger Electron Spectroscopy (AES) and were found to be free of carbonaceous residues. Only inorganic residues such as tin and chlorine were found. When a Pyrex filter was placed between the UV light and the films, or when a nitrogen atmosphere was used instead of oxygen, the depolymerization was hindered. Thus, Bolon and Kunz recognized that oxygen, and wavelengths shorter than 3,000 Å played a role in the depolymerization.

In 1974, Sowell et al. [2] described UV cleaning of adsorbed hydrocarbons, from glass and gold surfaces, in air and in a vacuum system. A clean glass surface was obtained after 15 hours of exposure to the UV radiation in air. Figure 1 shows the decrease in contact angle as a function of UV exposure time. In a vacuum system at 10^{-4} torr of oxygen, clean gold surfaces were produced after about 2 hours of UV exposure. During cleaning, the partial pressure of O_2 decreased, while that of CO_2 and H_2O increased. The UV also desorbed gases from the vacuum chamber walls. In air, gold surfaces which had been contaminated by adsorbed hydrocarbons could be cleaned by "several hours of exposure to the UV radiation". Sowell et. al. also noted that by storing clean surfaces under UV radiation it was possible to maintain the surface cleanliness indefinitely.

Starting in 1974, Vig et. al.[3-5] described a series of experiments aimed at determining the optimum conditions for pro-

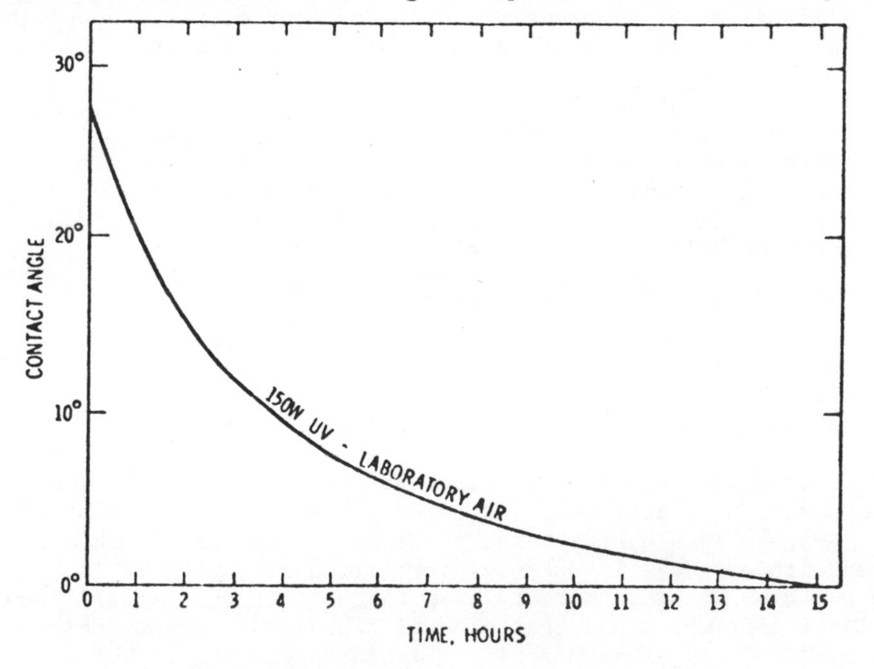

Figure 1. UV/ozone cleaning of a glass surface. (Ref. 2)

ducing clean surfaces by UV irradiation. The variables of cleaning by UV irradiation were defined, and it was shown that, under the proper conditions, UV/ozone cleaning was capable of producing clean surfaces in less than one minute.

THE VARIABLES OF UV/OZONE CLEANING

The Wavelengths Emitted by the UV Sources

To study the variables of the UV cleaning procedure, Vig and LeBus[5] constructed the two UV cleaning boxes shown in Figure 2. Both were made of aluminum and contained low-pressure mercury discharge lamps and an aluminum stand with Alzak[6] reflectors. The two lamps produced nearly equal intensities of short-wave UV light, about 1.6 mW per cm^2 for a sample 1 cm from the tube. Both boxes contained room air (in a clean room) throughout these experiments. The boxes were completely enclosed to reduce recontamination by air circulation.

Since only the light which is absorbed can be effective in producing photochemical changes, the wavelengths emitted by the UV sources are important variables. The low pressure mercury discharge tubes generate two wavelengths of interest, 1849 Å and 2537 Å. The 1849 Å wavelength is important because it is absorbed by oxygen, and it thus leads to the generation of ozone[7]. The 2537 Å radiation is not absorbed by oxygen; it therefore does not contribute to ozone generation. However, it is absorbed by most hydrocarbons[8,9] and also by ozone[7]. The absorption by ozone is principally responsible for the destruction of ozone in the UV box. Therefore, when both wavelengths are present, ozone is continuously being formed and destroyed. An intermediate product

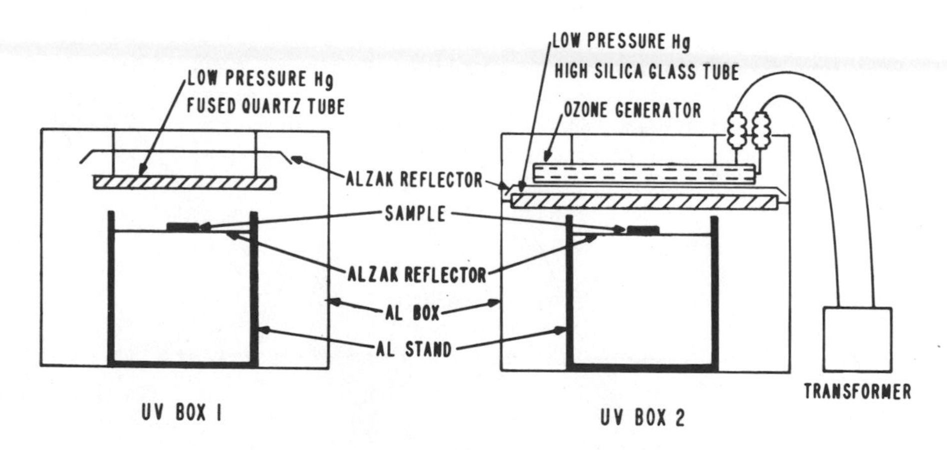

Figure 2. Apparatus for UV/ozone cleaning experiments.

of both the formation and destruction processes is atomic oxygen, which is a very strong oxidizing agent.

The tube of the UV lamp[10] in box 1 consisted of 91 cm of "hairpin-bent" fused quartz, which transmits both the 2537 Å and 1849 Å lines. The lamp emitted about 0.1mW per cm^2 of 1849 Å radiation, measured at 1 cm from the tube.

The lamp in box 2 had two straight and parallel, 46-cm-long, high-silica glass tubes. The glass was Corning UV Glass No. 9823 which transmits at 2537 Å but not at 1849 Å. Since this lamp generated no measurable ozone, a separate Siemens-type ozone generator[11] was built into box 2. This ozone generator did not emit UV light. Ozone was produced by a "silent" discharge when high-voltage ac was applied across a discharge gap formed by two concentric glass tubes, each of which was wrapped in aluminum-foil electrodes. The ozone-generating tubes were parallel to the UV tubes, approximately 6 cm away.

UV box 1 was used to expose samples simultaneously to 2537 Å, 1849 Å, and the ozone generated by the 1849 Å radiation. UV box 2 permitted the options of exposing samples to 2537 Å plus ozone, 2537 Å only, or ozone only.

Vig et. al. used contact angle measurements, wettability tests and Auger electron spectroscopy (AES) to evaluate the results of their cleaning experiments. Most of the experiments were conducted on polished quartz wafers whose cleanliness could be evaluated by the "steam test", a highly sensitive wettability test.[5,12,15]

A "black-light" long wavelength UV source, which emitted wavelengths above 3000 Å only, was also tried, however, it produced no noticable cleaning even after 24 hours of irradiation.

It was found early in the studies of Vig et. al. that samples could be cleaned consistently by UV irradiation only if gross contamination was first removed from the surfaces. Their precleaning procedure consisted of the following steps:

(1) Scrub the sample with a swab while it is immersed in ethyl alcohol,

(2) Agitate ultrasonically in fresh ethyl alcohol,

(3) Boil in fresh ethyl alcohol, and then agitate again ultrasonically,

(4) Rinse in running ultrapure (18 MΩ·cm) water,

(5) Spin dry immediately after the running water rinse.

Subsequent to this precleaning procedure, the steam test and contact angle measurements invariably indicated that the surfaces were contaminated. However, after exposure to UV/ozone in box 1, the same tests invariably indicated clean surfaces. The cleanliness of such UV/ozone cleaned surfaces have been verified in the author's laboratory and elsewhere by AES and Electron Spectroscopy for Chemical Analysis (ESCA) on numerous occasions[1,3,4,13,15] For example, Figures 3 and 4 are AES[14] and ESCA[15] results respectively, showing the cleaning effect of UV/ozone. The effectiveness of UV/ozone cleaning has also been confirmed by Ion Scattering Spectroscopy/Secondary Ion Mass Spectroscopy (ISS/SIMS).[16]

A number of quartz wafers were precleaned and exposed to the UV light in box 1 until clean surfaces were obtained. Each of the wafers was then thoroughly contaminated with human skin oil

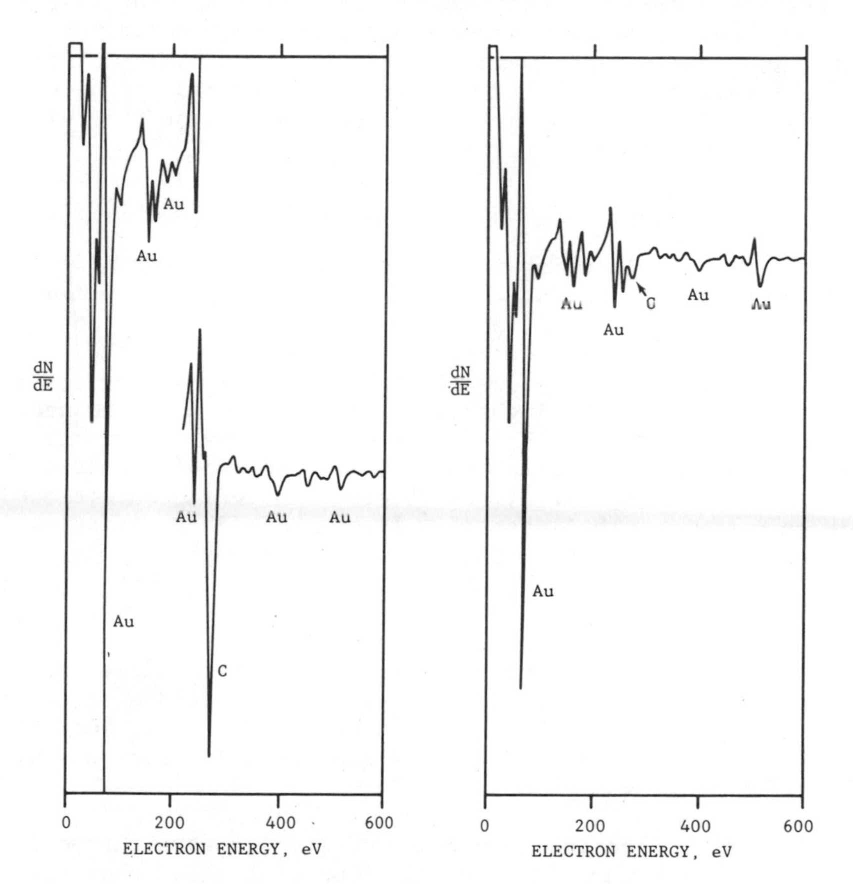

Figure 3. AES scans of contaminated (left) and UV/ozone cleaned (right) gold surface (Ref. 14).

which has been a difficult contaminant to remove. The wafers were precleaned again, groups were exposed to each of the four UV/ozone combinations mentioned earlier, and the time to attain a clean surface, as indicated by the steam test, was measured. In each UV box, the samples were placed within 5 mm of the UV source (where the temperature was about 70°C).

The wafers exposed to 2537 Å + 1849 Å + ozone in UV box 1 became clean in 20 s. The samples exposed to 2537 Å + ozone in UV box two reached the clean condition in 90 s. Samples exposed to 2537 Å without ozone and to ozone without UV light took about 1 h and 10 h, respectively, before clean surfaces were obtained.

The conclusion one can draw is that while both UV light without ozone and ozone without UV light can produce a slow cleaning effect in air, the combination of both short-wavelength UV light and ozone, such as is obtained from a quartz UV lamp, produces a clean surface orders of magnitude faster. Although the 1849 Å radiation is also absorbed by many hydrocarbons, it was not possible from these experiments to isolate the cleaning effect of the 1849 Å radiation. The ozone concentrations had not been measured. As is discussed below, the concentrations vary within each box with distance from the UV source.

Distance Between Sample and UV Source

Another variable which can greatly affect the cleaning rate is the distance between the sample and the UV source. Because of the shapes of the UV tubes and of the Alzak reflectors above the tubes and below the samples, the lamps in both boxes were essentially plane sources. One might, therefore, have expected that the intensity of UV light reaching a sample would be nearly independent of distance. This was not so, however, when ozone was present, because ozone has a broad absorption band[7,17,18] centered at about 2600 Å. At 2537 Å, the absorption coefficient of ozone is 130 cm^{-1} atm^{-1}. The intensity, I, of the 2537 Å radiation reaching a sample therefore decreases as

$$I = I_0 e^{-130pd}$$

where p is the average ozone pressure between the sample and the UV source in atmospheres at 0°C, and d is the distance to the sample in centimeters. When a quartz UV tube is used, both the ozone concentration and the UV radiation intensity decrease with distance from the UV source.

Two sets of identically precleaned samples were placed in UV box 2. One set was placed within 5 mm of the UV tube; the other, at the bottom of the box, about 8 cm from the tube. With the ozone generator off, there was less than a 30 percent differ-

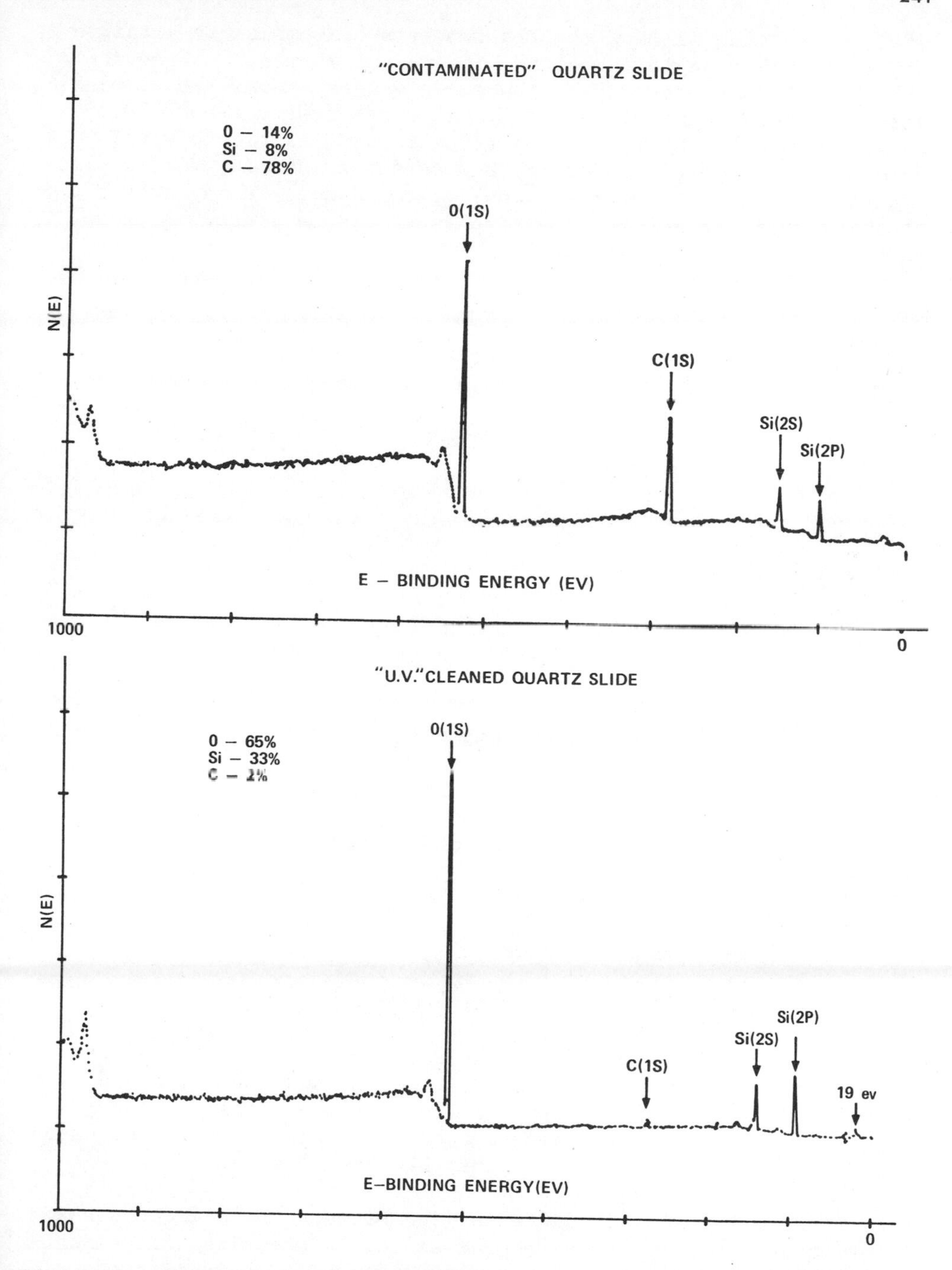

Figure 4. ESCA scans of contaminated (top) and UV/ozone cleaned (bottom) quartz surfaces. (Ref. 15).

ence in the time it took for the two sets of samples to attain a minimal contact angle (about 60 min versus 75 min). When the experiment was repeated with the ozone generator on, the samples near the tube became clean nearly ten times faster (about 90 s versus 13 min). Similarly in UV box 1, samples placed within 5 mm of the tube were cleaned in 20 s versus 20-30 min for samples placed near the bottom of the box, 13 cms away. To maximize the cleaning rate, therefore, the samples should be placed as close to the UV source as possible.

The Contaminants

Vig et. al. tested the effectiveness of the UV/ozone cleaning procedure on a variety of contaminants. Among the contaminants were:

(1) human skin oils,

(2) contamination adsorbed during prolonged exposure to air,

(3) a cutting oil,[19]

(4) a beeswax and rosin mixture,

(5) a lapping vehicle,[20]

(6) a mechanical vacuum pump oil,[21]

(7) DC 704 silicone diffusion pump oil,[22]

(8) DC 705 silicone diffusion pump oil,[22]

(9) a silicone vacuum grease,[22]

(10) an acid (solder) flux,[23]

(11) a rosin flux from a rosin core lead-tin solder.

The contaminants were applied to clean, polished quartz wafers. After contamination, the wafers were precleaned, then exposed to UV/ozone by being placed within a few millimeters of the tube in UV box 1. After a 60 s exposure, the steam test and AES indicated that all traces of the contaminants had been removed.

Since AES could not differentiate between the silicon peaks due to quartz and those due to the silicon containing contaminants, the removal of silicone diffusion pump fluids was also tested on Alzak, which normally has a silicon-free oxide surface, and on gold. AES examination of the Alzak and gold surfaces following UV/ozone cleaning showed no traces of silicon present.

During the course of their studies, Vig et. al. learned from colleagues working on ion implantation for integrated circuits that the usual wet cleaning procedures (with hot acids) failed to remove the photoresist from silicon wafers which had been exposed to radiation in an ion-implantation accelerator, presumably because of cross-linking of the photoresist. Ion-implanted silicon wafers, each with approximately a 1 μm coating of exposed Kodak Micro Resist 747[24], were placed within a few millimeters of the source in UV box 1. After an overnight (≈10h) exposure to UV/ozone, all traces of the photoresist were removed from the wafers, as confirmed by AES.

Films of carbon, vacuum deposited onto quartz to make the quartz surfaces conductive for study in an electron microscope, were also successfully removed by exposure to UV/ozone. Inorganic contaminants such as dust and salts cannot be removed by UV/ozone. Such contaminants must be removed in the precleaning procedure.

UV/ozone has also been used for wastewater treatment and for destruction of highly toxic compounds.[25-28] Experimental work in connection with these applications has shown that UV/ozone can convert a wide variety of organic and some inorganic species to harmless, mostly volatile products such as CO_2, CO, H_2O, N_2, etc. Compounds which have been successfully destroyed in water by UV/ozone include: ethanol, acetic acid, glycine, glycerol, palmitic acid; organic nitrogen, phosphorous and sulfur compounds; potassium cyanide; complexed Cd, Cu, Fe, and Ni cyanides; photographic wastes, medical wastes, secondary effluents; plus chlorinated organics and pesticides such as pentachlorophlenol, dichlorobenzene, dichlorbutane, chloroform, malathion, Baygon, Vapam and DDT. It has also been shown[29] that the combination of UV and ozone is more effective in destroying microbial contaminants in water, such as E. Coli and streptococcus faecalis, than UV or ozone alone.

The Precleaning

Although UV/ozone is able to remove contaminants such as thick photoresist coatings and carbon films without any precleaning, the procedure cannot, in general, remove gross contamination. For example, when a clean quartz wafer was coated thoroughly with human skin oils and placed in UV box 1 without any precleaning, even prolonged exposure to UV/ozone failed to produce a low-contact-angle surface. This is possibly due to the fact that human skin oils contain inorganic salts which cannot be removed by photosensitized oxidation.

The UV/ozone removed silicones from surfaces which had been precleaned, as described earlier, and also from surfaces which had been just wiped with a cloth to leave a thin film. However, when the removal of a thick film was attempted, the UV/ozone removed most of the film upon prolonged exposure; but it also left a hard cracked residue on the surface. This may be due to the fact that many chemicals respond to radiation differently depending on whether or not oxygen is present. In the presence of oxygen many polymers, for instance, degrade when irradiated; whereas, in the absence of oxygen (as would be the case for the bulk of a thick film) these same polymers crosslink. In the study of the radiation degradation of polymers in air, the "results obtained with thin films are often markedly different from those obtained using thick specimen ...".[30]

For the UV/ozone cleaning procedure to work reliably, the surfaces must be precleaned; first, to remove contaminants such as dust and salts which cannot be changed to volatile products by the oxidizing action of UV/ozone, and second, to remove thick films whose bulk could be transformed into an UV resistant film by the cross-linking action of the UV light which penetrates the surface.

The Substrate

The UV/ozone cleaning process has been successfully used on a variety of surfaces, including: glass, quartz, mica, sapphire, ceramics, metals, semiconductors* and a conductive polyimide cement.

Quartz is especially easy to clean with UV/ozone since it is transparent to short wavelength UV. For example, when a pile of thin quartz plates a couple of centimeters deep were cleaned by UV/ozone, both sides of all the blanks, even the ones at the bottom of the pile, were cleaned by the process. Since sapphire is even more transparent, it too could probably be cleaned the same way. When flat quartz plates were placed one on top of the other so that there could have been little or no ozone circulation between the plates, the UV/ozone cleaning was able to clean both sides of these plates also. (There is evidence in the literature[31] that photocatalytic oxidation of hydrocarbons, without the presence of gaseous oxygen, can occur on some oxide surfaces).

When white alumina ceramic substrates were cleaned by UV/ozone, the surfaces could be cleaned properly, however, the sides facing the UV turned yellow, probably due to the production of (UV induced) color centers. After several days at room temperature, or a few mintues at high temperatures, the white color returned.

Metal surfaces could be cleaned by UV/ozone without any problems as long as the UV exposure was limited to the time

*See note at the end of the references

required to produce a clean surface. (This time should be generally about one minute or less for surfaces which have been properly precleaned). However, prolonged exposure of oxide forming metals to UV light can produce rapid corrosion. Silver samples, for example, turned black in UV box 1 within 1 h. Experiments with sheets of Kovar, stainless steel (Type 302), gold, silver, and copper showed that upon extended UV irradiation the Kovar, the stainless steel and the gold appeared unchanged, the silver and copper oxidized on both sides, but the oxide layers were darker on the bottom sides than the top sides. When electroless gold plated nickel parts were stored under UV/ozone for several days, a black powdery coating gradually appeared on the parts. Apparently, nickel diffused to the surface through pinholes in the gold plating, and the oxidized nickel eventually covered the gold nearly completely. The corrosion was observed even in UV box 2 when no ozone was being generated. The rate of corrosion increased substantially when a beaker of water was placed in the UV boxes to increase the humidity. Even Kovar showed signs of corrosion under such conditions.

The corrosion may possibly be explained by the fact that, as is known in the science of air-pollution control, in the presence of short-wave UV light, impurities in air, such as oxides of nitrogen and sulfur, combine with water vapor to form a corrosive mist of nitric and sulfuric acids. The use of controlled atmospheres in the UV box should, therefore, minimize the corrosion problem.

Since UV/ozone dissociates organic molecules, it may be useful for cleaning some organic materials just as etching and electropolishing is sometimes useful for cleaning metals. The process has been used successfully to clean quartz crystal resonators which have been bonded with silver filled polyimide cement.[32] Teflon (TFE) tape exposed to UV/ozone in UV box 1 for 10 days experienced a weight loss of 2.5%.[33] Also, the contact angles measured on clean quartz plates increased after a piece of Teflon was placed next to the plates in a UV box.[34] Similarly, Viton shavings taken from an O-ring experienced a weight loss of 3.7% after 24 hours in UV box 1. At the end of the 24 hours, the Viton surfaces had become sticky.[33]

THE MECHANISM OF UV/OZONE CLEANING

The available evidence indicates that UV/ozone cleaning is the result of photosensitized oxidation processes, as is represented schematically in Figure 5. The contaminant molecules are excited and/or dissociated by the absorption of short wavelength UV light. Simultaneously, atomic oxygen and ozone[17,18] are produced when O_2 is dissociated by the absorption of UV with a wavelength less than 2454 Å. Atomic oxygen is also produced[17,18]

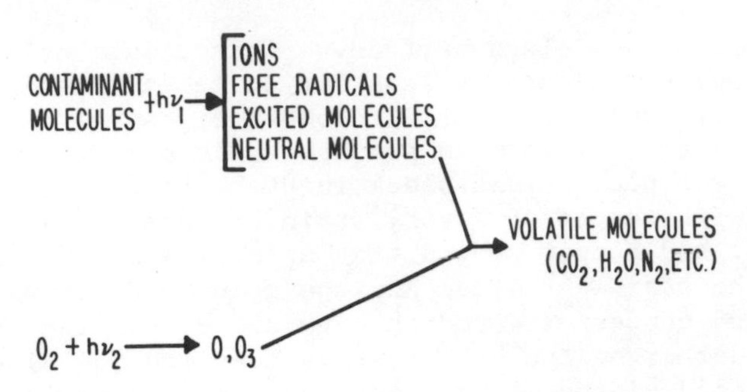

Figure 5. Simplified schematic representation of UV/ozone cleaning process.

when ozone is dissociated by the absorption of the UV and longer wavelengths of radiation. The excited contaminant molecules, and the free radicals produced by the dissociation of contaminant molecules react with atomic oxygen to form simpler, volatile molecules, such as CO_2, H_2O, N_2, etc.

The energy required to dissociate an O_2 molecule into two ground state O atoms corresponds to 2454 Å, however, at and just below 2454 Å the absorption by O_2 is very weak.[7,17,18] The absorption coefficient increases rapidly with decreasing wavelengths. A convenient wavelength for producing O_3 is the 1849 Å emitted by low pressure Hg discharge lamps in fused quartz envelopes. Similarly, since most hydrocarbons have a strong absorption band between 2000 Å and 3000 Å, the 2537 Å wavelength emitted by the same lamps are useful for exciting or dissociating contaminant molecules The energy required to dissociate ozone corresponds to 11,400 Å. The absorption by ozone reaches a maximum near the 2537 Å wavelength. The actual photochemical processes occurring during UV/ozone cleaning are more complex than shown in Figure 5 . For example, the rate of production of ozone by 1849 Å photons is promoted by the presence of other molecules, such as N_2 and CO_2.

As described earlier, the combination of short wavelength UV light with ozone produced clean surfaces about 200 to 2,000 times faster than UV light alone or ozone alone. Similarly, Prengle et. al.[25,28] had found in their studies of wastewater treatment that UV enhances the reaction with ozone 10^2 to 10^4 fold, and the products of the reactions are materials such as CO_2, H_2O and N_2. Increasing the temperature was found to increase the reaction rates. Mattox[35] has also found that mild heating increases the UV/ozone cleaning rates. Bolon and Kunz[1], on the other hand, had found that the rate of UV/ozone depolymerization of photoresists did not change significantly between 100°C and 300°C. The rate of destruction of microorganisms was similarly insensitive to a temperature increase from room temperature to 40°C.[29]

UV/OZONE CLEANING IN VACUUM SYSTEMS

Sowell et. al.[2] reported that when 10^{-4} torr pressure of oxygen was present in a vacuum system, short wavelength UV desorbed gasses from the walls of the system. During UV irradiation, the partial pressure of oxygen decreased, while that of CO_2 and H_2O increased.

One must exercise caution in using a mercury UV source in a vacuum system because should the lamp envelope break or leak, mercury would enter and ruin the usefulness of the system. Mercury has a high vapor pressure; its complete removal from a vacuum chamber is a difficult task. Other types of UV sources such as xenon or deuterium lamps may be safer to use in vacuum systems. The UV light can also be radiated into systems through sapphire or quartz windows. A small partial pressure of oxygen should be present during UV cleaning.

SAFETY CONSIDERATIONS

In the construction of a UV cleaning facility, one should be aware of the safety hazards associated with short-wave UV light. Exposure to intense short-wave UV light can cause serious skin and eye injury within a short time. For the UV boxes used in Vig, et. al.'s experiments, switches are attached to the doors in such a manner that when the doors are opened, the UV lamps are shut off automatically. If the application demands that the UV lamps be used without being completely enclosed (as might be the case, for example, if an UV cleaning facility is incorporated into a thermocompression bonder), then proper clothing and eye protection should be worn to prevent skin burns and eye injury.

Another safety hazard is ozone, which is highly toxic. In setting up an UV cleaning facility, one must assure that the ozone levels to which people are exposed do not exceed 0.1 ppm, the OSHA standard.[36]

UV/OZONE CLEANING FACILITY CONSTRUCTION

The material chosen for the construction of a UV/ozone cleaning facility should be one which is not corroded by extended exposure to UV/ozone. One material that can be used is polished aluminum with a relatively thick anodized oxide layer, such as Alzak.[6] Such materials are resistant to corrosion, have a high thermal conductivity which helps to prevent heat buildup, and are also good reflectors of short wavelength UV. Most other metals, including silver, are poor reflectors in this range.

Vig et. al. initially used ordinary, shop variety aluminum sheet for UV box construction. After a while, however, it was noticed that a thin coating of a white powder (probably aluminum oxide particles) appeared at the bottom of the boxes. Even in a UV box made of standard Alzak, after a couple of years' usage white spots started appearing on the Alzak. To avoid the possibility of particles being generated inside the UV/ozone cleaning facility, the facility should be inspected periodically for signs of corrosion. The use of "Class M" Alzak may also help, since this material has a much thicker oxide coating. It is made for "exterior marine service", vs. the "mild interior service" specified for standard Alzak.

Organic materials should not be present in the UV cleaning box. For example, the plastic insulation usually found on the leads of UV lamps should be replaced with inorganic insulation, such as glass or ceramic. The box should be enclosed so as to minimize recontamination by circulating air, and to prevent accidental UV exposure.

The most widely available sources of short-wave UV light are the mercury arc lamps. Low-pressure mercury lamps in pure fused quartz envelopes operate near room temperature, emit approximately 90 percent at 2537 Å, and generate sufficient ozone for effective surface cleaning. Approximately five percent of the output of these lamps is at 1849 Å. Medium and high pressure UV lamps[7] generally have a much higher output in the short-wave UV range. These lamps also emit a variety of additional wavelengths below 2537 Å, which may enhance their cleaning action. However, they operate at high temperatures (the envelopes are near red hot), have a shorter lifetime, a higher cost, and present a greater safety hazard. The mercury tubes can be fabricated in a variety of shapes to fit different applications.

Other good sources of short wavelength UV such as xenon lamps and deuterium lamps are also available. These lamps must also be in an envelope transparent to short wavelength UV, such as quartz.

In setting up an UV cleaning facility, one should choose an UV source which will generate enough UV/ozone to allow for rapid photosensitized oxidation of contaminants; however, too high an output at the ozone generating wavelengths can be counterproductive because a high concentration of ozone will absorb most of the UV light before it reaches the samples.

To maximize the intensity reaching the samples, the samples should be placed as close to the UV source as possible. In the UV cleaning box 1 of Vig et. al's, the parts to be cleaned are placed on an Alzak stand the height of which can be adjusted to bring the parts close to the UV lamp. Figure 6 shows a UV cleaning box in which the folded spiral UV tube is placed at the bottom of the box. The parts to be cleaned are placed directly onto the tube.[37]

APPLICATIONS

The UV/ozone cleaning procedure has found numerous applications during the past few years. A major use is substrate cleaning prior to thin film deposition as is widely used in the quartz crystal industry during the manufacture of quartz crystal resonators. There is probably no other device whose performance is so critically dependent on surface cleanliness. For example, the aging requirement for a 5-MHz resonator currently under development is that the frequency

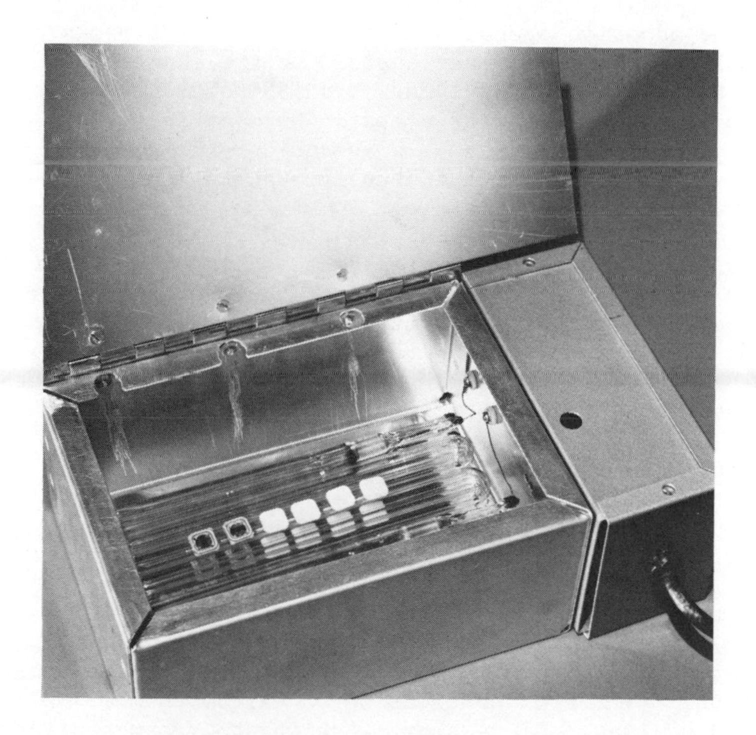

Figure 6. UV/ozone cleaning box with parts to be cleaned resting on UV tube.

change no more than two parts in 10^{10} per week, whereas adsorption or desorption of a monolayer of contamination from such a device changes the frequency by about one part in 10^6. The surface cleanliness must therefore be such that the rate of transfer of contamination within the (hermetically sealed) resonator enclosure is less than 10^{-4} monolayers per week! In the author's quartz resonator fabrication laboratory, UV/ozone is used at several points during the fabrication sequence, such as for cleaning and storing metal tools, masks, resonator parts, and storage containers.

The process is also being applied in a hermetic sealing method which relies on the adhesion between clean surfaces in an ultrahigh vacuum.[38-40,14] It has been shown that metal surfaces will weld together under near zero forces if the surfaces are atomically clean. A gold gasket between gold metallized, (UV/ozone cleaned) aluminum oxide sealing surfaces is currently providing excellent hermetic seals in the production of ceramic flatpack enclosed quartz resonators, The feasibility of achieving good hermetic seals by pressing a clean aluminum gasket between two clean, unmetallized aluminum oxide ceramic surfaces has also been shown.

The same adhesion phenomenon between clean (UV/ozone cleaned) gold surfaces has been applied to the construction of a novel surface contaminant detector.[41,42] The rate of decrease in the coefficient of adhesion between freshly cleaned gold contacts is used as a measure of the gaseous condensable contaminant level in the atmosphere.

The process has also been applied to improve the reliability of wire bonds, especially at reduced temperatures. It has been shown[43,44] for example, that the thermocompression bonding process is highly temperature dependent when organic contaminants are present on the bonding surfaces. The temperature dependence can be eliminated by UV/ozone cleaning of the surfaces just prior to bonding.

When the nonuniform appearance of thermal/flash protective electrooptic goggles was traced to organic contaminants on the electrooptic wafers, a number of cleaning methods were tested. UV/ozone proved to be the most effective method for removing these contaminants, and thus it was chosen for use in the production of the goggles.[45]

Other applications which have been described are photoresist removal[1,5,13] and the cleaning of vacuum chamber walls[2], as discussed earlier. Since short wavelength UV can generate radicals and ions, a side benefit of UV/ozone cleaning of insulator surfaces can be the neutralization of static charges.[46]

SUMMARY AND CONCLUSIONS

The UV/ozone cleaning procedure has been shown to be a highly effective method of removing a variety of contaminants from surfaces. It is a simple-to-use dry process which is inexpensive to set up and operate. It can produce clean surfaces at room temperature, in a room atmosphere.

The variables of the UV cleaning procedure are: the contaminants initially present, the precleaning procedure, the wavelengths emitted by the UV source, the atmosphere between the source and sample, the distance between the source and sample, and time of exposure. For surfaces which are properly precleaned and placed within a few millimeters of an ozone producing UV source, the process produces a clean surface in less than 1 min. The combination of shortwave UV light plus ozone produces a clean surface substantially faster than either short-wave UV light without ozone or ozone without UV light. Clean surfaces will remain clean indefinitely during storage under UV/ozone, but prolonged exposure of oxide forming metals to UV/ozone in room air produces rapid corrosion.

The cleaning mechanism seems to be a photosensitized oxidation process in which the contaminant molecules are excited and/or dissociated by the absorption of short-wave UV light. Simultaneously, atomic oxygen is generated when molecular oxygen is dissociated and when ozone is dissociated by the absorption of both short and long wavelengths of radiation. The products of the excitation of contaminant molecules react with atomic oxygen to form simpler molecules, such as CO_2 and H_2O, which desorb from the surfaces.

REFERENCES

1. D. A. Bolon and C. O. Kunz, Polymer Eng. Sci., 12, 109 (1972).
2. R. R. Sowell, R. E. Cuthrell, D. M. Mattox and R. D. Bland, J. Vac. Sci. Technol., 11, 474 (1974)
3. J. R. Vig, C. F. Cook, Jr., K. Schwidtal, J. W. LeBus and E. Hafner, in "Proc. 28th Annu. Symp. Frequency Control", US Army Electronics Command, Ft. Monmouth, NJ, AD 011113, pp. 96-108, 1974. Article reprinted as ECOM Tech. Rep. 4251, AD 785513, 1974.

4. J. R. Vig, J. W. LeBus and R. L. Filler, in "Proc. of 29th Annu. Symp. Frequency Control", US Army Electronics Command, Ft. Monmouth, NJ, AD A017466, pp. 220-229, 1975. Copies available from Electronics Industries Association, 2001 Eye St., NW, Washington, DC 20006.
5. J. R. Vig and J. W. LeBus, IEEE Trans. Parts, Hybrids and Packag., PHP-12, 365, (Dec. 1976).
6. Alzak is an aluminum reflector material with a corrosion resistant oxide coating. The alzak process is licensed to several manufacturers by the Aluminum Co. of America, Pittsburgh, PA 15219.
7. J. G. Clavert and J. N. Pitts, Jr., "Photochemistry", pp. 205-209, 687-705, John Wiley & Sons, New York, 1966.
8. V. S. Fikhtengolt's, R. V. Zolotareva, and Yu A. L'vov, "Ultraviolet Spectrum of Elastomers and Rubber Chemicals" Plenum Press Data Div., New York, 1966.
9. L. Lang, "Absorption Spectra in the Ultraviolet and Visible Region", Academic Press, New York, 1965.
10. Model No. R-52 Mineralight Lamp, Ultraviolet Products, Inc., San Gabriel, CA 91778.
11. See, e.g. "Encyclopaedic Dictionary of Physics", Vol. 5, p. 275, Pergamon Press, New York, 1962.
12. M. E. Schrader, in "Surface Contamination: Its Genesis, Detection and Control", K. L. Mittal, Editor, Vol. 2, pp. 541-555, Plenum Press, New York, 1979.
13. P. H. Holloway and D. W. Bushmire, in "Proc. of the 12th Annual Reliability Physics Symposium", pp. 180-186, Institute of Electrical and Electronic Engineers, Piscataway, NJ, 1974.
14. R. D. Peters, in "Proc. of the 30th Annual Symposium on Frequency Control", pp. 224-231, 1976. Copies available from Electronics Industries Assoc., 2001 Eye St., NW, Washington, DC 20006.
15. C. E. Bryson and L. J. Sharpen, in "Surface Contamination: Its Genesis, Detection and Control", K. L. Mittal, Editor Vol. 2, pp. 687-696, Plenum Press, New York, 1979.
16. W. L. Baun, (1978), personal communication.
17. J. R. McNesby and H. Okabe in "Advances in Photochemistry" W. A. Noyes, Jr., G. S. Hammond and J. N. Pitts, Editors, Vol. 3, pp. 166-174, Interscience Publishers, New York, 1964.
18. D. H. Volman, in "Advances in Photochemistry", W. A. Noyes G. S. Hammond and J. N. Pitts, Editors, Vol. 1, pp. 43-82 Interscience Publishers, New York, 1963.
19. P. R. Hoffman Co., Carlisle, PA.
20. John Crane Lapping Vehicle 3M, Crane Packing Co., Morton Grove, IL 60053.

21. Welch Duo-Seal, Sargent-Welch Scientific Co., Skokie, IL 60076.
22. Dow Corning Corp., Midland, MI 48640.
23. Dutch Boy No. 205, National Lead Co., New York, NY 10006.
24. Eastman Kodak Co., Rochester, NY 14650.
25. H. W. Prengle, C. E. Mauk, R. W. Legan and C. G. Hewes, Hydrocarbon Processing, Vol. 82, (Oct. 1975).
26. H. W. Prengle Jr. C. E. Mauk, and J. E. Payne, "Ozone/UV Oxidation of Chlorinated Compounds in Water", Forum on Ozone Disinfection, 1976; International Ozone Institute, Warren Bldg., Suite 206, 14805 Detriot Ave., Lakewood, OH 44107.
27. H. W. Prengle, Jr., and C. E. Mauk, "Ozone/UV Oxidation of Pesticides in Aqueous Solution", Workshop on Ozone/ Chlorine Dioxide Oxidation Products of Organic Materials, EPA/International Ozone Institute, Warren Bldg., Suite 206, 14805 Detriot Ave., Lakewood, OH 44107, Nov. 1976.
28. H. W. Prengle, Jr., in Proc. International Ozone Institute Ozone Symposium, Warren Bldg., Suite 206, 14805 Detriot Ave., Lakewood, OH 44107, 1978.
29. J. D. Zeff, R. R. Barton, B. Smiley and E. Alhadeff, "UV-Ozone Water Oxidation/Sterilization Process", US Army Medical Research and Development Command, Final Report, Contract No. DADA 17073-C-3138, Sept. 1974. Copies available from NTIS, AD A004205.
30. H. V. Boenig, "Structure and Properties of Polymers", p. 246, Wiley, New York, 1973.
31. V. N. Filimonov, in "Elementary Photoprecesses in Molecules", B. S. Neporent, Editor, pp. 248-259, Consultants Bureau, New York, 1968.
32. R. L. Filler, J. M. Frank, R. D. Peters and J. R. Vig, in Proc. of the 32nd Annual Symposium on Frequency Control, pp. 290-298, 1978. Copies available from Electronics Industries Assoc., 2001 Eye St., NW, Washington, DC 20006.
33. J. W. LeBus and J. R. Vig, (1976), unpublished information.
34. J. Kusters, (1977), personal communication.
35. D. M. Mattox, This proceedings volume, pp. 223-231.
36. "Occupational Safety and Health Standards", Vol. 1, General Industry Standards and Interpretations, Oct. 1972, Pt. 1910. 1000, Table Z-1, Air Contaminants, p. 642.4 as per change 10, June 26, 1975.
37. R. D. Peters, (1976), personal communication.

38. J. R. Vig and E. Hafner, "Packaging Precision Quartz Crystal Resonators", Technical Report ECOM-4134, US Army Electronics R&D Command, Fort Monmouth, NJ, July 1973. Copies available from NTIS, AD 763215.
39. E. Hafner and J. R. Vig, "Method of Processing Quartz Crystal Resonators", US Patent No. 3,914,836, Oct. 28, 1975.
40. P. D. Wilcox, G. S. Snow, E. Hafner and J. R. Vig, in "Proc. of the 29th Annual Symposium on Frequency Control", pp. 202-210, 1975. See ref. no. 4 above for availability information.
41. R. E. Cuthrell and D. W. Tipping, Rev. Sci. Instrum., 47, 555 (1976).
42. R. E. Cuthrell, in "Surface Contamination: Its Genesis, Detection and Control", K. L. Mittal, Editor, Vol. 2, pp. 831-841, Plenum Press, New York, 1979.
43. J. L. Jellison, IEEE Trans. Parts, Hybrids, Packag., PHP-11, 206 (Sept. 1975).
44. J. L. Jellison, in "Surface Contamination: Its Genesis, Detection and Control", K. L. Mittal, Editor, Vol. 2, pp. 899-923, Plenum Press, New York, 1979.
45. J. A. Wagner, in "Surface Contamination: Its Genesis, Detection and Control", K. L. Mittal, Editor, Vol. 2, pp. 769-783, Plenum Press, New York, 1979.
46. D. H. Baird, "Surface Charge Stability on Fused Silica", Final Technical Report, TR 76-807.1, Dec. 1976. Copies available from NTIS, AD A037463.

NOTE: Subsequent to the completion and editing of this paper, the author received a letter from E. Lasky, Aerofeed, Inc., Chalfont, PA, which stated in part: "There has been some concern about the effect of UV/ozone cleaning affecting the functioning of silicon devices. I thought you would be interested in knowing that we cleaned a 4K static RAM I.C. for 120 minutes in our Model 100 UV/Ozone Cleaning Station and after testing it was still found to be working and in specification. The IC was made using N-channel silicon gate technology with a junction depth of 1 to 1 1/2 microns which is shallower than most 4K RAMs, I am told. Although this junction depth makes the device more vulnerable to failure, the long exposure to UV/ozone did not affect this very sensitive device."

PLASMA CLEANING OF SURFACES

W. W. Balwanz

Mattox, Inc.

2905 Eastside Drive, Alexandria, Va. 22306

Surface cleaning is achieved by various techniques; such as washing with solvents, thermal heating, and ion bombardment. No one method has the desired features of simplicity, low cost, and effectiveness. Each has its range of applicability. Solvent cleaning has the greatest range of utility but is inadequate in many cases, particularly where the solvents themselves are contaminates. Thermal heating is useful up to the temperature limits of the surfaces to be cleaned. Ion bombardment with plasmas provide a method of cleaning where contaminant bond strengths exceed the temperature limits of the system. The plasma energy can be much higher than that achieved thermally, yet not damage surfaces because of the low thermal flux. The current techniques used in surface cleaning, their application and limitations, and the potential for future developments are reviewed. Emphasis is placed on plasma cleaning because of its versatility and its potential for cleaning of high bonding energy contaminants.

INTRODUCTION

Glow discharges and plasmas provide a method for removing monomolecular layers of the most adhesive of contaminants from surfaces. In common with other methods, plasma cleaning has advantages and limitations. In combination with other methods, it provides the ultimate in surface cleanliness. Basic principles common to other cleaning techniques also apply to plasma cleaning. A brief discussion of cleaning techniques aids in establishing the applicabilities of the plasma methods.

DISCUSSION OF METHODS

Surface cleaning is used for many different purposes. These range from the simplistic to the sublime, from the mother checking behind the ears of the small boy to the ultra-sophisticated requirements of many of our space probes.

The applications being reviewed are listed in Table 1. Cleaning for cosmetic purposes provides the majority industry in the surface cleaning area, having an impact on each of us individually and on the manufacture of an extremely wide range of consumer goods. Plasma cleaning plays a negligible role in this field of cleaning.

Cleaning for the purpose of improving the adhesion of coatings is the next in importance insofar as volume is concerned, and here plasma cleaning does play a role in the laboratory and potentially in industry. In the laboratory it enables the establishment of a standard surface for testing the adhesion of various coatings. In industry it is expected that it will be used for water removal on critical components prior to coating.

In the cleaning of laboratory vacuum systems, glow discharges have been used for years. The use of a direct current plasma provides a rapid, efficient method for cleaning large vacuum systems, more details of which are presented later in the discussion of plasma cleaning techniques.

The removal of corrosives, scale, and abrasives are problems of importance in many manufacturing processes. Plasma cleaning plays a negligible role in these areas and any future use is expected to be limited.

Toxic materials must be removed or neutralized in many areas of activity, such as systems in which a nontoxic material in one location migrates to another where it is toxic. One area of concern is in the small air bearings used in some space applications. Contaminants on the rotating parts are nonharmful until they are

Table 1. Cleaning Purposes

Cosmetic
Adhesion of Coatings
Attainment of High Vacuum
Removal of Corrosives
Removal of Toxic Materials
Removal of Scale
Removal of Abrasive Material
Friction Improvement

removed by mechanical forces and contaminate the air stream where they may be harmful. Plasma cleaning has been considered for such cases.

Friction improvement is another important area for surface cleaning, particularly the cleaning of roads and aircraft runways to remove loose material as well as oils and greases. Plasma cleaning is not of importance in such areas of activity.

A number of methods of cleaning are used; some are listed in Table II. Detergents, solvents, chemicals, and mechanical methods provide the major techniques for surface cleaning. In most cases, these methods are adequate and have the further advantage of relatively low cost, and plasma cleaning has little to offer. However all these techniques except plasma cleaning leave at least monomolecular layers of contaminants on the surfaces, usually leaving a range of contaminants. Where and when these must be removed, more sophisticated methods such as vacuum, thermal, and plasma techniques are utilized.

Thermal methods have long been used to drive off the more volatile, less tightly bound contaminants from surfaces, such as the use of heat to drive water vapor from stainless steel surfaces prior to welding; and they have been used to aid in outgassing high vacuum systems, where they are successful in driving off all but the most highly bound layers of contaminants. Mass spectrometer tests show that water molecules are the last to be removed when heating is used. The bonding energy of the water molecule to the metal surface is high, similar to that for hydrogen atoms, i.e., 1.5 electron volts or more.[1] This energy level is above the melting point of iron, hence, practical levels of thermal heating as a whole are very slow in removing the water molecules from the surface.

Table II. Surface Cleaning Methods

- Detergents
- Solvents
- Chemical
  - Photochemical
- Mechanical
  - Abrasive
  - Blasting
  - Ultrasonic
- Thermal
  - Laser
  - X-Ray
- Plasma

When the surfaces and the adjacent gas is heated to remove surface contaminants, the surface is bombarded by the adjacent gas. Where the energy of the impinging molecules is higher than the bonding energy of the contaminant molecules the bond is broken and the contaminant is released from the surface. The energy in the molecules of the adjacent gas has a maxwellian distribution, in which some few molecules have an energy high enough to break the surface bond. The distribution of the energy in a nitrogen gas at three different temperatures is shown in Figure 1. At room temperature (300° Kelvin) negligibly few molecules have sufficient energy to break a water molecule bond. The required 1.5 electron volts is off the plot, some 36 orders of magnitude below the peak. Practical values of heating, say 500 degrees Kelvin are much faster in removing water molecules but still much too slow in practical cases. The density of impinging molecules with the required energy is still more than 20 orders of magnitude below the peak. Even at 1000° Kelvin, an impractical temperature in most cases, only 1 in 10^{10} molecules has sufficient energy to dislodge a water molecule with bonding energy of 1.5 electron volts.

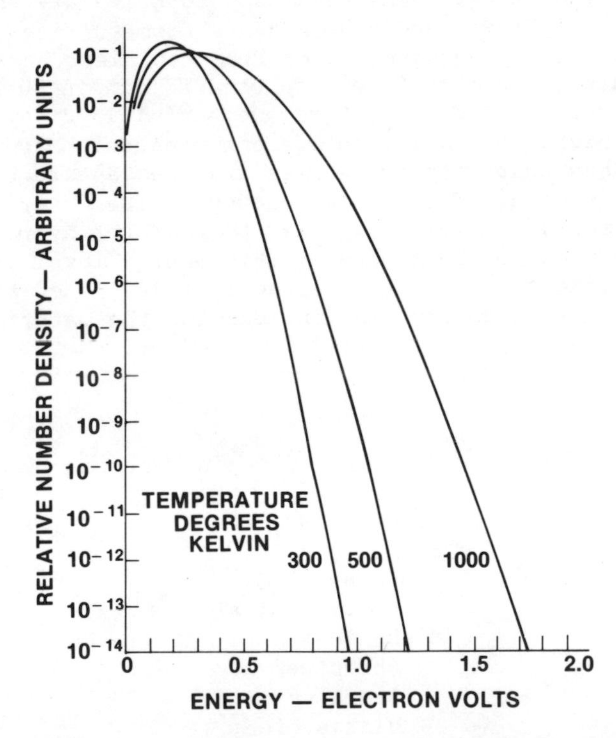

Figure 1. The number density of molecules drops rapidly with increasing energy. At normal permissible temperatures very few molecules have sufficient energy to dislodge water molecules from a surface. The energy distribution for nitrogen molecules is shown.

High energy bombardment at low flux rates so that only the surface layer gets to elevated temperatures is much faster in removing tightly bound contaminants. One method is to use high energy beams, such as lasers, ultraviolet, and x-rays which heat primarily only the surface layer. These methods bombard only a small area at a time, and are slow and inefficient. They also allow reattachment to the cleaned area when the beam is moved to adjacent areas. A more practical method is to use plasma bombardment because of simplicity and low cost, and, because of easy adaptability to any size and most shapes of surfaces to be cleaned.

In any cleaning method, two basic premises apply. One, the energy or forces in the cleaning flux must be sufficient to overcome the bonding energy of the contaminant and, two, the contaminant must be removed from the system before it can recontaminate the surface. If, as in many cases, the contaminant comes from the air or other media that the surface is exposed to after cleaning, protective coatings must be applied or the surface must be recleaned before every use.

EXPERIMENTAL RESULTS

Our first application of plasma cleaning was the outgassing of a large vacuum chamber, a chamber far too large to clean thermally 55 feet long and 12 feet in diameter. The surface had to be cleaned after every atmospheric exposure because of recontamination with atmospheric water vapor.

A few simple experiments showed that bombardment of the surface with positive nitrogen ions was quite effective. Nitrogen was used because of the low cost of relatively pure nitrogen, i.e., vaporized liquid nitrogen. Chamber pressure in the range from 1 to 10^{-3} torr was used during the cleaning with appreciable flow through the chamber to sweep out contaminants before they could recontaminate the chamber and its contents. The plasma discharge was maintained between a centrally placed wire electrode and the chamber walls. After initial breakdown, induced by an R. F. discharge, the glow discharge was maintained by D. C. source of the order of 10 amperes at about 20 volts, voltage and current varying with the ambient pressure in the system.

The result was: surface cleaning to remove water and other volatiles from the tank in an hour to an extent that would have required months of heating and pumping otherwise. In one test case, starting with walls contaminated with water and oil, one hour of plasma cleaning enabled tank evacuation to 10^{-9} torr within the following hour using a 36" oil diffusion vacuum pump, a phenomenally fast rate compared to that achieved with the usual methods for obtaining high vacuum in large systems.

Figure 2. A three-foot diameter by ten-foot long vacuum chamber was used for measurements on plasma cleaning. Instrument and controls used are shown.

The effectiveness of this technique lead to a more systematic study of the plasma method[2]. An existing vacuum system with a cold trap and a tank three feet diameter by ten feet long was modified for the investigation (Figure 2). A lucite plate over a twelve inch opening allowed easy visualization of the discharge, while a rod through a vacuum seal on the plate allowed sample manipulation. An easily opened and closed access port behind the plate permitted insertion and removal of small samples without the necessity of removing the plate.

Dry argon gas was used for the plasma cleaning since it is less chemically active as a plasma than nitrogen. (Based on previous experience nitrogen gas would have worked equally well.) Dry nitrogen gas was used for back filling the tank to atmospheric pressure because of its lower cost prior to tests on the cleaned surface. The sample was positioned in front of the viewing window at various distances from the center anode, grounded in some instances, ungrounded in others. Test samples were cleaned abrasively and washed with distilled water prior to mounting in the chamber.

The dried gas was flowed through the tank during the plasma discharge to sweep out contaminants before they could recontaminate the sample and the chamber walls. The flow rate and chamber pressure was adjusted by a controlled leak in the argon gas line. The plasma discharge was maintained at chamber pressures varied between 0.002 and 0.2 torr. The higher pressures are desired since they allow the dislodged contaminants to be swept from the system before they again come into contact with a cleaned surface. Higher pressures require larger power supplies for developing the plasma discharge. The pressure limits used here were dictated by the small size of the power supply available for the experiement. The polarity was selected to give positive ion bombardment of the sample.

The total plasma current to the sample and the chamber wall was varied between limits of 0.1 to 0.4 amperes. The plasma current density to the sample, when grounded, varied between 4 and 40 microamperes per square centimemeter. The relationship between the total current and the current to the test sample is shown in Figure 3.

The cleaning rates as a function of these variables were determined. Two tests were conducted to give a measure of the

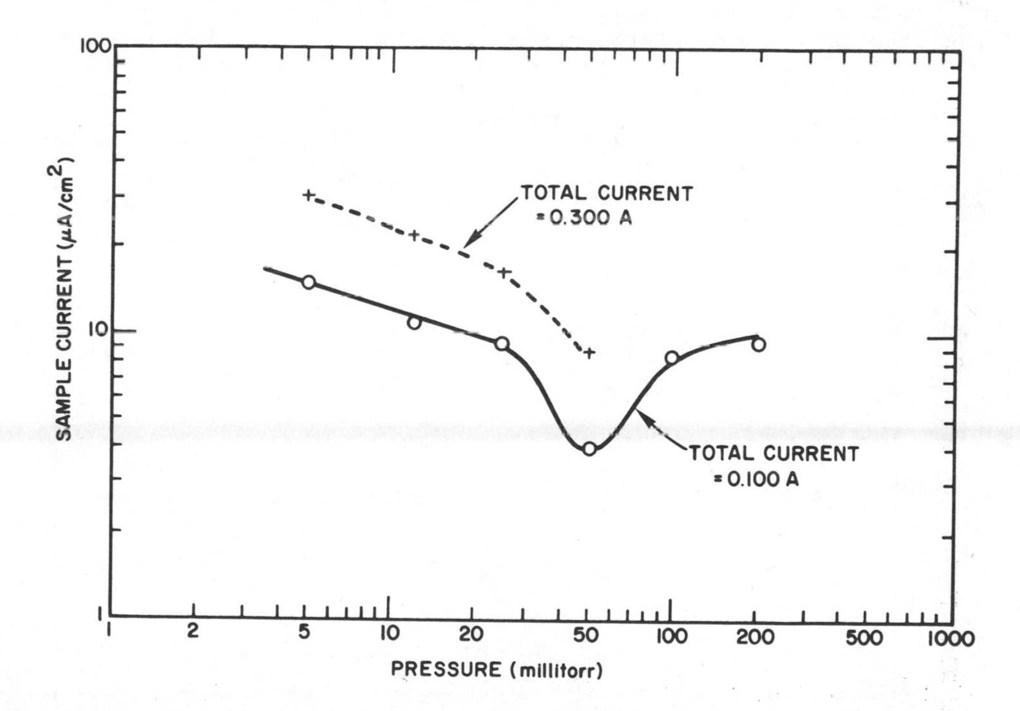

Figure 3. Only a small part of the total system current went to the sample to be cleaned. The major part of the current went to other surfaces of the system insuring that removed contaminants were swept out by the flowing gas instead of being left to recontaminate the sample.

relative cleanliness, oil drop spreading to give a measure of the remaining water and water drop spreading to give a measure of the remaining hydrophobic molecules.

The measurement method used initially was the standard contact angle technique wherein the angle between the surface of the test sample and the top surface of the drop at its edge gives a measure of drop spread[3-6]. (Allied work at the Naval Research Laboratory on surface cleaning is described in references 8 through 14.) The degree of cleanliness achieved with plasma cleaning was greater than any previously encountered, so that an improved technique relating the spreading diameter to the contact angle was developed[7]. The contact angle then provided a measure of the cleanliness of the sample as a function of plasma parameters and cleaning time. (Angles less than 5 degrees indicate less than one molecular thickness of the contaminant.)

The experimentally observed trends are illustrated in Figures 4 through 6. Figure 4 shows the removal of water vapor as function of time from an ungrounded stainless steel sample located 38 centimeters from the center electrode. Significant improvement in cleaning rate is achieved at the higher current levels. Similar cleaning

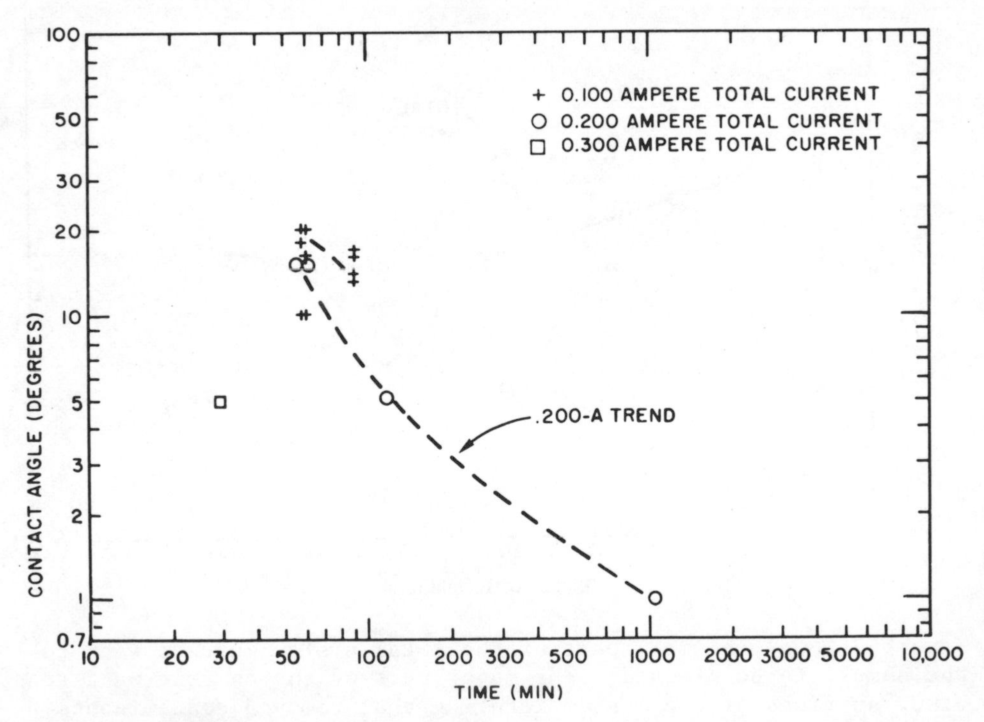

Figure 4, Surface cleaning increases with time. Faster cleaning is achieved at higher current rates.

is achieved on other materials as shown in Figure 5. The average impinging energy for the ionized argon atoms for these samples is 0.06 electron volts, much below the optimum value of 1.5 electron volts for water vapor.

Increasing the impinging energy resulted in a material increase in the cleaning rate as shown in Figure 6. Here the stainless steel sample was moved closer to the inner electrode to give a five fold increase in impingement energy, i.e., 0.3 electron volts. The cleaning time for a given contact angle decreased by the same factor of five to one. The methylene iodide $/\overline{I_2(CH2)}/$ drop spread gives a measure of the water vapor removal, whereas the water drop (H_2O) spread gives a measure of the hydrophobic material removed.

The impact energy of the individual ions on the test samples varies inversely with the pressure for any given current rate as shown in Figure 7. Continued reduction in pressure, however, places a limit on the current that can be maintained so that the curve bottoms out at about one millitorr for the particular system used.

Exposure of a test sample to the atmosphere or other contaminating environment results in recontamination of the sample. The

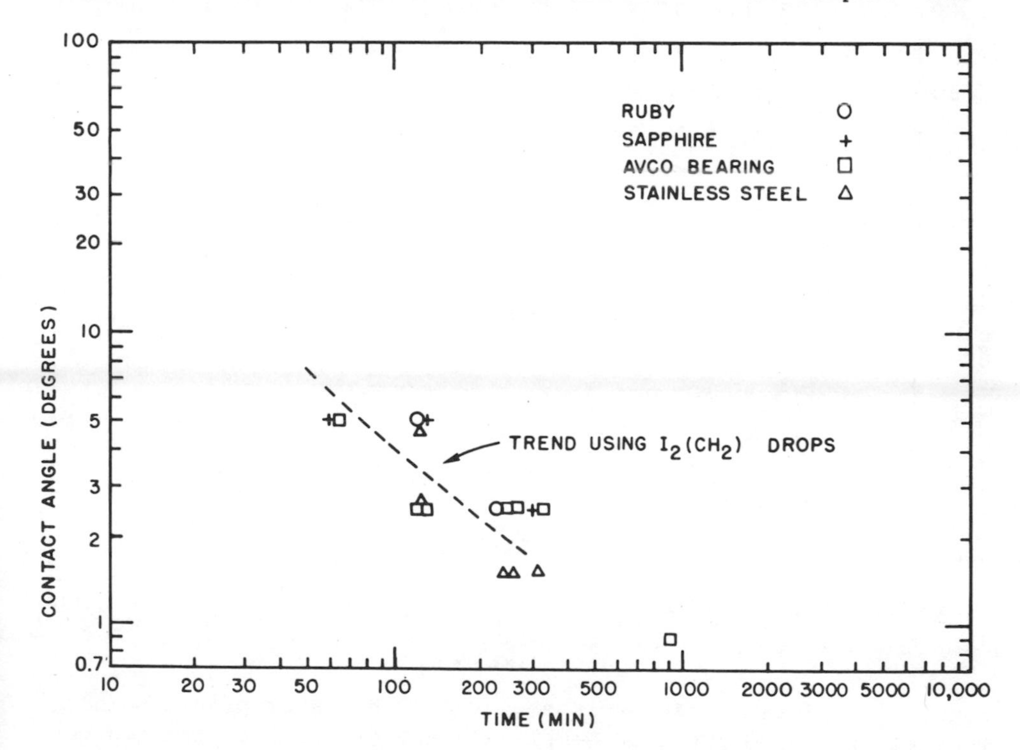

Figure 5. Cleaning rates for various materials do not vary significantly.

test results reported were achieved while maintaining the sample in an inert atmosphere. Measurements made after exposure of the sample to ambient air showed, a very rapid recontamination with water vapor and a slower recontamination with hydrophobic molecules as shown in Figure 8.

Rapid recontamination such as this emphasizes the importance of maintaining cleanliness, of preventing exposure to a contaminating environment until the purpose for cleaning is achieved. If coatings are to be applied, they must be applied while still in a noncontaminating environment. If tests are to be performed, they must be performed while still in a benign environment. These requirements apply not only to plasma cleaning but equally to all cleaning methods.

CONCLUSIONS

It has been demonstrated that plasma cleaning used in combination with other conventional methods to remove scale, dust, grit

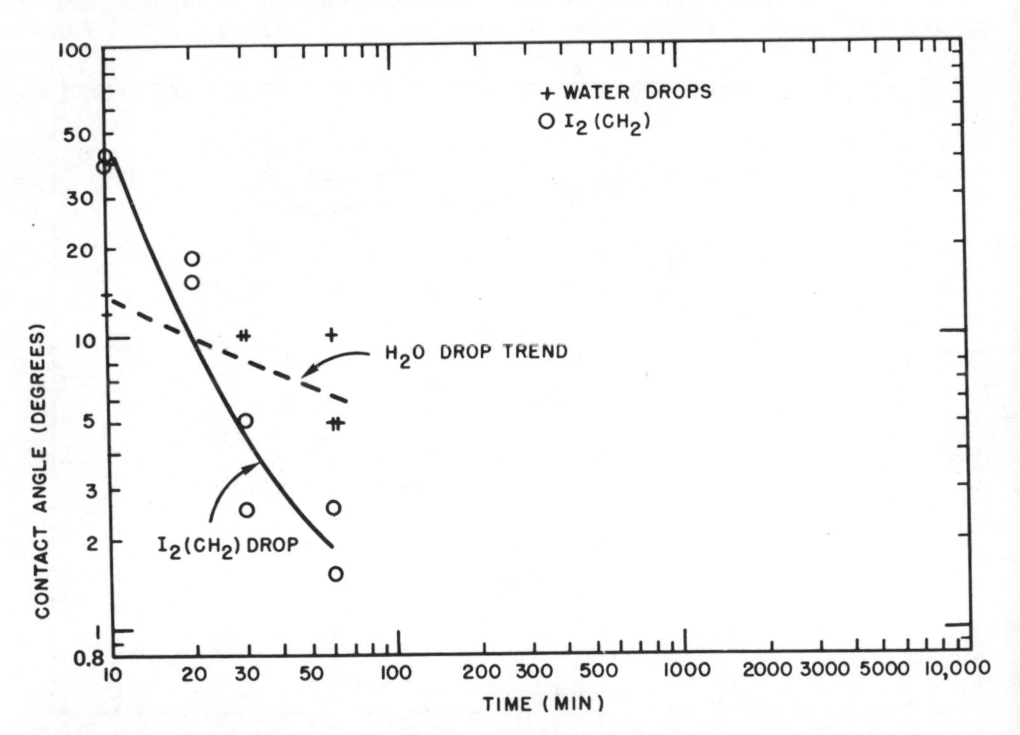

Figure 6. Faster cleaning is achieved with higher plasma bombardment energy. The grounded sample was moved closer to the source electrode to give a five-to-one increase in energy, resulting in a similar decrease in cleaning time.

and readily soluble contaminants provide a degree of cleanliness not achievable by other more conventional techniques. In common with other methods it requires continuous purging of the cleaning fluid to prevent recontamination by the contaminants being removed. Similarly, in common with other methods the surface must be protected until the purpose for the cleaning has been achieved. Application of the technique to the laboratory and to industry offers promise, particularly in the areas of coatings where contaminants may impact on the adhesion and characteristics of the coatings.

(Work by others in the plasma cleaning area is described in references 15 through 46.)

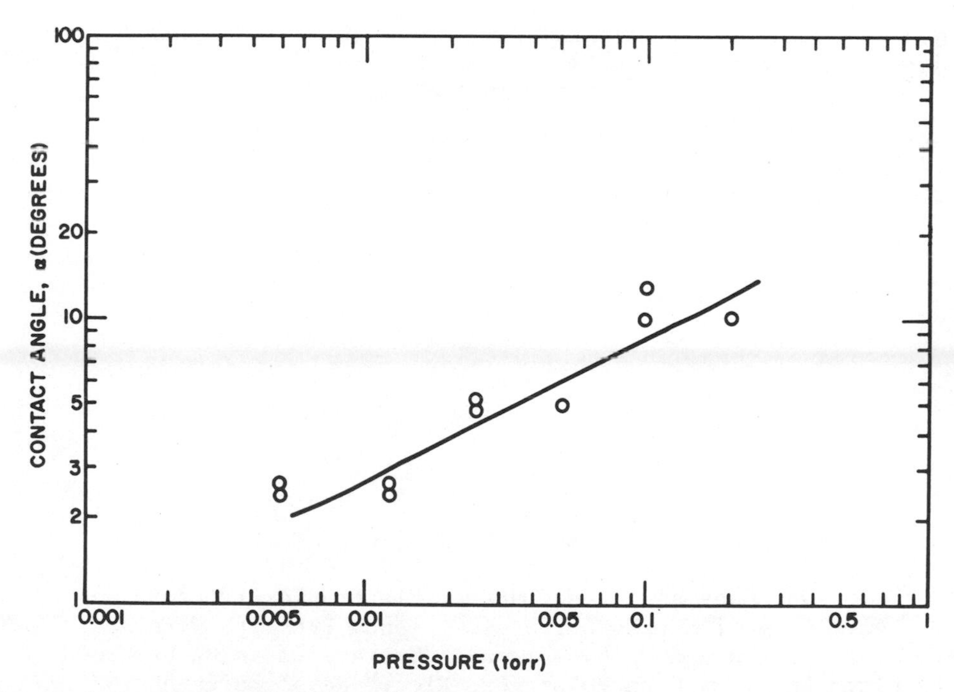

Figure 7. Reducing the pressure while holding the current (0.1 ampere) and cleaning time (60 minutes) constant increases the impingement energy and the cleaning achieved.

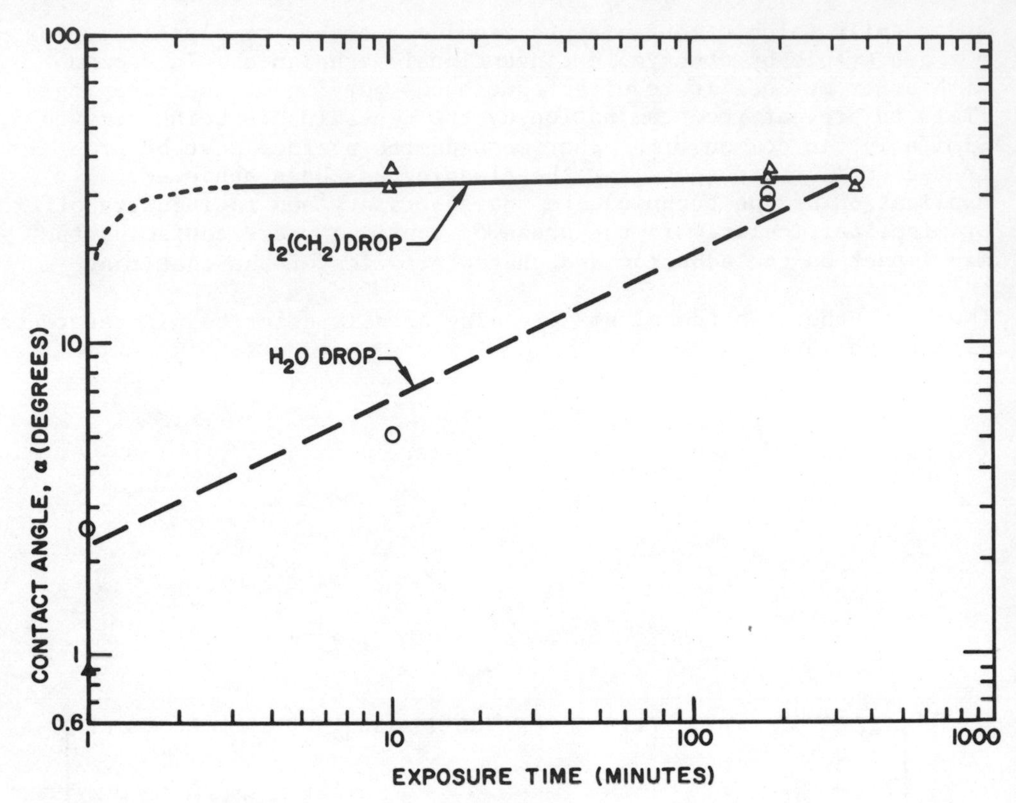

Figure 8. Recontamination with water vapor (upper curve) and hydrophobic molecules (lower curve) is rapid when cleaned samples are exposed to the ambient atmosphere.

REFERENCES

1. E. G. Derouane and A. A. Lucas, "Electronic Structure and Reactivity of Metals Surfaces," Plenum Press, New York, 1976.
2. J. P. Weston and W. W. Balwanz, "Surface Cleaning by Glow Discharge in a High Volume Gas Flow," Naval Research Laboratory (NRL) Report 7973 (April 7, 1976).

3. E. G. Shafrin and W. A. Zisman, "Upper Limits for the Contact Angles of Liquids on Solids," NRL Report 5985 (September 1963).
4. E. G. Shafrin and W. A. Zisman, J. Am. Ceram. Soc. 9 (1967).
5. M. K. Bernett and W. A. Zisman, J. Colloid Interface Science 28, 2 (1968).
6. M. K. Bernett and W. A. Zisman, J. Colloid Interface Science 29, 3 (1969).
7. J. P. Weston, "A Method to Estimate the Contact Angle of a Drop Spread Upon a Flat Surface When It Is Otherwise Too Flat to Measure," NRL Memorandum Report 3201 (January 1976).
8. E. G. Shafrin and J. S. Murday, J. Vac. Sci. Technol., 14, 246, (1977).
9. W. A. Zisman and M. K. Bernett, "Surface-Active Compositions and Method for Displacing Liquid Organic Films from Solids Surfaces," Patented March 3, 1970.
10. J. S. Murday, E. G. Shafrin, E. P. Kingsbury, and S. Allen, "Surface Chemistry of Ball Bearing Steels I: Interim Report," NRL (April 1975).
11. W. A. Zisman and M. K. Bernett, "Displacing Organic Liquids from Solid Surfaces," Patented June 23, 1970.
12. H. R. Baker and P. B. Leach, "Surface Chemical Methods of Displacing Water and/or Oils and Salvaging Flooded Equipment--Part 4--Aggressive Cleaner Formulations for Use on Corroded Equipment," Interim Rpt., NRL (June 15, 1965).
13. H. R. Baker, R. N. Bolster and P. B. Leach, "Surface Chemical Methods of Displacing Water and/or Oils and Salvaging Flooded Equipment: Part 6--Field Experience in Removing Seawater Salt Residues from Aircraft Cockpits and Avionics Equipment," NRL (December 12, 1968).
14. H. R. Baker and P. B. Leach, "Composition and Method for Cleaning Salt Residues from Metal Surfaces," Patented January 20, 1970.
15. J. B. Bradform, Jr., "The Adhesion of Nickel Using An Ion Bombardment and Heating Cleaning Technique," 1968 Vacuum Met. Conf., Beverly Hills, Calif., June 10-13, 1968.
16. J. M. Walls and H. N. Southworth, Surface Technol., 4 255, (1976).
17. Y. Aoki, J. Polym, Sci. Polym. Phys. Ed., 15, 199 (1977).
18. J. C. Bean, E. G. Becker, P. M. Petroff, and T. E. Seidel, J. Appl. Phys., 43, 907 (1977).
19. S. Naiditch, "Stable Dense Cold Plasma," Final Report, Unified Science Associates, Inc., Pasadena, Calif. (February 1966).
20. J. E. Houston and R. D. Bland, J. Appl. Phys., 45, 2504 (1973).
21. D. F. O'Kane and K. L. Mittal, J. Vac. Sci. Technol., 11, 567 (1974).
22. J. H. Greiner, J. Appl. Phys., p. 5151 (1971).

23. K. H. Tiefert and W. H. Legat, "Sputtering In the Manufacturing of Advanced Semiconductor Devices," Trans. Conf. and Sch. on Elements, Techniques and Applications of Sputtering, p. 21, Bringhton, England, November 7-9, 1971.
24. E. E. Gifford and R. T. Yoshino, "Plasma Treatment of Railway Rails to Improve Traction," 1971 Rail Transportation Proc. Joint IEEE/ASME Railroad Conference, New York, New York, April 20, 1971.
25. E. Suurmeijer, A. Boers and S. Begemann, Surface Sci., 4, 424 (1970).
26. N. Henzler, Surface Sci., 22, 12 (1970).
27. S. Naiditch, "Stable Dense Cold Plasma: Final Technical Report," Unified Science Associates, Inc., Pasadena, Calif. (February 1966).
28. S. K. Dolotov, S. A. Evstigneev, S. Y. Luk'yanov, Y. V. Martynenko, and V. M. Chicherov, "Removal of Foreign Atoms From a Metal Surface Bombarded with Fast Atomic Particles," International Conference on Plasma Physics and Controlled Nuclear Fusion Research, Berchtesgaden, Germany, October 6, 1976.
29. T. L. Thomas, "An Experimental Study of the Effects of Surface and Gas Contamination on Langmuir Probe Measurements in An Argon Glow Discharge," Air Force Inst. of Tech, Wright-Patterson Air Force Base, Ohio (June 1969).
30. R. L. Shannon and R. B. Gillette, "Plasma Cleaning Device," National Aeronautics and Space Administration (NASA), Huntsville, Alabama, Patent Filed May 10, 1976.
31. G. J. Kominiak and J. E. Uhl, "In Situ Investigation of Substrate Surface Recontamination During Glow Discharge Sputter Cleaning," Sandia Laboratory, Albuquerque, New Mexico (September 1975).
32. T. Farrell, "Cleaning of Specimens in the Electron Spectrometer Using an Argon Ion Gun," Electricity Council Research Center, Capenhurst, England (March 1972).
33. A. P. Dehamel and L. D. Nelson, "Improvement of Interlaminar Shear Strength in Organic Matrix Composites Utilizing Plasma Deposited Coupling Agents," Commonwealth Scientific Corp., Alexandria, Virginia (April 30, 1972).
34. "Evaluation of Plasma Cleaning and Electron Spectroscopy for Reduction of Organic Contamination," Boeing Co., Seattle, Washington (May 1972).
35. J. R. Ligenza and M. Kuhn, Solid State Tech., 13, 33 (1970).
36. V. V. Lyaginskov, Elektron Obrab Mater., 667, 46 (1976).
37. R. L. Bersin, Ind. Res., 17, No. 4, 60, (1975).
38. G. J. Kominiak and J. E. Uhl, J. Vac. Sci. Tech., 13, 170 (1976).
39. S. K. Sharma and R. Kushwaha, Surface Sci., 18, 449 (1969).
40. L. Smith, D. Hill, J. Hibbs, S. W. Kim, J. Andrade and D. Lyman, Am. Chem. Soc. Div. Polym. Chem. Prepr., 16, 186 (1975).

41. D. V. McCaughan, "Use of Plasmas in CCD Processing," University of Edinburgh, Center for Ind. Consult. and Liaison, Scotland (1974).
42. D. J. Dobbs, "Energy Transfer and Surface Cleaning with a D. C. Arc Plasma Generator," 9th International Conference on Phenomena in Ionized Gases, September 1-6, 1969.
43. F. M. Cobb, Sheet Metal Ind., 51, 280 (1974).
44. D. V. McCaughan and R. A. Kushner, Thin Solid Films, 22, 359 (1974).
45. A. W. Jones, E. Jones and E. M. Williams, Vacuum, 23, 227 (1973).
46. F. G. Koch, R. L. Meek and D. V. McCaughan, J. Electrochem. Soc., 121, 558 (1974).

SURFACE CONTAMINATION REMOVAL FROM SOLID STATE DEVICES BY DRY CHEMICAL PROCESSING

H. B. Bonham* and P. V. Plunkett

Sandia Laboratories

Albuquerque, NM 87185

Many authors have discussed the detrimental results of surface contamination upon thermocompression bondability. The problem is acute when gold-aluminum intraconnections are formed, since most wet chemical cleaning procedures degrade bondability further because of residual contamination. This paper discusses the dry chemical approach of plasma cleaning to remove organic contamination from aluminum surfaces.

Oxygen radicals produced by the plasma react with hydrocarbons to produce water vapor, carbon monoxide, and carbon dioxide, which are pumped from the system. Plasma cleaning has three fundamental parameters: rf power, time, and gas flow rate. Each of these parameters is varied, independently, resulting in a parametric envelope. The envelope illustrates the range of parameter variability over which sufficient cleaning results. Parametric settings outside this range result in poor cleaning or degradation of adjacent hybrid materials; i.e., epoxies, thin-film resistors, and silver capacitor metallizations. The evaluation shows that a large range of parameters may be used when cleaning only aluminum surfaces; however, when assembled hybrids are cleaned just prior to wire bonding, parameters must be more closely controlled.

* Electronic Systems Group, Rockwell International, Dallas, TX

Cleanliness was evaluated in terms of bondability and the ability of a bond to withstand a 300°C, 4-hour accelerated life test. In addition to the effect of plasma cleaning upon thermocompression bonding, data is presented to show that adequate cleaning of bonding surfaces may be achieved without significantly affecting silver capacitor terminations or thin-film tantalum nitride resistors.

Organic contamination was identified using Auger Electron Spectroscopy. The analysis indicated carbon contamination on uncleaned bond surfaces of 60 Å and on plasma cleaned surfaces of 15 Å. The high carbon contamination in conjunction with integrated circuit processing technology indicated a film of baked-on photoresist. Such baked-on photoresist typically is extremely hard to remove by liquid chemical processes.

INTRODUCTION

Previous authors have discussed the influence of surface contamination upon the bondability of hybrid metallizations.[1] These discussions show (1) that contamination reduces production throughout since repeated attempts may be required to achieve an intraconnection and (2) that contamination reduces bond reliability since bonding is only achieved between islands of contamination.

Traditionally, hybrid manufacturers use incoming tests, both visual and bondability, to certify component bondability prior to circuit assembly. The problem with this approach is what to do with parts that fail incoming, particularly when manufacturing schedules are short. Cleaning is obviously required. Wet chemical techniques, which are usually employed, are designed to lightly etch metallic bond pads or to dissolve inorganic contaminants. These techniques are expensive since hybrid components are small and difficult to handle; the techniques are also frequently ineffective.

As a last resort, many hybrid manufacturers use nondestructive pull tests to screen intraconnection in an attempt to identify and remove components which are contaminated. Nondestructive pull testing is only marginally successful since nondestructive force limits are low. Also for gold to aluminum thermocompression ball bonding, the as-made bond strength may be high but strength may be low after limited temperature storage. (The temperature storage referenced here is less than that required for Kirkendall failure.)

Dry chemical plasma cleaning is a method that offers a solution to these problems. The method does not require special fixturing, since completed hybrids may be cleaned just prior to wire bonding. The idea is to use rf power to generate free oxygen radicals. The radicals react with hydrocarbons to produce water vapor, carbon monoxide, and carbon dioxide, which are pumped from the system. Thus, bonding surfaces are cleaned of organic contamination by chemical oxidation. Since hybrids employ a number of oxidizable materials (epoxies, silver, and tantalum nitride resistors), the concentration of oxygen radicals and plasma clean time must be carefully controlled.

DEFINITIONS AND INITIAL EXPERIMENTATION

The objective of this study was to determine a parametric envelope for plasma cleaning of aluminum surfaces prior to thermocompression wire bonding. A parametric envelope illustrates the range of process parameters that yield acceptable cleaning from a bondability point of view without degrading adjacent hybrid materials.

A Model PDS-302 dry plasma stripper manufactured by LFE Corporation was used.[2] The unit has 2 parallel chambers, 4 inches in diameter and 8.5 inches long. Each chamber can easily accommodate 15 hybrids. Automatic control is built into the unit so that once the fundamental parameters are preset, the operator only loads and unloads the chambers. Since the procedure developed requires two cleaning steps, the plasma stripper was modified to handle two gases, oxygen and argon. Gases were switched using a manual valve.

Prior to performing the experiment, it was necessary to define bondability and to select samples. Bonding was accomplished using conventional thermocompression ball bonding techniques with the following parameters:

a. 0.007-inch diameter gold wire

b. Dwell time - 1.0 seconds

c. Base temperature - 220°C

d. Capillary temperature - 500°C

e. First bond force - 50 grams

f. Second bond force - 30 grams

Bondability was evaluated by monitoring bond yield and nondestructive pull test yield and by measuring intraconnection pull strength after an accelerated life test. Bond yield is the percentage of bonds achieved on the first attempt. Nondestructive pull testing was done with the loop technique to a

1.0-gram level. The accelerated life test consisted of storing bonds at 300°C for 4 hours. After the temperature stress, bonds were required to be stronger than the intraconnection wire; that is, only wire failures were accepted. When the 1.0-eV activation energy shown by Peck[3] was used, the 300°C stress is equivalent to approximately 20 years at 100°C ambient. Sample selection was difficult since the contamination levels of interest are typically not visible with standard light optics. In addition, the level of contamination throughout an incoming group of components typically varies. For most samples evaluated, no more than one bond in 100 would fail the bondability criteria. Although this failure rate is unacceptable for manufacturing, it is much too low for evaluating cleaning techniques, for which extremely large samples would be required. A large number of devices was found that bonded poorly. These devices were stored in air for a period of over 3 years and were initially stored because they demonstrated poor bondability. Visual inspection of the parts did not show obvious contamination, although 30 percent of the bonds failed bondability testing.

After bond samples were selected and bondability requirements were defined, samples were cleaned and bondability was evaluated. This initial study indicated that bondability of aluminum surfaces could be remarkably improved, but that silver metallization (epoxy filler and capacitor terminals) was discolored. The discoloration was only visual since capacitor terminals remained bondable. Samples of the discolored capacitors were baked at 300°C. This temperature should readily reduce silver oxide. It was found that the oxide ws reduced within a few minutes. Lower temperatures were also tried, but the time needed to restore the metallic color was excessive: approximately 1 hour at bonding temperatures of 220°C.

Finally, the plasma system that used argon in addition to oxygen was tried. The idea was to use the inherent heat generated in the cleaning to disassociate the silver oxide. Even though temperatures generated by this technique are lower than the temperatures evaluated above, the cleaner applies heating in a vacuum and bombards the oxide with argon. The study showed that the argon plasma easily restores the metallization color.

GENERATION OF PARAMETRIC ENVELOPE

As stated earlier, the parametric envelope for plasma cleaning is an illustration of the range of cleaning parameters over which adequate cleaning, from a bondability point of view, may be achieved. The envelope shows directly the interdependency of process parameters. Its boundaries indicate failure modes. Plasma cleaning as defined above has six variables, three for the oxygen cleaning cycle and three for the argon cycle. The

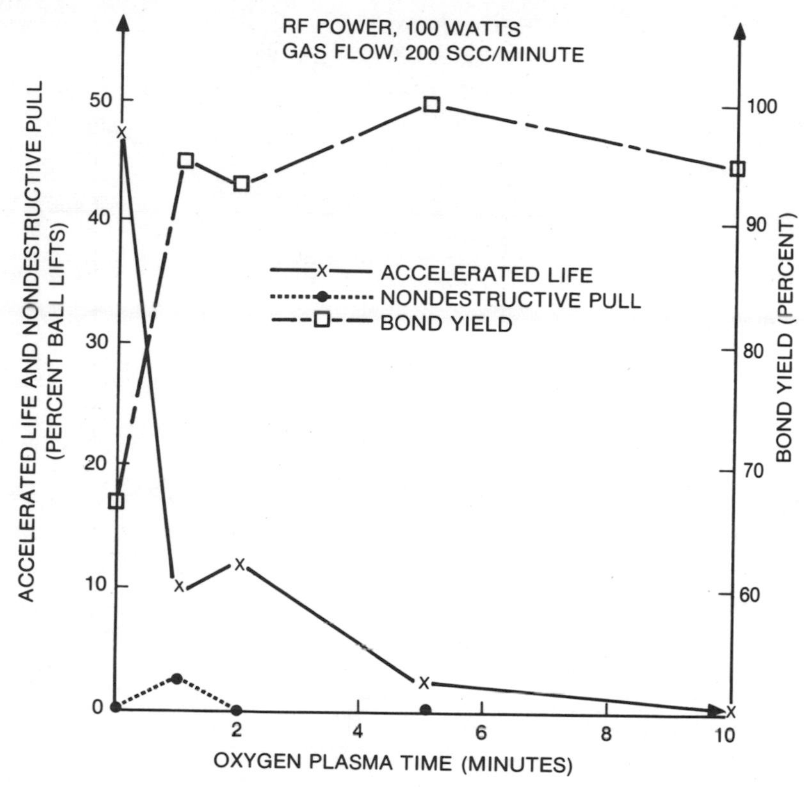

Figure 1. Influence of cleaning time.

variables are: (1) process time, (2) gas flow rate, and (3) rf power.

To simplify equipment operation, gas flow rate and rf power for oxygen were used for argon. In addition, argon plasma time was kept constant at 2 minutes throughout the evaluation. First, bondability was evaluated in terms of oxygen process time. Approximately 60 bonds were produced for each process time evaluated. Bonds were evaluated by recording bond yield in percent, percent of ball lifts during nondestructive pull, and percent of ball lifts after the accelerated life testing. In Figure 1, plasma cleaning was evaluated by increasing the cleaning time and keeping the rf power and gas flow rate constant. In Figure 2, the cleaning was evaluated by increasing the gas flow rate and keeping rf power and cleaning time constant. As both figures illustrate, plasma cleaning greatly improved bondability. For oxygen cleaning times of 9 minutes or greater, ceramic capacitor silver terminations were visually degraded.

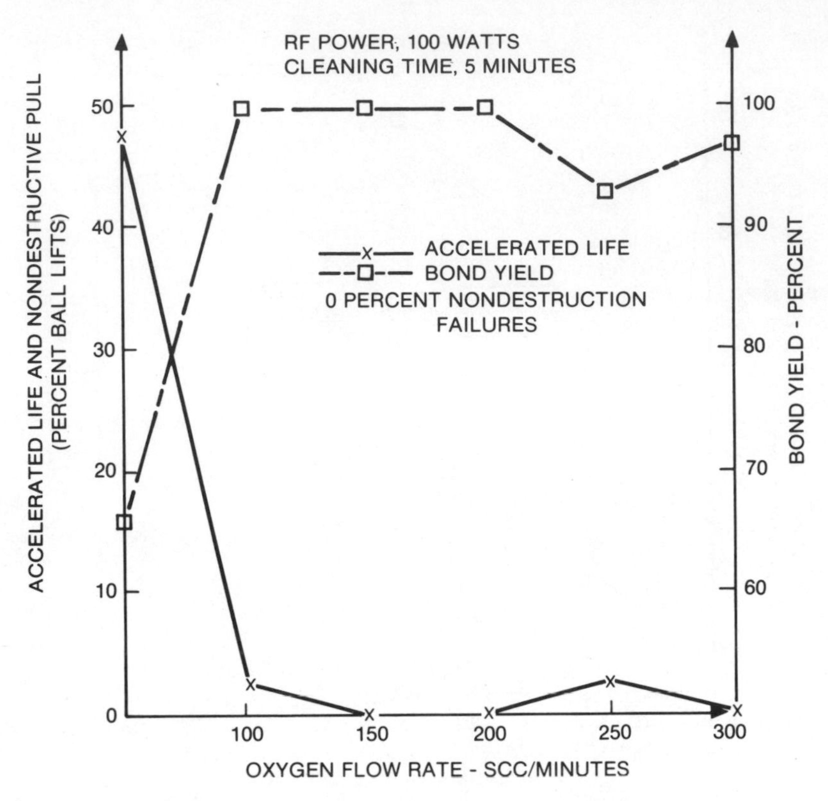

Figure 2. Influence of oxygen flow rate.

The first step in generating the envelope was to determine the cleanliness-completion relationship; that is, to determine the combination of rf power and cleaning time that will result in aluminum surfaces sufficiently clean for thermocompression bonding. Gas flow rate was kept constant at 130 scc/min for this evaluation. The procedure consists of selecting a cleaning time and increasing rf power until completion is obtained. Completion here is defined as 100 percent wire failures after the 300°C temperature stress, 0 percent nondstructive pull test failures, and 100 percent bond yield. After determining the completion relationship for 130 scc/min gas flow, the relationship was verified for gas flow of 150 and 170 scc/min. Figure 3 illustrates the results.

The rf powers above 150 W and below 50 W were not evaluated. Above 150 W, metallic films could be sputtered; below 50 W, cleaning times would reduce production throughout. Since plasma cleaning process time is dependent upon the number of parts or the level of part contamination, the data in Figure 3 was gathered in a full load condition. Full loading was simulated by placing a 2- by 2-inch substrate coated with photoresist in each chamber during cleaning. This quantity of organic contamination should be greater than the level of hybrid contamination in a cleaning cycle.

By employing the techniques discussed, minimum cleaning parameters required to achieve bondability were established. To generate a parametric envelope, maximum cleaning parameters are also required. Samples wre cleaned with a process time of 10 minutes and rf powers of 50, 100 V and 150 W. All samples evaluated demonstrated good bondability; however, silver surfaces were visually degraded (see photomicrograph, Figure 4). The envelope shown in Figure 3 illustrates the combinations of plasma cleaning parameters effective for cleaning aluminum and gold metallizations if silver is not present. A wide range of cleaning parameters may be used. The envelope applies to both gold and aluminum, since the bonding investigated consisted of a ball bond to aluminum and a wedge bond to gold.

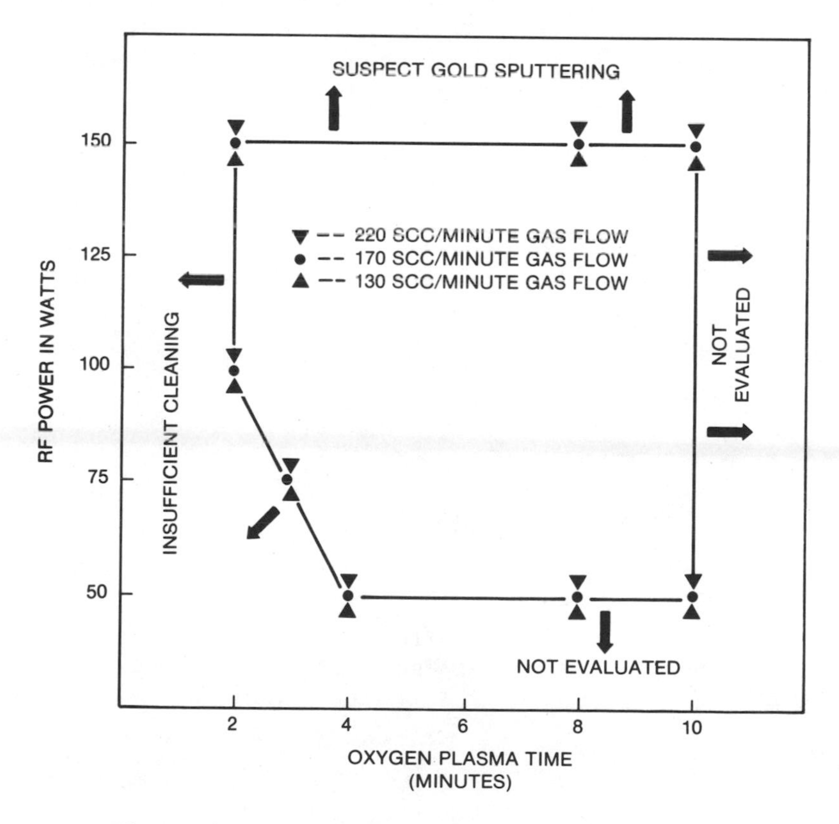

Figure 3. Parametric envelope for aluminum and gold metallizations.

Since silver metallizations are used in most hybrids, the parametric envelope of Figure 3 typically may not be employed. This restriction is a result of choosing to implement plasma cleaning at prebond for cleaning assembled hybrids. A silver degradation relationship is required. To obtain this relationship, samples were cleaned at a constant process time while the rf power was increased until visual damage appeared. Visual damage was determined using 1000X scanning electron micrographs. Any appearance of the cauliflower structure shown in Figure 4 was cause for rejection. All silver samples cleaned met the defined bondability criteria. The silver degradation relationship is shown along with the completion relationship in Figure 5. The data for this relationship was gathered under minimum loading conditions to present the worst case. Figure 5 illustrates the parametric envelope for plasma cleaning of assembled hybrids. Insufficient cleaning and overcleaning regions are illustrated, as is the region of usable parameters. Two additional curves are shown in the figure. These curves represent constant energy. The first (12,000 J) closely follows the completion curve and the second (24,000 J) corresponds to the over clean or silver degradation curve. The data, therefore, indicates that plasma cleaning may be evaluated only in terms of gas flow (130 scc/min to 220 scc/min) and the total energy applied to the system.

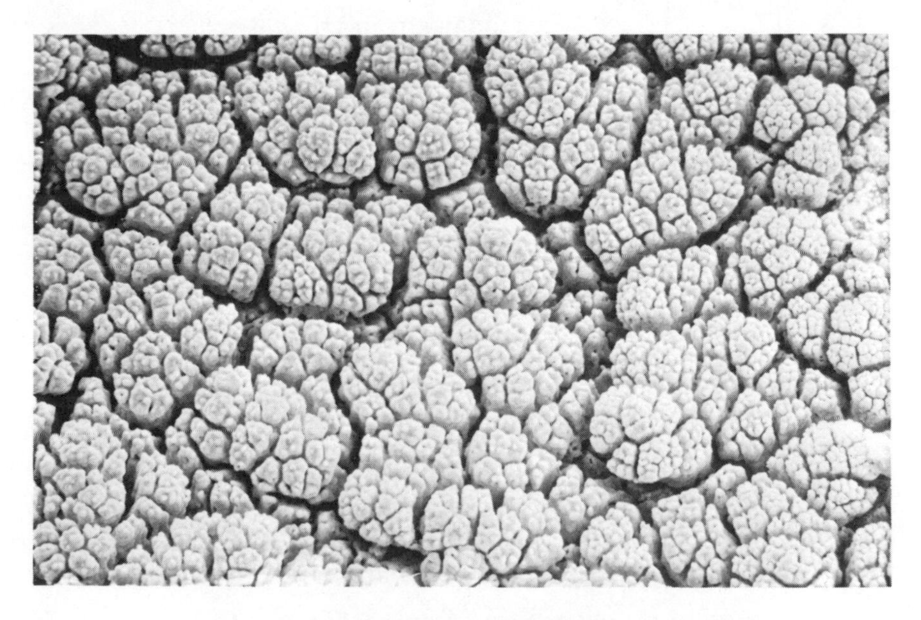

Figure 4. Cauliflower appearance - scanning electron micrograph.

SELECTION OF OPERATING PARAMETERS

After defining the parametric envelope, a set of operating parameters was selected to minimize the effects of parameter variations. This operating point would be the center of the envelope if both rf power and cleaning time could be controlled with equal tolerances. Since cleaning time is more easily controlled, the rf power (the point in the upper left, 3.5 min, 85 W, of Figure 5) was selected. Cleaning or process time may be controlled to 10 scc/min. Thus, the width, height, and depth of the envelope through the operating point are:

a. 20 clean time tolerances - wide.

b. 10 rf power tolerances - high.

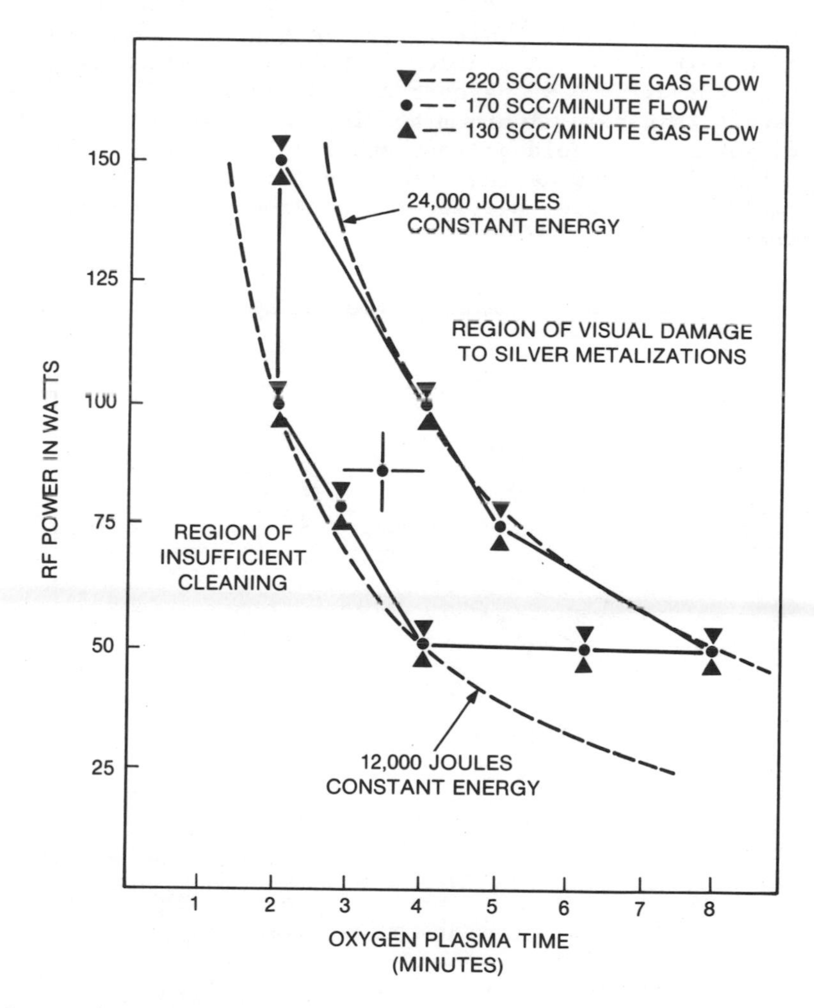

Figure 5. Parametric envelope for plasma hybrid cleaning.

c. 9 gas flow tolerances - deep.

Such an operating point should be easily controlled in manufacturing.

CERTIFICATION OF PARAMETRIC ENVELOPE

Certification of the envelope was accomplished by assembling and cleaning test samples at each envelope vertice. Also, a control group with no cleaning was tested. After cleaning, the samples were bonded and stressed at 300°C for 4 hours. The stress was applied only to gold-aluminum bonds, since temperature exposure of gold-silver bonds improved bond strength. Each test sample was pulled to destruction using the loop technique. Failure mode and strength were recorded. Figure 6 illustrates the results for aluminum. This figure is a probability plot in which normal distributions are straight lines. The envelope about these curves show 99 percent confidence limits. This data clearly shows that gold-aluminum bonding was greatly improved by plasma cleaning. The gold-silver bond sample predicts no more than one bond in 10,000 having strengths less than 1.2 grams. For the gold-silver case, all interconnections tested failed as wire breaks.

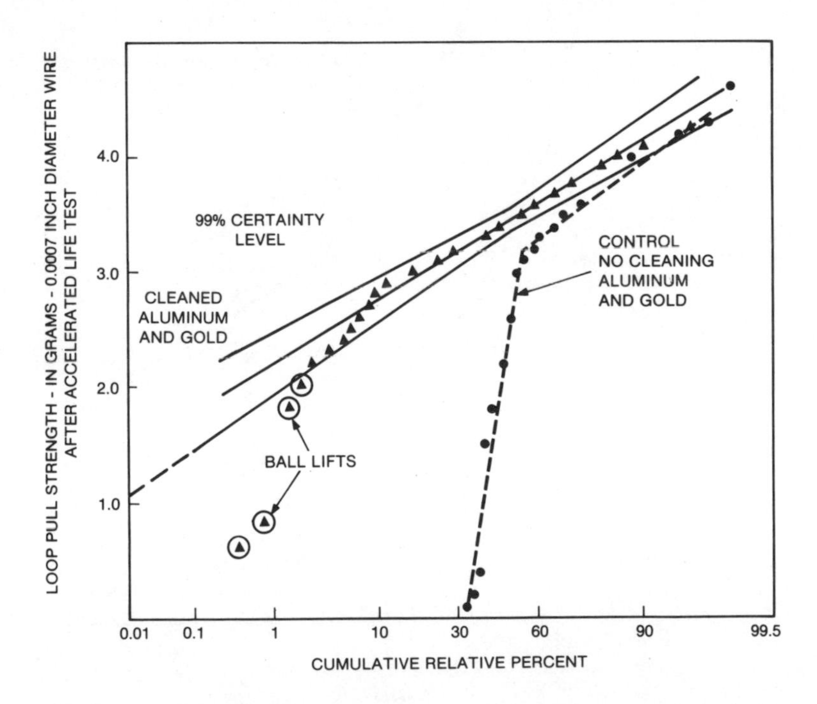

Figure 6. Influence of plasma cleaning on bondability of gold and aluminum metallizations.

INFLUENCE OF PLASMA CLEANING UPON TANTALUM NITRIDE RESISTORS

Tantalum nitride resistors are stabilized using a high-temperature bake in air. The temperature bake anneals and forms an oxide that reduces further oxidation during circuit application. When plasma cleaning is used, it is possible to increase the oxide layer on resistors, and thus force an increase in resistor value. Also, plasma cleaning may influence long-term resistor stability. Resistors were measured before and after plasma cleaning and after long term stability stressing. A 0.026 percent incease in resistance resulted from plasma cleaning.

Resistor stability was investigated using the standard 1,000 hour, 150°C stress in air. Percent change in resistance was determined after 160, 760, 1,000, and 2,000 hours. For each test, eight resistors were measured. Resistor stability was determined for resistors having no bake, one hour, and two hours of stabilization bake. A plasma-cleaned lot and a control lot (no cleaning) were evaluated. Plasma cleaning consisted of 100 W of rf power, 15 minutes of oxygen plasma time, 2 minutes of argon plasma time, and 350 scc/min gas flow rate. Thus, a worst-case condition was evaluated and the results are shown in Figure 7. Resistor stability was improved for the no-bake and one-hour stabilization bakes, but for the 2 hour stabilization bake, resistor stability decreased. Since resistor stability is specified at 0.5 percent after 1,000 hours at 150°C in air, plasma cleaning did not degrade the resistors.

IDENTIFICATION AND SOURCES OF CONTAMINATION

Auger Electron Spectroscopy was utilized to determine the type and level of contamination on cleaned and uncleaned IC bond pads. Table I summarizes the relative AES peak intensities for the plasma cleaned and uncleaned devices used in generating the parametric envelope. The oxygen peak was used as the reference peak because of its sensitivity to AES and its abundance of the samples analyzed. The table also shows the relative peak intensities after removing approximately 15 Å of the cleaned sample and 60 Å of the uncleaned samples.

The uncleaned samples revealed a large amount of carbon and a trace of nitrogen. This is indicative of photoresist since the by-products of photoresist that is bombarded by the high energy electron beam used to excite the Auger electrons are mostly carbon. As Table I illustrates, the plasma cleaning process was very effective in reducing this carbon contamination. Other authors[4] have shown similar difficulty in thermocompression bonding gold lead frames to a circuit with more than 15 Å of carbon present. The ion-sputtered surfaces are listed for reference and verify a clean interior metallization.

The small amount of carbon is probably residue from the ion sputtering.

Other features of this analysis was the AES data off the bonding pads. This data illustrated that all the contamination was restricted to the bond pad, typical of a photoresist process.

Since the devices used to construct the parametric envelope were contaminated to levels much greater than devices used in

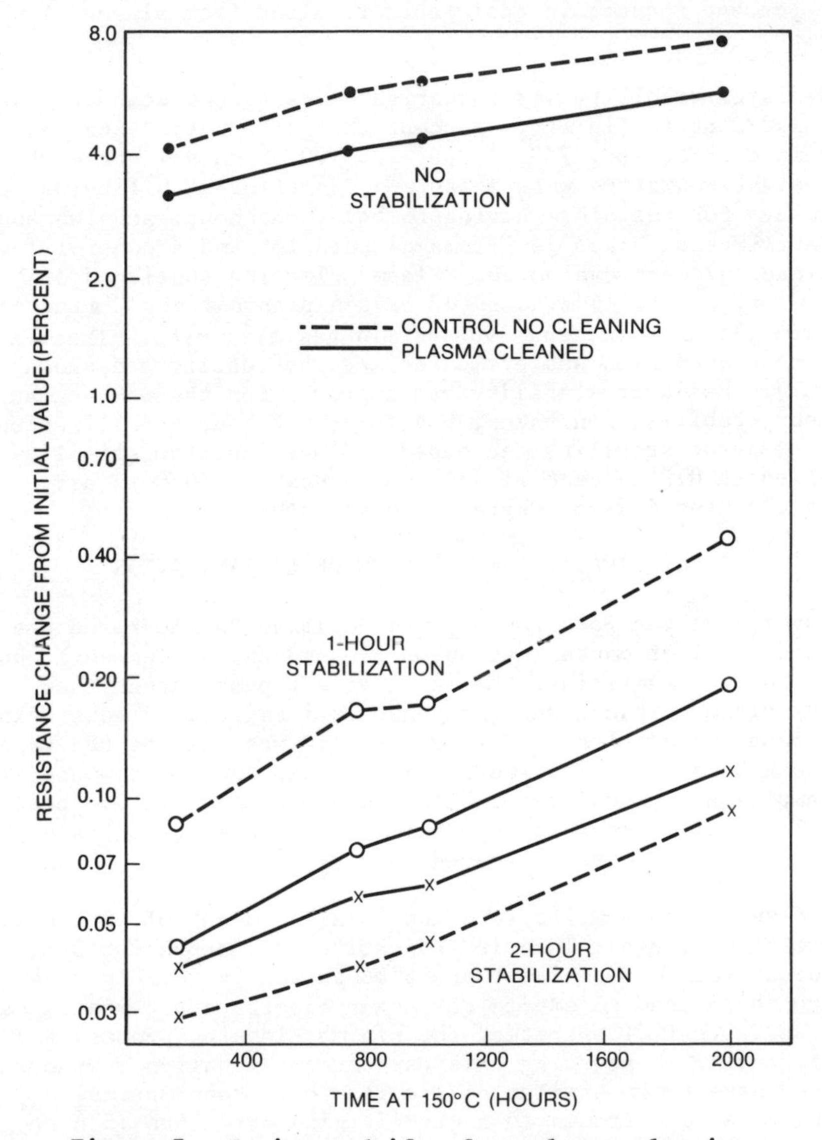

Figure 7. Resistor drift after plasma cleaning.

Table I. AES Peak Intensities for Test Devices

	O	Al	C	Si	S	Na	N
Uncleaned Bond Pads	100	33	64	8	5	3	Trace
Plasma Cleaned Bond Pads	100	27	10	4	3	4	
Ion Etched Uncleaned Sample	100	20	14	11			
Ion Etched Cleaned Sample	100	18	4	8			
Off Bonding Pads Typical of Cleaned and Uncleaned Samples	100		3	105		3	

Table II. Relative AES Peak Intensities for Manufacturing Processes.

	O	Al	C	Si	F	Cl	S	Na
Typical Manufacturing IC Chip Uncleaned	100	24	35	6			2	
Typical Manufacturing IC Chip After Plasma Cleaning	100	13	13	6			2	
Test Device After Plasma Cleaning and Before Epoxy Process	100	27	10	2	3		3	1
Test Device After Epoxy Process	100	26	43	4	10	3	7	

most hybrid manufacturing, Auger analysis was also performed on devices taken directly from the manufacturing area. This data is shown in Table II.

The level of contamination found on typical manufacturing samples was surprisingly high, approximately 50 percent of the level known to degrade bond quality. A second analysis was performed on a device that had previously been plasma cleaned and then subjected to an epoxy chip mounting process. As Table II shows, the epoxy process adds a significant level of contamination to the device. The level again was greater than 50 percent of the level known to create bond problems. Plasma cleaning reduced the carbon contamination in all cases tested to acceptable levels.

CONCLUSIONS

This study characterized a cleaning technique that is inexpensive and easy to employ as a prebond step. Specifically, the study concluded:

a. Plasma cleaning is an effective technique for removing the surface contaminations detrimental to thermocompression bonding.

b. A parametric envelope was generated that shows the acceptable range of cleaning parameters for gold, aluminum, and silver hybrid metallization.

c. Since the range of parameters applicable to cleaning hybrids is wide, the process should be easy to control in production.

d. Plasma cleaning may be characterized in terms of only two parameters--total energy input and gas flow rate.

e. Tantalum nitride resistors are not significantly influenced by plasma cleaning within the range of cleaning parameters evaluated.

f. Most organic contamination can be removed without affecting underlying inorganic materials.

REFERENCES

1. J. L. Jellison, "Proc. of the Electronic Components Conference," p. 271, IEEE, 1975.
2. LFE Data Sheet, Bulletin No. 8203-PBI, LFE Corporation, Process control Division, Waltham, MA, February 1974.
3. D. S. Peck, "Proc. 12th Reliability Physics Symposium," p. 253, IEEE, 1975.
4. P. A. Holloway, "Quantitative Analysis of the Influence of Contaminants on Thermocompression Bonding of Gold, " SAND73-1099, Sandia Laboratories, Albuquerque, NM, March 1974.
5. D. F. O'Kane and K. L. Mittal, J. Vac. Sci. and Tech., 567, 1974.

SURFACE CLEANING BY LOW-TEMPERATURE BOMBARDMENT WITH HYDROGEN PARTICLES: AN AES INVESTIGATION ON COPPER AND Fe-Cr-Ni STEEL SURFACES*

R. Bouwman, J.B. van Mechelen and A.A. Holscher

Koninklijke/Shell-Laboratorium, Amsterdam

(Shell Research B.V.), The Netherlands

Hydrogen has been used at a pressure of 6.6 x 10^{-3} Pa and at room temperature in an "ion-etching beam mode" for cleaning solid surfaces. Two systems, viz. solvent-cleaned and atmosphere-exposed copper and heat-resistant Fe-Cr-Ni steel, were analyzed by Auger spectroscopy before and after bombardment with hydrogen particles produced in a conventional ion gun. Copper is almost completely cleaned of carbon, sulfur, and oxygen within 30 min. A bulk-carbon-containing steel is completely freed from surface carbon while the bulk carbon is preserved in the subsurface zone at a satisfactory level. The proposed cleaning method seems to be very surface specific and nondestructive to subsurface layers.

INTRODUCTION

Surface cleanliness is a subject of considerable interest to all branches of surface science, especially those dealing with the investigation of the topmost one or more layers. Mainly because of the advances in ultrahigh vacuum technology, it is now possible to keep a solid surface clean during investigation over an appreciable period of time. However, the situation is different when samples are to be studied for their original state. Nearly all samples inevitably become covered with one or more monolayers of foreign material, either by adsorption from

* This paper was previously published in J. Vac. Sci. Technol., the publishers of which have no objections against the reproduction in this book.

the ambient atmosphere or as a result of specific pretreatments. With the widespread adoption of modern surface analysis techniques like x-ray photoelectron spectroscopy (XPS) and electron-excited Auger electron spectroscopy (AES)[1-3], it is now generally recognized that the elements commonly found in these monolayers are carbon, oxygen, sulfur and, to some extent, nitrogen.

As the removal of such "masking" layers is the subject of the present paper, we shall briefly touch upon the methods commonly used until now. Most typical are:

a) ion etching[4,5], i.e. removing the contaminants by bombarding the surface with mainly (noble) gas ions, which causes sputtering of exposed atoms as a result of the transfer of momentum from the incident ions to the target atoms;
b) chemical etching[6], a chemical reaction between particles and the surface e.g. the removal of carbon by exposing the surface to oxygen at roughly 500°C;
c) plasma etching[4], in which the sample is exposed to a reactive plasma, possibly at lower temperatures and at a pressure of about 133 Pa. This etching procedure is not a sputtering process but a chemical reaction that converts the etched material into a volatile compound;
d) thermal desorption[6], by which the sample is heated or flashed in vacuum.

Each of these methods has its inherent advantages and disadvantages. For instance, ion etching is quite a general cleaning method of relatively simple setup. Selective sputtering[5,7-14] however, may seriously affect the (sub)surface constitution of multicomponent systems, resulting in a difference between surface and bulk composition, and a possible erosion of the original state of the sample to be studied. Subsequent annealing of ion-etched samples is sometimes thought to restore the original state of the surface, but it may also cause the equilibration of the system, resulting in another difference between surface and bulk[10,11].

Chemical etching and thermal desorption, like sputtering, have a similar potential disadvantage for multicomponent systems. Both methods are generally carried out at temperatures sufficiently high to cause drastic changes of the surface composition[10,11,15,16]. Plasma etching has the advantage of the low-temperature approach. However, the partial pressure of contaminants in the treatment gas should be extremely low in order to avoid recontamination of the surface.

We presently report results obtained in the cleaning of copper and stainless steel surfaces by bombardment with accelerated hydrogen particles at low pressures (~10^{-3} Pa) and room temperature. It is, of course, well known that hydrogen can attack carbon; for instance, in microwave plasmas at elevated temperatures it has been employed to etch single crystals of graphite[17]. Rye used atomic hydrogen produced from a hot tungsten filament to remove adsorbed ethylene and decomposition fragments thereof from a tungsten surface[18]. Behrisch et al.[19] measured the erosion of stainless steel, carbon and SiC by hydrogen bombardment (0.5-7 keV) and Sone et al.[20], in their determination of sputter efficiency of 0.1-6 keV protons on graphite targets, found that a considerable amount of carbon is removed by "chemical sputtering" (methane production). To our knowledge, however, hydrogen has not been applied before in the "ion-etching mode" ("ion sputtering" mode[20]) for the purposes of very selective surface cleaning.

EXPERIMENTAL

We selected two sample materials: one single-component material (bulk-carbon free, pure copper) and one representative of frequently occurring multicomponent samples, viz. a centrifugally cast, heat-resistant Fe-Cr-Ni steel containing 25 wt. % Cr, 35 wt. % Ni, 1.7 wt. % C, the balance being iron.

The samples, both in ethanol-cleaned and atmosphere-exposed form, were mounted in an ultrahigh vacuum chamber equipped with a Physical Electronics Cylindrical Mirror Auger Analyzer, model 11-500 and a Physical Electronics ion gun, model 20-005. The background vacuum during the analysis was 6.6×10^{-9} Pa. Hydrogen, purified by passing through an on-line utrahigh vacuum (10^{-7} Pa) inlet system equipped with a Pd-diffusion membrane, was admitted at working pressures of up to 6.6×10^{-3} Pa. The samples were bombarded at normal incidence at an acceleration energy of 800 eV. Typical (uncorrected) target currents were 10 µA. cm^{-2}. The exact composition of the incident hydrogen particle beam is unknown since it was not mass analyzed for H, H_2 and H_3 particles.

Auger spectra were taken at a primary energy of 2 keV and a modulation amplitude of 4 V, both before bombardment and after specific periods of time. The hydrogen bombardment was discontinued during Auger analysis. The steel sample was also bombarded once with argon (from Messer and Griesheim GmbH, W. Germany, spec. pure quality) at 1.3×10^{-3} Pa and an acceleration energy of 800 eV.

RESULTS AND DISCUSSION

Copper

Figure 1 shows Auger spectra of copper before and after 10-

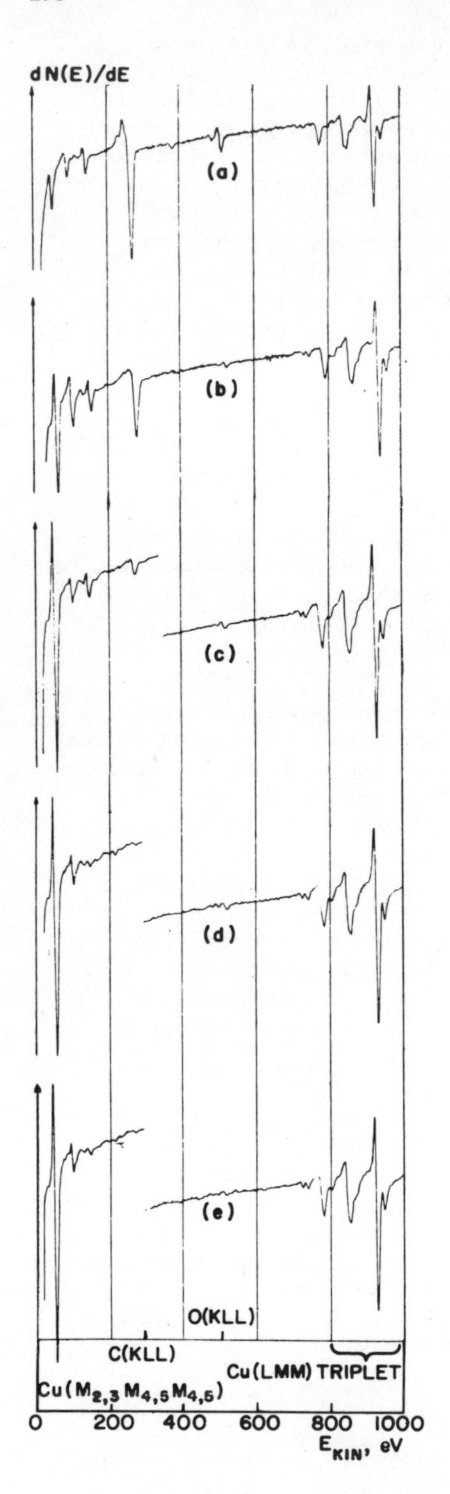

Figure 1. Auger spectra of copper: a) Solvent-cleaned and atmosphere-exposed; b) After 10-min hydrogen bombardment (6.6×10^{-3} Pa H_2 800 eV); c) After 25-min hydrogen bombardment (6.6×10^{-3} Pa H_2 800 eV); d) After 1-h hydrogen bombardment (6.6×10^{-3} Pa H_2 800 eV); e) After 16-h hydrogen bombardment (6.6×10^{-3} Pa H_2 800 eV).

min, 25-min, 1-h, and 16-h bombardments with hydrogen. The C(KLL) line intensity decreases in the course of this treatment and reaches nearly zero intensity after a 1-h bombardment. The O (KLL) intensity also decreases but some oxygen remains bound. The same holds for sulfur. The situation after 16 h is almost identical to that after 1 h. Note in Figure 1 the increase of the Cu ($M_{2,3}$ $M_{4,5}$ $M_{4,5}$) Auger line at about 60 eV kinetic energy and the Cu (LMM) triplet in the 800-1000 eV range, as carbon, oxygen and sulfur are removed.

We estimated the original thickness, d, of the contaminating layer to be about 6 Å from the increase of the [Cu(60 eV)/ Cu(925 eV)] line intensity ratio by assuming (1) the contaminant to be uniformly deposited on the copper surface and (2) specific values for the mean free paths of inelastic scattering, λ, of the appropriate Auger electrons[21]: $\lambda(60) = 0.4$ nm and $\lambda(925) = 1.3$ nm. By plotting the d-values calculated from the [Cu(60 eV)/Cu (925 eV)] line intensity ratio against time (Figure 2), it can be seen that the major proportion of contaminants is removed after some 30 minutes of hydrogen bombardment.

STEEL

Figure 3 shows spectra of the carbon-containing Fe-Cr-Ni steel is untreated form, after hydrogen bombardment for 1 and 16

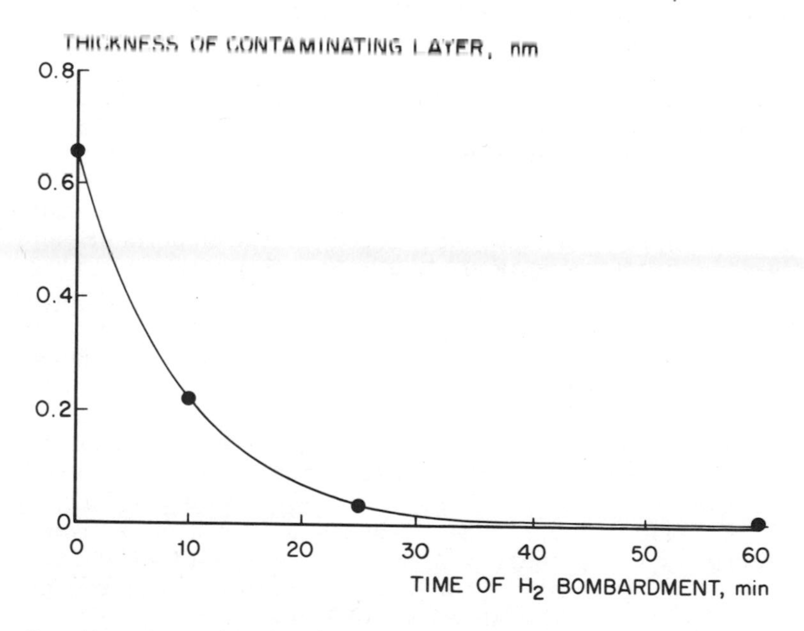

Figure 2. Decrease in thickness of contamination layer with time of hydrogen bombardment.

h, respectively, and after a mild argon bombardment for 15 min ($p = 1.3x10^{-3}$ Pa, ion energy = 800 eV, target current density (uncorrected) ≅ 15 μA cm^{-2}). As can be seen from Figure 3(d), the argon bombardment removes oxygen and sulfur. However, this treatment is ineffective in removing carbon; it even causes enrichment of carbon, possibly by selective sputtering of the metallic constituents[7-9].

The hydrogen bombardment results in removal of sulfur and some of the carbon, while oxygen is left in an apparently unaffected state. This latter observation could indicate that the hydrogen bombardment is rather selective towards the surface, because we expect that steel systems like the one under study contain oxygen not only in the topmost atomic layer but also in subsurface (superficial oxidation). Incidentally, the same argument would apply to copper, on which some sulfur and oxygen remain unaffected by hydrogen. It is well known that the thermodynamics favor the formation of bulk copper oxides and sulfides[22]. Exposure to the atmosphere is therefore likely to induce oxidation and/or sulfidation of more than the exposed surface layer.

The complete disappearance of carbon from copper, the partial disappearance of oxygen and sulfur from copper at the rate observed and the stability of oxygen all suggest that the process as observed from the spectra discontinues at the interface of the metal phase and the contaminating layer, i.e., as soon as the contaminant is removed. An explanation for this relatively rapid[23] "surface-selective" cleaning could be that the removal of carbon, sulfur, and oxygen takes place by a chemical process like in plasma etching rather than by sputtering, although the latter cannot be ruled out[19,20,23].

Now we are still left with the question whether the carbon composition of steel remains unaffected in the uppermost subsurface zone. To check this we have calculated the "bulk" composition of carbon in the steel from the Auger spectra [Figure 3(c)]. For this calculation we have to assume that our sample which is supposedly cleaned from spurious carbon is homogeneous throughout the analyzed depth. To be specific:

(a) The oxygen and carbon Auger signals are assumed to originate from oxygen and carbon associated with or dispersed throughout the steel matrix within a (sub)surface region with thickness at least equal to the escape depth of the relevant Auger electrons, i.e., roughly 4 nm depth, as estimated from the mean free path of the Auger line with highest kinetic energy, the Ni ($L_{2,3}M_{4,5}M_{4,5}$) line at about 840 eV.

(b) The analysed electrons are supposed to be scattered

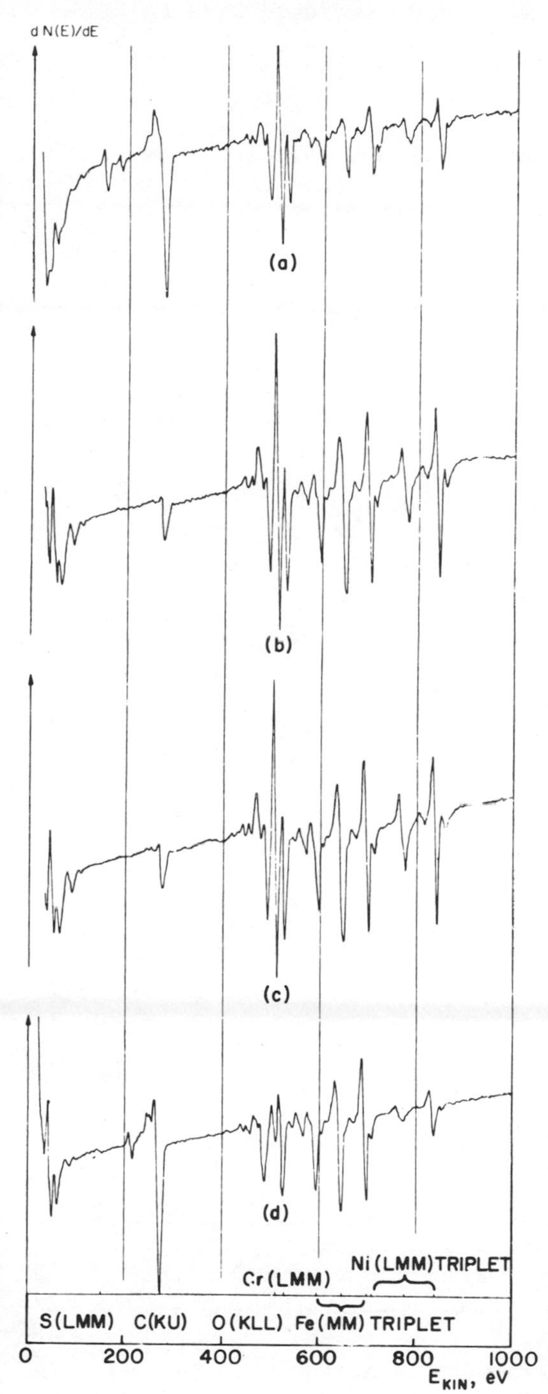

Figure 3. Auger spectra of a centrifugally cast, heat-resistant alloy of Fe (38.3 wt.%), Cr (25 wt.%), Ni (35 wt.%) and C (1.7 wt.%): a) Solvent-cleaned and atmosphere-exposed; b) After 60-min hydrogen bombardment (6.6 x 10^{-3} Pa H_2 800 eV); c) After 16-h hydrogen bombardment (6.6 x 10^{-3} Pa H_2 800 eV); d) After 15-min Argon bombardment (6.6 x 10^{-3} Pa H_2 800 eV).

isotropically throughout the steel. Furthermore we have to assume that the specific Auger electron yield from each of the elements in the steel is not different from that obtained from elemental reference systems[24].

On the basis of these assumptions we may calculate the carbon content as

$$x_c = \frac{i_c/s_c}{\sum_\alpha i_a/s_\alpha}$$

where x_c is the atomic fraction of carbon, i_c is the signal intensity of the C(KLL) Auger line, s_c is the atomic sensitivity of carbon at the prevailing experimental conditions and α stands for C, O, Fe, Cr and Ni. s_α values were taken from Ref. 24 by extrapolation of the available data to our experimental conditions. The x_c value thus calculated is 6.1 at. % (1.6 wt. %), which is in very satisfactory agreement with the bulk carbon content of the steel (1.7 wt. %). The significance of this agreement is further substantiated by analyzing the cross section of fractured piece of the same stainless steel characterized by a (known) carbon concentration profile. In Figure 4 the Auger-derived bulk carbon data obtained after cleaning the steel surface by means of the above hydrogen treatment are compared with carbon bulk analysis values obtained from independent methods.

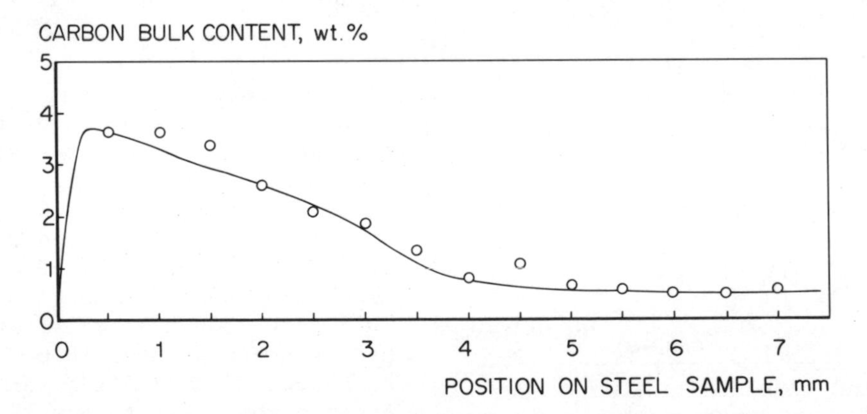

Figure 4. Carbon concentration profile in a sample of Fe-Cr-Ni steel[10,11]: o values calculated from Auger spectra, ___ curve obtained by bulk analysis.

CONCLUSIONS

(a) Hydrogen used at low pressures (10^{-3} Pa range) and room temperature in the "ion-etching mode", that is instead of a noble gas in a conventional ion gun, is capable of removing adsorbed contaminating layers (carbon, oxygen, sulfur) from copper and steel.

(b) This method might have a more general significance as a routine surface cleaning procedure for solids to be studied in ultrahigh vacuum. The results obtained so far strongly indicate that the mechanism of surface cleaning is based not on momentum transfer of hydrogen particles, but on a chemical process, possibly comparable to that in plasma etching. This would imply that surfaces of solids "as delivered" can be cleaned without appreciable deterioration of the original state masked by the contaminants.

REFERENCES

1. C.C. chang, in "Characterization of Solid Surfaces", P.F. Kane and G.B. Larrabee, Editors, Plenum Press, New York 1974.
2. A. Joshi, K.E. Davis and P.W. Palmberg, in "Methods of Surface Analysis", A.W. Czanderna, Editor, Elsevier, Amsterdam, 1975.
3. C.R. Brundle, J.Vac.Sci. Technol. 11, 212 (1974).
4. S. Somekh, J. Vac. Sci. Technol. 13 1003 (1976).
5. G.K. Wehner, in "Methods of Surface Analysis", A.W. Czanderna, Editor, Elsevier, Amsterdam, 1975.
6. J.M. Thomas and W.J. Thomas, "Introduction to the Principles of Heterogeneous Catalysis", Academic Press, London, 1967.
7. M.L. Tarng and G.K. Wehner, J. Appl. Phys. 42, 2449 (1971).
8. S.D. Dahlgren and E.D. McClanahan, J. Appl. Phys. 43, 1514 (1972).
9. M.L. Tarng and G.K. Wehner, J. Appl. Phys. 43. 2269 (1972).
10. R. Bouwman, L.H. Toneman and A.A. Holscher, Vacuum 23, 163 (1973).
11. R. Bouwman, L.H. Toneman and A.A. Holscher, Surface Sci. 35, 8 (1973).
12. L.A. West, J. Vac. Sci. Technol. 13, 198 (1976).
13. B. Navinsek, Prog, Surf. Sci. 7, 49 (1976).
14. W.D. Westwood, Prog, Surf, Sci. 7 71 (1976).
15. R. Bouwman and W.M.H. Sachtler, Surf. Sci. 24, 350 (1971).
16. R. Bouwman and P. Biloen, Surf. Sci. 41, 348 (1974).
17. B. McCarroll and D.W. McKee, Carbon 9, 301 (1971).
18. R.R. Rye, J. Vac. Sci. Technol. 13, 364 (1976).
19. R. Behrisch, J. Bohdansky, G.H. Oetjen, J. Roth, G. Schilling and H. Verbeek, J. Nucl. Mater. 60, 321 (1976).

20. K. Sone, H. Ohtsuka, T. Abe, R. Yamada, K. Obara, O. Tsukakoshi, T. Narusawa, T. Satake, M. Mizuno and S. Komiya, in "Proc. Int. Symp. on Plasma Wall Interaction", 1976, Julich (JAERI memo 6786).
21. R. Bouwman, Ned. Tijdsch, Vacuumtechn. (Eng. ed.) 11, 37 (1973).
22. National Bureau of Standards, Circular 500, "Selected Values of Chemical Thermodynamic Properties", February 1, 1952.
23. M. Kaminsky "Atomic and Ionic Impact Phenomena on Metal Surfaces", p. 176, Springer, Berlin, Heidelberg, 1965.
24. "Handbook of Auger Spectroscopy", 2nd ed., Physical Electronics Ind., Eden Prairie,MN, 1976.

SOME PROMISING APPLICATIONS OF ION MILLING IN SURFACE CLEANING

S. I. Petvai and R. H. Schnitzel

IBM Data Systems Division, East Fishkill

Hopewell Junction, New York 12533

Chemical (wet) etching or reverse sputtering processes have long been used in the semiconductor industry to remove contamination from the surfaces of silicon wafers in thin-film processing. By using ion milling to remove contamination, we offer an advantageous alternative process: removal rates greater than 300 Å/min were obtained at temperatures less than 160°C. We used a pure ion beam at 0.75 W cm^{-2} power density and 6 rpm sample rotations. The ion-beam angle of incidence determined the simultaneous removal rates of various materials, a uniform (nonselective) or nonuniform (selective) removal rate, and the degree of temperature. By selecting an appropriate angle of incidence, we can increase the uniformity of the removal rates, and thus obtain faster removal of the unwanted material or material in the wrong place, with minimum effect on the wanted material in the right place and the electrical parameters of the device. With this method, we have achieved faster removal rates at lower surface temperatures than previously possible, with uniform or nonuniform removal rate ratios of Al-Cu and SiO_2 at various angles of incidence.

INTRODUCTION

In a previous paper[1] we reported some advances in via cleaning of semiconductor devices by use of ion milling. In this paper we will address ion-beam kinetics and thermodynamics.

Chemical etching or reverse sputtering processes have long been used in the semiconductor industry to remove contamination from the surfaces of silicon wafers in thin-film processing. Generally device fabrication is practiced by masking an area where materials are to remain and then opening areas in the mask where materials are to be removed.

By using ion milling to remove contamination, we offer an advantageous alternative process to chemical etching and reverse sputtering: removal rates greater than 300 Å/min were obtained at temperatures less than 160°C. Using a pure argon ion beam at 0.75 W cm^{-2} power density and 6 rpm sample rotations, we engineered the removal rate of Al-Cu and SiO_2 by changing only the ion-beam angle of incidence, and thus obtained faster removal rates at lower temperatures than previously possible.

The purpose of this paper is to present the results obtained in our work and some viable extensions of the ion-beam processes.

THE SYSTEM OF ION-BEAM CLEANING IN A METAL EVAPORATOR

The system, comprising an ion gun with a two-inch exit orifice attached to a wafer-holding chamber, is shown in Figure 1.[1]

The evaporator[2] can comprise a source chamber under vacuum, evacuated by a diffusion pump, roots blower, and mechanical pump combination using a cold trap. During venting, the source chamber is isolated from the process chamber by a valve.

Metals are evaporated with e-guns. All the sensors and actuators are received and automatically driven through an interface package by an IBM System/7 process computer.

Process Cycle

Figure 2 is an illustration of a process cycle of a metal evaporator, using ion-beam cleaning immediately before metal deposition. Chains E and F demonstrate the ion-beam cleaning process cycle.

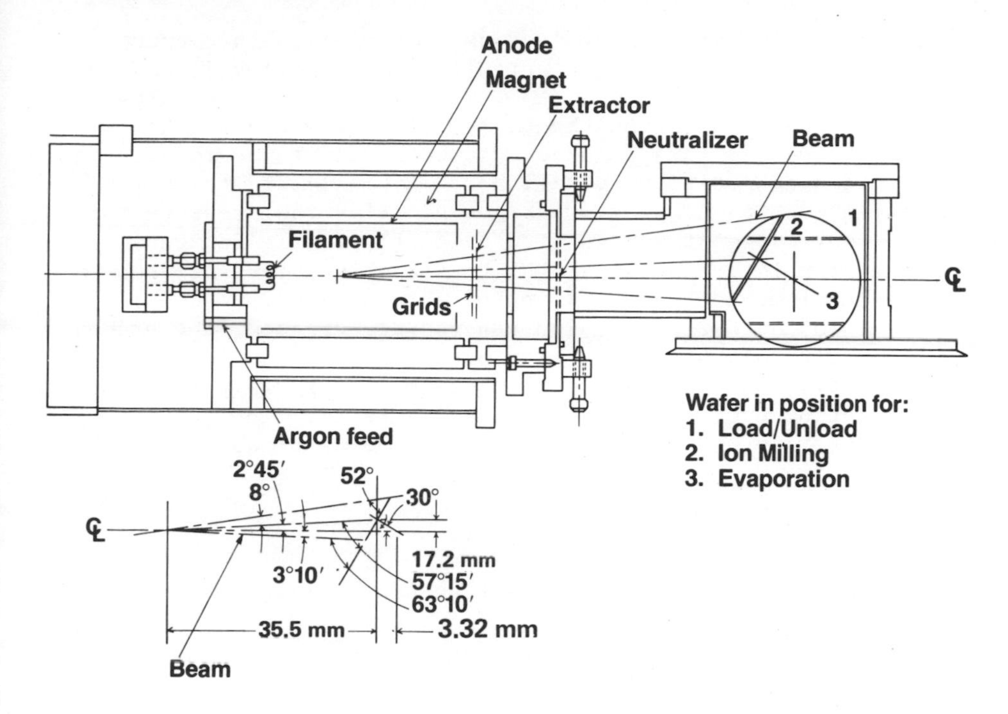

Figure 1. Ion-milling apparatus.

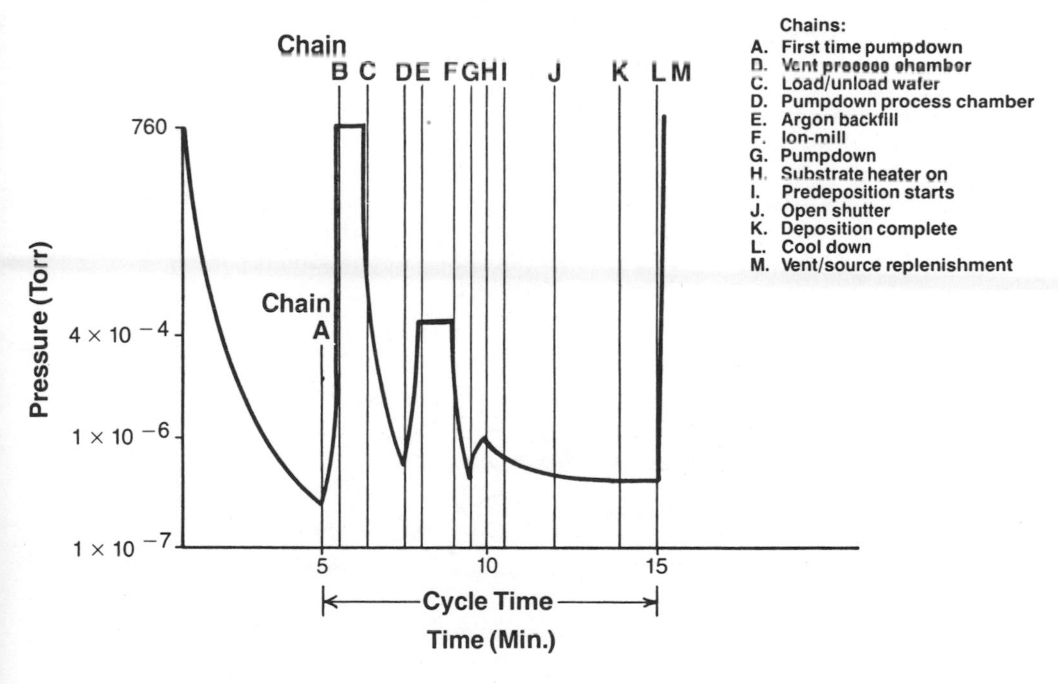

Figure 2. Typical ion-beam cycle in metal evaporation.

The cycle was started by loading a wafer into the vacuum process chamber (position 1, Figure 1), with the stencil facing upward, prior to deposition of the second layer of film. The chamber was then closed and the pressure was lowered to the 10^{-6} Torr range. During this pump-down time the wafer was rotated to the via cleaning position (position 2, Figure 1) and the chamber was backfilled with high purity argon to raise the pressure to the 10^{-4} Torr range at a regulated argon flow. The ion gun was energized at this pressure. The lift-off structure[3] was then subjected to 1 min of ion cleaning at a rate of 0.75 W cm^{-2}. After 1 min of operation the ion gun was deenergized, the argon gas flow was turned off and the chamber pressure was lowered to the 10^{-6} Torr range. During this time the wafer was rotated to the evaporation position (position 3, Figure 1). The second layer evaporation was then completed and the wafer was unloaded after venting.

EXPERIMENTS

Selective and Nonselective Ion Milling

The term "selective" means that the removal rates of all the materials are not uniform; "nonselective" means that they are uniform.

Figure 3 shows a lift-off photoresist stencil over a via prior to the deposition of a second layer of metal. In this structure, the top surface of the first layer of the metal exposed in the bottom of the via will be cleaned of any debris by ion beams, so that the second layer will make good contact with the first layer.

The ion-beam cleaning will also remove surface material all over the device, as shown in Figure 4. In this example, the exposed surfaces should be nonselective to the ion-beam cleaning to equalize the removal rates of all the materials.

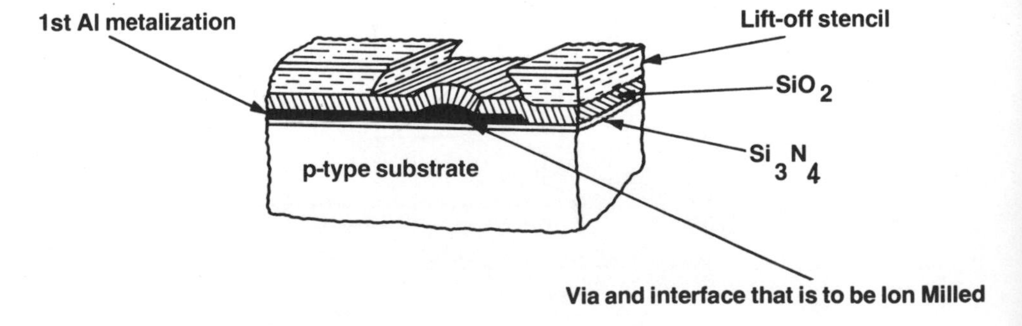

Figure 3. Photoresist stencil over via.

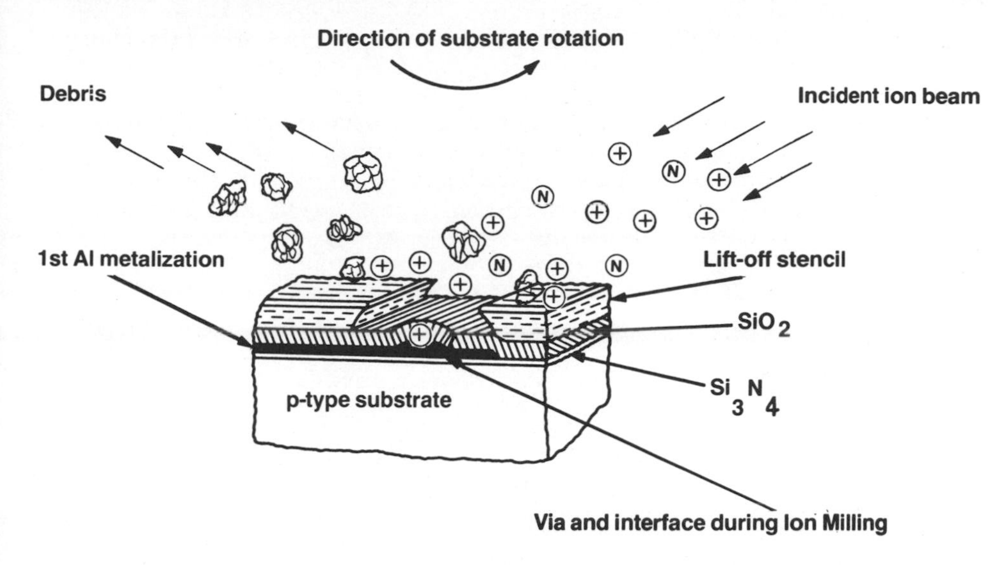

Figure 4. Direction of incident ion beam, debris (unwanted material), and substrate rotation.

Table I shows that at an incident angle of 30°, the removal rates for sputtered SiO_2 and Al-Cu are identical. The removal rate of the spun-on glass is higher. In this case, however, SiO_2 and Al-Cu are parts of the device, and the spun-on glass is part of the stencil to be removed. Therefore, we want the removal rates of the SiO_2 and the Al-Cu to be nonselective and the removal

Table 1. Rate of Removal as a Function of Incidence Angle for Various Materials.

		Removal Rate (Å/Min ±/10%)					
Angle (°)	Acc. (V)	Si	SiO_2	Al-Cu	Spun on Glass	Photo AZ-1350	Sputter Planar Glass
45	1000	273	330	342	620	625	400
30	1000	260	256	260	450	364	300
15	1000	185	156	260	ND[a]	ND	ND
0	1000	ND	141	208	ND	ND	ND

a) ND - Not Determined

rate of the spun-on glass to be selective. Figure 5 shows the completed device after the metal deposition has been completed.

The optical properties of structures exposed to ion beams may affect the integrity of the cleaned structures. If the reflective, absorbing, and transmission properties of a structure are not selected properly, a loss of the stencil may occur, as shown in Figure 6.

Selectivity of removal rates may be obtained, as shown in Figure 7, in which a uni-material, composite, geometric surface is exposed to an ion beam at different angles of incidence. The yields and temperature differences mapped in the figure indicate selectivity. On the right side of the figure an ion beam with a 60° angle of incidence impinges on the surfaces, and on the left

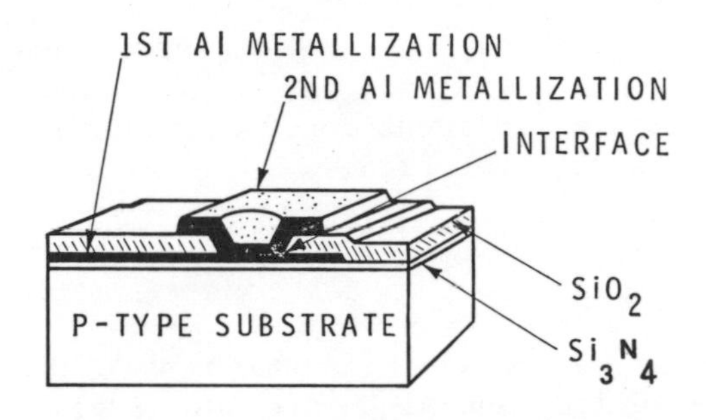

Figure 5. Via and via connection.

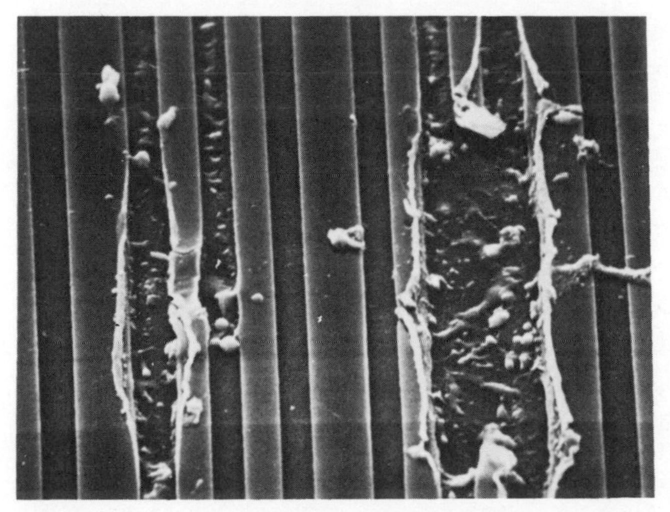

Figure 6. Evidence of photoresist running with Al-Cu substrate underlay at 1000 V acceleration, 30° incidence angle, 20 minutes.

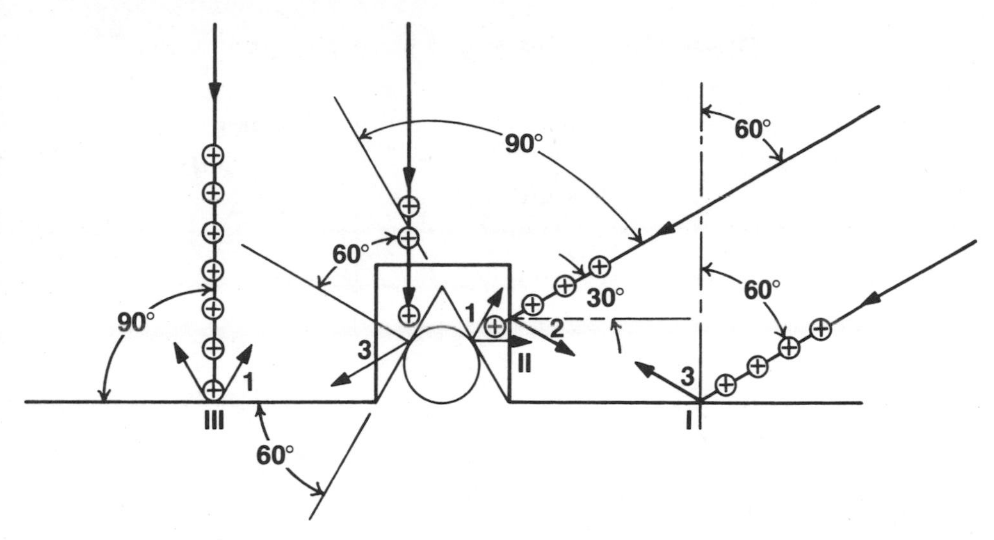

Figure 7. The function of material removal rates exposed to uniform ion milling on geometric surfaces. I < II < III-directions and induced surface temperatures, 1 < 2 < 3 - material removal rates, ⊕ ⊕ ⊕ - 90° < 30° < 60° angles of incidence.

side an ion beam with a normal angle of incidence (90°) impinges on the same surface.

For convenience, arrows indicating the direction of debris removal are also shown in Figure 7. The knowledge of the direction of debris removal is of major importance in preventing recontamination of the wafer by redeposition of the removed material.

Device Processing

Figure 8 is an illustration of a molybdenum mask with etched holes for material removal openings. If this mask is firmly attached to a wafer and exposed to ion milling, a hole with sharp edges will be etched into the wafer. This step into the wafer can then be probed with a stylus-type instrument or with an interferrometer to determine the removal rate vs. time.

In this fashion, we have accumulated removal rate data which is shown in Table I and in Figure 9. These measurements are the ion-milling removal rates of a p-type silicon wafer coated with approximately 10,000 Å of each material listed in the table.

Figure 10 is a SEM (scanning electron microscope) photo of a

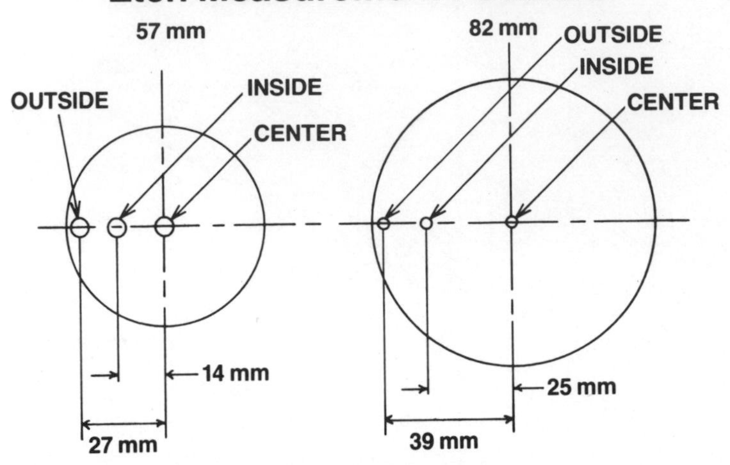

Figure 8. Molybdenum mask.

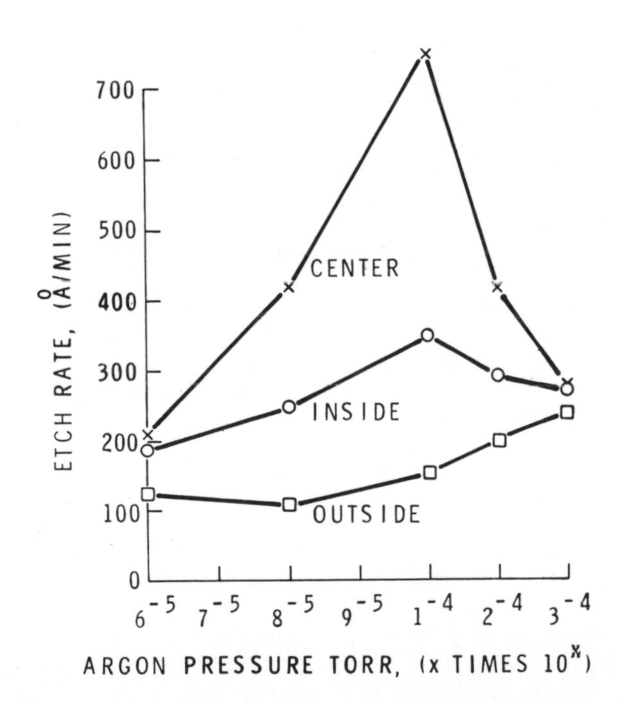

Figure 9. Uniformity function in ion milling (three points on a 57-mm surface). Δ etch rate (angstroms per minute) vs Δ argon pressure (Torr).

completed device prepared by the methods described in this paper and with the sequences shown in Figures 3-5.

Figure 11 is a SEM photo showing the lift-off structure after ion-beam cleaning. The photo clearly indicates the integrity of the stencil after ion milling. The stencil shows no detrimental effects of any kind, such as trenching, undercut, or photoresist flow.

Measurements

The effects of various operating parameters (temperature, argon pressure, and device parameters) on etching were investigated. The procedure consisted of varying a specific parameter, e.g., the argon bleed-in pressure, and then measuring the amount of material removed. We shall discuss only the pertinent parameters from which notable trends can be derived.

Figures 7, 9, 12, and 13 illustrate the parameter measurements of ion milling at a 30° angle of incidence. The incident angle of 30° and the offset of 17.2 mm with respect to the ion milling center line are shown in Figure 1, a schematic of the ion-milling apparatus. Figure 12 shows the uniformity function obtained with the setup shown in Figure 1 and with the metal mask shown in Figure 8. One can see that at a removal rate of 250 Å/min and an acceleration voltage of 1000 V, a ±10% uniformity is obtainable across a 57-mm wafer rotated at 6 rpm.

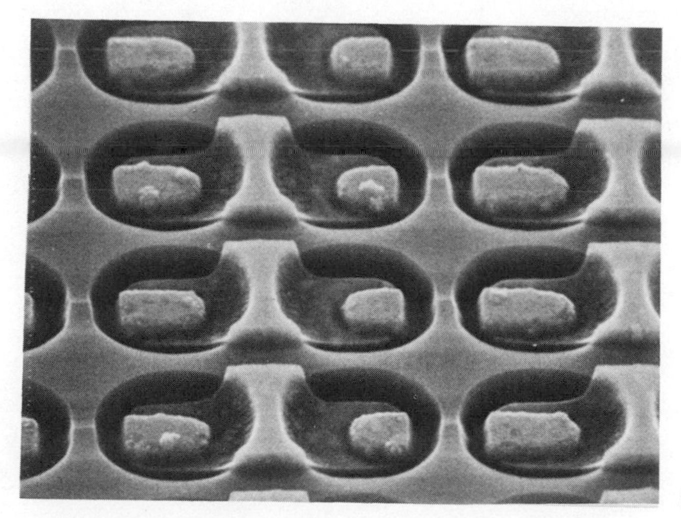

Figure 10. SEM photo showing ion-milled nonoverlapping vias after liftoff.

Figure 13 shows the temperature rise in an ion-milled wafer as a function of acceleration voltage and time. This figure clearly shows that in an ion-milling cycle of two minutes or less, the temperature will not exceed 160°C. The figure also indicates that thermocouple measurements are not practical.

Figure 9 is an illustration of the uniformity function vs argon pressure and etching rate. This figure shows that the most sensitive parameter and measurement is the argon pressure. We used 99.99995% pure argon to prevent surface oxidation during the cleaning process, and we sealed the cleaned surfaces immediately after ion-beam cleaning by evaporating the next level of thin film without breaking vacuum. Ion-beam cleaning proved to be very fast (see Table II).

Via Resistance Measurements

Low via resistance, $\leq$0.20 Ω/via, and ±5% repeatability is important for the electrical and thermal performance of LSI thin-film devices. We have obtained low via resistance and ±5% repeatability and tabulated our results in Table II. These via resistances were measured on multilayer thin-film devices such as that shown in Figure 5.

Figure 11. Representative SEM for the various angles, showing no effect of trenching.

Table II. Resistance of Vias After One Minute of Ion Milling.

	Acceleration Voltage, 1000 V						Acceleration Voltage, 1100 V					
	Mil angle (°)						Mil angle (°)					
	30		30		45		30		45		45	
	Pressure (x 10^{-4} Torr)						Pressure (x 10^{-4} Torr)					
	4		2		2		2		4		2	
Wafer	a	b	a	b	a	b	a	b	a	b	a	b
1	145	114	110	95	112	100	110	95	114	102	107	97
2	130	114	104	92	112	103	112	98	109	99	104	95
3	128	110	104	92	110	100	113	99	106	96	107	97
4	136	118	108	94	118	103	115	100	110	99	111	100
5	142	120	---	91	138	126	122	104	114	104	114	102
6	136	120	110	94	120	105	120	103	116	104	111	98
7	134	116	108	94	130	100	114	100	111	100	106	95
8	134	115	105	80	115	104	112	99	109	99	106	96
9	140	118	104	86	130	107	116	100	110	100	106	96
Minimum	128	110	104	80	110	100	110	95	106	96	106	95
Median	136	116	107	91	121	105	115	100	111	100	108	97
Maximum	145	120	110	95	138	126	122	105	114	102	114	102

a) Ohms per 1000 vias; vias annealed 1 hr at 400°C.
b) Ohms per 1000 vias; vias hand-probed. VRTS wafers, 57 mm diameter.

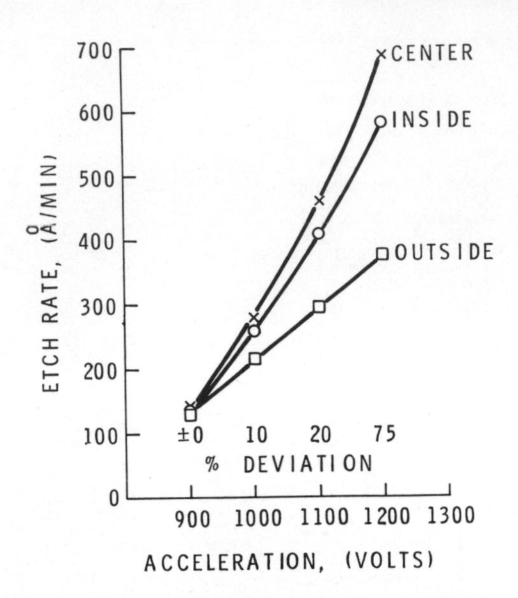

Figure 12. Uniformity function (percentage deviation on a 57-mm surface at 6 rpm). Δ etch rate (angstroms per minute) vs Δ acceleration (volts).

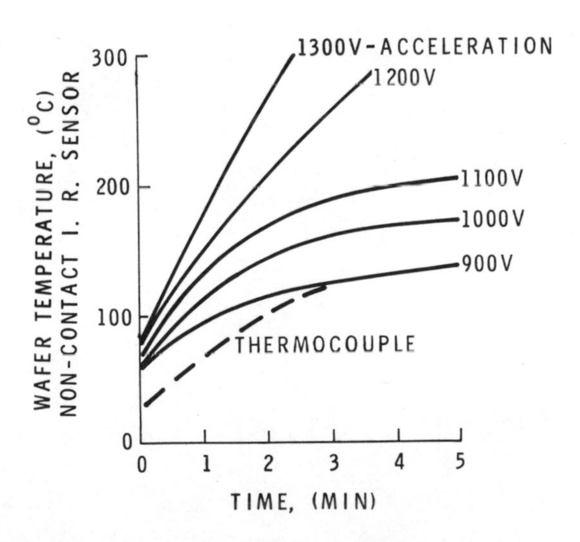

Figure 13. Function of acceleration volts in ion milling. Δ time (minutes) vs Δ temperature (° centigrade).

The SEM photo (Figure 10) shows the actual device that yielded the measurements for Table II. This device had via holes of 6.5 μm^2 and thin-film conductors 5 μm wide (lower conductor) and 10 μm wide (upper conductor). The vias were cleaned by an ion beam of 0.75 W cm^{-2} power density, with an argon pressure of 4 x 10^4 Torr in the vacuum chamber.

Ion-milling effects on charge-sensitive devices such as platinum Schottky barrier diodes (SBD) are tabulated in Table III. The forward and reverse characteristics of the SBD were not changed by ion milling.

CONCLUSIONS

Our work on ion-beam processing has proved that controlled transportation of materials can be obtained on advanced thin-film device surfaces. We were also able to control the transport of the unwanted materials and contaminations from the device surfaces to an easily removable target shield plate. Therefore, we were able to minimize recontamination of the device surfaces during processing and to maintain the cleanliness of the vacuum device after cleaning by ion beam.

The discovery that selective and nonselective removal rates of materials can be obtained merely by changing the ion-beam angle of incidence on geometric surfaces promises further useful extensions of ion beam material transportation processes.

In conclusion, we have designed an apparatus to effectively clean surfaces in a metal evaporator by ion beam. We have obtained a removal rate greater than 300 Å/min^{-1} on photoresist (stencil) lift-off wafers where the stencil temperature rise had to be maintained under 200°C during cleaning. The charge-sensitive device parameters of the SBDs were not affected by ion-beam cleaning.

Table III. SBD kerf data: 15 sites per wafer.

Wafer	Current	Minimum (mV)	Median (mV)	Maximum (mV)	Specification (mV)
1	10 μA	553	555	559	530–550
	100 μA	627	629	635	607–680
	1 mA	767	777	792	763–860
	BV 10 μA (V)	21	22.6	23.4	5.98
2	10 μA	564	567	574	530–550
	100 μA	641	647	652	607–680
	1 mA	808	821	838	763–860
	BV 10 μA (V)	21	21.5	22.2	5.98

REFERENCES

1. S. I. Petvai, R. H. Schnitzel, and R. Frank, Thin Solid Films, 53, 111 (1978).
2. R. H. Brunner, Q. J. Carbone, and W. C. Lester, "Vacuum Coating Apparatus," U. S. Patent 3,921,572, November 25, 1975.
3. A. N. Broers and M. Hatzakis, Scientific American, p. 42 (November 1972).

GLASS CLEANING AND CHARACTERIZATION OF CLEANLINESS

L. L. Hench and E. C. Ethridge

Ceramics Division
Department of Materials Science and Engineering
University of Florida
Gainesville, Florida 32611

The use of a number of different surface analytical techniques for surface characterization of glasses is outlined. Since glass surfaces react with the atmosphere producing surface films which are different from the bulk composition an understanding of the nature of these films is needed in order to understand the meaning of surface cleanliness. The five types of surfaces which can form on glasses are reviewed and examples are given. Glass durability tests require that the surface films be removed prior to testing. A procedure for characterizing the cleanliness of samples for durability tests is reported. The kinetics of silica rich film formation are presented as well as the effects of environmental solution variables on film formation and stability. The use of Auger electron spectroscopy to characterize the cleanliness of glass surfaces cleaned by various procedures is presented. It is demonstrated that sulfuric acid dichromate solution effectively removes surface contamination from glasses. Even for cleaned glasses that exhibit durable surfaces, interaction with water occurs within minutes and the resulting surface compositional gradients must be considered in subsequent use or study of the glass.

INTRODUCTION

The cleaning of glass surfaces is important to the food, pharmaceutical, health care, and electronic industries, for example. Glass cleanliness is also a vital topic from the viewpoint of the research community because of the common use of glass as a laboratory ware, containers, substrates for growth of cultures, and recently in potential applications for nuclear[1] and thermonuclear[2] sources of energy and as a replacement material for teeth, bones, joints and other parts of the body.[3,4] Also, the maintenance, cleaning and restoration of glass antiguities and medieval stained glass windows is a subject of considerable importance to the historical and art community.[5]

Because of the considerable interest in the production and characterization of clean glass surfaces a considerable volume of literature has developed.[6-8] However, many of the early efforts in this field were hampered by the lack of scientific tools for analyzing the structure and composition of the glass surface at sub-micrometer depths. Methods for surface characterization of glass are now available[9] and the objective of this paper is to review briefly the instruments required.

Much early work in the field of cleaning and contamination of glass treated the glass as an inert substrate with a composition equivalent to that of the average chemical analysis of the glass. By the use of surface characterization methods it is now well established that the glass surface at the μm level is usually of a composition significantly different from that of the bulk.[10,11] A compositional continuum often exists between the outer surface and the bulk, or discrete compositional layers may even be present. Thus, a second objective of this communication is to illustrate the five major categories of surfaces that are observed to be present on glass of various compositions[12,13] and indicate how the thickness of the surface films may be determined.

Cleaning and maintenance of glass cleanliness depends upon the type of surface present on the glass. If deep reaction zones of several micrometers are present, adsorbed species and contaminants may be contained within the microporosity associated with such deep layers. Cleaning such layers may require removing the glass reaction zone. Chemical removal of the contaminants or reaction zone is often difficult without producing further attack of the glass network which makes maintenance of the surface condition very difficult. Even glasses with very high chemical durability, such as soda borosilicate compositions, exhibit the problem of maintenance of a cleaned surface state to some degree. A third objective is to discuss the results of a study characterizing cleaning and post-cleaning reactions on a commerical soda borosilicate glass.[14] These results typify the type of fundamental glass surface characterization that is possible with the advanced instrumental tools now available.

ANALYTICAL TECHNIQUES

Compositional changes in the near surface of glass (0-2.0 nm) can be measured with Auger electron spectroscopy (AES), electron spectroscopy for chemical analysis (ESCA), ion-scattering spectroscopy (ISS), or secondary-ion mass spectroscopy (SIMS). Coupling these tools with ion-beam milling yields highly detailed compositional profiles of the intermediate glass surface (2.0-200 nm). Measurement of the average composition to the far surface (0-1000 nm) is now routinely available with electron-microprobe analysis (EMP), energy dispersive X-ray analysis (EDX) in the scanning electron microscope (SEM) or infrared reflection spectroscopy (IRRS). The related merits of these techniques in glass characterization have been recently reviewed.[9]

In our laboratory, AES coupled with Ar ion beam milling is routinely used for characterization of the near surface of the glass article and compositional profiles of the surface to depths of ∿1200 nm when necessary. Details of the experimental procedure are summarized elsewhere.

When reaction layers >0.2 μm are present, the surface characterization tool of choice is usually IRRS for several reasons.[10] From the viewpoint of characterization of cleaning procedures, a major advantage of IRRS is that it requires no vacuum so the surface remains in the as-received condition during analysis. By sequential polishing with accurate weighing of the layers removed, surface compositional profiles can also be obtained with IRRS.[17] When surface reaction layers are many μm deep this is an especially useful procedure since each polishing step removes contaminants as well as the surface active reaction zone which provides the adsorption sites for the contaminants. However, the compositional resolution is generally limited to ±3% and the calculated thickness of the removed layers is often only ±10% because of density variations within the surface layers.

An example of the large changes in IRRS spectra due to aqueous surface attack of a very simple glass are shown in Figure 1. Three spectra of a 33 Li_2O-67 SiO_2 mol % glass (33L) are compared in Figure 1: 1) a freshly polished (600 grit dry SiC) state; 2) after exposure to air containing 85% relative humidity (RH) for 72 h at 40°C; and 3) after exposure to a static aqueous solution which reduces the silicon-non-bridging oxygen-alkali (NSX) peak[18] as lithium ions are removed from the glass surface. A silica-rich film remains

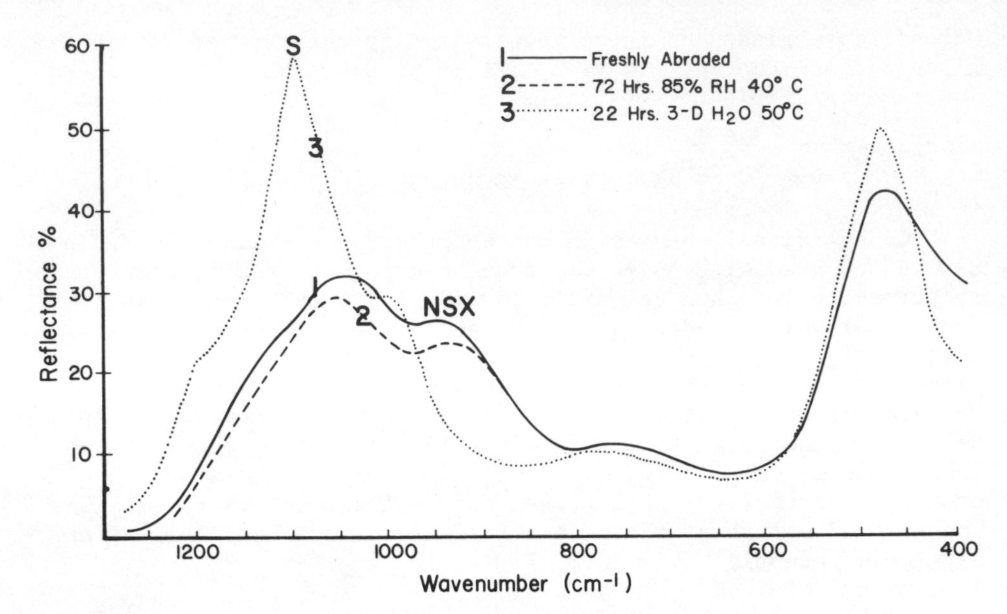

Figure 1. Comparison of corrosion reactions on a 30 Li_2O–70 SiO_2 mol % glass using IRRS.

as evidenced by growth of the S peak attributed to Si-O-Si stretching vibrations.[18] This vitreous silica-like film is very similar to that discussed for cleaned soda borosilicate chemical ware in a later section.

The 85% RH environment does not permit stable silica-rich film formation as shown by the decrease in intensity of both the S and NSX peaks caused by surface roughening. Because the alkali cannot be removed from the surface after proton exchange in the 85% RH, the mode of attack has to be one of stoichiometric or uniform dissolution (curve 2) rather than the mode of selective dissolution, shown in curve 3, when the alkali ion can diffuse from the surface. Thus IRRS can be used to detect the mode of attack, rate of dealkalization and rate of film formation. In addition, the composition of the film can be determined using IRRS by comparison of the frequency location of spectra features with compositional standards.

After removal of 10 μm of the silica-rich reaction film, curve 3, the original spectra of the glass, curve 1 is restored. Polishing sequences followed by IRRS until the original glass are obtained are used to obtain compositional profiles such as shown in Figure 3.

REPRODUCIBLE SURFACES FOR GLASS DURABILITY TESTS

In the characterization of the chemical durability of glass it is mandatory that the surface of glass be as uncontaminated as possible before kinetic studies are performed. Figure 1 demonstrates that leaving samples in air can lead to an attack of the surface by water vapor. If consistent reproducible results are to be obtained with glass durability tests precautions must be taken in order to remove such reaction films from the glass surface. In many glass durability studies, the larger number of samples needed from the same glass batch requires that samples be reused. One must insure, however, that the films which form on the glass surface are completely removed before additional tests are performed. A commonly used procedure is to repolish the corroded glass surfaces with SiC polishing papers. Analysis of the repolished glass surfaces by IRRS can insure that a new "bulk composition" glass surface is produced.

Figure 2 illustrates this technique. The reacted surface of a 14 Na_2O - 10 CaO - 86 mol % SiO_2 glass corroded for one week at 100°C has an IRRS spectrum demonstrating a silica rich surface layer.

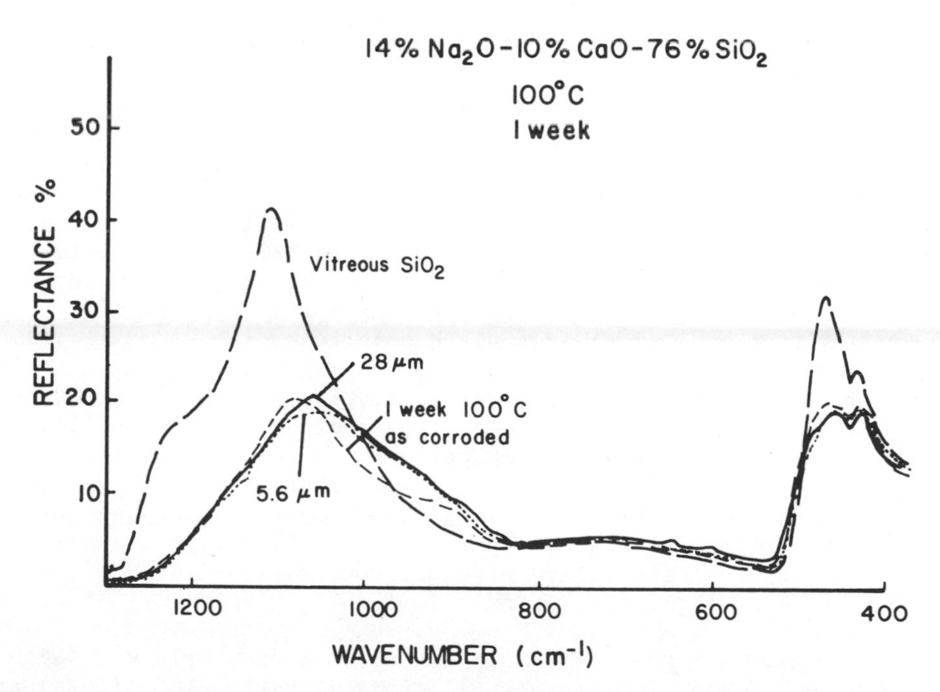

Figure 2. Illustration of IRRS spectral changes as one polishes through a glass surface exposed to deionized water at 100°C.

A systematic polishing through the surface removes successive layers until eventually a spectrum typical of the uncorroded surface is obtained. The thickness of glass removed, as calculated from the change in weight after polishing, is less than 28 μm. A standard polishing procedure for cleaning this glass uses 320 SiC paper and removes approximately 430 μm of glass while a polish with 600 grit paper removes approximately 140 μm of surface. Comparison of these values confirms that the standard polishing of the reacted glass surfaces removes sufficient material to insure that a bulk glass composition is obtained on the surface. Such an instrumental check for efficacy of a cleaning procedure should be performed before reusing any glass article in experiments where the glass is exposed to moisture.

TYPES OF GLASS SURFACE LAYERS

Experiments have shown that exposure of a pristine glass surface to even very small quantities (ppm) of water vapor result in proton-alkali ion and proton-alkaline earth exchange reactions on the glass surface within minutes or less.[16] A compositional gradient within the surface is the consequence. The depth of the gradient is a function of bulk glass composition, time, temperature, partial pressure of H_2O (vapor attack), solution pH (solution attack), electrolytes within the solution, and surface heterogeneities on the glass which tend to localize the attack preferentially.

In general, the surface reaction of different glass systems can be summarized by the five types of surfaces which develop upon exposure to moisture.[12,16] These surfaces are illustrated in Figure 3. One type of surface occurs when equilibrium conditions exist between the glass surface and aqueous solution. Hydration occurs at a very thin depth and minor network silica dissolution establishes the equilibrium concentration of silica in solution (Type I). This type of inert surface behavior is typical of vitreous silica exposed to neutral pH solutions.

A second type of surface that can develop during glass corrosion has a characteristic silica-rich protective film (Type II). This is the type of surface which develops on the binary alkali silicate glasses with higher SiO_2 concentrations, i.e., Li_2O-SiO_2 (<20%); Na_2O- and K_2O-SiO_2 (<15%). The hydrogen ion exchange of alkali ions produces a leached layer rich in silanol groups which polymerize to produce a silica rich vitreous silica-like surface which is stable to solution of $pH < 9$.[17]

Protective surface films form on multicomponent glasses (Type III). When CaO and Al_2O_3 (or other stabilizing glass modifiers) are present in the original glass composition calcium-silicate and alumino-silicate layers can develop on the glass surface by structural modifications within the surface region.[16] In some cases these ions can be precipitated onto the surface forming the stable film.[19] These protective films produce a diffusion barrier to ion exchange reactions and also make the glass network less susceptible to attack by the solution.

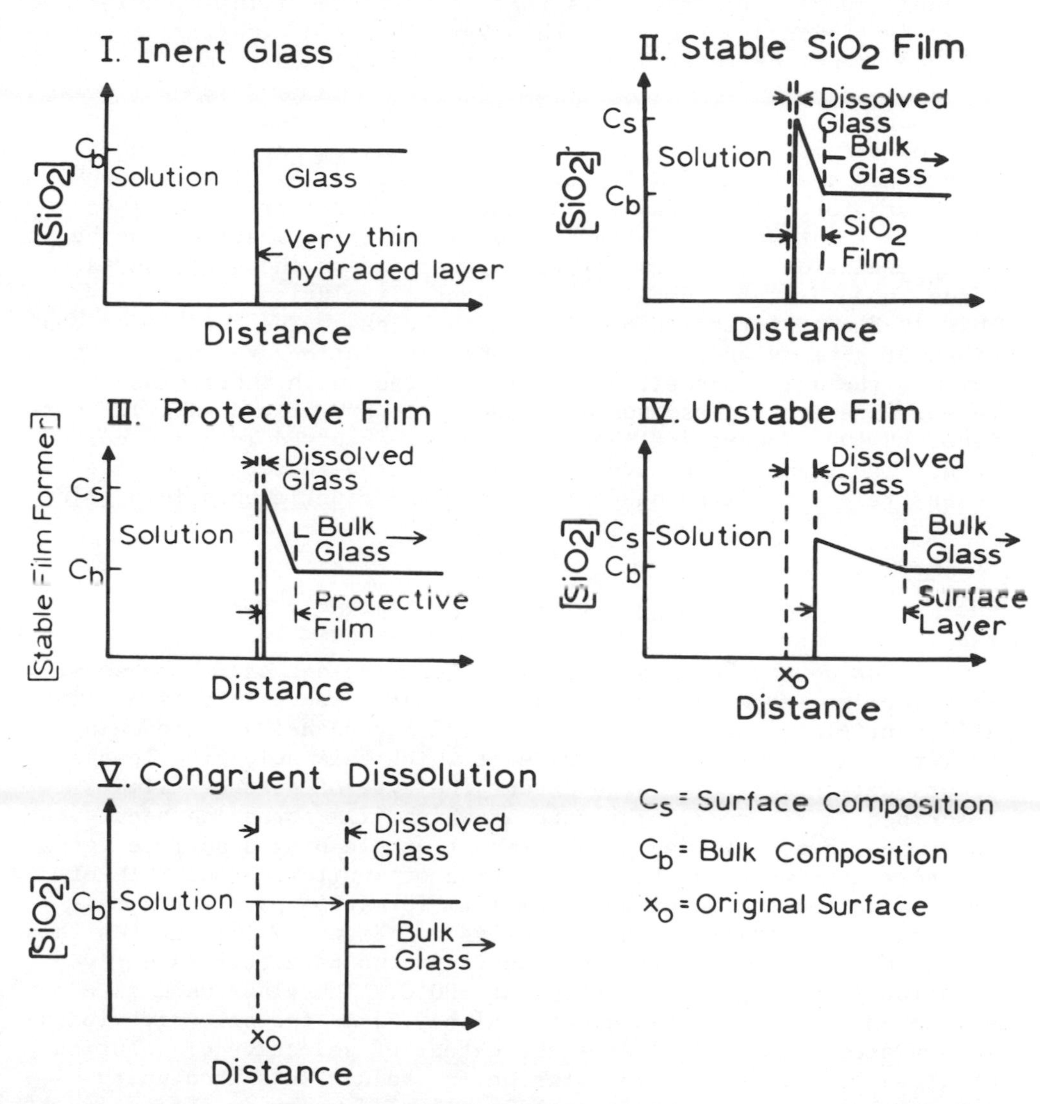

Figure 3. Schematic diagram of five types of corrosion processes observed with different glass systems. In each case the concentration of SiO_2 or stable film former is plotted vs. distance across the corrosion solution glass interface.

Other glasses also produce silica-rich leached layers, however, these glasses are very reactive. This is attributed to formation of unstable surface films (Type IV). Glasses with large concentrations of alkali ions and glasses with large alkali ions are most likely to form unstable films. The large intersticies left after the leaching of large numbers of alkali ions and/or by alkali ions of large radius are filled by water molecules which prevent the polymerization of silanol groups. The result is that an unstable silica-rich leached layer is produced. A glass with a Type IV surface easily degrades into congruent or total dissolution of the glass network.

Congruent dissolution of the glass surface results when the rate of network silica attack leading to silica extraction equals or exceeds the rate of selective leaching of alkali ions (Type V). It occurs for binary alkali silicate glasses at corrosion times when the solution pH > 10 and rapid silica network attack has begun. The presence of scratches or heterogeneities on the glass surface tends to accelerate localized congruent dissolution of the glass. This is due to formation of pits on the glass. The surface compositions of glasses undergoing congruent dissolution are nearly the same as the bulk composition, as is the case with inert glass surfaces. The inert glass surface and the congruently dissolving surface, however, represent the two extremes observed in glass corrosion. The former reacts very little, but the latter rapidly degrades releasing large quantities of glass constituents into solution.

FORMATION AND STABILITY OF SURFACE FILMS

Films on the surface of glasses form for one of two reasons: 1) selective leaching, or 2) dissolution and reprecipitation. The different glass constituents have varying tendencies of leaching from the glass surface. The rate at which this selective leaching occurs determines to a large extent the type of surface film described in Figure 3. The rate at which film formation occurs can determine to a large degree the amount of cleaning a surface needs in order to remove the film. A method of monitoring the rate of film formation can be gained from film formation parameters calculated from complete solution analysis. Figure 4 illustrates this approach. In this figure the changes in two parameters are given as a function of glass reactions at 100°C. The glass used is a simple 33 Li_2O - 67 mol % glass. Alpha, α, is the selective leaching parameter which indicates the extent of selective dissolution of glass constituents into solution.[21] Values less than unity indicate selective leaching while a value of unity is indicative of congruent dissolution. Initially the glass is selectively leached, to a large extent, of Li^+ ions producing the silica-rich leached layer. The formation of the developing leached layer is illustrated

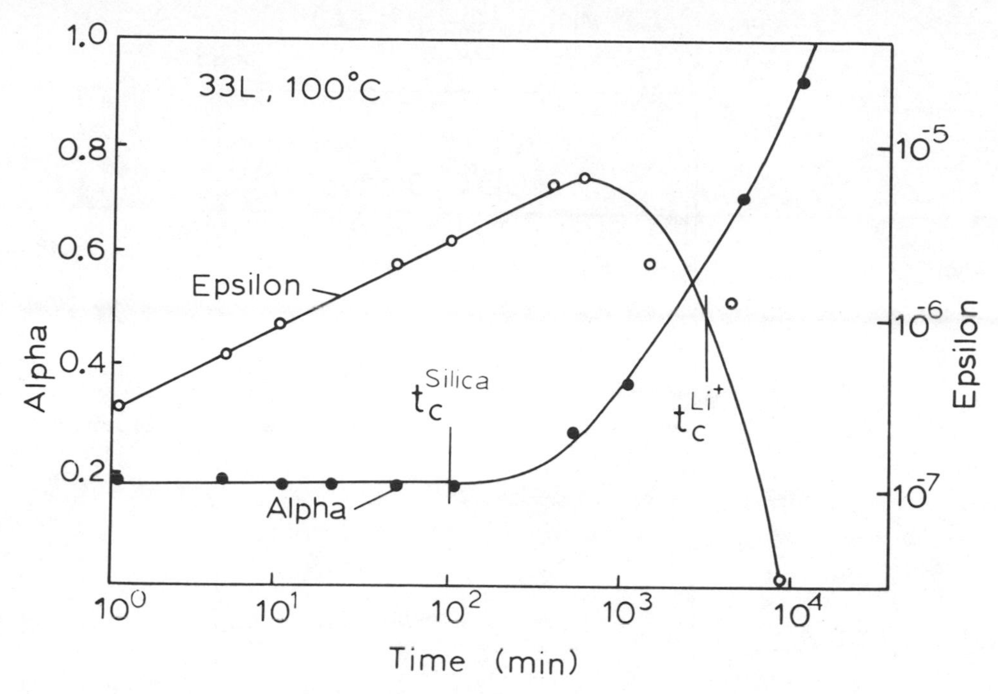

Figure 4. Glass corrosion parameters calculated from the concentration of glass constituents in solution for a 33 mol % Li_2O - 67% SiO_2 composition.

by epsilon, defined in terms of the "excess" silica available for surface film formation.[21] As epsilon increases the thickness of the leached layer also increases. For the initial stages of glass reacting to the environment, development of this type of leached layer predominates. From such calculations one can determine the extent of leaching from the glass surface and also obtain an approximation of the thickness of the reacted layer in the glass[17].

Long exposures of glasses to stagnant solutions, such as the test in Figure 4, results in a large increase in the pH of the corrosion solution. This increase in pH causes a change in corrosion kinetics from a leaching phenomenon to rapid network attack described as congruent dissolution in Figure 3, leaching to a Type V surface. The changes in the corrosion parameters in Figure 4 are typical of these conditions. As the exposure time increases, a point in time is reached when the selective leaching parameter α begins to increase. This is accompanied by a decrease in the value of epsilon, ε. Such changes in the kinetics indicate the initiation of leached layer dissolution and Type V corrosion.

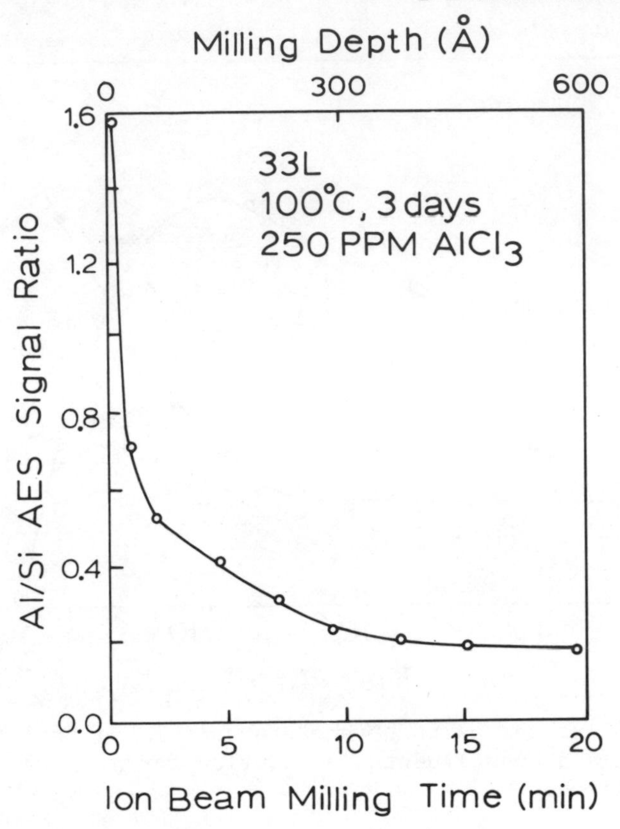

Figure 5. The incorporation of Al from solution into a glass surface as demonstrated by AES.

The presence of cations in solutions can affect film stability. Figure 5 illustrates how multivalent cations such as Al^{3+} become incorporated into the glass surface silica film producing a protective Type III film (Figure 3). In this case a binary 33 Li_2O - 67 mol % SiO_2 glass was placed in a solution containing 250 ppm $AlCl_3$ at 100°C for 3 days. The surface was then examined by AES and ion milling techniques. The Al to Si peak to peak heights are plotted against the depth to which the ion milling proceeded. As can be seen the surface is very rich in Al which is incorporated into the surface to a depth of several hundred Angstroms. Surface films such as these help to protect the glass surface from the environment. Without the presence of Al ions in solution the rate of attack of the glass was greatly increased.[19]

The addition of multivalent cations such as Al^{3+} and Cr^{3+} to bottle washing solutions has long been known to be useful in preventing iron staining.[20] The above results indicate how these cations effectively adsorb onto active centers in the surface silica gel, preventing the iron from complexing with the surface.

USE OF AUGER ELECTRON SPECTROSCOPY FOR GLASS CLEANING CHARACTERIZATION

A recent study was conducted to establish the type of surface resulting from the use of various cleaning procedures on commerical soda borosilicate tissue culture flasks.[14] The experimental procedure for the various cleaning steps examined, reagents, etc., and operation conditions of the AES are described in detail elsewhere.[14,15,16] Part of the results are reviewed herein to illustrate the complexity of both achieving and maintaining a cleaned and characterized glass surface. Maintenance of a cleaned glass surface that is a very stable Type II surface is difficult. Maintaining Type III, IV, and V surfaces in a characterized state is often nearly impossible under normal laboratory conditions.

Figure 6 summarizes some of the most important findings of the previous study.[14] AES spectrum #1 is that of the inside of the as-received borosilicate culture flask #1290 (774 Pyrex, Corning Glass Works). After an acetone rinse, spectrum #2, little change of the surface composition occurs other than removal of a slight amount of chlorine. In contrast, other solvents investigated showed either no effect on the surface chlorine contamination or significant changes to the composition of the glass surface.

The presence of the carbon, chlorine, and sodium on the glass surface was shown to be a very thin film of adsorbed species, C and Cl, and a thin layer of ion exchanged alkali by 1 minute of ion beam bombardment, spectrum #3. Removal of approximately 3.0 nm of the surface in producing spectrum #3 results in the presence of only Si and O on the surface. The Si/O peak ratio of spectrum #3 is similar to that of fused silica and therefore the alkali-proton exchange of the surface results in a stable Type II glass surface. The previous study[14] showed that the depth of this fused silica-like layer was approximately 15 nm.

A common cleaning procedure for glass surface is prolonged soak in sulfuric acid dichromate solution. The AES spectrum after 1 h soak in Chromerge, Manostat Corp., is shown in spectrum #4. The surface structure and composition is nearly equivalent to that obtained from ion beam milling in high vacuum, i.e., spectrum #3. The surface is devoid of contamination and consists only of silica and oxygen. Again, the Si/O ratio corresponds to that detected for a fused silica specimen.

In contrast to the acid cleaning procedure, prolonged exposure to triply distilled low conductivity (LC) water results in a glass

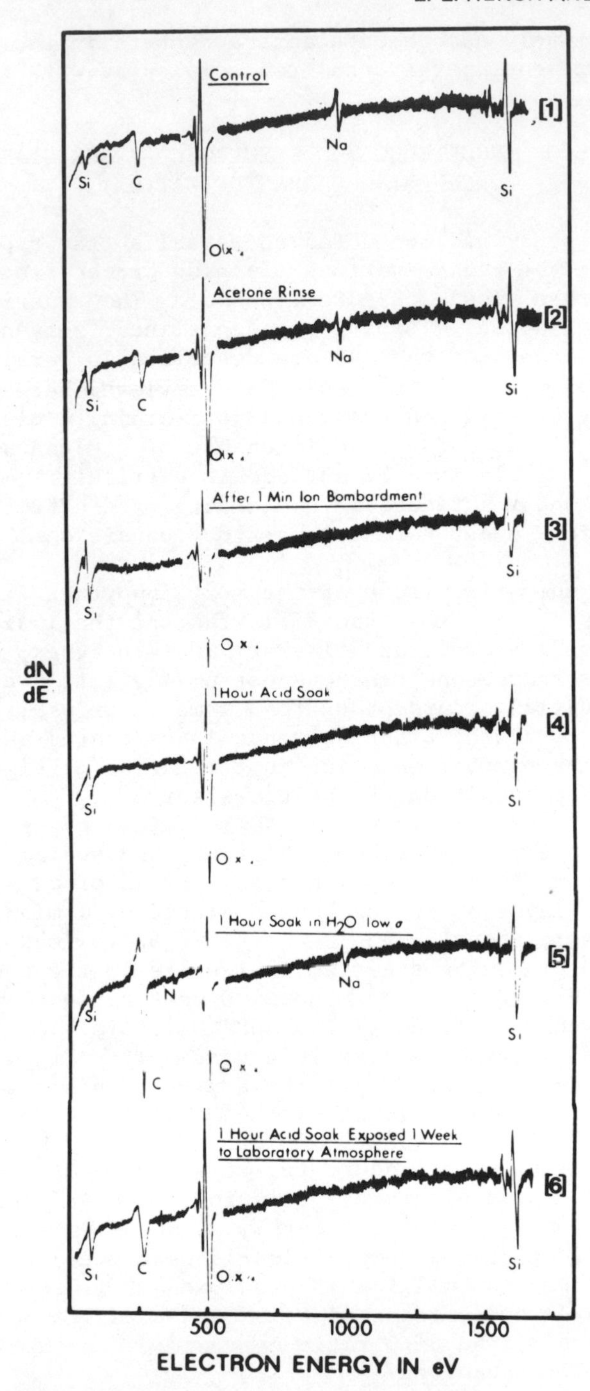

Figure 6. Summary of Auger data following sequence of change on surface of a soda borosilicate glass subjected to various cleaning procedures.

surface showing considerable surface contamination, spectrum #5. A high concentration of carbon is present in addition to a small concentration of chlorine. The sodium has not been removed, and probably has become even a thicker layer of reaction zone. Longer term soaks of up to 5 days[14] show progressive buildup of surface contamination in the LC water whereas the spectra for the acid cleaned sample remains constant.

Finally, stability, maintenance and reproducibility of a cleaned surface is often even more critical than the final surface state achieved after the cleaning process. Spectrum #6 in Figure 6 shows that even after 1 week exposure to laboratory atmosphere, uncovered, the primary contaminant is adsorbed carbon. The extent of the carbon coverage appears to be no greater than that of the as-received samples. More importantly, the oxygen concentration has increased dramatically in the surface indicating that the carbon has perhaps adsorbed as CO_2 vapor. Some of the increase in surface oxygen is also likely to be due to H_2O adsorption as well. The surface is still devoid of sodium and chlorine which indicates stability of the clean surface. Thus, the use of AES has shown that cleaning procedures involving water soaks may produce ion exchanged glass surfaces which can adsorb significant quantities of carbon, nitrogen, and chlorine. This is probably due to formation of a thicker, more porous Type IV glass surface. In contrast, exposure to the acid cleaning solution produces a thinner, less porous Type II surface which apparently is sufficiently oxidized where readsorption of contaminants is difficult. No concurrent contamination of sulfur or chromium from the cleaning solution was present. Confirmation of the vitreous silica like surface formed on tissue culture flasks cleaned in such acid solutions for many years has been obtained using the IRRS method as well.[14] Thus, this study has shown: 1) that cleaning of soda-borosilicate glass in a sulfuric acid dichromate solution followed by a 15-s rinse in LC water and acetone drying produces a stable cleaned glass surface structure similar to vitreous silica; 2) AES coupled with ion beam milling can be used to characterize such cleaning processes and their stability; 3) analysis of cleaning of thicker reaction layers can be followed with IRRS methods in a similar manner.

ACKNOWLEDGMENTS

The authors acknowledge the collaboration of C. G. Pantano, Jr. for the cleaning investigation and the National Institute of General Medical Sciences grant #GM 24854-01 for financial support of research in this area.

REFERENCES

1. J. E. Mendel, in Proceedings of ERDA Workshop on Ceramics and Glass Radioactive Waste Forms," Jan. 4-5, 1977, Germantown, U.S. Government Printing Office, CONF-770102.
2. G. F. Hurley, Ceramic Industry, 8 and 12, April-May (1968).
3. L. L. Hench, R. J. Splinter, T. K. Greenlee, and W. C. Allen, Jr., J. Biomed. Mater. Res. 2, 117 (1971).
4. L. L. Hench and H. A. Paschall, J. Biomed. Mater. Res. 4, 25 (1973).
5. R. G. Newton, Editor, "News Letters, Comite Technique die Vitrearum," University of York, England, Aug. 27, 1976.
6. L. Holland, "The Properties of Glass Surfaces," Chapt. 5, Chapman and Hall, London, 1964.
7. F. M. Ernsberger in "Annual Review of Materials Science," R. A. Huggins, Editor, Vol. 2, pp. 529-572, Palo Alto, Calif., 1972.
8. R. M. Tichane, Bull. Amer. Ceram. Soc. 42 (8), 441 (1963).
9. L. L. Hench, in "Characterization of Materials in Research Ceramics and Polymers," V. Weiss and J. Burke, Editors, pp. 211, Syracuse University Press, Syracuse, N. Y., 1975.
10. L. L. Hench, J. Non-Crystalline Solids 19, 27 (1975).
11. D. E. Clark, M. F. Dilmore, E. C. Ethridge and L. L. Hench, J. Amer. Ceram. Soc. 59, (1976).
12. "Proceedings of the International Congress of Glass," Prague, Czechoslovakia, July, 1977.
13. L. L. Hench and D. E. Clark, J. Non-Crystalline Solids 28, 83 (1978).
14. C. G. Pantano, Jr. and L. L. Hench, J. Testing and Evaluation 5 (1), 66 (1977).
15. G. Y. Onoda, Jr., D. B. Dove and C. G. Pantano, Jr., "Materials Science Research," pp. 39, Plenum, New York, 1974.
16. D. E. Clark, C. G. Pantano, Jr. and L. L. Hench, "Glass Corrosion," Glass Industry, N.Y., in press.
17. E. C. Ethridge, "Mechanisms and Kinetics of Binary Alkali-Silicate Glass Corrosion," Ph.D. Dissertation, University of Florida, 1977.
18. D. M. Sanders, W. B. Person and L. L. Hench, Appl. Spectroscopy 28 (3), 247 (1974).
19. M. F. Dilmore, "Chemical Durability of Multicomponent Silicate Glasses," Ph.D. Dissertation, University of Florida, 1977.
20. E. C. Marboe and W. A. Weyl, J. Amer. Ceram. Soc. 30, 320 (1947)
21. D. M. Sanders and L. L. Hench, J. Amer. Ceram. Soc. 57 (7), 373 (1973).

A SYSTEMATIC APPROACH TO GLASS CLEANING

P. B. Adams

Corning Glass Works

Corning, New York 14830

The rational design of a cleaning process for glass includes a definition of the soil to be removed, an awareness of the alternative methods available for removing dirt, and an understanding of the chemical and physical interactions between the soil, the glass and the cleaning materials. This discussion focuses on the interactions that occur between the glass and the cleaning materials. The glass surface must be understood; it is not necessarily characterized by the bulk glass. Aqueous cleansers react with the surface in much the same way as do other acqueous solvents. Chemical interactions are either of an etching character, as with alkaline cleansers; or leaching character, as with acidic cleansers. The action of detergents on glass can be severe since they are compounded from various aggressive chemicals; certain additives may mitigate some of the harmful effects on glass. In general, non-acqueous solvents affect glass the least; they may be difficult to remove however. Mechanical cleaning steps may damage the surface. With a full understanding of the effects of these various alternatives it is possible in most cases to design cleaning procedures that are simple, yet meet the particular "cleanliness" objective at hand.

INTRODUCTION

A great variety of situations give rise to the question of how best to clean a glass surface. The housewife wants her dishes and her windows "squeaky clean". The process engineer needs an efficient and cost effective method for cleaning an optical component. The scientist must have a surface of quantifiable cleanliness to perform his experiment. The demands, and thus the definition of "clean", are dependent on the end use.

A few years ago, a task group in the Glass Committee of the American Society for Testing Materials (ASTM C14.03) decided to try to organize a logical approach to the subject. It was recognized that many cleaning procedures are an art, but that technical principles could be applied. The first step was to publish a bibliography that provided access to the literature by contaminant type, evaluation technique, product and cleaning method.[1] The committee wrote two specific methods for evaluating cleanliness - by water vapor condensation[2] and by contact angle.[3] The committee then attacked the problem of giving general guidance to the person who wished to design a glass cleaning process.[4]

An approach to cleaning glass should include at least four elements: (1) a definition of the soil to be removed, (2) a knowledge of the alternative methods that are available for removing the soil, (3) an understanding of the interactions between the soil and the glass causing it to adhere and (4) an understanding of the effects that the cleaning procedure can have on the glass surface. This discussion will focus on item 4 above. A more complete description of alternatives available for cleaning is available.[1,4] A number of papers in these proceedings volumes relate specifically to glass[5,6,7,8]; and most of the other papers are applicable even though glass may not be specifically discussed.

THE NATURE OF GLASS AND THE GLASS SURFACE

The chemical composition of the glass to be cleaned is of extreme importance in selecting a process that does not damage the surface. Compositions of various silicate glasses differ widely as shown in Table I, although glass numbers 1 and 2 are certainly most common. Each cleaning process must be tailored to the glass composition and product application in question.

The structure of glass is represented schematically in Figure 1. The surface, however, may be quite different from the bulk in both chemical composition and structure.[8] The "normal" glass surface will be covered with various radicals and monolayers of molecular material as suggested in Figure 2. Hydroxyl ion, water molecules and organics are probably found most commonly.

Table I. Some Technical Glass Compositions

Oxide (wt. %)	Glass Number									
	1	2	3	4	5	6	7	8	9	10
SiO_2	73	81	16		27		30	49		
B_2O_3		13	10	20		7	10		29	
Al_2O_3	1	2				18	10	12	5	18
Na_2O	17	4			1	14		3		
K_2O			1		1					
Li_2O			2							
CaO	5		8					11		
MgO	4							10		
BaO			15				50			
La_2O_3			13	30						
ZrO_2			5							
TiO_2			19					3		
ThO_2				18						
Ta_2O_5				5						
Nb_2O_5				20						
WO_3				7						
ZnO			11						56	7
PbO					71	30				
Pb_2O_5						33		1		73
Fe_2O_3								11		
V_2O_5									10	
CdO										2

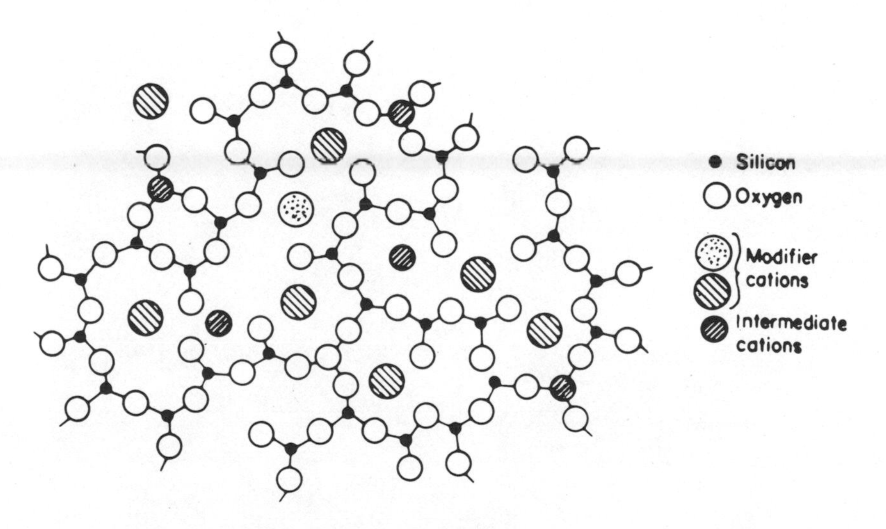

Figure 1. Two-dimensional representation of the atomic arrangement in bulk glass.

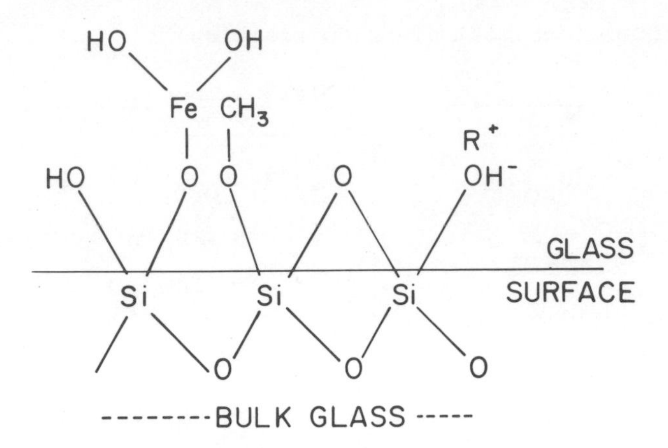

Figure 2. Schematic representation of a glass surface.

Glass is not necessarily smooth on a micro scale, although it may appear to be so to the eye. The surface may be porous as a result of leaching or other causes; this can result in microcracks. There may be surface deposits and irregular topography resulting from volatilization-recondensation processes that occur during glass forming. Micrographs of some glass surfaces are shown in Figure 3.

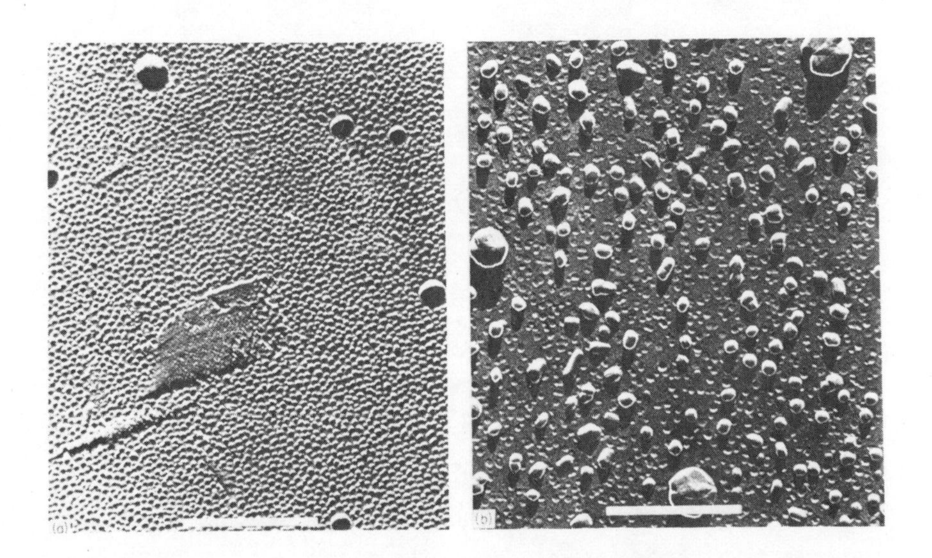

Figure 3. Structures of borosilicate glass surfaces (X27000) white bar = 1 μm: (a) spongy layer after chemical reaction; (b) crystallites on an "as-formed" surface. (from Adams, ref. 16)

For further and more detailed information on glass, see references on composition,[10,11], surfaces[9,12] and surface reactions.[9,12]

ALTERNATIVE APPROACHES TO CLEANING GLASS

As detailed in the ASTM procedure[4], there are four general alternatives for designing a systematic approach to glass cleaning. These involve use of aqueous solvents, detergents, non-aqueous solvents and/or mechanical techniques. Specific components of a process may be used sequentially or concurrently.

Aqueous solvents are perhaps most common. They range from plain water to strong chemicals.

Detergents are usually a mixture of inorganic and organic compounds. They are characterized by their Dr. Jekyl-Mr. Hyde characteristic to adsorb strongly with one end of the molecule, yet dissolve readily with the other end of the molecule. A good detergent will adsorb on both the soil and the glass via the reactive end of the molecule and dissolve via the water soluble end of the molecule. It will lower surface tension, emulsify and deflocculate.[13] The detergent may be either anionic, cationic or amphoteric. It will often contain other additives to enhance its effectiveness and diminish any harmful effect on the glass.

Non-aqueous solvents include hydrocarbons, and halocarbons, as well as mixtures of these with each other or in some cases with aqueous solvents. They are particularly useful for removing oils, non-polar compounds and particulates in the 0.1 to 1000 micron size range, especially if they are bonded principally by physical forces.

Mechanical methods of cleaning include a host of techniques. Scrubbing, wiping or brushing may be most logical in many instances. Polishing and buffing are employed when it is necessary to remove a small amount of surface glass in order to dislodge the dirt. Ultrasonics, high pressure sprays and air jets depend on fluid displacement to carry away dirt. Combustion and ion bombardment such as the plasma cleaning technique sometimes in combination with ozone or UV radiation,[14] are used to enhance the degradation of organic material. Often an entrapment material such as collodian or gummed tape is effective.

The effects of these methods on the glass surface will vary depending on the glass in question.

INTERACTIONS BETWEEN GLASS AND AQUEOUS SOLVENTS

An ideal aqueous solvent will either dissolve the soil (and

$$2x\ NaOH + (SiO_2)_X \rightarrow xNa_2SiO_3 + xH_2O$$

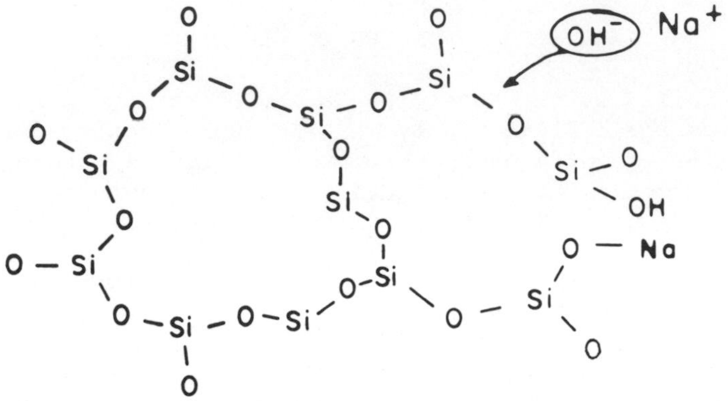

Figure 4. The etching process.

leave the substrate glass unaffected) or dissolve the substrate glass to undercut the soil (and leave the substrate glass smooth). Both processes occur in practice; the glass is always affected to a greater or lesser degree.

Basic solutions (and hydrofluoric acid) etch silicate glasses. Some acids may "etch" other glasses that are composed principally of acid soluble oxides. A schematic representation of the etching process is shown in Figure 4. This is a first order reaction of the type $Q = at$, where "Q" is the extent of corrosion, "t" is the time and "a" is a constant. Temperature dependence is generally in agreement with the Arrhenius expression. As a rule of thumb, corrosion rate doubles for every ten degrees centigrade temperature increase.

Acidic solutions leach silicate glasses. A schematic representation is shown in Figure 5. This is an ion exchange, diffusion controlled process, in which the mobile glass modifier elements are selectively removed leaving the silicate structure intact. It is a reaction of the type $Q = at^{\frac{1}{2}}$. As a rule of thumb, the temperature effect doubles for about every twenty degrees contigrade.

The solubility, in aqueous solvents, for a variety of oxides are listed in Table II. Now consider some glasses from Table I. Glass No. 2, a borosilicate, could be readily cleaned in hydrochloric acid without damage; although it contains a small amount of leachable modifier oxide - the 4% Na_2O is not enough to cause a problem. However, a long exposure to hot concentrated alkali would severely etch

Table II. Solubilities[a] of Oxides.

Oxides of	HF 49%	H_2SO_4 96%	HNO_3 70%	HCl 37%	H_3PO_4 85%	NaOH 50%
Al	s	s	s	s	i	s
Sb	i	i	i	s	i	s
As	s	s	s	s	s	s
Ba	i	i	s	s	s	s
Be, Bi	s	s	s	s	s	i
B, Cd	s	s	s	s	s	s
Ca	i	s	s	s	s	s
Ce	i	s	i	i	i	i
Cr	i	i	i	i	i	i
Co, Cu	s	s	s	s	s	i
Er, Eu, Gd	i	s	s	s	s	i
Ga	s	s	s	s	s	i
Ge	s	s	s	s	s	s
Au	i	i	i	i	i	i
Hf	s	i	i	i	i	i
Fe	s	s	s	s	s	i
La	i	s	s	s	s	i
Pb	i	i	s	i	s	s
Li	s	s	s	s	s	s
Mg	i	s	s	s	s	i
Mn	s	s	s	s	s	i
Mo	s	s	i	s	s	s
Nd	i	s	s	s	s	i
Ni	s	s	s	s	s	i
Nb	s	i	i	i	i	i
Pd	s	s	i	i	i	i
P	s	s	s	s	s	s
Pt	i	i	i	i	i	i
K	s	s	s	s	s	s
Pr, Pm, Rh Rb, Ru, Sm	i	s	s	s	s	i
Se	s	s	s	s	s	s
Si	s	i	i	i	i	s
Ag	s	s	s	i	s	i
Na	s	s	s	s	s	s
Sr	i	i	i	i	i	i
Ta	s	i	i	i	i	i
Te	s	s	s	s	s	s
Tl	s	s	s	s	s	i
Th	s	s	i	i	i	i
Sn	s	s	s	s	s	s
Ti	s	s	i	s	i	i
W	s	i	i	i	i	s
U	s	s	s	i	i	i

Table II. Solubilities[a] of Oxides (Cont'd)

V	s	s	s	s	s	s
Yb, Y	i	s	s	s	s	i
Zn	s	s	s	s	s	s
Zr	s	s	i	i	i	i

[a] s = soluble; i = insoluble.

the surface of this glass, somewhat in the manner shown in Figure 6. On the other hand, glass No. 4 could probably be cleaned in sodium hydroxide with little glass damage because a major share of the oxides, (of lanthanum, thorium, tantalum, niobium and tungsten) are virtually insoluble in this alkali. If sulfuric acid were used, the surface might be smoothly etched if the solution were slightly agitated to remove any small amount of precipitated reaction products as they formed. Glass 8 would probably be badly leached by sulfuric acid leaving an insufficient (49%) silica network which would crack on drying in a manner similar to that shown in Figure 7.

Even if all dirt is removed and the surface is visibly unaltered, it may not be suitable in some applications since a glass surface "getters" various radicals and molecules. Anions such as OH^-, Cl^- and CH_3^- react with surface hydroxyl groups. Cations also adhere, dependent on the free energy of the metal-surface bond.

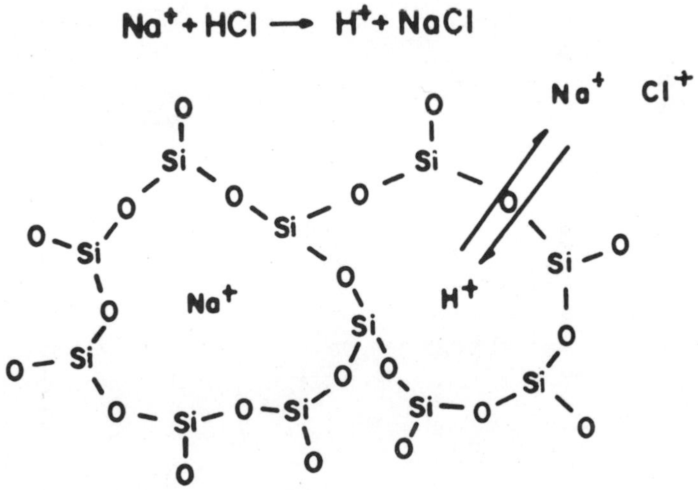

$$Na^+ + HCl \rightarrow H^+ + NaCl$$

Figure 5. The leaching process.

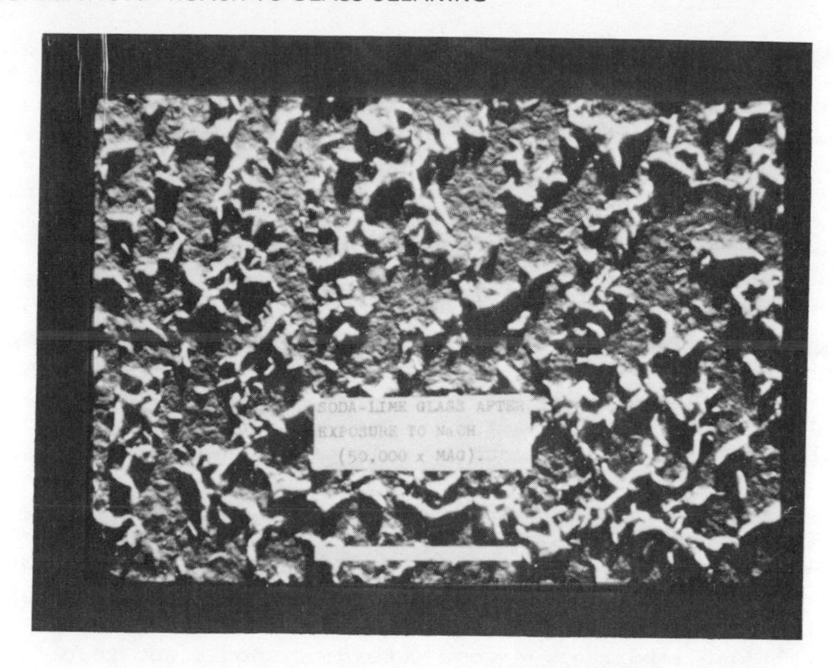

Figure 6. Electron micrograph of an etched glass surface. (white bar = 1 μm)

Figure 7. Optical micrograph of a leached glass surface showing cracked surface (30X magnification).

The strength of adhesion to a silicate surface decreases in the order Fe^{+++}, UO_2^{++}, Al^{+++}, Cr^{+++}, La^{+++}, Cu^{+++}, Co^{++}, Zn^{++}, Cd^{++}, Ag^{+}, Ni^{++}.[15] Generally speaking, a pure silica glass will adsorb ions weakly, a borosilicate more strongly and a soda-lime more strongly yet. Note that the commonly used dichromate cleaners leave a considerable amount of adsorbed chromium ion on the surface.

Water is not necessarily a good solvent for producing a clean glass surface. It is fairly effective in removing acid residues from prior cleaning steps but less useful in removing alkaline residues, whether they are dirt, contaminants or the remains of prior cleaning efforts. If it is used as the final cleaning step, evaporation is a poor way to remove the water since non-volatiles remain. A gas jet displacement or spin dry technique is preferable.

INTERACTIONS BETWEEN GLASS AND DETERGENTS

A good detergent should adsorb preferentially on the glass surface thus blocking soil adhesion, complex with and/or emulsify the soil, defluocculate, and lower the surface tension of the solvent.[13] Most important, a good detergent should not react chemically with the glass and should be readily desorbed. Since these criteria are somewhat contradictory, they are most difficult to fulfill.

Some detergent components react chemically; the oxide in the "dirt" and the oxide in the glass behave similarly. Phosphates and EDTA are especially corrosive to some glasses. Alkaline detergents which often have pH 12 or 13, are effective "etchants" of silicate glasses. The relationship between pH and corrosivity of a variety of detergents is shown in Figure 8. This effect can be lessened by the addition of inhibiting agents such as silicates which are often found in common detergents. When soft or deionized water is used as the detergent solvent, attack on the glass will usually be much more severe than when tap water is used; however, cleaning effectiveness is probably better.

In general, detergents can be considered to act in the same manner as "aqueous" solvents. Organic complexing components may add another dimension, but their effect is usually secondary.

INTERACTIONS BETWEEN NON-AQUEOUS SOLVENTS AND GLASS

The obvious reason for selecting a non-aqueous solvent is its relative chemical inactivity toward the glass surface. The principle disadvantage is its tendency to adsorb. Almost by definition, it adsorbs on the glass more tenaciously than the soil since its success usually lies in the fact that it can compete

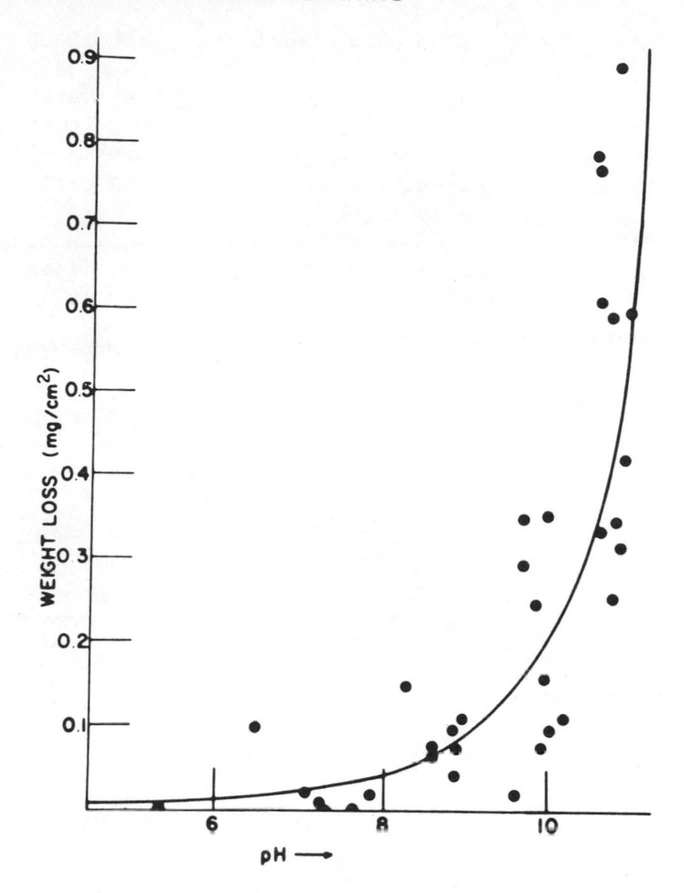

Figure 8. Attack on a borosilicate glass by various detergents as a function of pH. (after Ross, ref. 17)

preferentially for the adsorption sites on the glass surface.

Inert gas displacement or spin drying is preferred to evaporation for removing such a solvent in order to avoid the inorganic residue problem cited for water. Combustion usually removes the solvent residues but may condense the glass surface silanols to siloxanes rendering the surface more hydrophobic.

INTERACTION BETWEEN MECHANICAL CLEANING METHODS AND GLASS

If an abrasive technique is used to clean a glass surface, it will obviously be altered if the abrasive is harder than the glass. A slurry of a soft abrasive such as calcium carbonate may be useful in some applications. Polishing with a harder material such as cerium oxide may also be effective in certain cases.

Displacement techniques are useful since they have virtually no effect on the glass surface. A high pressure spray, a spinning technique or a gas jet method removes soil and solvent. In combination with an anti-static device, a clean part is obtained that stays clean for a finite time.

Combustion techniques alter the glass surface if inorganic residues remain, if silanols condense to siloxanes or if the glass softens. The latter can be an advantage if it "heals" a previously damaged surface or a disadvantage if residues fuse to the surface.

Ion bombardment may alter the surface by molecular ablation or localized heating. The latter may cause mobile glass constituents to migrate.

CONCLUSION

Scientific principles can be used to design optimum cleaning processes. To achieve this, it is essential to understand the effect of the cleaning process on the properties of the resultant surface.

ACKNOWLEDGEMENTS

The work and support of the Clean Glass Task Force in ASTM Committee C14 (on glasses), Subcommittee C14.03 (on Chemical Properties) has been an essential factor in the coalescence of the ideas expressed in this paper.

Mrs. T. S. Magliocca and Mr. W. R. Strzegowski, Corning Glass Works Technical Staff, prepared the solubility data presented in Table II.

REFERENCES

1. P. B. Adams, J. Testing Eval. 5, 53 (1977)
2. Method C-812-75 in "Annual ASTM Book of Standards", Part 17, American Society for Testing and Materials, Philadelphia, PA, 1978.
3. Method C-813-75 in "Annual ASTM Book of Standards", Part 17, American Society for Testing and Materials, Philadelphia, PA, 1978.
4. Proposed New Standard Method prepared by ASTM Committee C14.03 for "Annual ASTM Book of Standards", Part 17, American Society for Testing and Materials, Philadelphia, PA, 1978.
5. L. H. Hench and E. C. Ethridge, This proceedings volume, pp. 313-326.

6. B. D. Ratner, J. J. Rosen, A. S. Hoffman and L. H. Scharpen "Surface Contamination: Its Genesis, Detection and Control". K. L. Mittal, Editor, Vol. 2, pp. 669-686, Plenum Press, New York, 1979.
7. C. E. Bryson, L. H. Scharpen and P. L. Zajicek, "Surface Contamination: Its Genesis, Detection and Control". K. L. Mittal, Editor, Vol. 2, pp. 687-696, Plenum Press, New York, 1979.
8. I. Stowers and H. G. Patton, This proceedings volume, pp. 341-349.
9. P. B. Adams in "Ultrapurity", M. Zeif and R. Speights, Editors, pp 293-351, Marcel Dekker, Inc., New York, 1972.
10. J. P. Williams and D. E. Campbell in "Encyclopedia of Industrial Chemical Analysis" Vol. 13, pp. 403-487, John Wiley and Sons, Inc., 1971.
11. M. B. Volf, "Technical Glasses", Sir Isaac Putman and Sons, Ltd., London, 1964.
12. L. Holland, "The Properties of Glass Surfaces", Chapman and Hall, London, 1964.
13. A. W. Adamson, "Physical Chemistry of Surfaces" 3rd ed., pp 474-483, John Wiley and Sons, New York, 1976.
14. J. Vig, This proceedings volume, pp. 235-254.
15. D. L. Dugger, J. H. Stanton, B. N. Irby, B. L. McConnell, W. W. Cummings and R. W. Maatman, J. Phys. Chem. 68, 757 (1964).
16. P. B. Adams, "The Biology of Glass", New Scientist, 1, 25, (1969).
17. A. R. Ross, Corning Glass Works, (Personal Communication),

TECHNIQUES FOR REMOVING CONTAMINANTS FROM OPTICAL SURFACES*

I. F. Stowers and H. G. Patton

Solid State Laser Program
Lawrence Livermore Laboratory
Livermore, California 94550

The advent of very high power solid state lasers for fusion research has created a need for extremely clean large optical elements. The contaminants of concern are particles or agglomerations of particles exceeding 1 μm diameter. The primary concern is the elimination of these particulates from optical surfaces before the application of evaporated or sputtered coatings, before the optic is installed into the laser system, and during periodic maintenance.

Cleaning methods are needed which are extremely effective on small and large surface areas, relatively rapid, and easily monitored. In general detailed knowledge of the composition of contaminants is of little importance except as necessary to aid in determining their source. Since inadvertant recontamination is inevitable, only cleaning methods applicable to all types of contaminants have been considered.

Particle removal procedures such as plasma cleaning, ultrasonic agitation of solvents, detergents, solvent wiping, mild abrasives, vapor degreasing, high pressure solvent spraying and others have been evaluated and the results are reported here and elsewhere[4,5,9,12]. Wiping with a lens tissue

*Research performed under the auspices of the U.S. Department of Energy by the Lawrence Livermore Laboratory under contract number W-7405-ENG-48.

wetted with an organic fluid and high pressure liquid spraying are the only methods by which particles as small as 5 μm can be effectively removed. All of the other methods tested were found to be at least two orders of magnitude less effective at removing small insoluble particles.

INTRODUCTION

Several ultra-high power laser systems have been constructed by the University of California at the Lawrence Livermore Laboratory and the Los Alamos Scientific Laboratory to prove the feasibility of driving thermonuclear fusion reactions by the inertial confinement of a plasma. The Shiva[1] laser system at the Lawrence Livermore Laboratory is constructed of twenty individual beams arranged to impinge on a fusion target simultaneously. This Nd:Glass laser system contains 1600 coated and uncoated optics and Nova[2], the next generation laser system, will contain approximately 5000 optical elements. To extract maximum power and energy from the laser, the beam intensity is maintained only slightly below the damage threshold of the coated and uncoated optical surfaces. Contaminants on these surfaces therefore act as sites for the initiation of damage to the bare optics or the anti-reflection coatings. Optically absorbing films and nontransparent particulate contaminants attenuate the beam and add intensity variations to the beam profile. Both of these effects reduce the amount of focusable energy reaching the target[3] and therefore must be avoided. A very significant problem therefore is the cleaning and maintenance of literally thousands of components in the laser system[4,5].

A contaminant distributed as a uniform film on an optical surface can be tolerated if its net beam attenuation does not exceed 0.01% per surface. This is derived from a desire to keep the cumulative effect of films on the approximately 100 surfaces in each beam from exceeding 1% attenuation. A typical size distribution of contaminants at a concentration of only 15 particles/cm^2 >10 μm diameter obscures .01% of an optical surface (calculated by integration of the known size distribution). We are primarily concerned with reducing the surface particle concentration below this level. This is a very clean surface and is very difficult to verify in practice. Multiple pass attenuation measurements taken before and after cleaning have been reasonably successful in measuring such low losses.

We have investigated many surface cleaning procedures to quantify their cleaning speed, particulate removal efficiency, and the cleanliness levels that are ultimately attainable with each procedure. A search of the literature indicates that quantitative cleaning evaluations of this type have not been performed and in general particle removal mechanisms do not appear to be well understood. Although measurements of the binding forces between particles and surfaces have been reported[6,7,8], the work has not been extended to include the application of this information to overcoming the binding forces. Small particles are strongly bonded to surfaces by Van derWalls and electrostatic forces far exceeding the particle's weight, and binding forces exceeding 10,000 times the particles weight[8] have been measured.

EXPERIMENTAL PROCEDURE

Because of a general lack of information by which cleaning methods using vastly different particle removal mechanisms could be directly compared, the authors undertook such a research program. Briefly, an artificially contaminated surface was created by uniformly distributing 5 µm Al_2O_3 particles onto both flat optical surfaces and polished metal surfaces[9]. The contamination concentration was determined by microscopic counting both before and after cleaning. The contaminated samples were then cleaned by the seven methods listed in Table I and the efficiency of cleaning ((initial - final)/initial concentration) and lowest achieved particle concentration tabulated.

Aluminum oxide was chosen as the contaminant because it closely resembles airborne contaminants in that it 1) it is insoluble in common solvents and therefore tests the effectiveness of particle removal as opposed to removal by chemical mechanisms, 2) has a reproducible size distribution, and 3) is as difficult to remove as any naturally occurring mineral such as silica. The basic size of 5 µm represents a size which is relatively easy to count microscopically and represents the size range which is by far the most difficult to remove. Particles exceeding 200 µm are relatively easy to remove by all the methods listed in Table I and therefore are not of practical interest.

The choice of solvent or fluid for a cleaning procedure may have a profound effect on the results. This is particularly true if the contaminant composition is known, then a solvent should be selected in which the contaminant is soluble. Solvent extraction will not be discussed further because with mixed contaminants the extraction process will only aid in the removal of soluble contaminants leaving behind the insoluble species. The choice of a fluid also influences the cleaning efficiency based on its evaporation rate. Adequate flushing action is difficult to achieve with rapidly evaporating fluids unless large quantities are applied to the surface. Rapid evaporation has the effect of

Table I. Cleaning Methods Showing Cleaning Efficiency ((Initial-Final)/Initial Concentration) and Lowest Particle Concentration Achieved with 5 μm Alumina Artificially Contaminated Glass and Metal Surfaces.

Cleaning Method			Removal Efficiency for Particles >5 μm	Lowest Achieved Particle Concentration/cm² >5 μm
1. Scrubbing and dragging with lens tissue using ethanol and acetone			99.6 – 99.98%	2 – 40
2. Spraying with liquid solvent, 5-30 s duration	Trichlorotrifluoroethane (Freon TF*)	345 kPa (50 psi)	97% 3% for >1 μm particles	1500
	" "	6.9 MPa (1000 psi)	99.7 – 99.9% 81% for >1 μm particles	10 – 35
	Water	17 MPa (2500 psi)	98 – 99.5%	2 – 60
3. Adhesive strippable coating Scotch 2253[†]			95 – 98%	500
4. Ultrasonic agitation of Freon TF, 1–2 min duration			24 – 92% 1% for >1 μm particles	9000 – 70000
5. Sequential cleaning operation Ultrasonic agitated TWD-602*/followed by Freon TF low pressure spray/vapor degrease/immersion in boiling solvent/ultrasonic agitation in rinse tank			92%	4000
6. Compressed gas jet for 10 s duration Micro-duster** 690 kPa (100 psi)			50 – 61%	5000 – 5800
7. Vapor degreasing in Freon TF			11 – 28%	65000 – 80000

* E. I. duPont de Nemours & Co. Inc., Wilmington, DE
† 3 M Co., St. Paul, MN
** Texwipe Co., Inc., Hillsdale, NJ

redepositing those suspended or soluble contaminants as small concentrated spots of contaminants when an isolated drop evaporates. These concentrated contaminant spots are particularly noticeable on optical surfaces.

EXPERIMENTAL RESULTS

Vapor Degreasing

Vapor degreasing creates a gentle flushing action on an optical surface when immersed in the fluid vapors. Soluble contaminants can be expected to be effectively flushed from the surface but small insoluble particles are simply redistributed and very few are ultimately removed as shown in Table I. The flushing time is also comparatively short in duration because of the relatively rapid heating of the glass surface.

We have observed the reentrainment of particles from the liquid solvent to the vapor and their deposition onto clean surfaces in a vapor degreaser. This occurred in a vapor degreaser with a high boiling rate created by electric immersion heaters. This effect can be reduced by preventing rolling boiling and by continuous filtration of the degreaser fluid.

In general, vapor degreasing is not needed nor recommended for optics cleaning unless it can be experimentally shown that the process indeed removes something not removed by one of the procedures to be discussed below.

Ultrasonic Cleaning

The ultrasonic agitation of fluids has long been used as a means of improving the effectiveness of the simple immersion cleaning and solvent extraction process. Simple solvent immersion is completely ineffective for removing insoluble 5 μm particles, but when aided by ultrasonic agitation can approach 60% removal as shown in Table I. The length of immersion time or the choice of a fluid appears to have little effect on improving these experimental results[9]. In fact, long immersion times (30-60 minutes) appear to reverse the cleaning effect and surfaces generally show more contaminants than after 15 seconds cleaning.

Ultrasonic cleaners with which the authors are familiar are totally inadequate in their filtration rate. Their pumps are generally sized to circulate the reservoir once per hour or less. Since very little filtered fluid comes into contact with the surface during immersion, it can be expected that particles are removed and then redeposited because of the lack of flushing action. Many of the ultrasonic cleaners we have investigated extract their fluid only from the bottom of the tank thereby accumulating buoyant contaminants on the fluid surface. The cleaned part must then be extracted through this surface layer and will again become contaminated. Ultrasonic cleaning therefore can only be recommended for gross cleaning and is not recommended for the precision cleaning of optics.

Strippable Adhesive Coatings

The use of a strippable adhesive coating to remove particles is a somewhat unconventional cleaning procedure but has proven to be quite effective at removing even small particles. The coating may be applied to the surface with a brush or by dipping. The coating is removed by peeling from the surface after thorough drying. Coatings such as collodion (cellulose tetranitrate in ether and alcohol) and commercial coatings such as Scotch 2253* Strip-Coat have been evaluated with results shown in Table I. A drawback to these adhesive coatings is that a thin organic film remains on the surface after the thick plastic coating is removed. This film may be subsequently removed by using an organic solvent, as in a vapor degreaser. If the organic film is composed of a low molecular weight constituent of the coating formulation, it may be possible to reformulate the strippable coating to minimize this undesirable effect.

A clean strippable coating could prove to be very helpful for cleaning mounted optics. Substantial effort is expended in large laser systems recleaning large optics such as turning mirrors exposed to airborne contaminants. Disassembly of most components is time consuming and exposes the optic to unnecessary physical abuse. In-place cleaning is particularly difficult inside inherently long and complex hardware such as disk amplifiers and polarizer-Faraday rotator assemblies. A strippable coating that could be applied by any laser technician and easily removed when dry would greatly improve system reliability and overall system cleanliness.

Cleaning by Viscous Shear Forces

The shear force exerted by a moving fluid on a small particle on a surface has been reported[7,10] in the literature. If the shear force exceeds the adhesion force holding the particle

*Minnesota Mining and Manufacturing, St. Paul, Minnesota.

to the surface, the particle will be removed (reentrained) and suspended in the turbulent fluid. This has the dual advantage of both dislodging small particles and applying a flushing action which carries the particle away from the vicinity of the surface. For a given free stream fluid velocity, the more viscous and dense the fluid, the more momentum will be imparted to the attached particle. This is verified by comparing the cleaning efficiency of compressed fluorocarbon "dusters" with liquid solvent spraying. Table I indicates that liquid spraying at 345 kPa (50 psi) is ten times more effective than a gaseous spray at a similar pressure. Increasing the pressure, and corresponding liquid velocity, results in an increase in removal efficiency. Using a narrow fan-spray nozzle, the nozzle-to-surface distance should not exceed one hundred nozzle diameters to retain the quoted removal efficiencies.

An unfortunate result of high pressure spraying with trichlorotrifluoroethane is the cooling of the fluid that results from its rapid expansion and the subsequent chilling of the glass surface. If the surrounding atmosphere contains water vapor, the surface can be cooled below the dew point causing water condensation. The evaporation of the cold water droplets can leave the surface spotted as described earlier. Ongoing experiments will determine if surrounding the surface being cleaned with dry nitrogen or saturated cleaning fluid vapors will prevent the observed water spotting.

An alternate method for removing water droplets that results in spotless drying is the use of centrifugal forces generated by spinning the optic while it is being sprayed. A commerically available integrated circuit silicon substrate cleaner was recently evaluated for particle removal efficiency. The cleaner uses a water spray at 17 MPa (2500 psi). The results are similar to those obtained using the high pressure organic liquid spray and did not result in water spotting. This process is the most viable alternative to the currently used lens tissue scrubbing procedure described below.

Cleaning by Scrubbing and Swiping

Rubbing an optical surface with a solvent saturated lens tissue is currently our standard cleaning procedure. It takes advantage of solvent extraction and imparts a high liquid shear force to attached particles when the fibers in the tissue pass over the surface. Final cleanliness is sensitive to the presence of contaminants in the solvent and in the lens tissue. Recontamination is avoided by discarding each tissue after one pass over the surface. Although this method achieves the surface cleanliness levels necessary for laser systems, it has the following limitations: 1) slow laborious process, and 2) results are subject to technician skill and attention to detail. Because of these inherent limitations, faster and more operator independent techniques such as the high pressure solvent spraying technique are being developed. Such techniques will be needed to efficiently maintain glass laser systems of the future.

CONCLUSIONS

Large glass laser systems have created a need to clean and maintain clean thousands of optical surfaces in an ultra clean condition for periods of time exceeding a year.

Measurements of the particle removal effectiveness of seven cleaning processes has shown that scrubbing with solvent saturated lens tissues is the most effective particle removal process leaving only 2-40 particles/cm^2 (> 5 μm) followed closely by spraying with high velocity liquid jets 2-60/cm^2 and strippable adhesive coatings 500/cm^2. Either of these techniques greatly exceeds the cleaning effectiveness of the more conventional production cleaning techniques of either ultrasonic agitation of a solvent or vapor degreasing.

High pressure liquid spraying using an organic fluid or water may evolve into a very rapid and efficient cleaning procedure for unmounted optics when problems of spotless drying are overcome. Strippable adhesive coatings hold great promise for the cleaning of mounted optics. Their single drawback, that of leaving an organic film when stripped from the optical surface, may be overcome by advances in strip-coating formulations.

REFERENCES

1. J. A. Glaze, R. O. Godwin, Laser Focus, August 1977.
2. "Laser Program Annual Report", Lawrence Livermore Laboratory, Livermore, CA, UCRL-50021-76.
3. W. W. Simmons, J. T. Hunt, App. Optics 17 (7), 999 (1978).
4. H. G. Patton, I. F. Stowers, W. A. Jones, D. E. Wentworth, "Status Report on Cleaning and Maintaining Laser Disk Amplifiers", Lawrence Livermore Laboratory, Livermore, CA, 1978, UCRL-52412.
5. I. F. Stowers, H. G. Patton, et al in "Proc. Electro-Optics/ Laser 77 Conference", pp. 75-80, Industrial and Scientific Conference Management, Inc., Chicago, Ill., 1977, or LLL report UCRL-78923.
6. R. I. Larsen, Ind. Hygiene J., Aug. 265 (1958).

7. J. M. Corn, J. Air Pollution Control Assoc., 11 (11) 523 and 11 (12) 566 (1961).
8. D. W. Jordan, Brit., J. Appl. Phys., Supp. 3 5, S194 (1954).
9. I. F. Stowers, J. Vac. Sci. Technol 15 (2), 751 (1978).
10. A. D. Zimon, "Adhesion of Dust and Powder," Plenum Press, New York 1969.
11. B. R. Fish, Editor, "Surface Contamination", pp. 45-54, Pergamon Press, N.Y. 1964.
12. I. F. Stowers, H. G. Patton in "Proc. SPIE Optical Coatings II", March 28-29, 1978, Washington, D.C., pp. 16-31, Soc. of Photo-Optical Engineers, Bellingham, Wash. or UCRL-80731.

THE CLEANING OF POLYMER SURFACES

James Koutsky

Chemical Engineering Department
University of Wisconsin
Madison, Wisconsin 53706

A short review of typical cleaning procedures for a variety of polymers is presented. Most common cleaning procedures involve liquid chemical etching with associated pre and post treatments. Included are typical aids to achieve uniform cleaning and etching of the polymer surfaces. Also, plasma and mechanical cleaning of polymer surfaces are tersely discussed.

INTRODUCTION

Many investigators have been concerned with obtaining "clean" surfaces of polymers using a variety of techniques and recipes each designed for particular polymer system and for a specific purpose. In the processing and handling of polymers, surfaces can be easily contaminated with processing impurities such as mold release agents, oils, segregation of polymer impurities and low molecular weight species. In the handling and storage of polymers freshly processed surfaces can "age" with time due to migration of low molecular weight species, such as plasticizers. The segregation of impurities on polymer surfaces often leads to poor adhesion with coatings or adhesives applied to the surface at some later time.

Another purpose for cleaning polymer surfaces is to obtain a more chemically reactive surface for bonding and to sometimes roughen the surface to increase the surface area; this process is commonly referred to as etching. A succinct review has been recently published on the various methods which have been employed to clean and activate polymer surfaces [1].

The cleaning of polymer surfaces has primarily involved liquid etching and plasma etching techniques. Mechanical abrasion and etching methods using laser beams and ion beams have also been investigated [2-5]. This paper focuses on liquid and plasma etching since these methods have been commonly used with most synthetic polymers.

LIQUID ETCHING

Liquid etchants can be used for both chemical modification or dissolving surface contamination in the form of low molecular weight species or unwanted foreign species. First, the liquid must spread and wet the contaminated surfaces and efficiently and uniformly remove impurities in order to perform this function. In order to achieve spreading, the surface tension of the liquid must be smaller than the critical degree of wetting of the contaminated surface, i.e. the contact angle of the liquid on the surface must vanish. The viscosity of liquid etchants must be low in order to achieve spreading in a reasonable time.

To facilitate this etching process, removal of surface oils, fingerprints, etc., is first accomplished by an alkaline solution treatment followed by a neutralization by a dilute acid [6,7].

Often pretreatments are used to aid the etching process. Dispersed organic solvent - water carrier systems are used extensively for many commercial polymers [7-11]. The organic solvent is used to soften and swell the polymer surface which subsequently gives a more uniform cleaning in the final liquid etch. The organic is dispersed in water to insure a gentle swelling.

Various aids are used in facilitating uniform surface etching. Some of these aids are indicated in Table I.

Incorporation of oxidizable phases or compounds which are soluble in the parent polymer controls surface pit size and density and generally allows more gentle etching conditions.

If the polymer is of semicrystalline nature, the control of surface crystallization is essential. Molding conditions can profoundly alter the surface crystalline content when compared to the bulk of the polymer. Normally crystalline regions are much less susceptible to etching than amorphous regions.

An interesting approach for improved etching is to incorporate solid fillers such as TiO_2, talc, etc. which can be etched out of the polymer surface leaving a well controlled pit size. The size of the filler is crucial. Generally, best performances for adhesive bonding to the etched polymer surface occur when the filler used has a size range of about 10 microns.

Table I. Some Aids Used In Etching.

1. INCORPORATION OF OXIDATION PHASES OR COMPOUNDS
   a) rubber particles (10 μ)
   b) ketones
   c) olefins
   d) phosphorus compounds
   e) unsaturated fatty acids
   f) others

2. CONTROL OF SURFACE CRYSTALLIZATION
   a) molding conditions
   b) nucleating agents
   c) additives which give amorphous regions or segregation

3. INCREASE OF SURFACE IRREGULARITIES BY INCORPORATION OF FILLERS (~ 10 μ)

4. COPOLYMERIZATION WITH REACTIVE MONOMER

Table II. Common Polyolefin Etching Pretreatments For Metallization.

1. Annealing to remove internal stress $T > T_g$

2. Rinse

3. Solvent pretreatment causing polymer surfaces to swell

4. Alkali cleaning (soap solution 65-70°C)

5. Dilute acid neutralization

6. Application of etchant

Copolymerization of monomers which readily react with the etchant are also used. These can be block or random copolymers. The reaction with the etchant should change chemical groups on the monomer units but should not attack the main backbone chain. The indiscriminate chemical attack on the main backbone chain would lower the molecular weight and create a "brittle" surface layer.

Common pretreatments for polyolefins used for metallization are shown in Table II.

The choice of the final cleaning liquid etchant depends on the polymer system. Polyolefins are usually treated by oxidizing acids such as chromic, sulfuric, nitric or mixtures of these acids. The temperature of these treatments is generally above 50° to 60°C to insure rapid etching. These acids chemically alter these polymers as well as producing roughened or sponge-like surfaces. This increased surface roughness aids in adhesion of subsequent coatings. Extensive etching of the surface can also lead to a weakened surface structure which is undesirable for bonding purposes, therefore, processing time using these acids is generally kept to minutes or even seconds.

Polyamides have been cleaned by formic acid solutions [12], chromic-sulfuric acid [13], or iodine solutions for improved adhesion [14].

Polyesters have been etched by amines, water hydrolysis at high temperatures 130-180°C and elevated pressures [15] and aluminum and titanium halides in nitrobenzene [16].

Polycarbonates have been cleaned by strong oxidizing acids [17], strong bases [18] and various organics [19, 12].

Fluorcarbon polymers, being quite chemically inert, must be cleaned by dissolved sodium in organic solvents such as tetrahydrofuran [21, 22].

Polyvinyls are normally etched by mixtures of oxidizing acids and potassium hydroxide in isopropanol specifically for polyvinyl chloride [23, 24].

ABS systems are generally treated by strong oxidizing acids in combination with pretreatments [20, 27]. Generally, the polybutadiene phase is removed preferentially from the surface by these cleaning procedures.

Epoxy thermosets are generally pretreated by swelling agents and subsequently etched with strong caustic etchants or oxidizing acids [28, 29]. Phenolic and modified phenolics have been cleaned and etched by strong acids with oxidizing agents, fuming sulfuric as well as methanol.

An important consideration of choosing an appropriate liquid etchant is the ease of removal by a rinse step or simple evaporation. It is especially important that highly reactive etchants be thoroughly removed in order to insure chemical and physical stability of the surface. This insures long term durability of the surface for bonding to other materials. For example, the cleaning of fiber surfaces has been essential for improved bonding. It has been found that carbon fibers, when etched with sodium hypochlorite solutions obtain higher epoxy-bond strengths in composite materials [30]. However, it has also been found that the traces of sodium could be detrimental to bond durability, and rigorous cleaning procedures after etching should be employed to remove the final traces of sodium.

PLASMA ETCHING

The main advantage of plasma etching over liquid etching is that it can be done quickly and without the handling and disposal problems of liquid etchants. It's disadvantages are that more expensive equipment is needed and to achieve uniform etching the design and control of the processes are more stringent. Another important factor is that plasma etching will alter only the top surface layers within 500Å of the surface when applied at short times.

Plasma cleaning can be done at atmospheric pressures where a high temperature arc subjects the polymer surface to rapid melting and degradation. If the current is limited, then corona discharges are obtained which are localized near the electrode surfaces, and if the pressure is also reduced a glow discharge is produced which is observable over a larger volume between the electrodes. In fact glow discharges can be maintained by electroless processes.

The type of gas used in the plasma can be of an inert type, usually argon or helium, which physically etches and crosslinks the surface, or a reactive gas such as oxygen which is widely used for producing oxidized sites on the surface as well as physical etching. Arc or corona discharge methods normally employ air in commercial treatment processes. An excellent review of plasma etching procedures has been recently published [31].

Often plasma etching changes the surface topography creating small pits which can vary with the type of gas, pressure and electric field. Normally, high frequency discharges are used to rapidly etch the surfaces of insulating materials. This prevents surface charge accumulation and aids in efficient cleaning. Some interesting recent articles on plasma treatments of polyester fabrics [32] and polyethylene [33] indicate major surface chemical changes with short treatment times (~ 10 sec) when using nitrogen, oxygen, air and carbon dioxide. However, the use of argon for polyethylene and ammonia for the polyester fabric took longer times for significant surface changes under the same reaction conditions.

Surface cleaning by mechanical abrasion has not been widely used since contamination and surface stresses generally give a poor surface for subsequent applications. An interesting discussion of abrasion of thermoplastics and the attendant problems has been recently published [34]. Usually, stressed surface layers produced by the local shearing action of the abrasive cause bonding problems and the chemical contamination and degradation during abrasion usually are additional factors. Subsequent annealing and etching of the abraded polymer surface is usually necessary to achieve uniform, reproducible, bondable surfaces.

REFERENCES

1. J. S. Mijovic and J. A. Koutsky, Polym.-Plast. Technol. Eng., 9(2), 139 (1977).
2. F. Hasko and R. Fath, Galvanotechnik, 59(1), 32 (1968).
3. Y. Yatsui, N. Toshitake, and U. Kazuo, J. Electrochem. Soc., 116(1), 94 (1969).
4. T. G. Vladkova, I. N. Kolev, P. Tsvetkov, and I. T. Mladenov, Dokl. Bolg. Akad. Nauk. 26(9), 1189 (1973); Chem. Abstr., 80, 71303p.
5. J. Rost, H. J. Erler, H. Giegengack, O. Fiedler, and Chr. Weissmantel, Thin Solid Films, 20(1), S15 (1974).
6. D. C. Simpkins, Electroplat. Met. Finish., 23(10), 22 (1970).
7. I. A. Abu Isa, Polym. Plast. Technol. Eng., 2(1), 29 (1973).
8. D. R. Fitchmun, S. Newman, and R. Wiggle, J. Appl. Polym. Sci., 14(10), 2441 (1970).
9. I. W. Rose, Electroplat. Met. Finish., 23(8), 24 (1970).
10. R. Aries, French Patent Demande 2,199,012 (1974); Chem. Abstr., 82, 87253z.
11. S. Sivaram and K. M. Naik, Polym.-Plast. Technol. Eng., 10(2), 145 (1978).
12. J. S. Mijovic, "Morphology Changes of Nylon Film and Fiber Surfaces during Etching, Sensitization, Activation and Electroless Metal Deposition," M. S. Thesis, Univ. of Wis. Madison, 1974.
13. V. Baeva and V. Ts. Toshkow, Dokl. Yubileinata Nauch. Seryia, Vissh. Mach. Elektrotekh. Inst., 5, 89 (1970); Chem. Abstr., 74, 134087v.
14. I. A. Abu Isa, J. Polym. Sci., A-1, 9, 199 (1971).
15. A. Miyagi and B. Wunderlich, J. Polym. Sci., A-2, 10(10), 2073 (1972).
16. I. H. Skoog, U.S. Patent 3,497,406 (1970).
17. D. Raudiene, M. Salkauskas, and R. Baltenas, Liet. TSR Mokslu Akad. Darb., Ser. B., 4, 179 (1970).
18. G. C. Eastmond and E. G. Smith, Polymer, 14(10), 509 (1973).

19. Donlop Rubber Co., Belgium Patent 633,471 (1963); Chem. Abstr., 61, 3268a.
20. H. Mueller, German Patent Offen. 2,336,585 (1974); Chem. Abstr., 81, 14403t.
21. H. Brecht, F. Mayer, and H. Binder, Angew. Makromol. Chem., 33, 89 (1973).
22. W. M. Riggs and D. W. Dwight, J. Electron Spectrosc. Relat. Phenom., 5 447 (1974).
23. V. Baeva, Wiss. Z. Tech. Hochsch. Ilmenau, 19(2), 81 (1973); Chem. Abstr., 80, 96895w.
24. M. Mikhailov, V. Baeva, and K. Petrov, Gummi, Asbest, Kunstst., 26(6), 458 (1973).
25. N. B. Rainer, South African Patent 7,300,323 (1973); Chem. Abstr., 81, 154017v.
26. E. Maguire, German Patent Offen. 2,046,689 (1971); Chem. Abstr., 75, 7572v.
27. G. V. Elmore and K. C. Davis, J. Electrochem. Soc., 116(10), 1455 (1969).
28. G. V. Elmore, U.S. Patent 3,668,082 (1972).
29. D. J. Lando, German Patent Offen. 2,203,949 (1973); Chem. Abstr., 79, 6357z.
30. N. C. W. Judd, British Polymer Journal, 9, 272 (1977).
31. R. S. Thomas, in "Techniques and Applications of Plasma Chemistry," J. R. Hollahan and A. T. Bell, Editors, Wiley, New York, 1974.
32. A. M. Wröbel and M. Kryszewski, Polymer, 19, 908 (1978).
33. A. R. Blythe, D. Briggs, C. R. Kendall, D. G. Rance, and V. J. I. Zichy, Polymer 19, 1273 (1978).
34. C. C. Lawrence, British Polymer Journal, 10, 93 (1978).

REMOVAL OF FLUID CONTAMINANTS BY SURFACE CHEMICAL DISPLACEMENT

Robert N. Bolster

Naval Research Laboratory

Washington, D. C. 20375

Water and oils can be removed from solid surfaces by displacing them with another fluid, the displacing action being driven by differences in surface tension. The displacing fluid must have a surface tension well below that of the contaminant, and have a high equilibrium spreading pressure. Moderate mutual solubility is necessary to aid penetration of the contaminant film and formation of a surface tension gradient. The volatility of the displacing fluid affects the rate of displacement and the respreading of the contaminant. For water displacement, butyl and amyl alcohols are most effective. Certain fluorinated compounds and low-molecular-weight silicones have been found particularly effective in the displacement of organic fluids. Additives capable of adsorbing on the surface to prevent the displaced contaminant from respreading are sometimes desirable to enhance the effectiveness of the process. Emulsion systems employing the displacement mechanism are useful, especially for the removal of gross oil contamination and mixed oily and inorganic contamination. These techniques and conventional detergents have been applied to the salvage of electronic and electrical equipment contaminated by water, seawater, fuels and lubricants, and smoke deposits.

INTRODUCTION

When fluid contaminants are to be removed from a non-porous solid surface, cleaning may be accomplished or at least aided by the action of a displacing fluid driven by surface tension differences. The contaminant may be water or an aqueous solution, or it may be an organic fluid, hereafter referred to as an oil. Solid contaminants, either particulates or films, are not amenable to removal by this technique, unless they are first dissolved or suspended in a liquid. Frequently the displacing action is not recognized as a distinct phenomenon, but occurs during common washing or vapor degreasing operations. Knowledge of the mechanism involved can aid in selecting or optimizing the cleaning process.

This paper is a summary of the results of research conducted by the Surface Chemistry Branch of the Naval Research Laboratory, begun over 30 years ago by Baker and Zisman, and continued by them, Bernett, Singleterry, and others.

DISPLACEMENT MECHANISM

As a model, we will assume that a flat solid surface is covered by a film of the contaminating fluid, a. A drop of the liquid displacing agent, b, hereafter referred to as the agent, is applied. An effective agent will spread on the contaminant, driven by a two-dimensional pressure, the spreading pressure. This pressure, which is measurable with a film balance, has been shown[1] to be related to the surface and interfacial tensions as shown in Equation (1), where F_{ba} is the spreading pressure, γ_a is

$$F_{ba} = (\gamma_a - \gamma_b) - \gamma_{ab} \quad (1)$$

the surface tension of the contaminant, γ_b is the surface tension of the displacing agent, and γ_{ab} is the interfacial tension between the liquids. For reasons to be described later, the agent selected should be substantially soluble in the contaminant. The interfacial tension decreases to zero with increased solubility[2,3], leaving the relationship shown in Equation (2) as a special case.

$$F_{ba} = \gamma_a - \gamma_b \quad (2)$$

High spreading pressure is desired, as this is the source of the displacing action. Baker and Zisman[4] recognized this in their development of water-displacing formulations, and it is equally applicable to the displacement of oils[5]. As the surface tension of the contaminant, a, is fixed, it is necessary to choose an agent, b, with a low surface tension.

If a drop of an insoluble agent is placed on the contaminant, it will spread radially, carrying the contaminant along by viscous drag as shown in Figure 1. Under these conditions the agent does not usually penetrate to the solid surface until the contaminant has been largely displaced. Therefore, the displacing action may be incomplete, leaving islands of contaminant behind, and emulsification, rather than true displacement, may occur.

When the agent is soluble in the contaminant, the mechanism shown in Figure 2 applies. Radial spreading occurs as before, but now the agent is also able to penetrate directly to the solid surface to provide more thorough displacement. Around the periphery of the clean area, the agent dissolving in the contaminant produces a surface tension gradient which pulls the contaminant outward as the displacing agent spreads to claim the territory. An annular ridge of contaminant forms as displacement proceeds, and, if the spreading is too rapid, this ridge may be overrun as shown in Figure 2c. Although these islands will contract and become mobile, additional agent may have to be applied to remove them. Agents which spread more temperately and which adsorb on the solid surface are more likely to produce the complete displacement shown in Figure 2d. The rate of displacement is proportional to the spreading pressure and inversely proportional to the viscosities of the fluids. The nature of the solid surface has

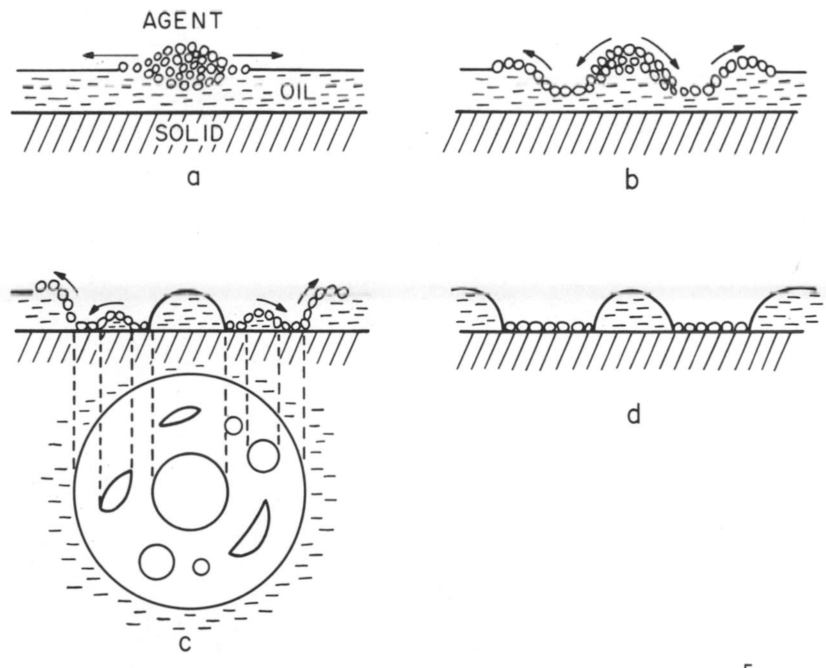

Figure 1. Displacement by an insoluble agent.[5]

little effect on the process, so long as it is wetted by both fluids. Chemisorption of the contaminant on the solid, however, may create an unfavorable competitive adsorption situation.

If the displacing agent is a pure, volatile liquid which does not adsorb strongly to the surface, it will soon evaporate and the contaminant can spread back over the cleaned area. This is usually prevented in practice by the continued application of the agent, by spray or vapor condensation, until the contaminant is completely displaced and flushed from the item being cleaned. Other techniques to prevent respreading include the addition to the agent of a non-volatile component, or the use of a strongly adsorbing agent, which modifies the surface so that it cannot be wetted by the contaminant. The adsorption of the agent onto the surface as a monomolecular layer is often sufficient to prevent respreading.

Displacing agents of low molecular weight generally act rapidly, as they have high spreading pressures and low viscosities. Figure 3, for example, shows how the spreading pressure of a series of alcohols on water increases with decreasing molecular size. In this case, increased solubility has a large effect, the three lowest alcohols being completely soluble. Plots of a

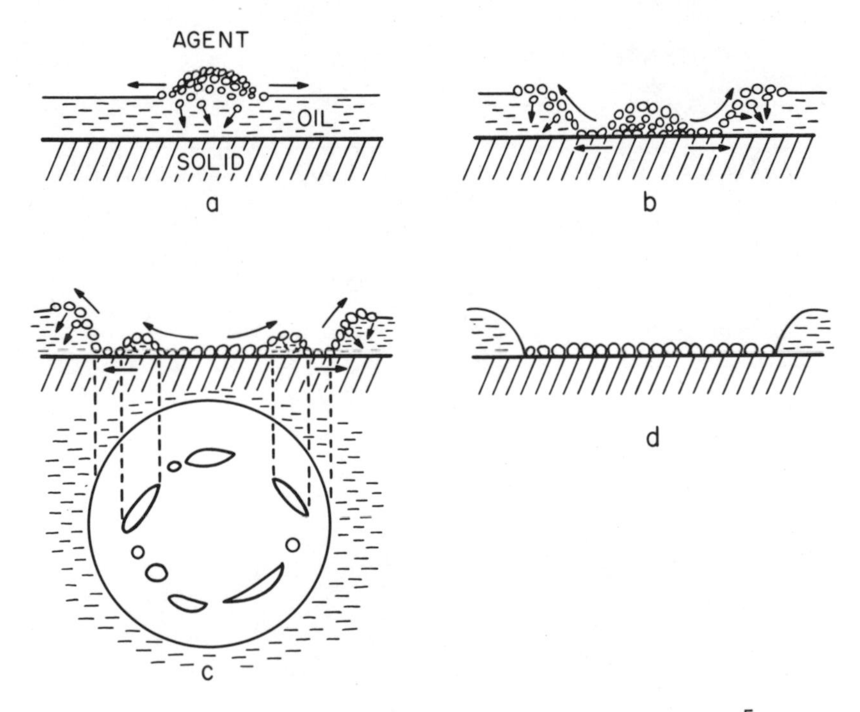

Figure 2. Displacement by a soluble agent.[5]

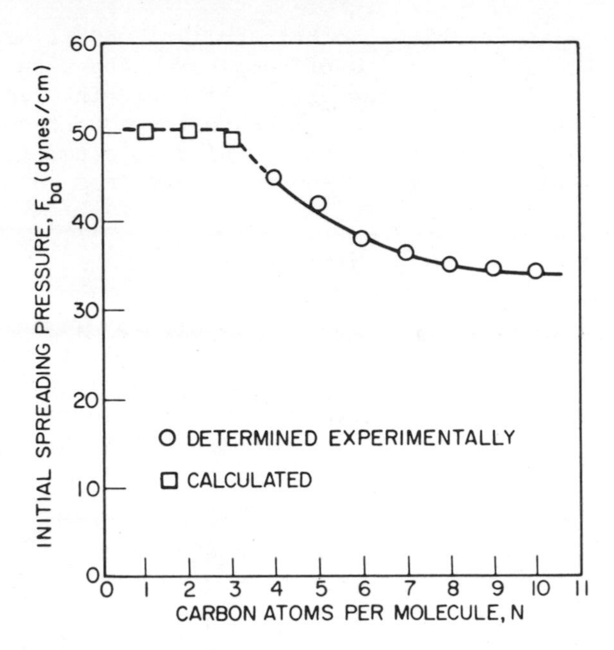

Figure 3. Spreading pressures of normal alcohols on water.[3]

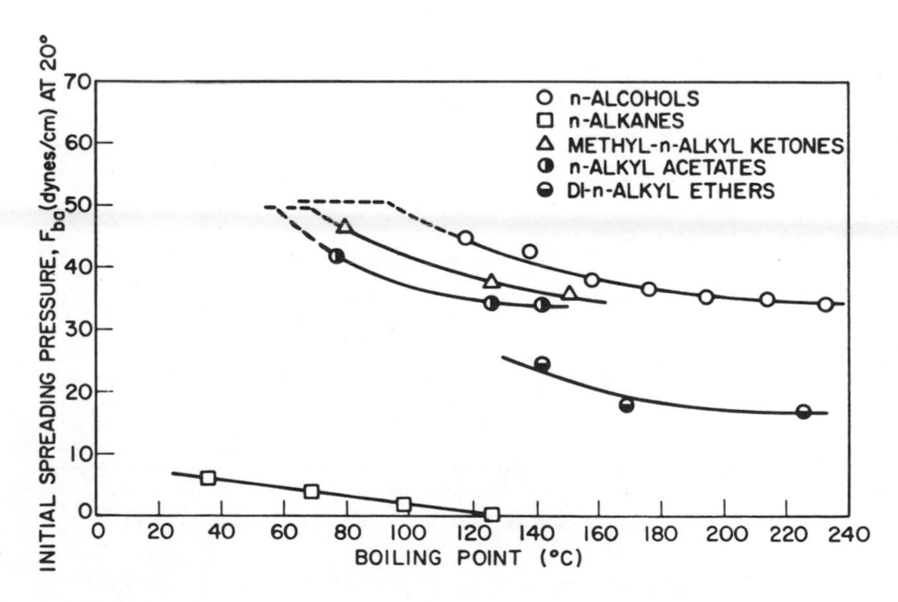

Figure 4. Spreading pressures of some organic liquids on water vs boiling point.[3]

number of agents in relation to their boiling points are shown in Figure 4. Within each class of compounds, the more volatile ones have higher spreading pressures. However, the less volatile compounds of higher molecular weight have the advantage of greater persistence and therefore thoroughness of displacement. They are also less flammable and may be less toxic and less injurious to plastics and coatings.

WATER DISPLACEMENT

The displacement of water is facilitated by its high surface tension and low viscosity. As shown in Figure 4, the alcohols have relatively high spreading pressures on water, due to their amphipathic or polar-nonpolar structure. The hydroxyl group is attracted to water (hydrophilic) while the hydrocarbon portion is repelled (hydrophobic)[3,4]. Among the aliphatic alcohols, the

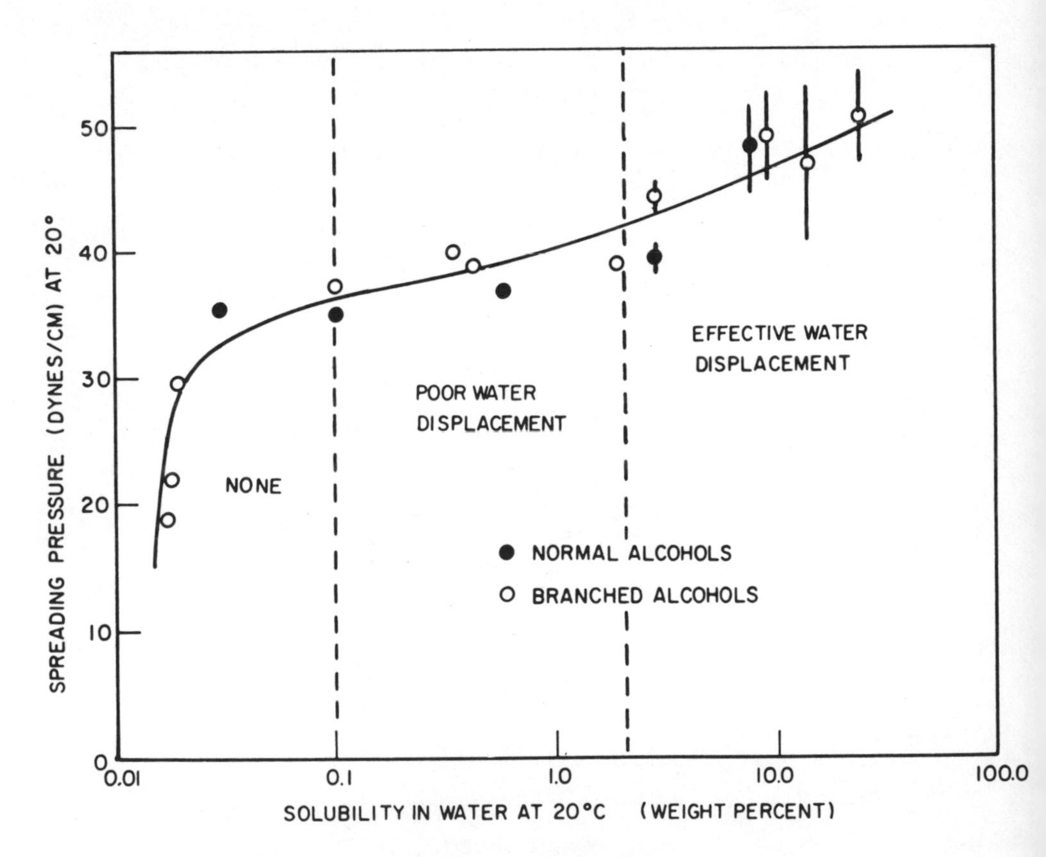

Figure 5. Relation of spreading pressure to water solubility for alcohols. Lengths of the vertical bars are proportional to water-displacing ability.[6,7]

effect of solubility in water is shown in Figure 5. The spreading pressure and displacing effectiveness increase with solubility, the optimum solubility range being from 2 to 25%[6,7]. Although their spreading pressures are high, the completely soluble alcohols are not persistent enough to provide effective displacement because of their solubility and volatility. Shown in Table I are several water-displacing chemicals, listed in order of their ability to displace 2-mm-thick water films. Their effective water-displacing properties made these fluids candidates for use in a formulation to restore flooded electrical equipment[4]. After consideration of their flash points, evaporation rates, and effects on insulating materials and coatings, 1-butanol was selected as the most suitable. A barium petroleum sulfonate rust inhibitor was added to prevent respreading of the water and to protect metal surfaces against corrosion, and an oxidation inhibitor to ensure stability of the alcohol during storage[7].

OIL DISPLACEMENT

Oils are more difficult to displace than water because of their lower surface tensions and higher viscosities. Silicone oils, for example, are notoriously difficult to remove from surfaces, largely because of their very low surface tensions, which make them immune to displacement by most solvents and detergents. It is helpful to know the nature of the contaminant so that an agent with a lower surface tension and sufficient solubility can be selected. In Table II a number of alcohols and fluorinated alcohols found to be effective displacing agents[5] are listed in decreasing order of their ability to displace propylene carbonate (γ_a = 41.4) from steel surfaces. Note that the ranking is approximately in order of increasing surface tension. Variations may result from solubility or volatility effects. Non-volatile compounds may also adsorb at the solid surface to prevent respreading. The fluorinated agents are more effective in this regard, as they form monolayers of lower wettability. Also listed in Table II are some non-polar agents in decreasing order of their ability to displace a fuel oil. As they are all soluble in the oil, surface tension effects dominate. Since they are not adsorbed, they evaporate completely and respreading occurs. Dimethyl silicones of low molecular weight are also effective in displacing hydrocarbon oils because of their low surface tensions[5]. Nonvolatile silicones can be added to prevent respreading if their presence is not objectionable. The addition of strongly adsorbing compounds such as fluorinated alcohols and acids has similarly been found very effective in enhancing the oil-displacing abilities of non-polar solvents such as naphtha and fluorocarbons[8]. Concentrations of only a few hundredths of a percent of the additive are sufficient to form an adsorbed monolayer which prevents respreading of the oil and helps the displacement proceed to completion. Such techniques,

Table I. Properties of Some Water-Displacing Fluids[6]

Compound	Equil. Spread. Press. (dynes/cm)	Solubil. in Water (wt-%)	Area of Water Displ. (cm^2)
2-Methyl-2-butanol	47.1	12.5	20
1-Butanol[3]	44.8	7.9	12
2-Butanol	50.4	12.5	12
2-Methyl-1-propanol	49.3	9.5	12
3-Pentanol	44.0	Sl.Sol.	5
Furfuryl alcohol	-	Complete	3
Monobutyl ether	39.6	Complete	3
Methyl n-amyl ketone	35.8	V.Sl.Sol.	1
Butyl lactate	43.5	Sl.Sol.	1
Diethyl carbonate	33.1	Insol.	0
Diisobutyl ketone	26.8	Insol.	0

Table II. Some Oil-Displacing Fluids and Their Surface Tensions

Compound	Surface Tension (dynes/cm)
Compounds with polar groups[5] (listed in the order of decreasing ability to displace propylene carbonate from steel)	
1-Perfluorooctanol	17.0
1-Perfluorohexanol	17.4
1-Perfluorobutanol	17.2
Perfluoroethanol	19.5
1-Butanol	23.9
Ethanol	22.3
1-Pentanol	24.9
1-Heptanol	26.2
1-Hexanol	24.9
1-Octanol	26.9
Non-polar compounds[7] (listed in the order of decreasing ability to displace fuel oil from steel)	
n-Hexane	18.4
n-Decane	23.9
Aliphatic naphtha	25
1,1,1-Trichloroethane	25.7

of course, cannot be used when the surface must be left completely free of organic matter, or where a wettable surface is desired for the application of an adhesive, coating, or lubricant.

APPLICATIONS

The principal motivation for the study at NRL of these techniques was their applicability to the cleaning of electrical and electronic equipment contaminated by seawater, fuels, lubricants, or smoke deposits as a result of flooding or fire. Research and experience have led to established procedures[9] for dealing with these contaminants, and a wide variety of equipment has been successfully salvaged, with savings of millions of dollars. Salts from seawater flooding are removed with an aqueous detergent solution, preferably by immersion in an ultrasonic tank. A flush and thorough rinse with pure water remove the detergent, and the bulk water is then blown off with compressed air. Drying is completed in a warm oven or vacuum oven, or with the aid of the butyl alcohol water-displacing fluid. When cleaning cannot be done immediately, a fresh water flush and application of the water-displacing fluid prevents deterioration pending cleaning, and may temporarily restore the unit to operation.

For the removal of fuel oils, an emulsion cleaner employing the displacement mechanism was developed[7,10]. Naphtha having a flash point of 60°C is emulsified in water with an oil-soluble nonionic surfactant. A small amount of an oil, such as diesel fuel, which is more soluble in the contaminant is included to aid penetration. When the emulsion is applied to an oil-covered surface, the solvent phase is released, penetrates the oil film, and displaces it. The displaced oil, mixed with solvent and surfactant, is emulsified and carried away. Any seawater salts remaining are dissolved by the water phase. Spray application of the emulsion is often adequate, but immersion in an ultrasonic tank is more effective. Flushing, rinsing, and drying are then carried out as before.

In practice, we have encountered fire contamination most often. These solid contaminants are outside the scope of this paper, but the procedures for dealing with them are similar. Alkaline solutions are often used to loosen or dissolve films of combustion products.

SUMMARY

In summary, an effective displacing fluid should have the following properties:

1) A low surface tension; must be lower than that of the contaminant.

2) A high spreading pressure on the contaminant.

3) Moderate solubility in the contaminant.

4) Moderate volatility.

5) Low viscosity.

6) If permissible, the ability to modify the surface so that it is not wettable by the contaminant.

In numerous applications from the cleaning of spacecraft components, miniature bearings, and electronic equipment to everyday housekeeping, an appreciation of the displacement mechanism can aid in the selection of effective and efficient cleaning procedures.

REFERENCES

1. P. Pomerantz, W. C. Clinton, and W. A. Zisman, J. Colloid Interface Sci., 24, 16 (1967).
2. J. R. Pound, J. Phys. Chem., 30, 791 (1926).
3. C. O. Timmons and W. A. Zisman, J. Colloid Interface Sci., 28, 106 (1968).
4. H. R. Baker and W. A. Zisman, "Water-Displacing Fluids and Their Application to Reconditioning and Protecting Equipment," Naval Research Laboratory Report C-3364, Washington, D.C., 20375, October 4, 1948.
5. M. K. Bernett and W. A. Zisman, J. Phys. Chem., 70, 1064 (1966).
6. H. R. Baker, C. R. Singleterry, and W. A. Zisman, "Factors Affecting the Surface-Chemical Displacement of Bulk Water from Solid Surfaces," Naval Research Laboratory Report 6368, Washington, D.C., 20375, Feb. 24, 1966.
7. H. R. Baker, P. B. Leach, C. R. Singleterry, and W. A. Zisman, Ind. Eng. Chem., 59 (No. 6), 29 (1967).
8. M. K. Bernett and W. A. Zisman, Ind. Eng. Chem. Prod. Res. Develop., 11 (No. 1), 83 (1972).
9. H. R. Baker and R. N. Bolster, "Guide to NRL Cleaning and Salvaging Techniques for Reclaiming Equipment Contaminated with Seawater, Oil, and Smoke Deposits," Naval Research Laboratory Report 7563, Washington, D.C. 20375, March 27, 1973.
10. H. R. Baker, P. B. Leach, C. R. Singleterry, and W. A. Zisman, "Surface Chemical Methods of Displacing Water and/or Oils and Salvaging Flooded Equipment, Part 1 - Practical Applications," Naval Research Laboratory Report 5606, Washington D.C., 20375, Feb. 23, 1961

CLEANING ISSUES RELATED TO SOLDERING

Howard H. Manko

Solder Consulting Services
751 Bayard Street
Teaneck, New Jersey 07666

Cleaning has become a major issue in conjunction with the soldering process. Two major areas of cleanliness are discussed; pre-soldering and post soldering. The solder process itself can be simplified into a sequence of operations starting with surface preparation by the flux, heating, melting and wetting by the solder, and subsequent freezing. The efficiency of the flux thus depends on the amount of foreign materials present on the surface which may interfere with the intimate contact of the flux with the surface to be soldered. In this respect, contamination of the surfaces is often associated with solderability problems and the inability of fluxes to perform their function properly.

The chemical corrosion, and electrical conductivity of residues left after the soldering operation is the second area of concern. In many critical assemblies flux removal is mandatory. The paper deals with additional sources of contamination beside the flux which may cause malfunctions in equipment after soldering, and points out the latest trends in cleaning technology for this industry as well as some other dangers to assembly because of flux residues.

INTRODUCTION

The industry has always recognized the importance of cleanliness as it is associated with soldering. While it is possible through careful planning to avoid the need of flux residue removal, other sources of contamination must be considered throughout the process. The efficiency of the flux, for instance, depends on the amount of foreign materials present on the surface and the ability of the specific formulation used to penetrate through these interference layers and react with the surfaces to be soldered. Thus, cleanliness prior to soldering has become synonymous with the solderability of the surfaces and the success of the soldering operation. Neglecting this important area may result in unnecessary touch-up which increases cost without necessarily improving reliability.

The nature of flux removal techniques have changed with recent environmental and safety pressures. Most of the dangers of contamination left after soldering are well understood, but there is a divergence of opinion as to the best methods of their removal. The volatile organic solvents, normally referred to as degreasing fluids, have been traditionally used for rosin flux removal. Our concern with the environment, as well as legislative pressures, have pushed the industry into considering water-based systems which are finding wide-spread acceptance these days. This is closely tied in with methods of measuring cleanliness which have become quite sophisticated. Unfortunately, the interpretation of cleanliness information and the preservation of cleanliness are subjects which are often neglected by the same organizations.

PRESOLDER CLEANING FOR FLUX EFFECTIVENESS

Fluxes serve as a catalyst for the soldering process, and their intimate contact with the metalic surfaces to be bonded is indispensible.[1] Like a true catalyst, the presence of the flux is needed before and during the wetting process, in order to achieve reliability in the bond formation. Any interference layer between the flux and the metallic surface would render the catalyst useless. A reduction of the effectiveness of the flux by surface contamination is one of the major aspects of "solderability problems".

Provided then that the flux has been properly selected for the base metals to be soldered, solderability testing becomes a useful tool to assess the cleanliness of the surfaces. One must remember that the average flux has only limited ability to penetrate under gross amounts of contamination. Thus it becomes necessary to identify the interference layers and preclean the parts prior to soldering. Proper handling, storage and assembly techniques minimize this undesirable condition and therefore make a better economical alternative. When gross contamination is the cause for loss of solderability, the move to a stronger flux is not always the right solution. The contaminating materials, furthermore, cause other solder-related problems like blow-holes, and should be eliminated wherever possible.

Presolder cleaning is used in one other major important section of the industry. It is often impossible to clean the soldered assembly after the process is completed because of the type of components and hardware which are assembled. In this case it might be necessary to clean all parts prior to assembly with major emphasis on maintaining high cleanliness levels. This is coupled with the use of mild (safe) fluxes, of known contamination content, in order to guarantee acceptable quality. Under these conditions the reintroduction of contamination during assembly, soldering and final inspection is kept to a minimum, and clean-room conditions may be required.

POST SOLDER CLEANING

The need for cleaning after soldering is no longer under debate and there is a strong trend within the electronic industry in this direction.[2] There are four basic considerations which are the reason behind this need for cleanliness; unfortunately, they are often misunderstood and confused with one another.

A. The Functional Needs

Post solder contaminants, including flux residues, can interfere with the proper behavior of mechanical contacts in relays, switches, variable resistors, etc. This constitutes the most difficult area of cleaning; however, since many switches, rotary-contacts and similar devices are often lubricated with materials that are easily removed by the cleaning process. Unless relubrication is feasible, they may well fall in the category of "uncleanable". They may have to be installed as a separate "add-on" item just prior to final assembly and after post solder cleaning. Furthermore, some shielded or covered components are not easily rinsed and cleaning solutions may be trapped inside aggravating the functional problems. These components too fall into the "uncleanable" category.

B. The Insulation Requirements

Exposed ionic residues, when operating in humid atmospheres, may lower the surface insulation properties of circuitry below their acceptable limits posing current leaking problems.[3] More about this problem later on in this paper.

C. The Corrosion Dangers

The chemical attack of metallic surfaces by post solder residues cannot be overlooked as an industrial hazard. A large portion of the industry is moving toward acidic and non-rosin fluxes.[2] This problem is especially severe if dissimilar metals are exposed to the residues, and the assembly is not protected from humidity. The corrosive contamination obviously does not originate only in the flux, but may be due to residual etching and plating solutions, body salts from human perspiration in handling, etc.[1]

D. The Cosmetic Appearance

Customer acceptance and the visual asthetics of the assembly are also an important parameter associated with cleaning. Shiny solder joints, bright metallic surfaces and similar characteristics are unfortunately associated with quality assemblies. While the author objects to cosmetics per se, and maintains that the technical accuracy of an assembly is a more important characteristic, many novices still make buying decisions on the basis of external appearance alone. Suffice it to say, that for cosmetic reasons, cleaning cannot be delayed after soldering. This is because of the danger of attacking (etching) hot surfaces which may result in somewhat duller appearance. Delayed cleaning often results in a larger degree of difficulty when removing residues, especially in the rosin family.

WATER BASE SYSTEMS VS. VOLATILE ORGANIC SOLVENTS

One major current issue revolves about the choice of the cleaning system. The growing trend of water usage is being fought by the proponents of the solvent systems. From an engineering point of view, both offer materials and equipment suitable to achieve identical levels of cleanliness. More conventional flux removal techniques in the rosin family have always relied on non-flammable volatile organic chemicals. Their effectiveness in the removal of ionic contamination has been assured through the design of azeotropic blends containing polar additives. Cold cleaning and vapor degreasing have both been utilized, and sophiscitated inline equipment is available. To date, these systems, however, are not suitable for the removal of the more aggressive intermediate group of fluxes like organic acids, amines, and non-rosin base materials.

On the other hand, the use of water base systems has become more popular, since the introduction of water additives which enable the removal of rosin fluxes through a chemical saponification process. Thus, water base systems are more universal in their applications, since they can remove a larger variety of fluxing formulations. While water base systems use a cheaper solvent - the water - they are much more energy intensive and thus there is no appreciable economical difference between them and the solvent systems. This statement is based on a current comparison of inline cleaning systems for the removal of fluxes from printed circuit boards after wave soldering. The economic comparison includes the amortization of equipment (in 5 years), the cost of the materials, and the energy requirements for identical quantities of printed circuit boards.

One important issue in this comparison between water and solvent systems is based on the effect they have on the environment and personal safety. While the solvents used presently are not considered pollutants[4], they are not bio-degradable, and their effect on the human system has posed restrictions on their level permitted in the air (measured in ppm per 8 hours exposure and quoted as TLV or MAC). This serves as a basis for fear that solvent systems may not be permissible in their present form for the near future. Legislation may require stricter limitations on their emission in the work area and into the environment. This would impose additional expenses such as the installation of activated charcoal beds, etc. The use of water base systems, on the other hand, is not free of problems since the effluent must be controlled before it is discarded. While all solder chemicals can be bio-degradable, other contaminants may well require treatment before they are returned into the sewer system. At present, however, this seems to be the lesser of the two dangers.

MEASURING CLEANLINESS

Measuring the extent of surface contamination removal is another vital issue. The best contemporary method relies on the determination of the ionic contamination picked up from the surface[5,6]. This is based on the assumption that the worst actors are ionizable materials, which in the presence of humidity would cause electrical current leakage and possibly galvanic corrosion damage. Specific discussion of equipment and methods of detection are handled by others in these proceedings volumes 7,8. Let us briefly examine some of the problems that exist with this measurement concept.

a) The method allows only the overall assessment of total ionizables on the surface without yielding any specific information on the distribution of the contamination. It is obvious that contamination is not uniformally distributed and that higher concentrations of dangerous materials may exist in crevices and difficult to wash places. Obviously, this method could be made more meaningful if specific areas are physically isolated from the rest of the assembly; unfortunately, in most cases this would mean destructive testing techniques. Test results would be reliable if the board was made for easy and total cleanability, something which is badly neglected as a design consideration.

b) The character of that solvent used (a mixture of alcohol and water) and its volume are such that the solubility and ionization potential of most materials become obliterated. Obviously just measuring quantity without considering the quality of the residues leaves much to be desired.

c) Masking by non-ionic, non-soluble materials. The measuring method is admitedly only sensitive to surface contamination. It also detects ionizable materials which are coated or imbeded in alcohol soluble non-ionic materials. However, in the electronic industry, deliberate and accidental coatings are deposited on the surface which are neither soluble nor removable by the test solution. These coatings may trap ionic contaminations which later bleed-out in the presence of humidity through natural crevices and capillaries. In addition, these coatings may in themselves craze, creating cracks which will enable humidity to penetrate below. The seepage of aggressive soldering fluxes under solder resists applied over tin lead is a typical example of such a problem. While the solder resist may have good adhesion to the tin lead surface when first applied, this bond is disturbed during the soldering process when the alloy melts due to the heat of the soldering operation. Furthermore, fresh molten solder often flows below the resist over the board traces, pusing flux ahead of it to become trapped under the resist.

Even though the objections raised to the test are formidable, it is an extremely valuable and indispensable tool for the industry. It allows the manufacturer to establish the levels of cleanliness which he achieves during his processing and to monitor them continuously. It permits specific adjustments in materials, equipment and process to obtain the type of results set as a goal.

TEST RESULT INTERPRETATION PROBLEMS

One of the more difficult tasks to be discussed is the correlation between the test results and the requirements of a specific assembly. Once the problem of contamination distribution on the surface is understood, and the quantitative evaluation of ionizables is made, we still have to assess the insulation properties of the surface. For this we require some more specific information as follows:

a) The other parameters which affect surface insulating properties, such as surface finish, roughness, and the type of materials involved.

b) The worst case of voltage gradient vs, spacing of conductors, and the sensitivity of the circuitry to minute current leakage.

c) The ambient in which the equipment will be operating; here too the worst case of temperature and relative humidity are important.

The task of sorting out this information and relating it to surface cleanliness levels is not as difficult as it may sound. Past field experience coupled with judicious testing of more conventional surface insulation properties under time, temperature and humidity conditions can easily establish acceptable levels. Obviously, no generalizations can be made which are or will be suitable for all segments of the industry.

PROTECTING CLEANLINESS

With planning and careful handling procedures, it is possible to avoid recontamination of the surfaces during the manufacturing cycle. Much of the detrimental materials can be airborne in our polluted air and environment and may accumulate on sensitive surfaces after they have left the manufacturing facility. The use of some preventive measures are, therefore, well worth considering. There are simple options as follows:

a) Sealing the clean assemblies in hermetical containers.

b) The use of conformal-coatings to act as a total envelope around the assembly. Two areas here need additional attention: first, if the surfaces are not free of non-ionic contaminants, such as silicone oils, the conformal coatings will not adhere to the surface. A phenomenon which cannot always be recognized through blistering. Second, a conformal coating applied over trapped ionizables may cause corrosion cells near sharp surfaces where conformal coating adhesion is minimal due to the natural wetting mechanism that prevents coverage of these geometric locations.

c) Total encapsulation which is often required also for mechanical and vibration reasons.

d) If none of the above protective methods are feasible, it is well to consider the use of a solder-resist as a partial coating over the surfaces. While it is true that the solder nodules and exposed component leads are still subject to surface shorting, much of the internal circuitry is covered. Properly applied insulation material will change surface insulation to volume insulation, thus making closer spacing and higher voltages entirely feasible.

The need for protective techniques is obvious, but surface cleanliness prior to their coating application and the permeability of these materials must be considered in order to get the degree of reliability required.

SUMMARY

While cleaning before soldering is vital to enhance flux action, post solder cleaning assures the function of the circuitry and prolongs its expected life. Solderability testing helps assess pre-solder cleanliness for specific flux systems. Water extract resistivity establishes the quantity of potentially harmful materials in post-solder stages. This test method has some shortcomings which requires intelligent interpretation vital. The preservation of cleanliness after processing must also be planned.

REFERENCE

1. H. H. Manko, "Solders and Soldering," 2nd edition, McGraw Hill Book Company, New York, to be published beginning 1979.
2. H. H. Manko, "Wave Soldering: A Look Into the Future," First International Printed Circuit Convention Proceedings, April, 1978.
3. H. H. Manko, Electronic Design, 17, No. 24, (November 22, 1969).
4. Air Quality-Recommended Policy on Control of Volatile Organic Compounds, Federal Register 42 No. 131, (July 8,1977).
5. H. H. Manko, Quality Management and Engineering, 11 No. 11, (November, 1972).
6. MIL-P-28809, "Printed Wiring Assemblies," Latest Revision.
7. J. Brous, in "Surface Contamination: Its Genesis, Detection and Control," K. L. Mittal, Editor, Vol. 2, pp. 843-855, Plenum Press, New York, 1979.
8. W. B. Wargotz, in "Surface Contamination: Its Genesis, Detection and Control," K. L. Mittal, Editor, Vol. 2, pp. 877-895, Plenum Press, New York, 1979.

NONIONIC CONTAMINATION PREVENTION - THE NEXT STEP TOWARD GUARANTEED PRINTED WIRING ASSEMBLY RELIABILITY

W. G. Kenyon

Freon[R] Products Division
E. I. du Pont De Nemours & Co.
Wilmington, Delaware 19898

Recognition of the devastating effects of contamination on printed wiring assembly reliability, identification of contaminant classes and control of certain ionic contaminants by post-defluxing testing were important first steps toward assured assembly reliability. After briefly summarizing these steps, the discussion will concentrate on nonionic contamination prevention, the next step the electronics industry must take to achieve guaranteed assembly reliability.

INTRODUCTION

Printed wiring assembly contaminants fall into three general classes: polar or "ionic", nonpolar or "nonionic" and particulate (Table I).

The solvent extract conductivity or resistivity test has rapidly gained popular acceptance for measuring ionic contamination levels on printed wiring assemblies. Government research at the Naval Avionics Center, Indianapolis, resulted in a unique military specification, MIL-P-28809, in 1975.[1] The specification was unique because, for the first time, a Printed Wiring Assembly specification called out a preconformal coating ionic contamination inspection method that included a test method, test solution composition, and an acceptable cleanliness level. This acceptable cleanliness level was based on the reliability history accumulated by the military over many years of soldering with mildly activated (RMA) fluxes.[2] The use of activated (RA) fluxes conforming to MIL F-14256D[3] was permitted, provided the same cleanliness level was met.

Table I. Printed Wiring Assembly Contaminant Classes.

"Ionic"	"Nonionic"	Particulate
Flux activators and activator residues	Rosin	Laminate resin & Fiberglass (drilling or punching debris)
Solder salts	Oils, Grease and Waxes	Dust
Etching salts	Soldering oils	Handling soils
Plating salts	Soldering salts	
Handling soils	Handling soils	

Other government agencies have followed the Navy's lead, and included the MIL P-28809 test procedure, applicable only to organic solvent defluxed rosin flux residues,[4] in soldering specifications such as MIL-S-45743E, MIL-S-46844C and MIL-S-46860B.[5] Recently the test was written into MIL-P-55110/MIL STD 275 to ensure printed wiring board cleanliness prior to permanent solder mask application.[6] In this case, the test is used to measure rosin flux residues remaining after the solder coating step.

Any solvent extract conductivity resistivity test is required to perform two functions. First, it must provide a mechanism for removing the post-defluxing residues from the printed wiring assembly surface and dissolving the ionic contaminants in the test solution. Second, the test must provide for accurate and reproducible measurement of the dissolved ionic residues.

Obviously, the test solution composition chosen should not suppress contaminant dissociation from molecules to ions (Equation 1).

Test Solution Composition

$$\underset{\text{(molecule)}}{AB} \underset{\text{Wrong}}{\overset{\text{Right}}{\rightleftharpoons}} \underset{\text{(ions)}}{A^{+} + B^{-}} \qquad (1)$$

If it does, the tester will assume the assembly is clean (no conductivity or resistivity change) but may later find assembly failure caused by the humidity encountered in the actual operating environment where contaminant dissociation to conductive ions can occur.

Historically, the first attempts to measure post-defluxing ionic soil levels used the technique developed at Bell Laboratories to measure the contamination level on gold-plated parts.[7] The procedure involved agitating a representative sample of plated parts with a specified volume of deionized water, previously equilibrated with air to eliminate the effect of CO_2 uptake as a source of error, for a specified time and determining the conductivity change. If the conductivity change (ΔC) did not exceed 1 micromho/cm per 5 in^2 of gold-plated piece parts surface area in 100 ml of water at 25°C, the parts were considered acceptable for spot welding and further processing (Equation 2).

$$\Delta C = C_E - C_W \qquad (2)$$

where

ΔC = Conductivity increase in μmho/cm

C_E = Final conductivity of sample extract in μmho/cm

C_W = Initial conductivity of water in μmho/cm

The Naval Avionics Center (Indianapolis) recognized that this method, although useful for plating residue levels, was not applicable for measuring ionic contamination levels in the post-soldering rosin flux residues remaining after solvent defluxing. Thus, the first requirement was to find a solvent system capable of dissolving away any post-defluxing rosin residues and liberating any incorporated ionic activator residues for solvation and subsequent measurement. The test solution chosen for MIL-P-28809 is 75/25 (vol./vol.) isopropyl alcohol/deionized water. This mixture was found to have excellent solvating power for the rosin residues while providing adequate dissociation and sensitivity for meaningful ionic conductivity/resistivity measurements. After matching the test solution and flux residues, the Naval Avionics team completed their work by setting an ionic cleanliness acceptance limit derived from an extensive reliability history data base.[1,2] Key points regarding MIL-P-28809 testing are summarized in Table II.

Detailed discussions of ionic contamination and appropriate testing are readily available.[8,9,10]

NONIONIC CONTAMINATION

Nonionic contaminants present a much greater challenge to production and quality assurance personnel, since the contaminants and their sources are often very difficult to recognize and identify.

Table II. Post-Defluxing Ionic Contamination Test: MIL-P-28809 Printed-Wiring Assembly Pre-Conformal Coating Cleanliness Test Key Points

- Solvent Extract Conductivity/Resistivity Method
- Isopropyl Alcohol/Deionized Water (75/25 V/V) Test Fluid
- Reliability History Used to Set Pass/Fail Limit
- Applicable Only to Post-Solvent Defluxed Rosin Flux Residues

The classic example of this problem involved the discovery of a tenaciously adherent, gray, nonionic contaminant on printed wiring board dielectric surfaces capable of severely degrading insulation resistance when exposed to high humidity.[11] The contaminant's appearance was viewed without optical aid, at 1000x magnification and at 3000x magnification is illustrated in Figures 1-3.

Figure 1. Appearance of the nonionic contaminant without optical aid.

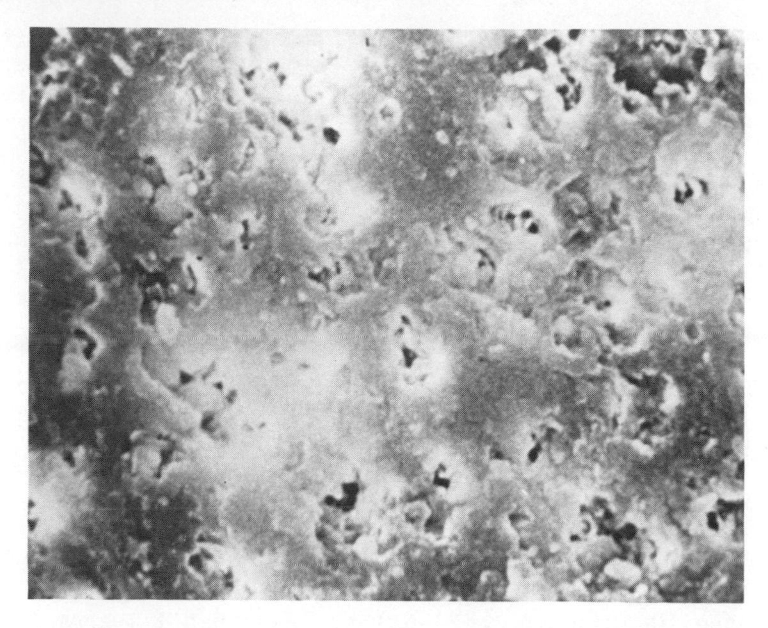

Figure 2. Appearance of the nonionic contaminant at 1000x magnification.

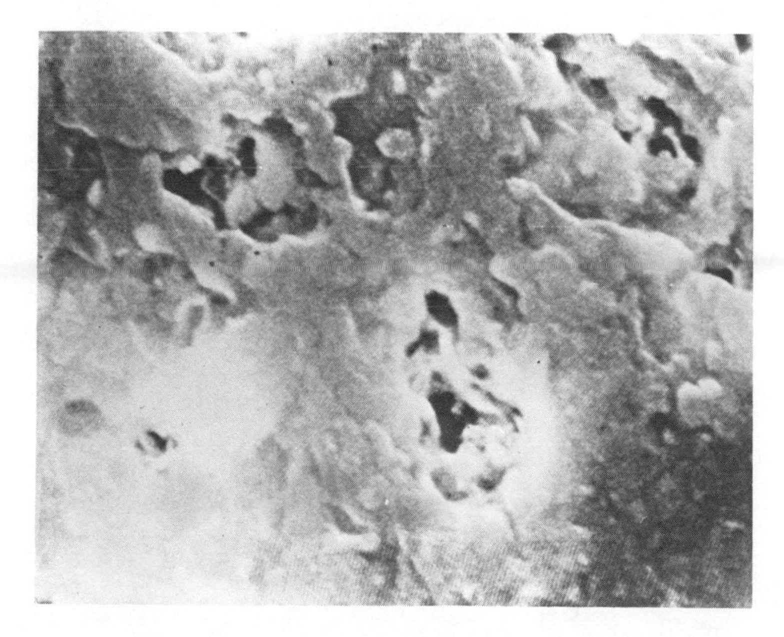

Figure 3. Appearance of the nonionic contaminant at 3000x magnification.

X-Ray fluorescence spectrometry, microprobe analysis (Figure 4) and ESCA (Electron Spectroscopy for Chemical Analysis) technique were used to identify the contaminant composition as a mixture of tin and lead oxides.

This unusual contaminant

- was first observed after final assembly and cleaning.
- was neither present on all printed wiring assemblies nor evenly distributed on the affected assemblies.
- was not removed by conventional "white residue" removal techniques such as a second application of flux followed by defluxing in clean solvent.
- had no effect on insulation resistance under ambient conditions.
- caused the insulation resistance to drop from greater than 10 gigaohms (ambient) to less than 1 megohm (87% relative humidity).
- could cause assembly failure at levels undetected by standard visual inspection methods.
- was designated as nonionic since it
    1. had no effect under ambient humidity conditions on the insulation resistance of printed wiring assemblies cleaned using the standard defluxing procedure.
    2. caused assembly failure under high humidity conditions.
    3. did not show any conductivity increase when subjected to the solvent extract resistivity/conductivity test for assembly cleanliness.

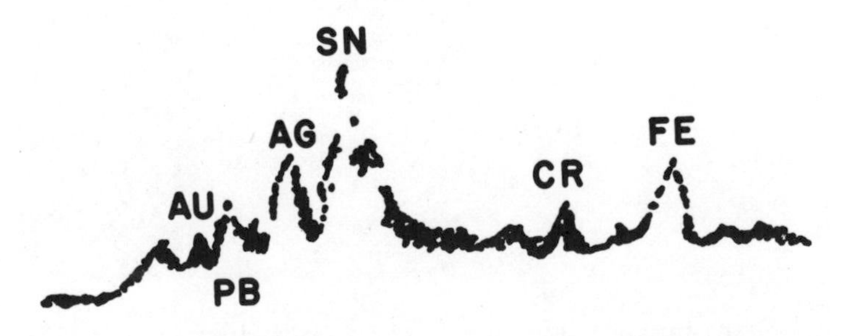

Figure 4. Microprobe analysis of nonionic contaminant.

Since standard solvent extract resistivity/conductivity tests[1,7] could not predict assembly failure, a new test was urgently needed. This need was fulfilled by the development of a novel humidity cycling test using a pair of clock circuits. This Dual Clock Circuit (DCC) test established that the observed insulation resistance degradation in CMOS digital switching circuitry was a function of both the contaminant and humidity.

The complete humidity cycling test system, consisting of a 300 Hz oscillator, humidity chamber, two identical clock circuits, test printed wiring assembly and a dual trace oscilloscope is shown in Figure 5. A pair of adjacent parallel traces on the test assembly and the dielectric surface between them, equivalent to a resistor, was electrically connected in parallel with the one megohm resistor in the test clock circuit. The control circuit lacks the "parallel resistor," but is identical in all other respects.

After connecting the test assembly to the test clock and placing the assembly and both clocks in the humidity chamber, the test is performed by impressing a 300 Hz square wave signal on the clock circuits while varying the humidity and observing the clock circuit outputs which are simultaneously displayed on the dual trace oscilloscope.

The resulting pulse width of the control circuit remained at approximately 450 microseconds throughout the relative humidity range of 50 to 95% while the identical circuit with one pair of parallel traces from a known contaminant bearing printed wiring assembly attached in parallel electrically with the one megohm resistor exhibited a decrease in pulse width from approximately 450 microseconds at 50% relative humidty to 150 microseconds at 70% relative humidity (Figure 6). The dramatic decrease in insulation resistance observed in this test established that inservice assembly failure would occur under moderate-to-high humidity conditions if the contaminant was present!

A second striking observation was that the initial high ambient insulation resistance of the test assembly could be readily restored by opening the humidity chamber door. The multiple tracings observed in the test clock portion of the oscilloscope display as the pulse width was restored to 450 microseconds from 150 microseconds are clearly shown in Figure 7.

To confirm that the test results displayed on the oscilloscope were due to the presence of the test printed wiring assembly and not to any electrical anomalies in the test equipment, the test printed wiring assembly was connected in parallel with one megohm resistor of the control circuit and the test was repeated.

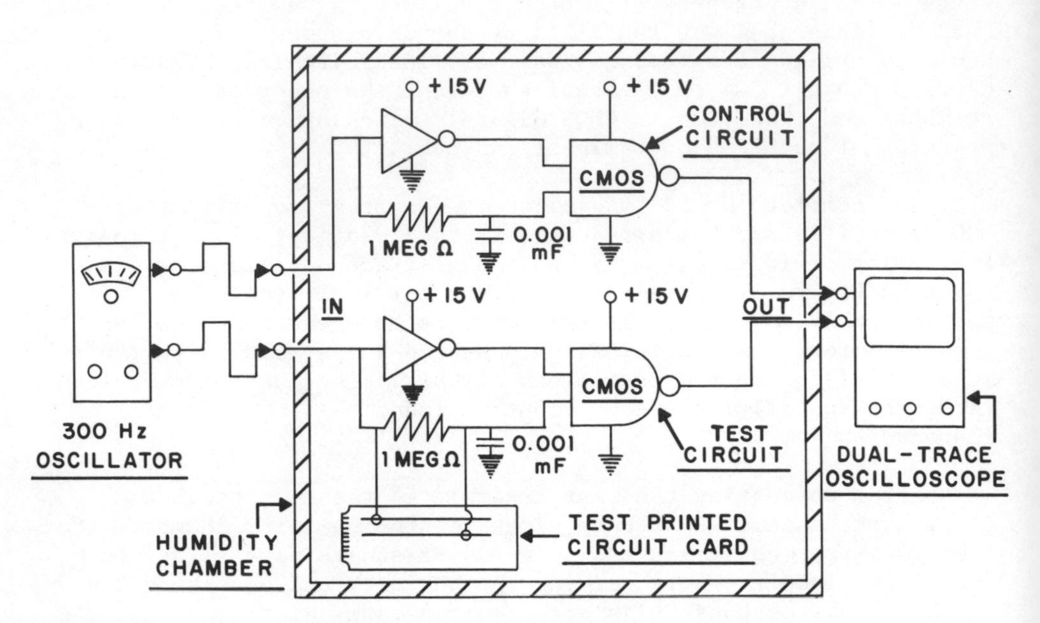

Figure 5. Novel test equipment for measuring the effect of humidity on printed wiring assembly insulation resistance.

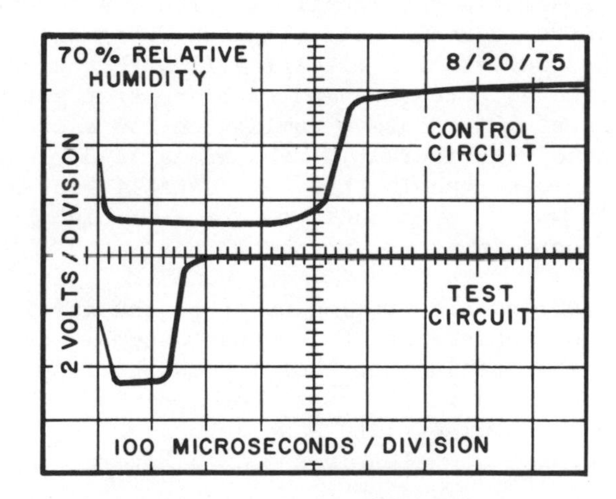

Figure 6. Insulation resistance degradation caused by nonionic contaminant and humidity resulting in decreased CMOS switching circuitry pulse width.

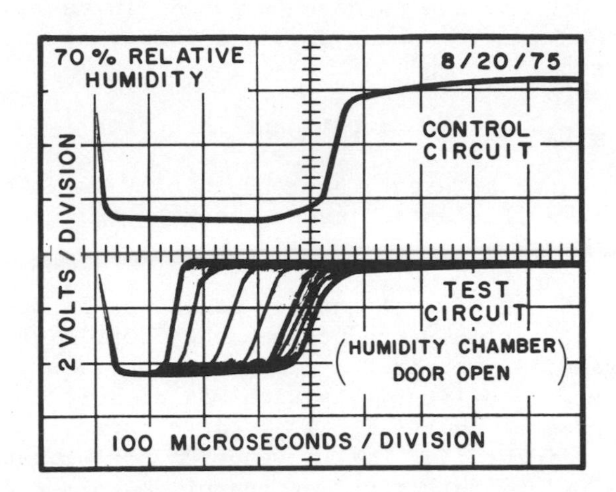

Figure 7. Recovering of high initial insulation resistance upon lowering the relative humidity from 70% (test level) to 50% (initial level).

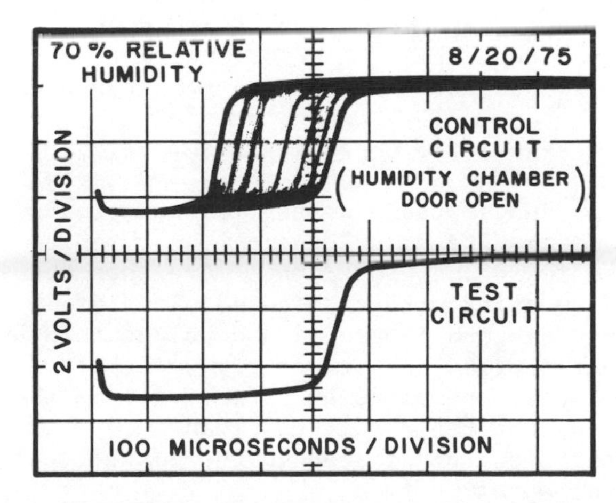

Figure 8. Control circuit-test circuit equivalence check. Recovery of high initial insulation resistance upon lowering the relative humidity from 70% (test level) to 50% (initial level).

Again, the decrease in pulse width was directly related only to the presence of contaminant and increased humidity. Reversal of the insulation degradation phenomenon was again easily demonstrated by opening the humidity chamber door and observing the oscilloscope display (Figure 8).

Nonionic Contaminant Detection

It later became apparent that essentially invisible levels of contaminant were sufficient to cause assembly failure under high humidity conditions.

Since Dual Clock Circuit testing could not be performed at the same speed as assembly production, the identification of unacceptable assemblies required a dielectric surface test system capable of reacting qualitatively with one or more contaminant components to produce an easily observed response. The detection system of choice identifies the presence of contaminant by reacting with lead, the major contaminant component, at a specific pH. A distinct yellow-to-red detector solution color change accompanies the reaction.

Performing the test requires separate buffer and detector solutions. The buffer solution is stable, but the detector solution has a limited shelf life (8 hours). To perform the test, place two drops of buffer on a clean cotton swab and then rub the dielectric test area with the swab. Next, place three drops of detector solution on a second, clean cotton swab, and rub it on the buffer-swabbed test area. A color change from yellow to red on the detector swab denotes a positive test for contaminant.

Nonionic Contaminant Removal

Electrically unacceptable assemblies were coated with a proprietary thermosetting resin solution[12] and cured at 150°F for 1 hour. The cured resin and contaminant were then readily removed by brushing and spraying with a patented blend of 1,1,2-trichloro-1,2,2-trifluoroethane and isopropyl alcohol containing a trace of nitromethane inhibitor.[13] Interestingly, neither of the pure solvents cleaned the assemblies as effectively as the blend. The removal of lead by the resin/solvent cleaning system was confirmed by positive identification of lead in the concentrated solvent washings by X-ray spectrometry.

The cleaned assemblies retained the high ambient insulation resistance level when tested at high humidity using the dual clock circuit and gave a negative response to the lead detector test. The assemblies were now judged to be electrically acceptable based on in-house life tests. Careful monitoring and follow-up

Table IIIa. Nonionic Contamination Sources, Problems & Preventive Measures.

Printed Wiring Board Fabrication Materials & Processes

Source	Problem	Preventive Action
Laminate	Incomplete Cure	Use only fully cured laminate (Check with "Pressure Cooker" test or by differential scanning calorimetry)
Plating	Acid	Minimize acid usage, espcially hydrochloric acid
Etching	Acid	Neutralize and rinse highly alkaline etchant per manufacturer's instructions
Reflow	Acid, Abrasive	Minimize acid usage, rinse completely, avoid abrasive brushing during rinse cycle
"Cosmetic Polishing"	Abrasive	Don't "pumice scrub" tin/lead traces to brighten their appearance
Solder Mask (UV Cure)	Incomplete Cure	Monitor UV output of curing lamps Use adequate UV exposure time
Solder Mask (Epoxy)	Incomplete Cure	Follow measuring, mixing and curing directions exactly

Table IIIb. Nonionic Contamination Sources, Problems & Preventive Measures.

Printed Wiring Assembly Materials & Processes

Source	Problem	Preventive Action
Water Soluble Fluxes	Acid, Glycols	Upgrade component & printed wiring board solderability specifications & storage conditions to permit adoption of milder fluxing and defluxing materials.
Alkaline "Saponification" Materials for Rosin Flux Residue Removal	Alkali	Recommend using alternative defluxing materials.
Unstabilized Chlorocarbon or Chlorocarbon/Alcohol Mixtures and Active Metals	Acid	Recommend using stablized fluorocarbon/alcohol mixtures when cleaning active metals.
Chlorocarbons or Chlorocarbon/Alcohol Mixtures with Spent Inhibitor Systems & Active Metals	Acid	Recommend using stabilized fluorocarbon/alcohol mixtures when cleaning active metals.
Unstabilized Fluorocarbon/Alcohol Mixtures and Active Metals	Acid	Recommend using stabilized fluorocarbon/alcohol mixtures when cleaning active metals.

analysis showed that no known in-use failures could be attributed to this contaminant removal method.

Nonionic Contamination Sources & Prevention

Nonionic contamination prevention is simpler and cheaper than contaminant removal, which requires additional processing steps and testing. Sources of nonionic contamination and corresponding preventive measures are summarized in Tables IIIa & b. Acid, alkali, abrasion and solder mask are usually involved in any nonionic contamination problem. Glycols, glycol esters and glycol ethers have recently been identified as the water-soluble flux component responsible for the insulation resistance degradation which make such materials unacceptable for use on long-term high reliability requirement products.[14]

Partially cured solder mask offered a chance to measure and compare the effects of ionic and nonionic contaminants on printed wiring assembly performance. Having established that partially cured solder masked assemblies would fail the dual clock circuit test unless they were post-cleaned using the D-7L thermosetting resin decontamination treatment described above,[12] an unpopulated, solder-masked printed wiring board was fluxed, soldered, defluxed and sheared in half. After one-half has received the D-7L resin treatment, both halves were sheared into test coupons for ionic contamination measurement. These data (Table IV) show that the ionic soil levels on the two halves are indistinguishable.[15] This is an excellent example of the inability of the solvent extract conductivity/resistivity test to predict assembly reliability.

Table IV. Ionic Contamination Levels on Solder Masked (UV Cure) Printed Wiring Boards.

Nonionic Contaminant Present?	Ionic Soil Levels, Equivalent µg NaCl/cm^2	
	Mean[a]	Std. Dev.
Yes	7.99	4.53
No	9.42	1.23

[a]Mean of three replicates, measured as described in Reference 9.

The use of stabilized fluorocarbon/alcohol combinations are strongly recommended to replace unstabilized halogenated solvents for defluxing in the presence of active metals, such as aluminum, magnesium and zinc. It is especially important to understand how the stabilizer action differs when used with either chlorocarbons or fluorocarbon/alcohol combinations. The stabilizer mixture used with chlorocarbons merely "accepts" or "neutralizes" any acid formed--until the stabilizer capacity is exceeded. In contrast, the nitromethane stabilizer used in modern fluorocarbon/alcohol combinations interferes with the acid generation mechanism to prevent acid formation (Table V). Laboratory studies which determined that the nitromethane level required for effective protection of fluorocarbon/alcohol combinations in the presence of aluminum 7075 alloy, moisture and air should be greater than 500 ppm have been reported elsewhere.[9]

Table V. Stablizers and Inhibitors for Protection of Halogenated Solvents in Contact with Active Metals.

Solvent Type	Stabilizer Type	Stabilizer Level, Wt.%[a]	Practical Degree of Protection
Chlorocarbon	Acid Acceptor	5 - 10	Limited
Fluorocarbon/ Alcohol	Acid Generation	0.25 - 1	Unlimited

[a]Based on weight % of halogenated defluxing solvent

SUMMARY

Nonionic contamination is now recognized as a significant roadblock to guaranteed printed wiring assembly reliability. Methods for nonionic contaminant detection, measurement and removal have been developed. The key factors for contaminant generation in many printed wiring assembly manufacturing steps have been identified, accompanied by appropriate contamination prevention measures.

ACKNOWLEDGEMENT

I would like to thank Mr. Garland J. Engelland, Wescom, Inc., Downers Grove, IL, Mr. John Zachariah, Hi-Grade Alloy Corporation, East Hazelcrest, IL, and Mr. Rene Guerra, Digital Equipment Corporation, Westfield, MA for their significant contributions to this paper.

REFERENCES

1. Military Specification MIL-P-28809: "Printed-Wiring Assembly."
2. W. T. Hobson, Materials Research Report 3-72, "Printed-Wiring Assemblies."
3. Military Specification MIL-F-14256D: "Flux, Soldering, Liquid (Rosin Base)."
4. W. T. Hobson, Materials Research Report 3-78, "Review of Data Generated with Instruments Used to Detect and Measure Ionic Contaminants on Printed-Wiring Board Assemblies."
5. Military Specifications MIL-S-45743E: "Soldering, Manual Type, High Reliability, Electrical and Electronic Equipment;" MIL-S-46844C: "Solder Bath Soldering of Printed-Wiring Assemblies;" MIL-S-46860B: "Soldering of Metallic Ribbon Lead Materials to Solder Coated Terminals, Process for Reflow."
6. Military Specification MIL-P-55110C: "Printed Wiring Boards;" Military Standard MIL-STD-275D: "Printed-Wiring for Electronic Equipment."
7. T. F. Egan, Plating, p. 350 (April, 1973).
8. C. J. Tautscher, "The Contamination of Printed Wiring Boards and Assemblies," Omega Scientific Services, Bothell, WA, 1976.
9. W. G. Kenyon, in "Technical Proceedings of NEPCON-WEST/EAST, 1977.
10. W. G. Kenyon, The Institute for Interconnecting and Packaging Electronic Circuits, Sept. 1977.
11. G. J. Engelland, and W. G. Kenyon, Technical Review, Feb., 1977; Circuits Manufacturing, July, 1977.
12. Hi-Grade Alloy Corp., Technical Bulletin: D-7L Thermosetting Resin.
13. E. I. du Pont de Nemours & Co. (Inc.) Freon® Products Division Technical Bulletin No. FST-5F: Freon T-P 35.
14. F. M. Zado, paper presented at the Institute for Interconnecting and Packaging Electronic Circuits, San Diego, CA, September, 1978.
15. M. G. Natrella, "Experimental Statistics," National Bureau of Standard Handbook, No. 91, 1966.

TECHNIQUES FOR REMOVING SURFACE CONTAMINANTS IN THIN FILM DEPOSITION

Bharat Bhushan

Mechanical Technology Incorporated
968 Albany-Shaker Road
Latham, New York 12110

This paper describes the way to clean bearing surfaces so coatings can be applied to increase wear life. The paper describes the origin of contamination and the techniques of removing them. Typically, bearing surfaces are cleaned in a vapor degreaser consisting of trichlorethylene solvent or by alcohol to remove any oil and grease and then they are pretreated by phosphating, sand blasting, vapor honing, hand polishing, acid etching and electro-etching. Mechanical processes such as sand blasting, vapor honing, and hand polishing cannot be used on a substrate thinner than approximately 0.2 mm, but the acid etching and electro-etching can be used on thin foil substrates and in other specialized applications. In vacuum deposition processes, the surfaces are cleaned in a vacuum prior to deposition so the cleaned surface is not exposed to air before coating and, therefore, the bond is very good.

INTRODUCTION

Surface contaminants are defined as undesirable material with chemical characteristics different from the surface of the basis material. The surface contamination could be present either in the film form or as particulates. What is purposefully added to one part of a system can be considered as an impurity or contaminant in another part. No manufactured material is without some degree of contamination. The contaminants often are introduced during processing of the materials such as shaping (machining, grinding, etc.), joining (soldering, brazing, etc.), heat treatment, and exposure to a reactive atmosphere. Typical contaminants are machining oils and greases, thin oxide layers from heat treatment or hot working, adsorbed layers, dust, lint, fingerprints, metallic contaminants, and the residues from reactions of outgassing products in assembly of metals and nonmetals (plastics), etc. Contamination may also be accidental (rust, corrosion). The problem in process control is to reduce the degree of contamination to a level conforming to specifications.

A clean surface is one that contains only a small fraction of a monolayer of foreign material. The reasons for cleaning depend on the application [1 to 3]. Solid lubricant coatings are applied on the bearing surfaces to reduce friction and wear during its operation. The cleaning of the bearing surfaces is carried out to improve the coating adhesion to the substrate [4 to 6]. Sometimes, however, deliberate addition of the foreign matter may be helpful in improving the solid-solid bonding. The adhesion between deposited film and metal surfaces depends on the interfacial morphology, the nucleation, chemical reaction, and diffusion in the interfacial region. Very thin oxide layers may provide effective barrier layers which prevent good bonding or adhesion. Since surface oxides and adsorbed hydrocarbons are commonly found on the surfaces, the removal of these materials is often of concern in thin film deposition technology.

If the mechanism of coating adhesion is mechanical in nature, then preroughening of the surface after cleaning of foreign matter is beneficial. Surface roughening forms micro-grooves, film material fills the grooves, and mechanical interlocking provides the adhesion. No chemical reaction between the materials is necessary, and the strength of the joint depends on the physical properties of the materials.

This paper will be limited to the description of cleaning processes and other surface preparations relevant to thin film deposition in bearings. Experiments on the pre-treatment of foil bearing surfaces are discussed in detail. Specifically, the sand blasting, vapor honing, acid etching, electro-etching, and sputter-etching are discussed.

SELECTION OF CLEANING PROCESSES

Cleaning processes may be categorized as specific or general. Specific cleaning processes are designed to remove specific types of contaminants without changing the surface of interest, for example, solvent cleaning techniques which dissolve or emulsify contaminants without attacking the surface. General cleaning techniques often involve removing some of the surface material, as well as contaminants, for example, sputter cleaning, electropolishing, chemical etching, and mechanical abrasion [5]. Often a cleaning procedure will involve a combination of procedures used in sequence. Unless the cleaning procedure is designed to take advantage of the strong points and weak points of each step, one cleaning step may defeat another.

The following cleaning techniques are commonly used: vapor degreasing, solvent cleaning, acid cleaning, emulsion cleaning, alkaline cleaning, salt bath descaling, pickling, abrasive blast cleaning, barrel finishing, mechanical polishing, chemical etching, electro-etching, plasma cleaning, sputter etching, sand blasting and vapor honing (for details for each technique see, e.g., References 5 and 7).

Since numerous cleaning techniques are available, selection of the proper technique becomes difficult. In many instances, a technique is selected by trial and error. To design a cleaning process, one must consider the following factors:

- Mechanical, chemical and geometrical properties of the base material to be processed
- Mechanical, chemical and geometrical properties of the contaminants
- Relevant cleaning techniques available
- Required degree of cleanliness
- Effect of cleaning on the mechanical and chemical properties of the base material
- Cost/advantage effectiveness of cleaning techniques
- Quality control on the procedure

The most important factor in designing a cleaning process is identifying the type of contaminant and its level of adherence to the substrate. A water-soluble contaminant and some organic contaminants can be removed with a detergent, by scrubbing with an abrasive cleaner, or by ultrasonic agitation. The common shop oils and greases can be effectively removed by several different cleaners: emulsion (organic solvent), alkaline, and solvent cleaners and vapor degreasers (employing hot vapors of a chlorinated solvent). Vapor degreasing is an effective and widely used method for removing a variety of oils and greases and is well adapted to cleaning oil-impregnated parts such as bearings. Small quantities of oil and greases can be removed by rubbing the surface with methyl alcohol or acetone.

Metal surfaces often have oxide layers which can be removed by mechanical scrubbing, blasting with abrasive particles or etching. Surfaces can be prepared by polishing with a silica paper or sand blasting. Thin surfaces can be prepared by vapor honing, chemical etching or electro-etching. If subsequently, the coating has to be put on by the vapor deposition process, the surface can be cleaned by sputter etching. Sputter cleaning or sputter etching has the advantage that it may be done in situ in the deposition system just prior to the deposition process.

EXAMINATION OF CLEANED SURFACES

After a surface has been prepared, it should be checked for its cleanliness and proper surface roughness. Visual examination with a microscope is one of the most effective and universal tests for examination of substantial contaminant films. With proper use of reflected or refracted light from a suitable light source, the presence of contaminant films can be detected (Reference 2). Very gross estimation of cleanliness is possible by rubbing the cleaned surface with white tissue paper and examining for grease on the paper.

To identify the surface contaminants not easily visible by a microscope, several direct and indirect monitoring techniques can be used. Direct monitoring involves identifying the surface species directly such as Auger Electron Spectroscopy (AES), Appearance Potential Spectroscopy (APS), Ion Scattering Spectroscopy (ISS), and multi-reflectance infrared analysis, etc. Indirect monitoring techniques utilize the behavior of the surface to imply the presence of contaminants such as contact angle, wetting, adsorption, and adhesion, etc. (for details see Reference 2.)

Once the contaminants have been detected, they can be identified using one of the metallurgical examination techniques such as X-ray diffraction, electron diffraction, electron microprobe and mass spectroscopy, etc. The surface roughness measurements can be made by a surface profilometer. Microscope photographs also help study the surface profile qualitatively.

GENERAL REMARKS ON PRETREATMENT OF BEARING SURFACES

The bearing surfaces have to be cleaned and roughened prior to coating for improved coating adherence. Typically, the bearing surfaces to be coated consist of oil, grease and thin oxide layers, and their removal is carried out in two steps:

- Clean oil and grease, if any, from the surface in vapor degreaser or by mechanical scrubbing and subsequent rubbing with solvent cleaners.
- Use mechanical or chemical means to remove any oxide layer and roughen the surface.

If the coating has to be applied by plasma gun, air spraying or the fusion process, the surfaces usually are prepared by chemical etching, electro-etching, phosphating, abrasive blast cleaning, vapor honing or hand polishing. If the coating has to be applied by sputtering, the sputter-etching technique is employed. The surface roughness of a thick surface so obtained should be about 1.5 μm (60 μ in.) CLA. If the substrate is very thin, this roughness cannot be obtained without altering the mechanical properties unfavorably, and a compromise has to be made between the desired roughness and acceptable mechanical properties of the treated surfaces.

PRETREATMENT OF FOIL BEARING SURFACES: EXPERIMENTS AND RESULTS

Research was conducted on the surface pretreatment technique employed in foil bearings. In a hydrodynamic resilient foil bearing construction, the bearing is comprised of a smooth top foil and a "bump" or convoluted foil. The bump foil gives distributed elastic support to the top foil which provides the bearing surface (see Figure 1). Both foil members are very thin, typically 0.069 to 0.127 mm (0.0027 to 0.005 in.), and made of Inconel X-750. Solid lubricant coatings are applied to the top foil to reduce friction and wear during starting and stopping of the bearing. The mating journal is usually made of stainless steel, eg., A286 and a hard coating is applied on it to minimize wear during sliding contacts. The top foil and the journal are heat treated in an air or argon atmosphere and this treatment results in an oxide layer (1000-3000Å thick) on the surfaces. The room temperature air exposure also results in a thin oxide layer (20-30Å thick) buildup within minutes. The oxide layer, oil and grease, and any fingerprints have to be removed prior to coating.

Cleaning of the Foil Surfaces

Experiments showed that oil, grease and other loose dirt could be removed from the foil members by cleaning in a vapor degreaser. To remove both the oxide layer and metallic contaminants and to roughen the surface, a procedure was developed to prepare the surface to be coated by an air spraying technique.

First of all, attempts were made to prepare 0.102 and 0.127 mm (0.004 and 0.005 in.) thick Inconel X-750 foils with mechanical processing. Sand blasting was done at varied pressures and grit sizes. Very fine grit and pressures as small as 35 kPa (5 psi) were used to reduce the distortion, but the foil was badly distorted after treatment. Vapor honing also distorted the foil. The foil was unsuccessfully treated on both sides to avoid curling. The surfaces were then prepared using 600 grit size polishing paper

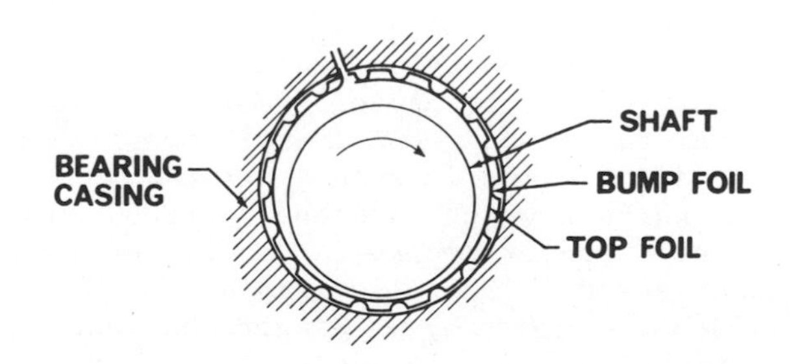

Figure 1. Construction of a foil bearing

on both sides of the foil. The foil warped slightly, and it was difficult to ensure that the foil would remain flat after pretreatment.

Because mechanical means of surface treatment seemed too severe, acid etching and electro-etching were used. One of the etchants successfully used to chemically etch an annealed foil was nine parts hydrochloric acid and one part hydrogen peroxide. An Inconel foil in the annealed condition dipped for 15 seconds in this etchant roughened the surface uniformly from 0.15 μm (6 μ in.) CLA to 0.4 μm (15 μ in.) CLA. The surface roughness of 0.4 μm on thin foils was considered adequate since any more roughness would reduce the thickness of the foil considerably and reduce its strength. The same etchant on foil heat treated in an argon atmosphere attacked the foil preferentially and the oxide layer came off at some spots earlier than others. The foil lost about 25 percent of its thickness before all the oxide layer could be removed. Microscopic examination revealed quite deep pits on the surface which would be detrimental to the mechanical strength of the foil.

The acid pickle for Inconel is usually an HF-HNO_3 solution. This solution attacks the basis metal and, therefore, was considered undesirable for thin foil due to danger of excessive metal loss. Other etchants unsuccessfully tried were: nine parts hydrochloric acid and one part hydrogen peroxide at 82°C (180°F), ten percent hydrofloric acid at room temperature, ten percent hydrofloric acid at 60°C (140°F), thirty percent sulfuric acid at 82°C (180°F), forty percent nitric acid at room temperature, forty percent nitric acid at 66°C (150°F), and five parts hydrochloric acid, one part nitric acid and five parts water at 77°C (170°F).

Next, an electro-etching technique was tried. Several etching solutions can be used. Murray and Calabrese [8] used one part concentrated hydrochloric acid, one part concentrated nitric acid and three parts water for etching of thin Inconel foils. They were able to achieve a surface roughness of about 3.3 μm (130 μ in.) CLA by etching at a current density of 1085 amps/m^2 (0.7 amp/sq in.) for 2.5 minutes. The examinations of profilometer traces of the etched foils showed that the etch pits were too deep and their effect on mechanical properties would be undesirable.

To retain the strength of the foil, a less potent solution was selected for electro-etching study. The solution consisted of 25 gms of sodium fluoride, 95 cc sulphuric acid, and 1000 cc water. The etching was carried out by making the Inconel foil as anode and any other metal as cathode. A six-volt power supply was used. The etchant removed the oxide layer and attacked the grain-boundaries of the surface which made the surface rough. The foil was etched until the oxide layer came off entirely and the metal phase was attacked. The foils warped due to mechanical processes but did not distort by etching techniques; thus, the electro-etching technique was considered to be a suitable choice.

The current density and etching time were varied to optimize the etching parameters. The surface roughness was measured and metallurgical examinations were conducted to study the condition of surfaces after treatment. Figure 2 shows typical optical micrographs of the etched foils. From metallurgical examinations, it was found that high current density ($\geq$ 4650 amp/m^2) attacked the grains as well as grain boundaries, and there was significant loss of the material. At lower current density, ($\sim$2325 amp/m^2), only grain boundaries were attacked; there was not as much loss of the material, and the surface was much rougher. The depth of grain boundary attack could be varied by varying the etching time. Very small current density, ($\leq$ 775 amp/m^2), resulted in preferentially etching and attacking of some grains since it took much longer to strip off the oxide layer. An optimum current density would provide more or less uniform grain boundary attack and virtually no attack at the grains.

Figure 3 shows typical Talysurf traces of the etched foils. The average surface roughness of the virgin foil was 0.15 μm (6 μ in.). The maximum roughness in etched foils, 0.28 μm (11 μ in.) was achieved at an etching condition of 2325 amp/m^2 (1.5 amp/in.^{2}) and two minutes. The surface roughness could be increased to 0.53 μm (21 μ in.) by etching at 2325 amp/m^2 for six minutes. Mechanical property tests showed that there was some reduction in strength of the foil etched at 2325 amp/m^2 for six minutes.

Based on the parametric study, an etching condition with an etching current of 2325 amp/m^2 and etching time of two minutes was selected. Since the solid lubricant coating needs to be applied on only one side of the foil; the other side of the foil was masked with stop-off lacquer (323 red, supplied by M&T Chemicals) during etching. The etched foils were thoroughly washed in ultrasonic cleaner and then cleaned with alcohol to remove all the chemicals from the surface.

Cleaning of the Journal Surfaces

The journal surfaces to be coated by air spray or flame spray techniques were treated by sand blasting subsequent to vapor degreasing. With silica of 100 micron grit size and an air pressure of 275 kPa (40 psi), the surface could be roughened to 1.5 μm-2 μm (60 - 80 μ in.) CLA in about ten minutes. If the journal surfaces were steel, a manganese-iron phosphate pretreatment resulted in long wear life of the surface. Close control of bath conditions was necessary to obtain consistent results. The wear life of a lubricant film applied over a surface which had been treated in a manganese-iron phosphate bath was affected by a number of factors; in particular, thickness of phosphate coating and the phosphate crystal pattern. If the coating has to be applied by electroplating, the surfaces can be adequately prepared by blasting with fine beads of glass. The blasting provides a surgically-clean

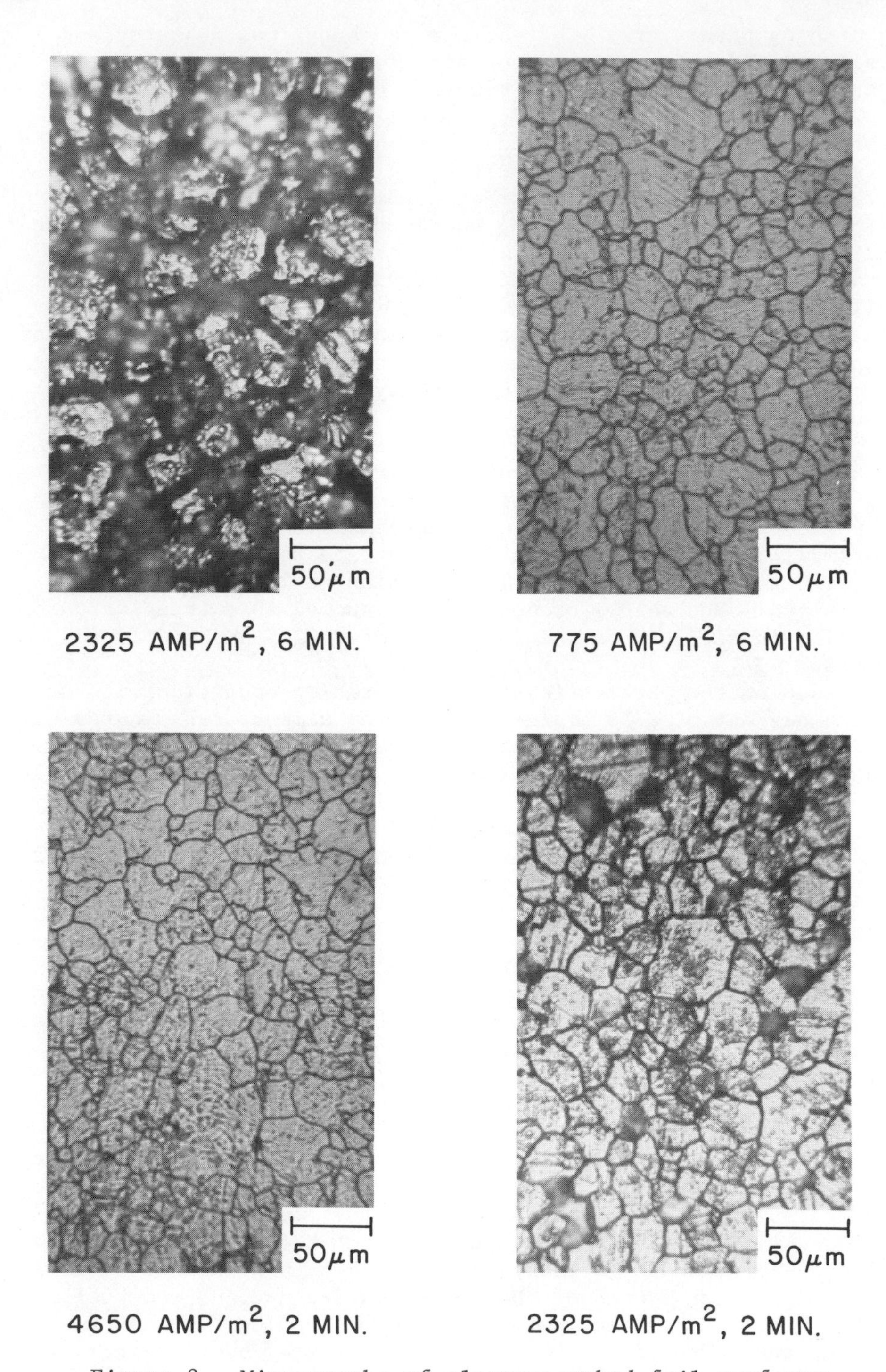

Figure 2. Micrographs of electro-etched foil surfaces

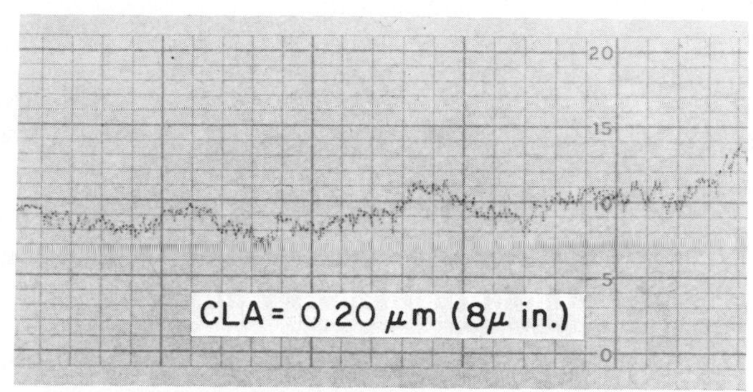

775 AMP/in.^{2} (0.5 AMP/in.^{2}), 6 MIN.

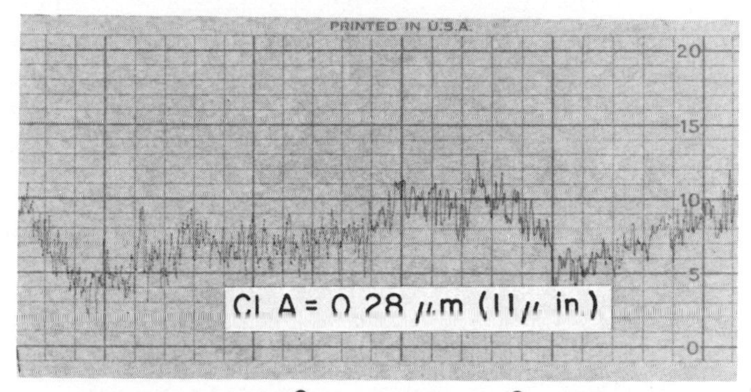

2325 AMP/m^2 (1.5 AMP/in.^{2}), 2 MIN.

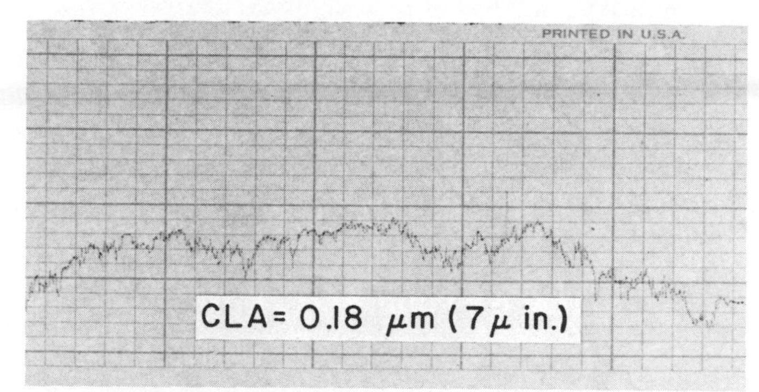

7750 AMP/m^2 (5 AMP/in.^{2}), 30 SEC.

Figure 3. Typical Talysurf traces of electro-etched foil surfaces
Vertical mag: Each small div. = 0.25 μm (10 μ in.)
Horizontal mag: Each small div. = 250 μm (0.010 in.)

surface; it removes practically no material from the surface, and it does not put any stresses on the surface.

Cleaning of the Bearing Surfaces for Sputtered Coatings

In the case of applying sputtered coatings on the journal and foil surfaces, the surfaces were cleaned in a vacuum chamber (sputter etching) prior to deposition. During sputter etching, the substrate had negative polarity with respect to the target. A shutter was inserted between the target and the substrate. Material from the substrate was removed by bombardment with positive ions of the argon plasma and deposited on the shutter. The etching was carried out for thirty minutes or until all the impurities appeared to have been removed. In this technique the cleaned surface was not exposed to air before coating and, as a result, the bond was usually very good.

The target was also cleaned prior to deposition. The target was installed and left under a vacuum for 24 hours before sputtering to insure that the target had outgassed. Sputtering was carried out for about thirty minutes with a shutter in between the target and the substrate. If the target was new, sputtering was carried out for at least one hour, since it was believed that there could be considerable contamination on the target surface during preparation and in handling. Sometimes hot pressing used in preparation of the target oxidizes the surface and changes its chemical formulation. This surface layer was removed before film deposition was carried out.

Preheating of the substrate prior to deposition has been shown to significantly increase the adhesion of rf sputtered TiC films on high speed steel substrates (Greene et al. [9]). The increased adhesion of the films on the preheated substrates was believed to be directly related to outgassing and desorption providing a cleaner deposition surface. In addition, it was possible that the heating might aid the formation of an Fe-Ti intermetallic interface layer which facilitated adhesion. The substrate was also heated during deposition which helped to degas and desorb any trapped gases during sputtering. In some cases, the substrate heating continuously anneals the coating and relieves the internal stresses developed during the deposition process.

A bias voltage, negative potential with respect to plasma, was also applied to lead to films of higher purity and good adhesion. The basis for bias sputtering is that during resputtering (due to bias voltage), most impurities should be preferentially removed relative to the atoms of the main film. Whether or not impurities are removed depends on the relative strengths of the metal-to-impurity and the metal-to-metal bonds.

Research on coating type 440C steel surfaces by several refractory coatings conducted by Brainard (Reference 10) showed that substrate bias, while producing good bulk film properties can adversely affect adhesion by removing interfacial oxide layers. Sputter cleaning of substrates prior to deposition produced poor adhesion for the same reason. Deliberately preoxidizing the substrate prior to coating produced significant gains in coating adherence. The improved adherence was believed to be related to the formation of a mixed oxide interface. It seems that pre-cleaning of the substrate is not always helpful and it depends on the substrate, the coating and influence of the environment on the substrate.

It was difficult to obtain good coating adherence on heat treated Inconel substrate due to its high hardness. If the Inconel X-750 was coated in the annealed condition, then heat treated, the coating adherence was improved. A bend test with 180° bend was conducted to study adhesion of the coating. The process sequence of coating the substrate in an annealed condition can be used for coatings which can withstand the heat treatment temperature without any adverse effect.

Coating of the Cleaned Surfaces

The foil surface and journal surfaces were coated with solid lubricant coating and hard coating, respectively, subsequent to pretreatment, and the mechanical bond of the coating was found to be significantly better. In one instance, etched and unetched Inconel foils were coated with CdO-graphite coating (MTI-HL-800). The foils were bent back upon themselves with the use of pliers and a vise and the coatings were on the outside of the bend. Each bend sample was then unfolded and examined for any cracks. The photographs of the specimens after tests are shown in Figure 4. In the case of the unetched foil as the heat treated foil, the coating flaked off the foil substrate in the area of the reverse bend. In the case of the etched foil, no evidence of cracking was found.

In another instance, a chemically adherent coating was applied successfully to the etched foils where the coating did not bond at all to untreated foils. The wear tests on bearings made with coated foils and coated journals were conducted using the test apparatus described by Bhushan, et.al. [11]. Generally, the life of the coating could be increased at least 50 percent with the use of surface pretreatment.

CONCLUSIONS

When surfaces are found to have contaminants, they must be cleaned in order to improve the coating adherence. The cleaning process depends on the type of contaminants, the method of coating and the substrate. In the foil bearing application, if the coating is to be applied by air spray, plasma spray or electroplating, the

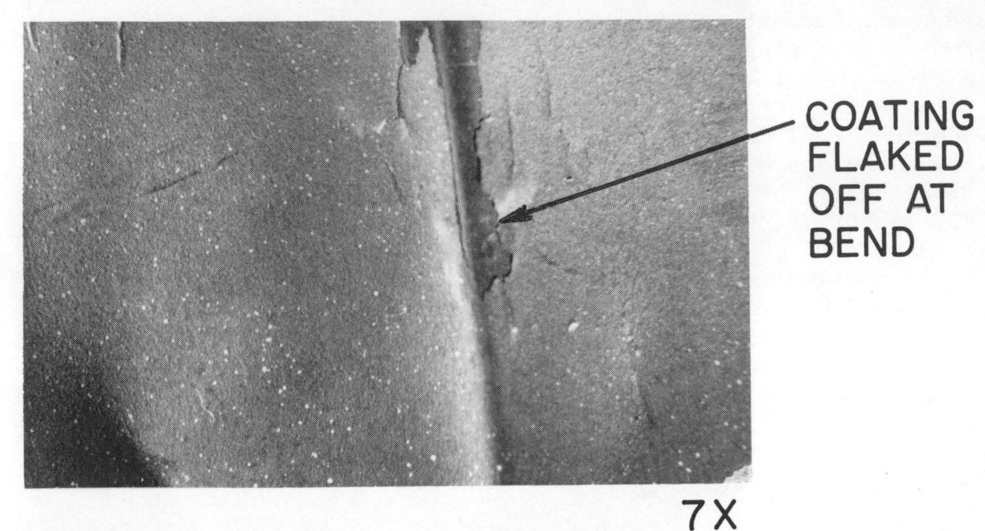

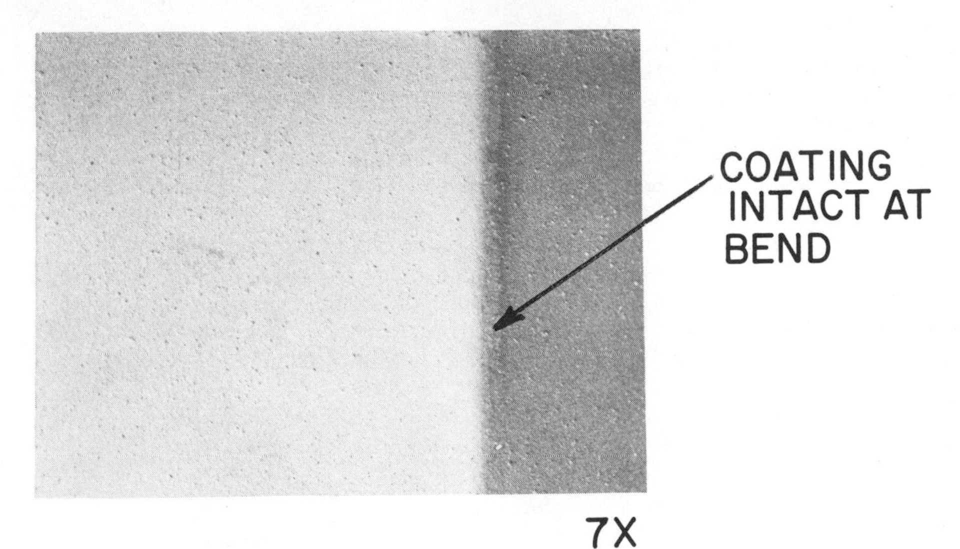

Figure 4. Photographs of foils coated with CdO-graphite after bend test.

thin foils can be successfully cleaned by subjecting them to a vapor degreaser and, subsequently, electro-etching; and the journal surfaces can be pretreated by subjecting to vapor degreaser and, subsequently, abrasive blast cleaning or phosphating, etc. If the coating is to be applied by sputtering, the foil and journal surfaces can be prepared by sputter-etching. Substrate heating and bias sputtering can be employed during the sputtering process to improve the coating material adhesion. Generally the coating adherence could be increased significantly with the use of a proper cleaning technique.

REFERENCES

1. D. M. Mattox, Thin Solid Films, 53, 81 (1978)
2. "A Survey of Contamination - Its Nature, Detection and Control", Sandia Laboratories Report SC-M-70-303, March 1971.
3. S. E. Beal, Thin Solid Films, 53 (1978).
4. B. C. Stupp, presented at the Joint Air Force - Navy - Industry Conference on Lubricants, Dayton, Ohio, February 1959.
5. B. Bhushan, Thin Solid Films, 53, 99 (1978).
6. H. E. Farnsworth, "Clean Surfaces - The Preparation and Characterization for Interfacial Studies", George Goldfinger, Editor, pp. 77-95, Marcel Dekker, New York, 1970.
7. "Heat Treating, Cleaning and Finishing", Metals Handbook, Vol. 2, American Society for Metals, Metals Park, Ohio, 1964.
8. S. F. Murray and S. J. Calabrese, "Selection and Evaluation of Foil Bearing Coatings for Use in Freon-11", Mechanical Technology Inc., Latham, N.Y., Report, June 1977.
9. J. E. Greene, J. Woodhouse, and M. Pestes, Rev. Sci. Instrum., 45, (6), 747 (1974).
10. W. A. Brainard, Proceedings of the 2nd International Conference on Solid Lubrication, pp. 139-147. American Society of Lubrication Engineers, Park Ridge, Illinois, August 1978.
11. B. Bhushan and S. Gray, ASLE Trans. (in press) paper presented at ASME/ASLE Joint Lubrication Conference, Minneapolis, Minnesota, October 1978.

CLEANING AND SURFACE PREPARATION TECHNOLOGY AND OTHER FACTORS

RELATED TO COATINGS PRODUCED BY FLAME SPRAYING

Y. A. Kharlamov,* M. S. Hassan,+ and R. N. Anderson++

*Voroshilovgrad Machine Building Institute,
Voroshilovgrad, 348034, USSR,
+Advanced Micro Devices, Sunnyvale, California 94086
++San Jose State University, Dept. of Materials Engr.
San Jose, California 95192

The physical characteristics of coatings produced by high temperature spraying is dependent on the equipment used and the condition of the surface. An experimental investigation was carried out using two types of thermal spraying equipment: induction plasma torch and detonation gun. The target surface parameters studied included velocity of the surface and the effect of thin oxide and water films and other contamination on the surface. Scanning electron microscope pictures clearly show the pronounced effect of surface contamination on particle formation on the target surface.

INTRODUCTION

Thermal spraying methods such as plasma-arc or flame-spray are used widely for making metal, ceramic and refractory coatings. Since its introduction in 1905, the development of thermal spraying techniques has increased the variety of materials processed and made possible improvements in the properties of spray coatings. A wide range of carbides, oxides, aluminides, etc. are now available for producing coatings resistant to high-temperature wear, abrasion and corrosion. Thermal spraying processes are also used for producing metallic coatings and coatings with special magnetic, electrical or optical properties.

HIGH TEMPERATURE SPRAYING METHODS

The most common methods for applying protective coatings are: 1) flame spraying; 2) plasma spraying; 3) detonation gun spraying; 4) electric arc metallizing.

The flame spraying process utilizes a wire, powder or rod of the material being sprayed which is fed into a chamber where it is heated to a high temperature in an oxyacetylene flame. The melted material in the wire and rod systems is atomized by a compressed air blast which transports the particles to a previously prepared target surface.

The plasma spraying process for applying metal and ceramic coatings uses a torch of high velocity inert gas which can be maintained at temperatures over 20,000^oF (11,366^oK). The hot gas stream can melt, and accelerate to a high velocity, particles of any solid inorganic material which is capable of melting without decomposition. When the molten particles strike the target substrate, they impact to form a dense, high purity coating. The coating material is introduced into the torch in the form of powder, wire or rod.

The detonation gun spraying process is similar to the flame powder spraying method. In this process, measured quantities of fuel gas and oxygen and suspended particles of a metal or nonmetallic coating material are pressure fed into the chamber of a specially constructed gun. A spark ignites the gaseous mixture and creates a detonation which hurls the coating particles, now heated to a temperature that makes the material plastic, out of the gun barrel at a speed of 800 metres per second and onto the prepared surface.

The electric arc spraying process uses a motor generator to provide direct current to two consumable wire electrodes. When the wire feed is started, the tips make contact, strike an arc and melt. The molten metal particles are propelled toward the substrate by a blast of compressed air.

VARIABLES INFLUENCING THE CHARACTERISTICS OF THERMAL-SPRAYED COATINGS

The microstructures of materials produced by thermal spraying generally consist of thin lenticular particles with a very fine grained structure. The particles are rapidly quenched from the molten, or near molten state and may have a high degree of residual stress. The physical characteristics of the as-deposited coatings depend on the type of equipment used, the operating parameters, and the chemical composition of the material. A bibliography of the literature on variables relating to the operation of the equipment, selection, and applications is given in references 1-104.

MECHANISMS OF FORMATION OF SPRAYED COATINGS

The mechanism of the coating formation by thermal spraying is not fully understood. A discussion on adhesion is complicated by the fact that, depending on the spraying process used and on the combination of materials involved, one or more bonding mechanisms may be operative. According to A. R. Moss and W. J. Young (105) the three main possible mechanisms in spraying are:

(i) mechanical interlocking
(ii) metal/metal bonds-
  liquid/liquid as in fusion welding,
  liquid/solid as in soldering and brazing,
  solid/solid as in solid phase pressure welding,
  solid/liquid as in hot dip galvanizing,
(iii) chemical bonds, for example spinel formation.

Industrial experience and special research confirm that significantly higher bond strengths are achieved on roughened target surfaces than on smooth ones of the same material. Although this would apparently support the interlocking hypothesis of bonding, it is unlikely that a simple mechanical concept is possible because of four observations (105):

(i) thermal isolation of the projections on roughened surfaces,
(ii) the existence of a critical temperature for adhesion,
(iii) penetration of the oxide film on roughened aluminum by high velocity spray particles,
(iv) reduction of bond strength at shot blasted surfaces by a film of contaminant.

The process of bonding between plasma sprayed particles and substrate was investigated by V. V. Kudinov (1). The interaction between material of the substrate and material of the particles is divided into three stages:

(1) physical contact of materials,
(2) activation of contact surfaces and chemical interaction of boundary materials,
(3) development of interaction effects.

The change of relative strength of attachment of particles in contact with the substrate is given by Equation (1).

$$\frac{N(t)}{N_o} = 1 - \exp\ (\nu t)\ \frac{-\ 1}{\exp(E_a/kt)} \qquad (1)$$

where No is the quantity of substrate surface atoms or particles in physical contact; N(t) is that portion of atoms or particles in

N_o which have reacted in time t; ν is frequency of self oscillations of atoms; E_a is the activation energy; T_k is the absolute temperature of contact; k is Boltzmann's constant.

In accordance with this equation, the formation of strongly adhesive spray coatings could be accomplished in three ways:

(1) increasing the time the particles are in contact with the substrate at elevated temperature T_k;
(2) decreasing activation energy E_a;
(3) increasing contact temperature T_k.

The first approach is limited because the particles are rapidly cooled on the substrate surface. Decreasing the activation energy E_a, is achieved by preliminary deposition on the substrate surface of thin layers of material with low melting point or by using powder particles covered with a low melting material. E_a can also be decreased by increasing the particle velocity.

Increasing contact temperature can be achieved by increasing the temperature of substrate surface and/or the particles as well as by modification of thermo-physical properties of surface of the substrate or the particles.

PARTICLE INTERACTION WITH A HOMOGENEOUS SUBSTRATE MATERIAL

The task of determining the temperature and pressure in the contact zone between particle and substrate by experimental methods is very difficult.

According to experimental and analytical investigations (107-109) the following parameters may influence the plastic deformation of substrate:

A. Parameters describing the substrate material
(1) density
(2) velocities of the compression and shear waves
(3) modulus of elasticity
(4) Poisson's ratio
(5) ultimate tensile strength
(6) compressive and shear strength
(7) fracture toughness
(8) hardness
(9) grain size
(10) surface roughness
(11) curvature of surface
(12) thickness
(13) temperature

B. Parameters describing the solid particles
(14) density
(15) velocities of the compression and shear waves
(16) modulus of elasticity

(17) Poisson's ratio
(18) compressive and shear strength
(19) hardness
(20) grain size
(21) shape of particles
(22) sizes of particles
(23) temperature
(24) size distribution of the particles

C. Parameters describing the melted particles
(25) density
(26) velocity of a compressive wave
(27) viscosity
(28) temperature
(29) surface tension
(30) shape of the particles
(31) size distribution of the particles

When a particle impinges upon a substrate surface, the particle comes to a sudden stop. The sudden deceleration causes a pressure build-up at the particle-substrate interface. The high pressure inside the particle then forces the ductile material to misshape. In the case of melted particles this high pressure forces the melted material to flow laterally. The plastic deformation of the substrate is produced both by the effects of the high pressure at the particle-substrate interface and by the lateral outflow of the particle material. The pressure acting across the particle-substrate inter-surface results in a force normal to the substrate surface. This force generates stress waves traveling along the substrate surface and inside of its material. The pressure can be given Equation (2):[107]

$$P = \frac{Z_p V \cos\theta}{1 + Z_p/Z_s} \quad (2)$$

where $Z_p = \rho_p C_p$ and $Z_s = \rho_s C_s$ are dynamic impedances, ρ_p and ρ_s are densities of particle and substrate materials, respectively;

C_p and C_s are the velocities of sound in the particle and substrate materials; V is the velocity of impact; and θ is the angle of impact. In the case of solid particles, it is possible to use the following expression to calculate the depth of particle penetration h [110-112]:

$$h = \left(\frac{m_o}{2\pi\rho_s K_o r_o^2}\right) \ln\left(1 + \frac{\rho_s V_o^2}{H_s}\right) \quad (3)$$

where m_o is mass of the particle; V_o is the velocity of impact; ρ_s is the density of the substrate material; r_o is the radius of the spherical particle; K_o is coefficient of penetration and H_s is the dynamic hardness of the substrate material.

Assuming the penetration of particles into the substrate is proportional to the impulse pressure, and taking into consideration the influence of heating on the particles, we may write for the penetration of a particle at temperature T_p

$$h_{T_p} = \left(\frac{2\ \rho\rho_p}{3K_o\rho_s}\right) \ln\left[\left(1 + \frac{\rho_s V_o^2}{H_s}\right)\left(\frac{C_p T_p(1 + Z_p/Z_s)}{C_p(1 + Z_p T_p/Z_s)}\right)\right] \quad (4)$$

The influence of increasing particle temperature on the intensity of substrate activation at the zone of impact may be expressed in terms of:

(1) A decrease in the "mechanical part" of activation because, increasing the temperature of material decreases the speed of sound therein and, consequently, decreases the impulse pressure on the contact zone of impact,

(2) An increase in the intensity of thermal activation of the substrate,

(3) An increase in the intensity of particle plastic deformation and its spreading on substrate surface which imposes high shear stresses along the surface, and hence, increases the "mechanical part" of the substrate activation.

In the case of melted particles, a sharp decrease in impulse pressure and a correspondingly sharp increase in substrate surface temperature could take place. Calculations of intensity of substrate surface thermal activation are given in several references.[1,2]

EXPERIMENTAL INVESTIGATION OF SPRAYED PARTICLES ON SUBSTRATE

Experimental Apparatus

Two energy sources for heating and accelerating of powder were used in this study: (a) induction plasma torch and, (b) detonation of oxygen-acetylene gaseous mixture.

Design and parameters of plasma-gun are described in dissertation of Hassan[116]. A schematic diagram of the plasma-gun with some modifications is shown in Figure 1. A Lepel induction heating generator T-2.5-1-MC-BW(J) was used in this work. The generator operated in the range 2.5 MHz to 8 MHz to deliver 2.5 KW of high frequency energy to an appropriate inductively coupled load in accordance with IEEE Standard No. 54.

The essential features of the basic torch are shown in Figure 2. The copper torch head is provided with a single entry port where gas is admitted tangentially with respect to the center of the head. Another central hole in the torch head serves as a guide to an axial 20 mm diameter baffle tube as well as an inlet

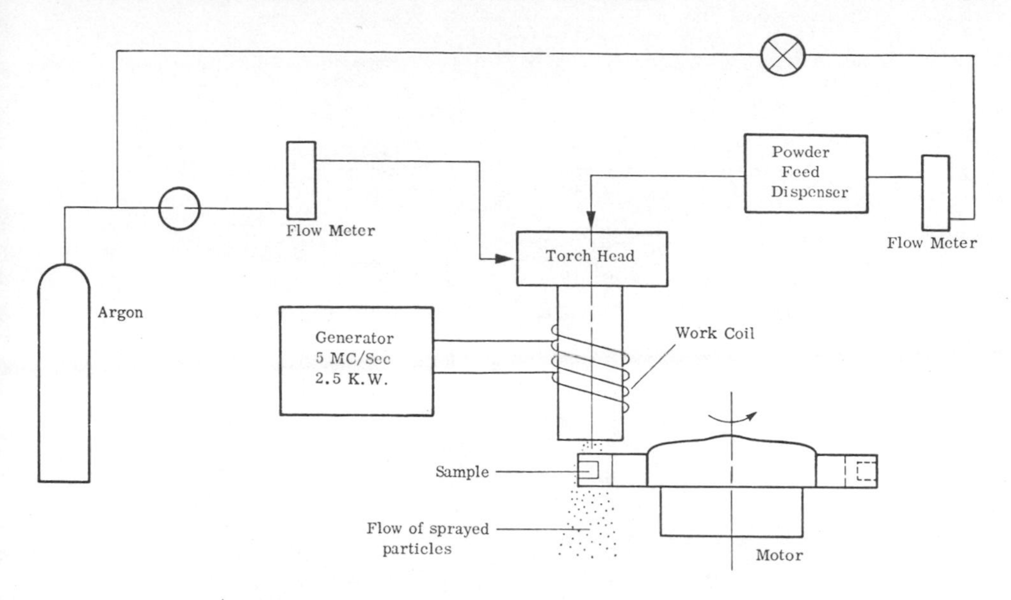

Figure 1. Schematic diagram of plasma-gun

passage for reactants. The confining discharge tube (54 mm O.D., 50 mm I.D., 29 cm long) of quartz is connected to the torch head and sealed to it by a teflon washer and a silicon rubber 'O' ring. The portion of the gas that strikes the baffle tube becomes turbulent thereby facilitating the initiation and sustaining of the plasma. The remainder of the gas swirling down the walls of the confining discharge tube proves a means for cooling the walls and also contributes to the creation of a low pressure region in the center.

A 3-turn 3/16" copper work coil is closely wound (clockwise when viewed from the top) round the confining tube with the high potential turn at the top and extends horizontally for 20 cm beyond the confining tube wall to the r.f. generator to which the two ends are connected.

A Tesla coil was placed external to the confining tube some distance downstream from the work coil as depicted in Figure 2. The Tesla-coil used was a Model BD-10 from Electro Technic Products, capable of producing 50,000 volts potential and situated with its tip slightly pointed up toward the axis of the tube.

With the r.f. generator turned on, and the argon flow rate at an appropriate level, the flux lines of the field will concentrate in the vicinity of the Tesla-coil tip, giving rise to a magnified local voltage gradient and a spark-type breakdown followed by the formation of the pilot discharge. Once the plasma has been initiated, the Tesla-coil is turned off.

The design of this torch employs the so-called aerodynamic or

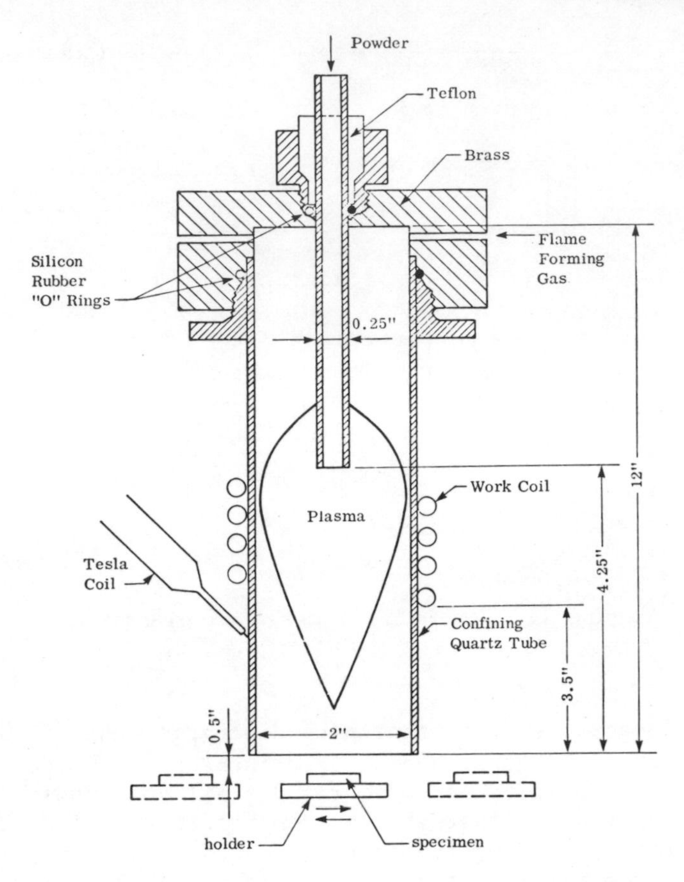

Figure 2. Essential features of the basic torch

vortex mode of stabilization. The vortex created due to the tangential introduction of the plasma forming gas acts to hold the discharge on the axis. The aforementioned baffle tube must, of course, be concentric with the confining tube. This is necessary to prevent attack of the confining tube by the plasma. The majority of the gas that swirls down the walls is essentially flowing faster and in the opposite direction to that in the central region of the confining tube. This feature of aerodynamically stabilized columns allows low-powered torches to be operated without the need for external cooling, as the high velocity gas between the plasma and the walls fulfills this purpose.

Design and parameters of detonation-gun are described in references[96]. Oxygen-acetylene mixtures of $C_2H_2:O_2=1:1.25$ were used. The length of barrel was 1800 mm and internal diameter was 20 mm, the frequency of explosions in the barrel was one detonation per second.

SPRAYED MATERIALS

The sprayed powders were: a) aluminum (-325 mesh), Alfa Products; b) nickel (-100 mesh), Alfa Products; c) tungsten (-100 mesh), Alfa Products.

SUBSTRATE MATERIAL

Only silicon (111) substrates were used in this research. The choice was based on the ease of investigating substrate plastic deformation by particle impact on such samples. Different methods of surface preparation were used:

1. Mechanical lapping and chemical treatment involving final HF etching.
2. Oxidation of silicon in dry O_2 at a temperature of 1100°C.
3. Deposition of (Al + 3% Si) coatings (thickness 1.3μ) on Si substrates by sputtering.

All samples were freshly prepared before use.

SPRAYING CONDITIONS

The spraying conditions are given in Table I. Spraying was done by placing the specimens on a disc and manually or mechanically traversing them under the stationary plasma-gun. The substrate surface in these experiments can be considered to be cold and the velocity of impact approximately equal to the linear speed of the disc.

Table I. Spraying Parameters

Plate Power (KW)	2.5
Argon flow rate (l min^{-1})	35
Power feed rate (g min^{-1})	1-2
Particle velocity, m sec^{-1}	3-8
Velocity of specimens (rotation), m sec^{-1}	11.4-27
Spraying distance (cm)	
a) manual traversing	1-2
b) rotation disc	2.5-3
Speed 1 - 11.4 m sec^{-1}	
Speed 6 - 27 m sec^{-1}	
Speed 3 - 19.0 m sec^{-1}	

EXPERIMENTAL RESULTS

Scanning electron microscopy (SEM) has been used extensively in this study to characterize the behavior of the substrate and the particles at the time of collision.

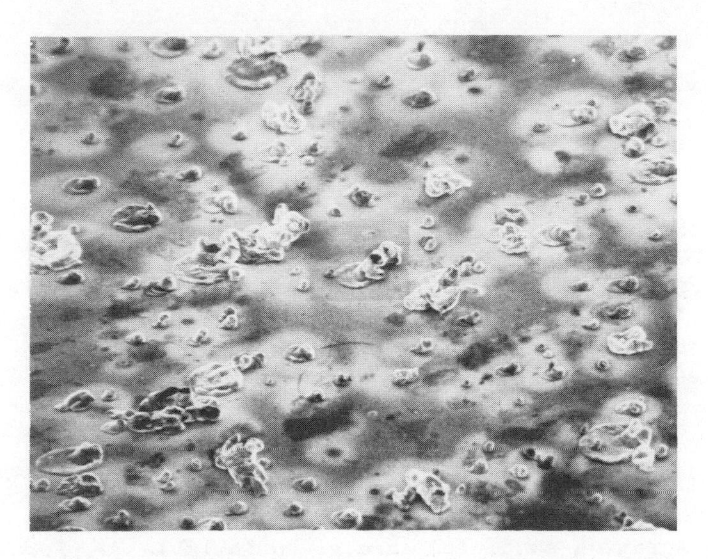

Figure 3(a) Velocity 0.1 m/sec. Shown at 50X

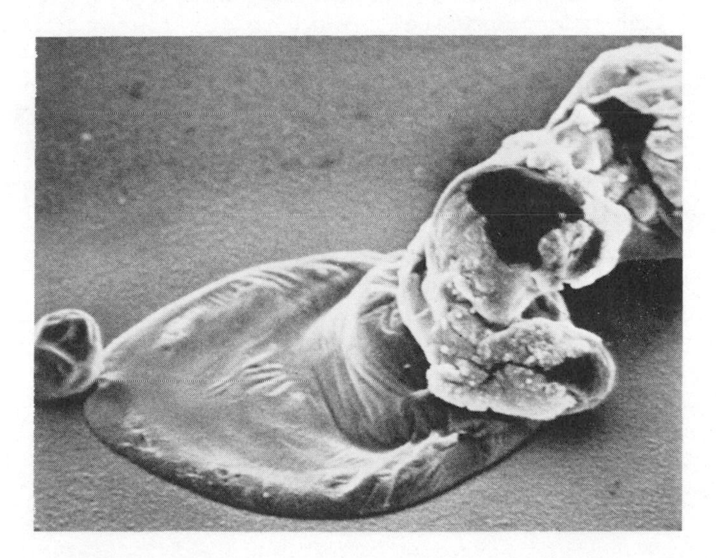

Figure 3(b) Velocity 27 m/sec. Al particles on Si substrates at 500X

PLASMA SPRAY RESULTS

Pictures of Al particles after impinging onto a flat silicon surface, prepared in the manner previously described, are shown in Figure 3. The figure shows particles on flat surface with different shapes. Some of them took the shape of a thin disc, others that of a half sphere. There is, however, a definite distinction between shapes of particles which were sprayed on specimen moving at 0.1 m sec^{-1} (manual transportation) - Figure 3a and those spraying on specimen mounted on rotation disc whose velocity of transportation was 27 m sec^{-1}. In the latter case we have a high velocity of impact, and, as a result, more particles have flattened shapes. The temperature of particles would be decreased somewhat due to the cooling influence of the rotating disc and the larger spraying distance.

The impact behavior of sprayed particles depends on their temperature and velocity. Low velocity promotes a small change in particle shape as shown by Figure 4(a). Greater changes in the shapes of particles at higher speeds are shown in Figure 4 (b) and Figure 3 (b).

Comparatively low velocity of particle impact is enough for plastic deformation of the silicon substrate. This is evident from Figure 5 which shows the silicon substrate after removing of the particles by etching in an aluminum etchant composed of phosphoric acid, nitric acid, acetic acid, and deionized water in the ratios 16:1:1:2 by volume respectively. Subsequent etching was done to reveal dislocations in the silicon crystals. The etch had the following composition: 0.15 molar solution of $K_2Cr_2O_7$ (or $Na_2Cr_2O_7$) in distilled H_2O: 1 part by volume: and HF: 2 parts by volume.

The intensity of plastic deformation of substrate surface depends on the energy state of particles as revealed by the quantity of circular or elliptical pits on zone of impact as shown in Figure 5. The more intensive plastic deformation takes place on periphery of contact zone, Figure 5(b). In case of very low speed of impacting particles only a small quantity of dislocations is formed on contact surface, Figure 5(c).

The tungsten particles did not undergo significant changes in their shape as shown in Figure 6. This is because of their higher melting point and the requirement for longer residence time in the plasma for their melting.

The external appearance of Al particles sprayed on a silicon substrate coated with Al (thickness 1.3μ) is shown in Figure 7 and is similar in appearance to the uncoated silicon surface.

The complexity of the hydrodynamic phenomena that affect melted particles in the process of collision with a substrate is shown in Figure 8, which was obtained after spraying nickel particles on a silicon substrate with 1.3μ coating of aluminum. The nickel is completely melted and essentially "splat cools" on contact with the substrate.

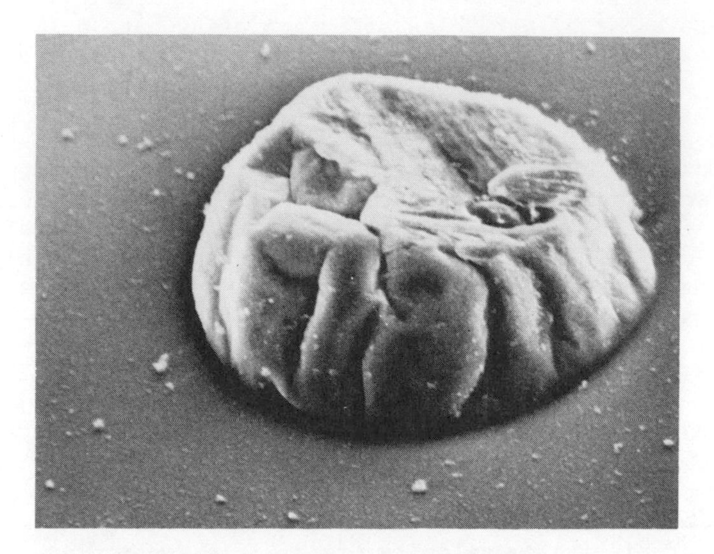

Figure 4(a) Low velocity

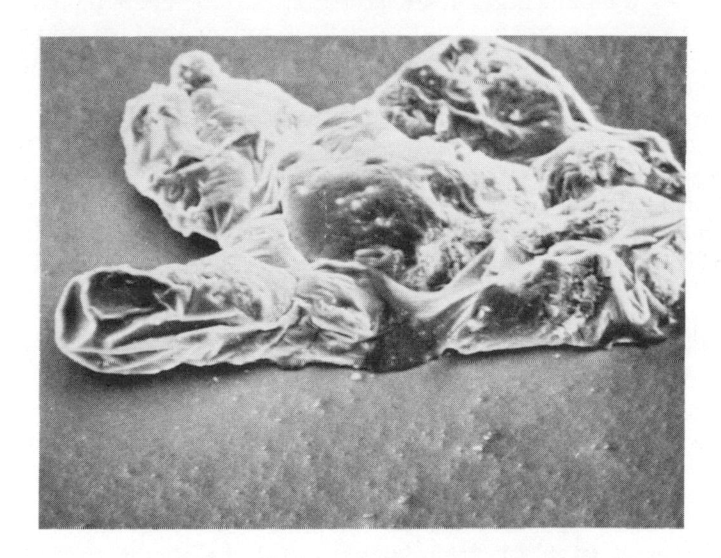

Figure 4 (b) Moderate velocity

Figure 4 (a) and (b) Aluminum particle on silicon substrate shown at 500x

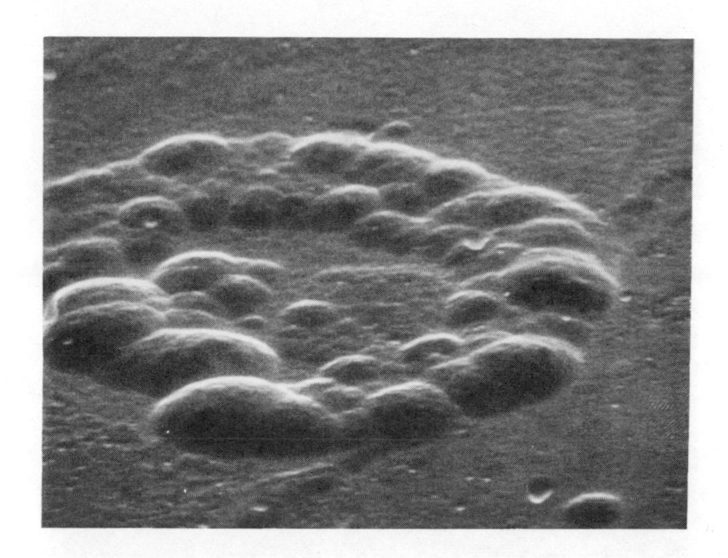

Figure 5(a) Velocity 0.1 m/s. Shown at 2500X.

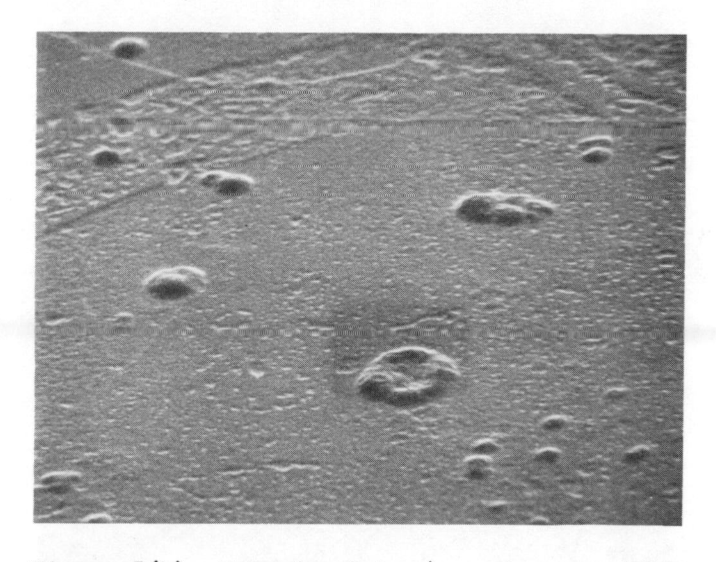

Figure 5(b) Velocity 0.1 m/s. Shown at 500X.

Figure 5 Etched silicon substrate showing deformation caused by aluminum particle.

Figure 5(c) Velocity 11.4 m/s. Shown at 1500X.

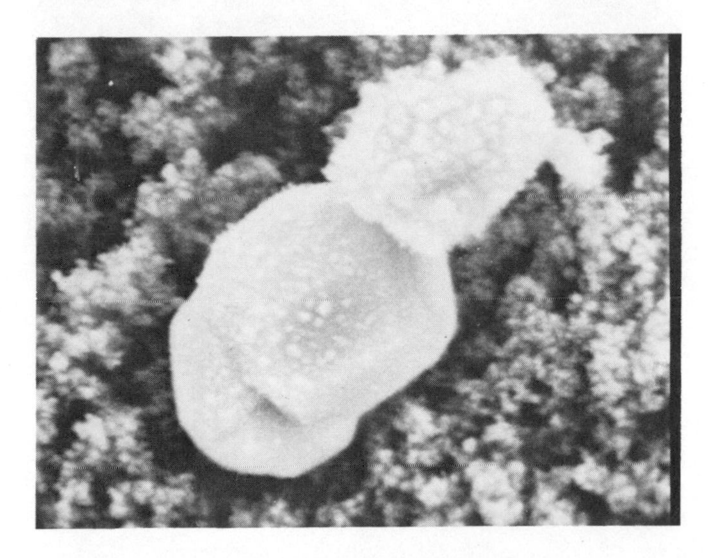

Figure 6. Tungsten particle on silicon substrate. 25,000X.

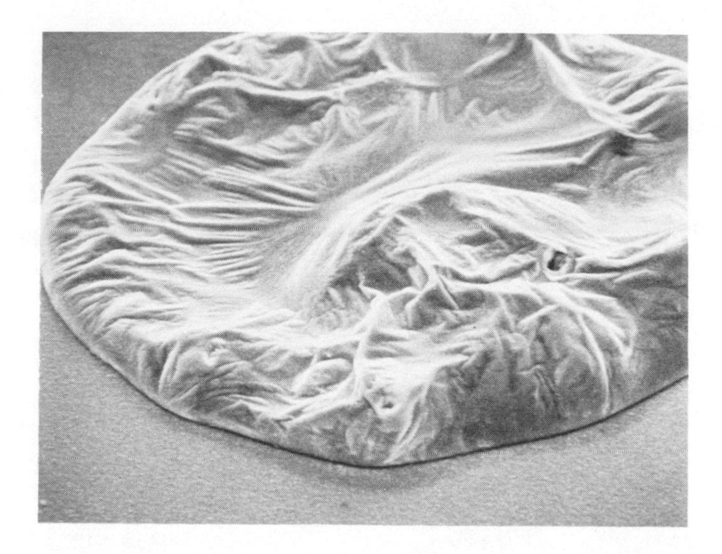

Figure 7(a) Velocity 11.4 m/s. Shown at 25,000X.

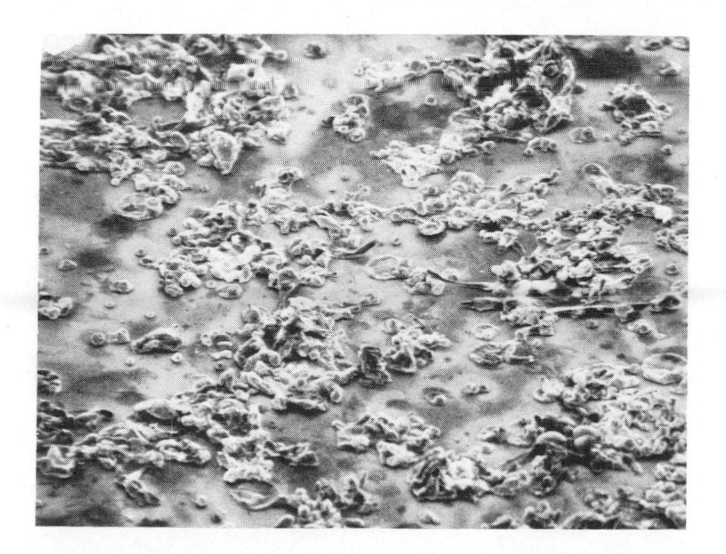

Figure 7(b) Velocity at 27 m/s. Shown at 50X.

Figure 7. Aluminum particles on aluminum coated silicon substrate.

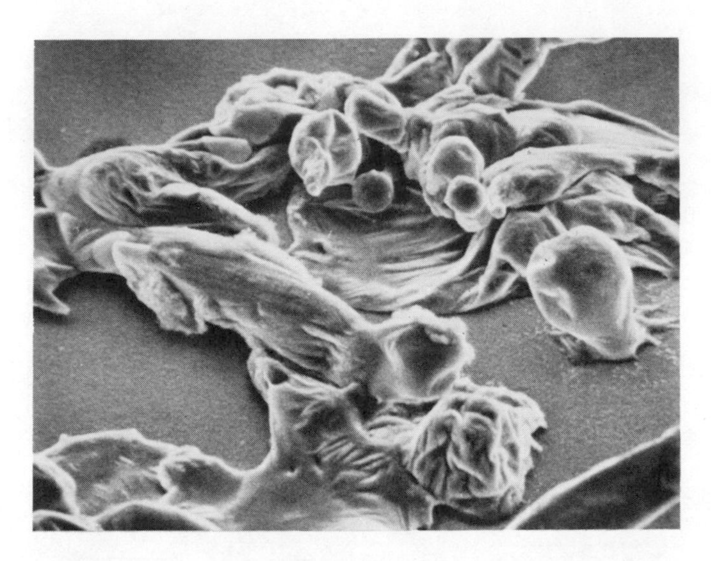

Figure 7(c) Velocity 27 m/s. Shown at 600X.

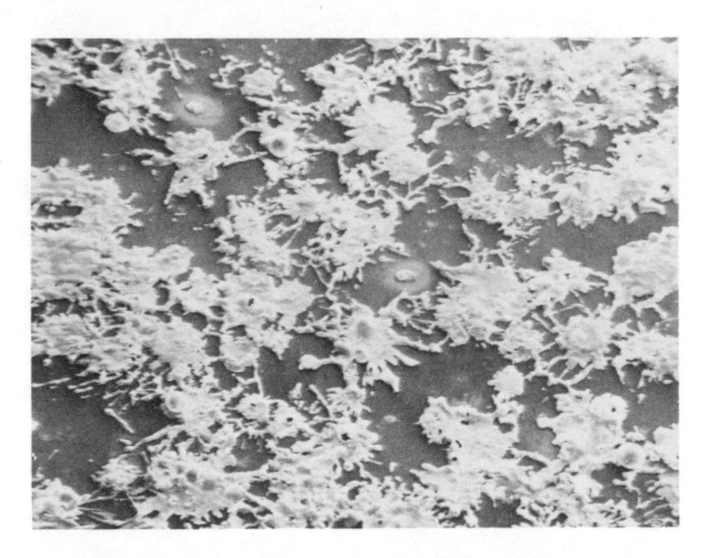

Figure 8(a) Nickel particles on aluminum coated silicon substrate. Velocity 27 m/s. Shown at 50X.

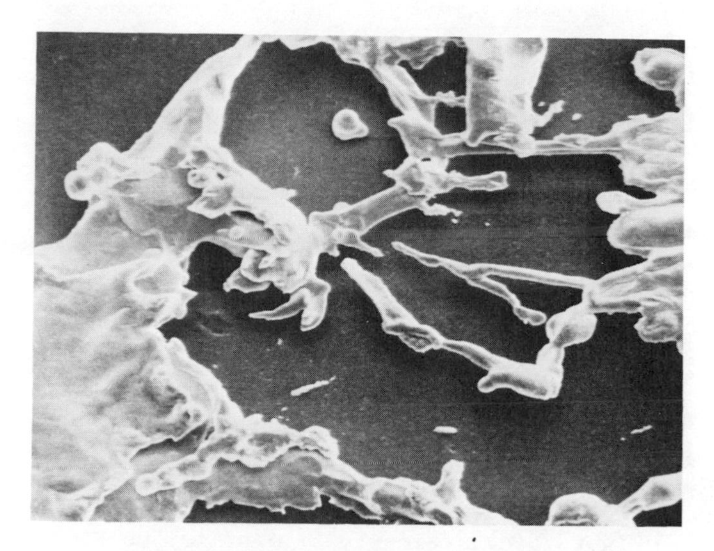

Figure 8(b) Velocity 27 m/s. 500X

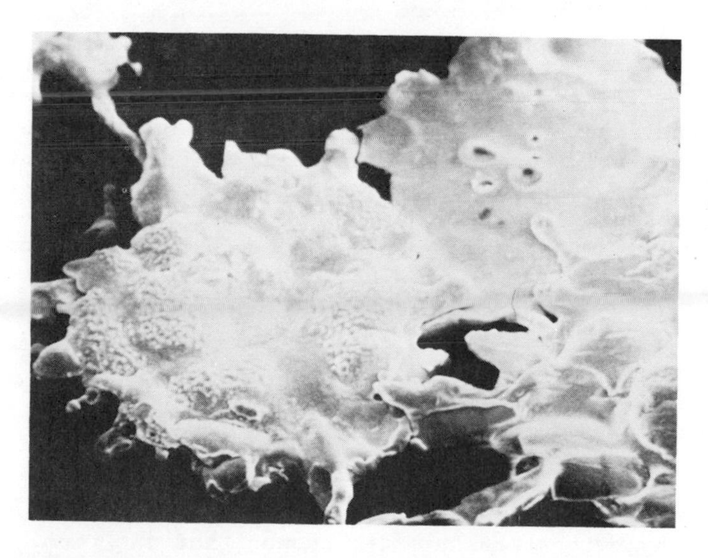

Figure 8(c) Velocity 27 m/s. Shown at 500X.
Nickel particles on Al coated silicon substrate.

Figure 8(d) Velocity 27 m/s. 2,500 X.

Under impact conditions the thin layer of Al coating on the silicon substrate could be destroyed or changed in structure and composition. It is possible to see the extent of these effects on the substrate after removing the Ni particles by a suitable etchant such as nitric acid followed by treatment in the aforementioned Al etchant as shown in Figure 9. At zones of particle impact the Al coating is absent. However the coating still remained in the areas surrounding the impact zones probably due to the formation of an alloy from the original Al + 3% Si that is more resistant to the etchant.

DETONATION GUN RESULTS

Detonation spray coatings have a much higher bond strength than plasma coatings and frequently require no special surface roughening of the substrate. Qualitative estimation of materials interaction by detonation spraying showed that the formation of high bond strength deteriorated with increase in vaporization energy, melting point, hardness, and brittleness of the substrate material[90]. The same functional relationships, but not so distinct, are found for plasma spraying. The influence of the parameters of detonation spraying on bond strength of the coating on a substrate is summarized by the curves in Figure 10.

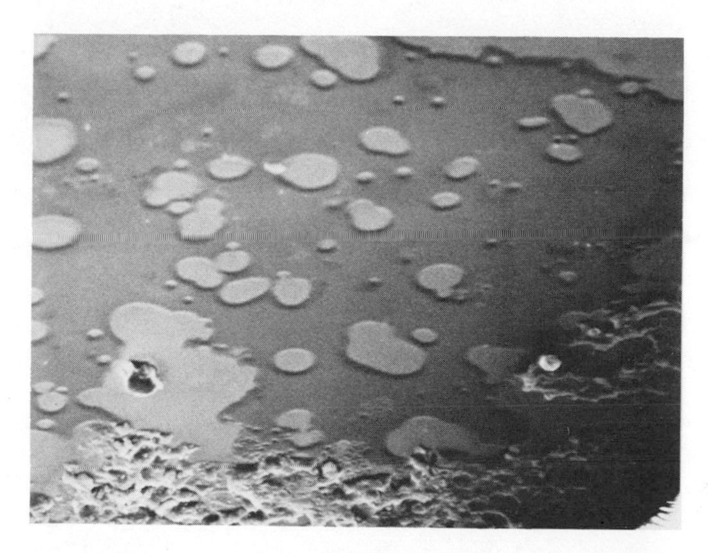

Figure 9(a) Etched silicon substrate after impact with nickel particles. Velocity 0.1 m/s. Shown at 150X.

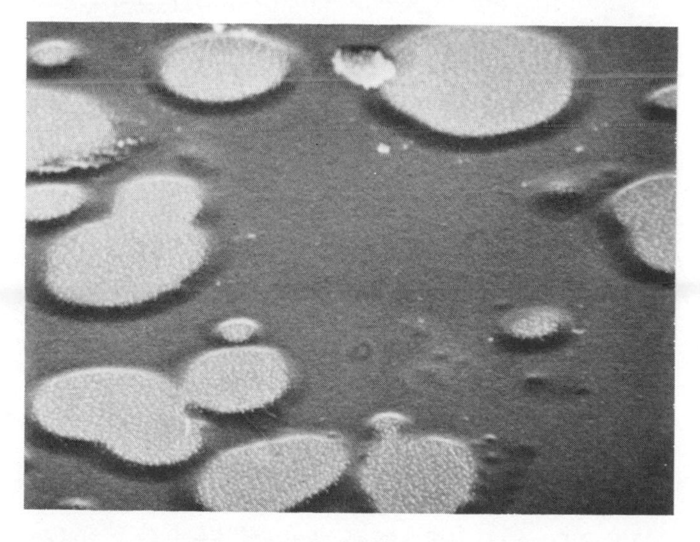

Figure 9(b) Velocity 0.1 m/s. Shown at 70X.

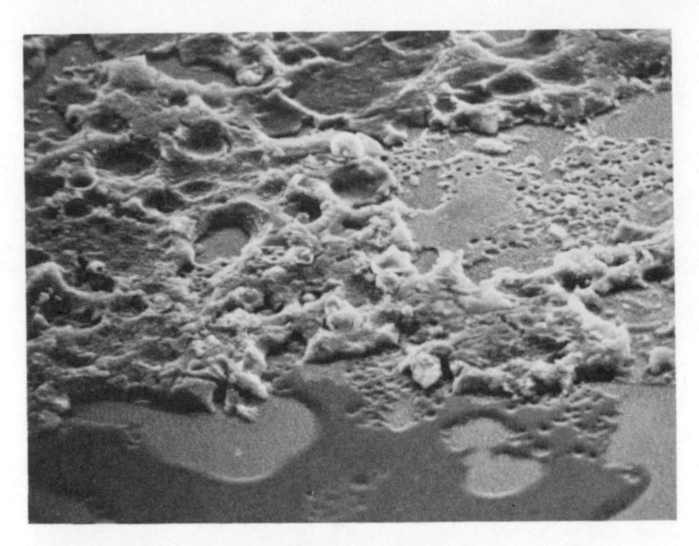

Figure 9(c) Velocity 0.1 m/s. Shown at 450X.
Etched silicon substrate after impact with Ni particles.

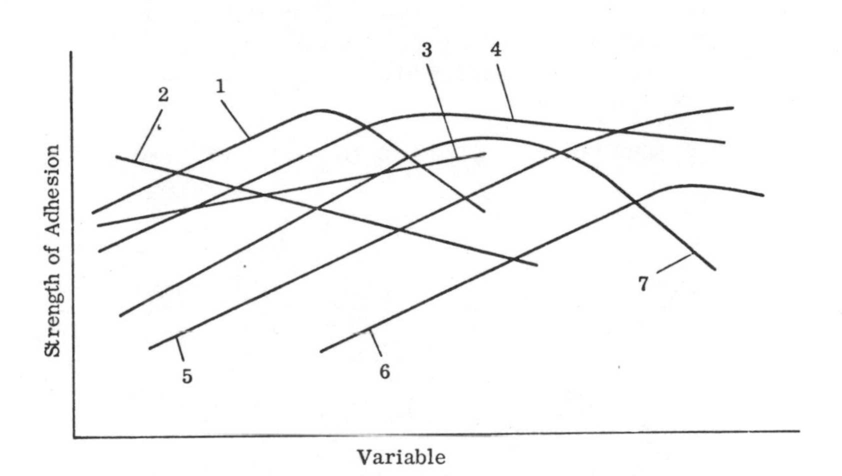

Figure 10. The effects of selected variables on the adhesion strength of detonation sprayed coatings.[90]

1- distance (S) from powder at moment of ignition to open end of barrel
2- spraying distance (l) for large values of S
3- spraying distance (l) for small values of S
4- particle temperature (T) for large values of particle velocity (v_p
5- particle temperature (T_p) for small values of v_p
6- particle velocity (v_p) for small values of T_p
7- particle velocity (v_p) for large values of T_p

The principle differences between detonation coatings and plasma coatings are:

(1) higher velocity of particle impact on the substrate;
(2) broader possibilities for regulation of the intensity of the thermal and mechanical interaction of the particles with the substrate by change in the ratio of the kinetic to thermal energy of the particles;
(3) increase in the activation of substrate surface by collision of a shock wave with the substrate.

The shock wave phenomenon plays an important role in the thermal and mechanical activation of the substrate surface at the initial moment of formation of a single layer of the coating. A sharp rise in pressure, temperature and density takes place when a shock wave (which is formed after the detonation wave leaves the barrel) meets the substrate.

The thermal energy is absorbed comparatively uniformly by the solid, but, as the absorption of mechanical energy is not uniform at the time of impact, micro-regions arise with a local stress higher than the average on the contact surface between the substrate and particle. These micro-regions are active centers where chemical interaction between materials takes place. With increasing impulse pressure the number of these centers increase and hence the bond strength increases.

In the case of a ductile metal substrate, impact with particles is characterized by intensive plastic deformation of sub-surface layers and the generation of fresh lattice defects in the contact zone which play the role of active centers. The intensity of plastic deformation, which it is possible to present as the depth of particle penetration into substrate, increases with increasing particle velocity and decreases with increasing particle temperature. The depth of craters with increasing section length of the powder acceleration in the barrel of the detonation gun decreases, but the ratio between the diameter of craters and their depth increases. The particles with higher temperatures create lower impulses of pressure and flatten more easily on the substrate surface and, therefore, have a larger crater diameter. As a result, the intensity of the mechanical activation of the substrate surface and bond strength usually decreases. It is possible to consider the penetration of particles into the substrate as a process of the sub-surface layer plastic deformation and determine the frequency of the generation of fresh lattice defects.

EFFECTS OF SURFACE CONTAMINATIONS ON BOND FORMATION BETWEEN PARTICLES AND SUBSTRATE

A substrate surface is normally coated with a film of oxide, sulphide or carbide, whose thickness lies within the range of 10 to 1,000 Å. It may also have a layer of adsorbed gas, and it may be contaminated by oil, grease or other non-metallic substances. This contamination interferes with adhesion between substrate and particles. The contamination is created as a result of: 1. surface preparation before spraying; 2. contamination due to handling and storage; 3. contamination by the spraying process itself.

The adsorbed atoms may or may not affect the quality of a bond but will affect the rate of formation of oxide layers on the surfaces [118]. These layers form instantly in air and grow in thickness through the transparent range and the tarnish range, to a thick oxide layer visible to the naked eye. Oxide layer growth is dependent on diffusion of one or more chemical species. Thus a metal that is serviceable for long periods of time at low temperatures may oxidize rapidly if the temperature is increased. Oxide films on some metals grow linearly with time whereas other films on metals such as nickel and aluminum grow at a decreasing rate (parabolic rate law) with time.

Surfaces can be cleansed by chemical etching, fluxing, degreasing, abrasion, and heating the metal to melt or decompose the oxide. If an oxide layer is present, pressure may also help to fracture the brittle oxide layer, thus exposing clean metallic surfaces.

Joining may be defined formally as the process of bringing two or more surfaces into intimate contact in order to establish continuity of a field across the resulting interface - electrical joining denotes establishment of continuity of an electric field across an interface; mechanical joining denotes establishment of continuity of a stress field across an interface. The surfaces of the substrate and the particles to be joined must be in such conditions that they are compatible with one another, both chemically and mechanically.

Oil films inhibit adhesion between a substrate and a particle when the temperature is such that they can persist during spraying[119] Oxide films, on the other hand hinder, but do not prevent adhesion. It is believed that during deformation of the surface the oxide film (or the hardened surface layer produced by scratch-brushing) fractures and exposes areas of clean metal, which bond to the opposite surface wherever two clean areas come into contact. At elevated temperatures the oxide is dispersed partly mechanically and partly by solution in the metal or agglomeration.

Obtaining a surface which is uncontaminated by foreign substances (materials different from that of the pure surface) which may be adsorbed on the surface, such as water and the hydrocarbons common to the atmosphere, is difficult. Chemical cleaning leaves a film of the final rinse material. Sputtering a layer of material off the surface to leave a clean layer, changes the structure of the surface. Sputtering on a new layer covering the old surface with all its contamination (sweeping it under the rug) also changes the structure with no assurance that the new layer will not peel off. A "bakeout" may "bake on" the unknown substances initially contaminating the surface, on the contaminating substances and the nature of the material of the surface itself.

Grit blasting is the most common method of surface roughening. Other methods include machining and chemical etching. Numerous microscopic irregularities exist on a grit-blasted surface. Therefore, the molten aluminum droplets are constrained by the surface irregularities which limit radial flow of the particle[117]. The particles striking a rough surface are smaller in diameter and generally thicker than the particles sprayed onto a polished surface. Transmission Electron Microscope (TEM) observations indicate that the sprayed aluminum particles form columnar grains oriented parallel to the direction in which heat flows, in other words, perpendicular to substrate surface[117].

Angular aluminum oxide is a typical abrasive material used in grit blasting. To achieve a surface roughness compatible with good adhesive strength using this type of material, parameters must be chosen which unavoidably result in a significant amount of aluminum oxide abrasive mechanically imbedded in the surface[7]. Spherical alumina appears to minimize this kind of surface contamination[120].

Kudinov[1] assumes that oxide films on a substrate limit bond formation in plasma spraying because an oxide has a higher activation energy than metals. He also assumes that destruction of oxide films can take place only during the high speed impact of particles with a substrate. In the present work, experiments were carried out to determine the behavior of oxide films on substrate under impact with sprayed particles. Sprayed particles on a silicon substrate have a flattened shape. After spraying these particles onto a silicon substrate with an SiO_2 film (thickness 1,083 A) the physical contact between substrate and particle was reduced. Although the velocity of the particles was relatively low (approximately 5-10 m/sec) one can observe the destruction of oxide films as shown in Figure 11. The form of particle penetration into the substrate had changed and they were observed to have more of a spherical than a flattened shape. The destruction of thinner oxide films takes place in larger areas. The flattening particles pull the oxide films from the substrate. In the case of spraying Ni powder, destruction of oxide films takes place in smaller areas, because Ni has a significantly high melting temperature and consequently is more difficult to flatten on the substrate surface.

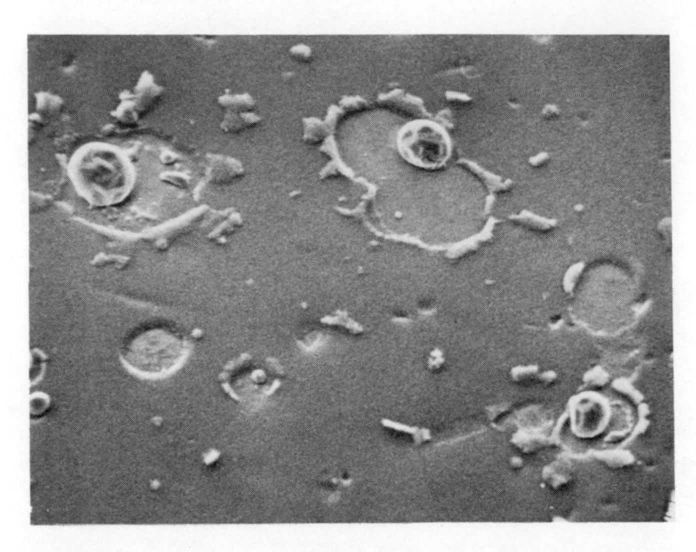

Figure 11(a) Aluminum particles on 1,083 A SiO_2. 200X

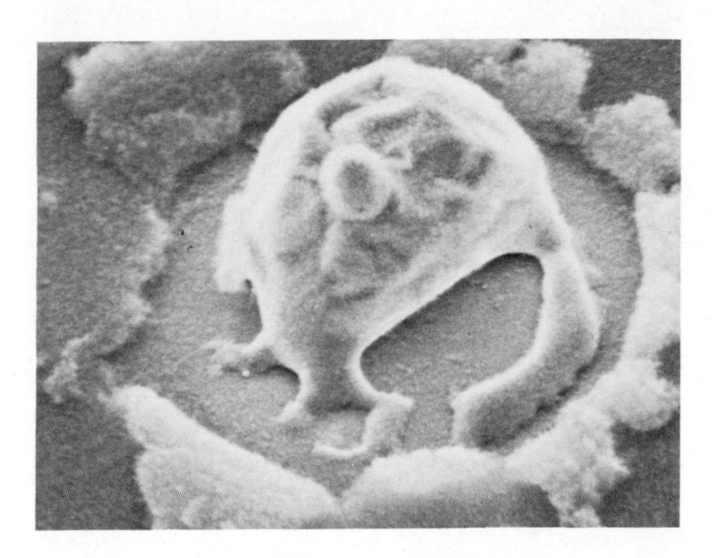

Figure 11(b) Al particle on 1,083 A SiO_2. 1,000 X.

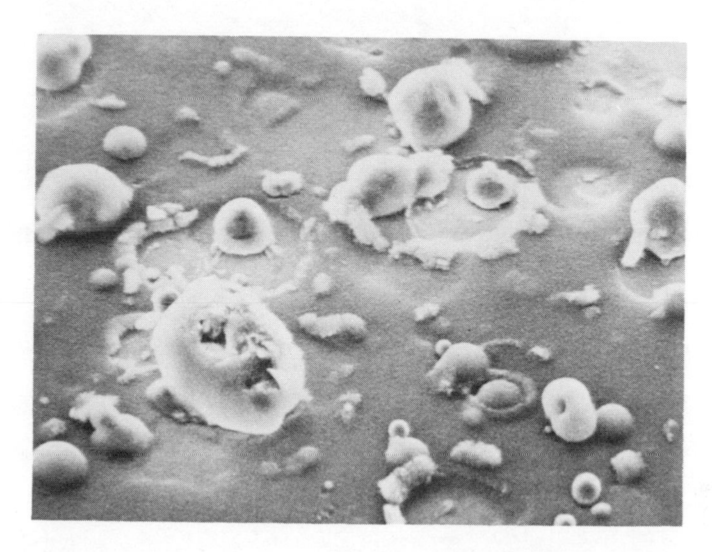

Figure 11(c) Al particles on 3000 A SiO_2. 300 X.

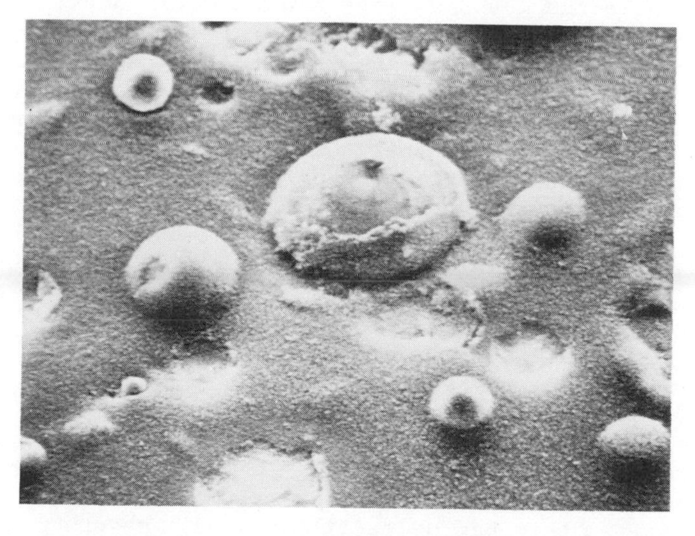

Figure 11(d) Ni particles on 3000 A SiO_2. 500 X

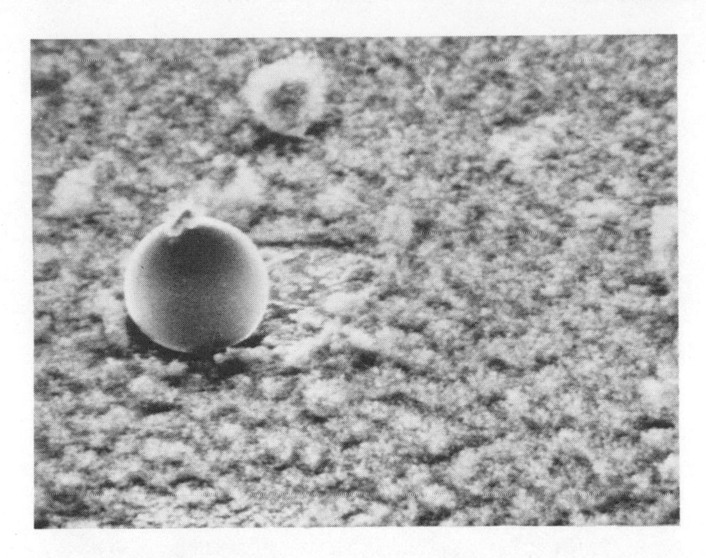

Figure 11(e) Ni particles and H_2O on silicon substrate. Shown at 2,500 X.

As shown in Figure 12, a small number of Ni particles adhered to the silicon substrate coated with an SiO_2 film (thickness 3,000 A). These results show that formation of adhesion between the substrate and the sprayed coating through oxide films is difficult. It is possible only if the velocity of the particle is very low and if there exists a strong adhesive force between the oxide films and the substrate. With increasing thickness of the oxide and higher adhesive strength between oxide films and substrate, more thermal and kinetic energy of particles are required for physical contact between the particles and the substrate. The oxide film at the initial moment of impact prevents the base metal from thermally activating and, in the process of its removal from substrate, oxide splinters take away thermal energy from contact zone. Therefore, it is better to spray particles at high velocities as this would promote faster deformation and removal of oxide films from the contact zone thereby decreasing thermal energy losses to the oxide films.

Special experiments were done with detonation-gun spraying. Different metal substrates were used. The surfaces of these substrates were polished and left in open air for 24 hours, and, therefore, all of them had oxide films as well as other forms of contamination. In the detonation-gun spraying powder particles were heated and accelerated to 500-600 m sec^{-1}, but the particles were still in the solid state before collision with the substrate surface. The spraying was done by one blast of the detonation-gun (the time of interaction of the hot gases with the substrate was about 5×10^{-3} sec) and, therefore, the temperature of substrate surface was about 20° C (no preliminary heating).

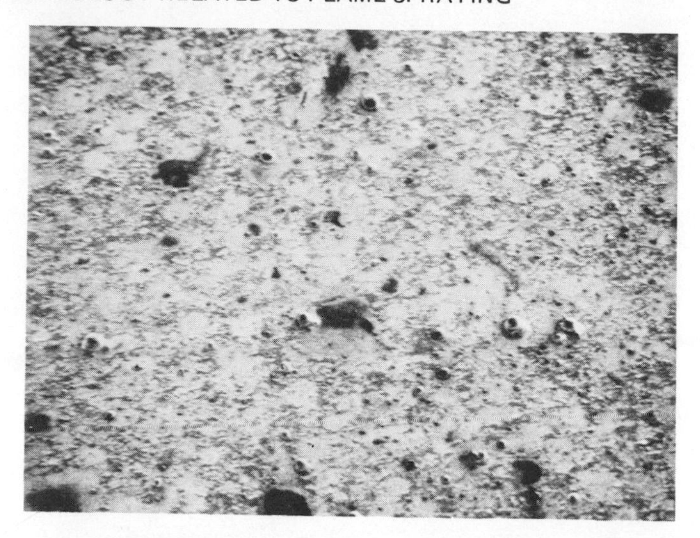

Figure 12(a) Ni particles on 3000 A SiO_2. Velocity 0.1 m/s. Shown at 50 X.

Figure 12(b) Ni particles on 3000 A SiO_2. Velocity 0.1 m/s. Shown at 500 X.

The presence of other contaminant films such as oils and greases on the substrate surface increase the requirement for energy to overcome these films. Very often in the case of rather thick films the formation of a physical bond between substrate and particles becomes impossible. This is shown in Figure 13 for H_2O

vapor contamination on a Si substrate and in Figure 14 for a thin film of oil on a Si surface. In both cases, Ni powder was used for spraying.

Figure 13(a) Ni particles on H_2O film on Si substrate. Shown at 50X

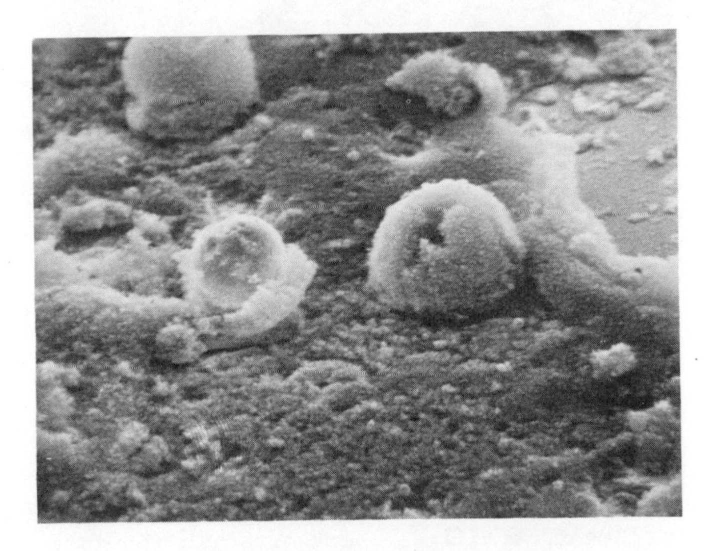

Figure 13(b) Ni particles on H_2O contaminated Si substrate. Shown at 500 X.

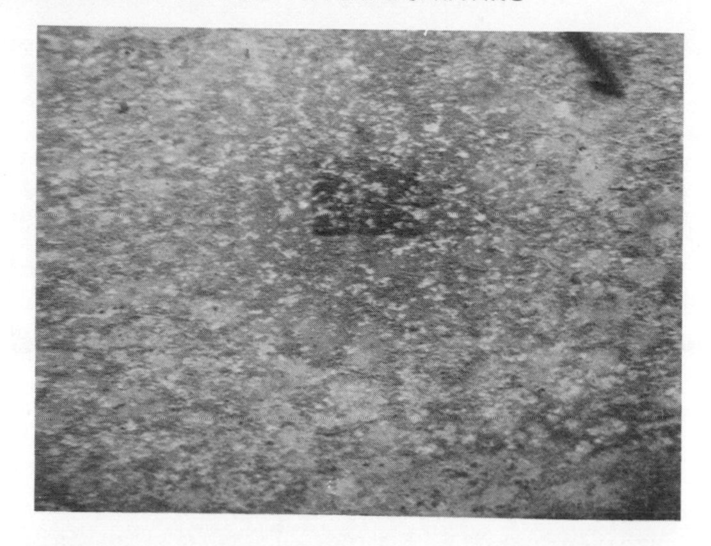

Figure 14(a) Ni particles on oil film. Shown at 50 X.

Figure 14(b) Ni particles on surface contaminated with oil. Shown at 2,500 X.

CONCLUSIONS

Plasma and detonation spraying for coating materials is sensitive to the surface contamination present on the work piece. Our studies have shown that the physical character of the sprayed particles can be very different for different forms of surface contamination. Some flexibility exists in the spraying process with respect to process variables such as temperature, velocity and pressure, and suitable selection of these can help to ameliorate the deleterious effects of surface contamination.

This study has included an extensive list of the pertinent literature associated with plasma and detonation spraying. These references on the subject are as thorough and as up to date as it was possible to assemble. With the many new applications that are being found, it appears that the number of investigators in the field are rapidly increasing and we forecast much more interaction between surface scientists and spray scientists in the future.

REFERENCES

1. V.V. Kudinov, "Plasma-Sprayed Coatings," Nauka, Moscow, 1977.
2. L.K. Druzhinin and V.V. Kudinov, Editors, "Obtaining of Coatings by High Thermal Spraying," Atomizdat, Moscow, 1973.
3. L.I. Gotlib, "Plasma Spraying," Moscow, 1970.
4. S.R. Levine and J.S. Clark, "Thermal Barrier Coatings--A Near Term, High Payoff Technology," NASA TM X-73586, 1977.
5. S. Stecura, "Two-Layer Thermal Barrier Coating for Turbine Airofoils--Furnace and Burner Rig Test Results," NASA TM X-3425, 1976.
6. F.S. Stepka, C.H. Liebert and S. Stecura, "Summary of NASA Research on Thermal-Barrier Coatings," NASA TM X-73584, 1977.
7. H.F. Butze and C.H. Liebert,"Effect of Ceramic Coating of JT8D Combustor Liner on Maximum Liner temperatures and Other Combustor Performance Parameters," NASA TM X-73581, 1976.
8. I. Preece and C.W.D. Andrews, J. Materials Sci., 8, No. 7, 964 (1973).
9. P. Predecki, B.A. Auslaender, J.E. Stefan, et al, J. Bio-Medical Materials Res., 6, No. 5, 401, (1972).
10. H. Fitzer and G. Karlsruhe, Werkst Korros, 24, No. 4, 274 (1973).
11. J. Lammer and F. Benesovsky, Planseeber Pulvermetal, 21, No. 1, 18, (1973).
12. F. Bardal, P. Molde and T.G. Eggen, Br. Corros J., 8, No. 1, 15, (1973).
13. H.D. Steffens, K.N. Mueller and H. Kayser, Schweissen Schneiden, 25, No. 2, 45, (1973).
14. S. Kimura, Electron Commun Japan, 54, No. 6, 104, (1971).
15. H.D. Steffens, Arch Eisenhuebtenwes, 43, No. 10, 775, (1972).
16. L.J. Toth, W.D. Brentnall and G.D. Menke, J. Metals, 24, No. 9, 19, (1972).
17. L.J. Walters, Anti-Corros Methods Materials, 19, No. 1, 8, (1972).
18. H.D. Steffens, H. Kayser and K.N. Mueller, Schweissen Schneiden, 24, No. 4, 118 (1972).

19. S. Kimura and S. Uchida, in "Sixth International Metal Spraying Conference," Vol. 1, paper D-3, 18p, Paris, France, Sept. 14-19, 1970.
20. "New Plasma-Sputtering Process Metal Plates Various Substrates on Mass Production Basis," Ind. Heat, 38, No. 11, 2228, (1971).
21. L.G. Davies, W.M. Powers and R.G. Shaver, SAMPE Quart, 3, No. 1, 32, (1971).
22. A. Faussat, Soudage Tech Connexes, 25, No. 3-4, 139, (1971).
23. F. Eichhorn and J. Metzler, Metalloberflaeche, 24, No. 4, 135, (1971).
24. H. Hantzche and K. Zabel, Schweisstechnik (Berlin), 20, No. 12, 549, (1970).
25. R.L. Brown, in "Sixth International Metal Spraying Conference," Vol. 2, paper E-6, Paris, France, Sept. 14-19, 1970.
26. E. Kretzschmar, in "Sixth International Metal Spraying Conference," Vol. 2, paper H-5, Paris, France, Sept. 14-19, 1970.
27. H.D. Steffens, in "Sixth International Metal Spraying Conference," Vol. 1, paper D-2, Paris, France, Sept. 14-19, 1970.
28. H. Hahn and W. Palich, J. Bio-Medical Materials Res., 4, No.4, 571, (1970).
29. R. Glaelzle and F. Benesovsky, Metall., 24, No. 8, 823, (1970).
30. "New Methods for Coating PTFE on Metals and Ceramics," Plating, 57, No. 10, 987, (1970).
31. F. Blume and L. Sieth, Schweisstechnik, (Berlin), 20, No. 3, 127, (1970).
32. J.W. Cywinski, in "Proc. Int. Symp. on Powder Coatings," paper 11/11, London, England, February 13-15, 1968.
33. J.E. Hauck, Mod. Plast. 47, No. 10, 164, (1970).
34. "Plasma Gun Spray-Coats PTFE Commercially," SPE J., 26, No. 10, 37, (1970).
35. J.D. Buckley, Amer. Ceramic Soc. Bull. 49, No. 6, 588, (1970).
36. K.G. Kreider and E.M. Breinan, Metals Progress, 97, No. 5, 104, (1970).
37. S. Kimura, IEEE Trans Parts, Material Packag., PMP-6, No. 1, 3, (1970).
38. D.J. Godfrey, J. Brit Interplanetary Soc., 22, No. 5, 353, (1969).
39. R.L. Brown, Metal Construction and Brit Welding J., 1, No. 7, 317, (1969).
40. R. Kreienbuehl, Oberflaeche Surf., 17, No. 4, 73, (1976).
41. O.B. Chevela, A.S. Nagin, A.G. Cherdakli and V.N. Gadalov, Sov Powder Met Ceramic, 15, No. 2, 108, (1976).
42. R.W. Babbitt, Amer. Ceramics Soc. Bull., 55, No. 6, 556, (1976).
43. E.S. Lysenko, I. Ya. Chernyavskii and V.F. Turov, Sov Powder Metall Met Ceram, 15, No. 1, 30, (1976).

44. F. Eichhorn and D. Boehme, Elektrowaerme Int., 31, No. 34, 3160, (1973).
45. T. Parmley, Europlast Mon., 47, No. 2, 59, (1974).
46. H.E. Sliney, ASLE Annual Meeting, 29th Tech. Paper, Cleveland, Ohio, April 28-May 2, 1974, Pap. 74AM-7A-3, published by ASLE, Park Ridge, Illinois, 1974.
47. S.A. Bortz and E.J. Onesto, Composites, 5, No. 4, 151, (1974).
48. H.D. Steffens, K.H. Mueller and G. Wittig, Materialpruefung, 16, No. 5, 125, (1974).
49. H. Kayser, in "Int. Metal Spraying Conference, 7th, London, England, Sept. 10-14, 1973," Vol. 1, Pap. 20, 5-10, published by Weld Intl, Abington, Cambridge, England, 1974.
50. H. Kayser, Schweissen Schneiden, 28, No. 3, 101, (1976).
51. H.D. Steffens, K.N. Mueller and H. Gruetzner, Schweissen Schneiden, 28, No. 3, 92, (1976).
52. H. Maeda, M. Hijikata, S. Kitahara and L. Morimoto, J. Japan Inst Metals, 40, No. 2, 180, (1976).
53. V.S. Terent'eva, E.V. Vasil'eva and Yu. E. Ugaste, Prot. Met., 11, No. 4, 463, (1975).
54. C.A. Elyard, D.J. Connoly and R. Lambert, Symp. on Spee Ceram, Sixth Proc., Pap. and Discuss, Penkhull Stoke-on-Trent, England, July 9-11, 1974, 79-90, published by Brit Ceram Res Assoc, Penkhull, Stoke-on-Trent, England, 1975.
55. K. Kirner, Ber Dtsch Keram Ges, 52, No. 4, 77, (1975).
56. H. Nagasaka, in "Int. Plastic Powder Coating Conference, 4th, London, England, March 5-6, 1974," published by Wheatland J. Ltd., Watford, Herts, England, 1974.
57. G. Perugini, High Temp, High Pressures, 6, No. 5, 565, (1974).
58. V.B. Raitses, M.M. Dzhons, V.P. Rutberg, Yu. P. Gavrilenko, V. Ya. Cheltsov and A.G. Shcherbakova, Russ Cast Prod, No. 3, 90, (1974).
59. M.A. Clegg, V. Silins and D.J.L. Evans, in "Int. Met. Spraying Conference, 7th, London, England, Sept. 10-14, 1973," Vol. 1, Pap. 4, 62-71, published by Weld Inst. Abington, Cambridge, U.K., 1974.
60. F.E. Montbrun, J. Richelmi, R. Liesenborghs and V. Leroy, in "Int. Metal Spraying Conference, 7th, London, England, Sept. 10-14, 1973," Vol. 1, Pap. 2, 102-116, published by Weld Inst, Abington, Cambridge, England, 1974.
61. B. Bourgoin, in "Int. Met. Spraying Conference, 7th, London, England, Sept. 10-14, 1973," Vol. 1, Pap. 11, 185-189, published by Weld Inst, Abington, Cambridge, England, 1974.
62. C.W.D. Andrews and L. Preece, IERE Conference Proc (London), No. 26, 1973, for Meet, Univ of Birmingham, England, July 10-12, 1973, 295-302.
63. L.J. Toth, W.D. Brentnall and G.D. Menke, J. Metals, p. 19, (Sept. 1972).
64. R.E. Cleary and C. Amman, Testing of High Emittance Coatings, NASA CR-1413, 1969.
65. G. Durmann and F.N. Longo, Cer. Bulletin, 48, No. 9, 221, (1967).

66. J.R. Rairden, Electrochemical Technol, 5, No. 7-8, 407, (1967).
67. T.A. Wolfla, J. Vac. Sci. Technol, 12, No. 4, 777 (1975).
68, R.S. McClocklin and B.A. Teal, J. Vac. Sci. Technol, 12, No. 4, 784 (1975).
69. T.A. Taylor, J. Vac. Sci. Technol, 12, No. 4, 790 (1975).
70. M.C. Willson and R.J. Janowiecki, "Planar High Energy Permanent Magnets," Report AD-771 956, 1973; AD-780 085 1974; AD/A-005 080, 1975, Monsanto Research Corporation.
71. G.A. Whitlow, R.I. Miller, S.L. Schrock and P.C.S. Wu, Corrosion, 30, No. 12, 420, (1974).
72. J.W. Van Wyk, "Ceramic Airframe Bearings," Report AD/A-020 170, Boeing Aerospace Company, 1975.
73. R.W. Babbitt, "Preparation and Reproducibility of Temperature Stable RF Ferrites," Report AD 734810, 1971.
74. R.W. Babbitt, "Arc Plasma Deposition of Nickel Zinc Ferrites," Report AD 747771, 1972.
75. D.H. Harris, "Tailored Dielectrics," Report AD-777808, Monsanto Research Corporation, 1974.
76. I. Ahmad and J. Vasilakis, "Application of High Strength Filament Reinforced Metal Matrix Composites to Advanced Rapid Fire and High Temperature Barrels," Report AD-A015 460, Watervliet Arsenal, 1975.
77. D.P. Ferriss and C.B. Cameron, J. Vac. Sci Technol., 12, No. 4, 795, (1975).
78. A.R. Stetson and C.A. Hauck, J. Metals, p. 479, (July 1961).
79. C.H. Liebert, R.E. Jacobs, S. Stecura and C.R. Morse, "Durability of Zirconia Thermal-Barrier Ceramic Coatings on Air-Cooled Turbine Blades in Cyclic Jet Engine Operation," NASA TM X-3410, 1976.
80. R.C. Bill, "Selected Fretting-Wear-Resistant Coating for Titanium-6-percent-aluminum-4 percent-vanadium alloy," NASA TN D-8214, 1976.
81. T.E. Strangman, E.J. Felton and N.E. Vlion, Ceramic Bull 56, No. 8, 700, (1977).
82. G.D. Moore, J. Vac. Sci. Technol., 11, No. 4, 754, (1974).
83. R.H. Johnson, S.L. Schrock and G.A. Whitlow, J. Vac. Sci. Technol., 11, No. 4, 759, (1974).
84. R.H. Singleton, E.L. Bolin and F.W. Carl, J. Metals, p.483, (July 1961).
85. M. Donovan, British Welding J., p. 490, (August, 1966).
86. R.F. Belt and G.C. Florio, Ceramic Bull., 56, No. 6, 523, (1972).
87. R.C. Tucker, J. Vac. Sci. Technol., 11, No. 4, 725, (1974).
88. D.R. Mash, N.E. Weare and D.L. Walker, J. Met., 13, 473, (1961).
89. I.A. Fisher, Int. Metall. Rev. 17, 117, (1972).
90. M. Kh. Shorshorov and Y.A. Kharlamov, "Physical and Chemical Fundamentals of Detonation-Gun Spraying," Moscow, Nauka, 1978, (t.b.p.)

91. "UCAR Metal and Ceramic Coatings," Physical Characteristics, Union Carbide Coatings Service, 1977.
92. The Metco Thermospray Process, Metco Inc., Bull. 153C, 1976.
93. "Problem Solving Technology," Union Carbide, Linde Division, Coating Service, Bulletin, 1975.
94. D.A. Gerdeman and N.L. Hecht, "Arc Plasma Technology in Materials Science," Springer-Verlag, New York, Wien, 1972.
95. Y.A. Kharlamov and B.L. Ryaboshapko, Powder Metallurgy (USSR), No. 2, 33 (1975).
96. Y.A. Kharlamov, M. Kh. Shorshorov, B.L. Ryaboshapko and O.V. Gusev, Combustion Process in Chemistry and Metallurgy, Chernogolovka, (USSR), 156, (1975).
97. Y.A. Kharlamov, M. Kh Shorshorov, V.N. Kudinov, O.V. Gusev and B.L. Ryaboshapko, Physics of Combustion and Explosion, (USSR), No. 1, 88 (1975).
98. Y.A. Kharlamov, B.L. Ryaboshapko and O.G. Ignatenko, in "Protective Coatings on Metals," Vol. 11, p. 31, Naykova Doomka, Kiev, 1977.
99. Y. A. Kharlamov and P.S. Banatov, Powder Metallurgy (USSR), No. 1, 40 (1974).
100. "Development of Thin Organic Rolled Film Capacitor," Progress Report AD-615, 753, 1964.
101. J.K. Totten, Ramjet Technology Programs, 1963, Vol. 10, "High Temperature Coated Tungsten Structures," Report AD-602049 1964.
102. W.A. Reaves and E.J. Chapin, "An Investigation of Barrier Coatings on Graphite Molds for Casting Titanium," Report AD-623 992,1965.
103. H.J. Van Hook and R. Maher, "Manufacturing Methods and Technology Measure for Arc-Plasma-Sprayed Phase Shifter Elements," Report No. AD-A040709, 1977.
104. D. H. Harris, "Arc Plasma Sprayed Microwave Circuits," Report No. AD-A015 077, 1975.
105. A.R. Moss and W.J. Young, "Science and Technology of Surface Coating," p. 287, Academic Press, London, New York, 1974.
106. P.M. Bartle, Metal Construction, 1, 241 (1969).
107. G.S. Springer, "Erosion by Liquid Impact," Scripta Publishing Co., a division of Scripta Technica, Inc., Wash. D.C., 1976.
108. O.G. Engel, "Basic Research on Liquid-Drop-Impact Erosion," Report NASA CR-1559, 1970.
109. Y.A. Kharlamov, Powder Metallurgy (USSR), No. 10, 55 (1974).
110. Ph. Ph. Vitman and N.A. Zlatin, J. Technology Physics (USSR), XXXIII, No. 8 (1963).
111. M. Kh. Shorshorov and Y.A. Kharlamov, The Porblems of Strength (USSR), No. 2, 70, (1975).
112. Eu. I. Andriankin and Yu. S. Stepanov, Artificial Satellites of Earth, Vol. 15, Moscow, publishing house of USSR Academy of Sciences, 1963.
113. S.J. Grisaffe and W.A. Spitzig, "Preliminary Investigation of Particle-Substrate Bonding of Plasma-Sprayed Materials," NASA TN-D-1705, 1963.

114. D.G. Moore, et. al., "Studies of the Particle-Impact Process for Applying Ceramic and Cement Coating," U.S. Air Force, Report No. ARL-59 (AD-266 381), 1961.
115. W.A. Spitzig and S.J. Grisaffe, "Analysis of Bonding Mechanism Between Plasma-Sprayed Tungsten and a Stainless Steel Substrate" NASA TN-D-2461, 1964.
116. M.S. Hassan, "Use of Induction Plasma in the Extraction of Gallium from Coal Fly Ash," PhD thesis, 1977, Stanford University.
117. S. Safai and H. Herman, Thin Solid Films, 295 (1977).
118. C. L. Bauer, G. G. Lessmann, Ann. Rev. Materials Sci. 6, 361 (1976).
119. J.E. Lancaster, "Metallurgy of Welding, Brazing and Soldering," American Elsevier Publishing Co., Inc., N.Y. 1965.
120. T.A. Wolfla, R.N. Johnson, J. Vac. Sci. Technol., 12, No. 4, 777, (1975).

CONTAMINATED CONCRETE SURFACE LAYER REMOVAL

J. M. Halter and R. G. Sullivan

Pacific Northwest Laboratory
Battelle Memorial Institute
Richland, Washington 99352

Equipment is being developed to economically remove contaminated concrete surfaces in nuclear facilities. To be effective this equipment should minimize personnel radiation exposure, minimize the volume of material removed, and perform the operation quickly with the least amount of energy. Several methods for removing concrete surfaces are evaluated for use in decontaminating such facilities. Two unique methods especially suited for decontamination are described: one, the water cannon, is a device that fires a high-velocity jet of fluid causing spallation of the concrete surface; the other, a concrete spaller, is a tool that exerts radial pressure against the sides of a pre-drilled shallow cylindrical hole causing spallation to occur. Each method includes a means for containing airborne contamination. Results of tests show that these techniques can rapidly and economically remove surfaces, and leave minimal rubble for controlled disposal.

INTRODUCTION

The concrete walls, floors and ceilings of many nuclear facilities have become contaminated with radioactive particles because of accidental spills or releases of vapors and fine particles. It is desirable to reduce the quantity of contaminated rubble that must be handled during the decommissioning of surplus nuclear facilities and to provide an easier method for cleaning smaller contaminated areas. The amount of contaminated material can be reduced by removing the contaminated concrete surface. Several contaminated surface removal methods are available.

This paper presents criteria for selecting a suitable removal technique. Currently used concrete surface removal techniques are summarized and detailed descriptions of two new techniques are presented. These techniques, the water cannon and the concrete spaller, have been developed or adapted at Pacific Northwest Laboratory.

BACKGROUND

The choice of a contamination removal method depends on the type of contamination, the depth of contamination and the type of surface.

If contamination has not penetrated the concrete significantly, vacuuming the surface and then scrubbing it with mild soap and water or solvents sometimes successfully decontaminates the concrete.[1] For facilities which are to be used again, paint is an acceptable short-term measure used to fix low-level contamination in place.[2] However, when the contamination has penetrated to the extent that it cannot be cleaned as described above, and it emits excessive radiation, it is necessary to remove the surface layer of the concrete.

In some clean-up operations of the past, whole walls were removed because this was faster than removing only the contaminated surface. But the high cost of storing contaminated materials requires economical surface removal techniques to minimize the quantity of rubble which must be placed in controlled storage facilities.

There are many ways to remove contaminated concrete surfaces and these methods have been used with varying degrees of success. A suitable removal method should satisfy the following criteria. It should:

- minimize personnel exposure to harmful radiation or toxic materials (reduce both the number of personnel involved and their individual dosage),
- minimize the recontamination possibility of previously cleaned surfaces,
- minimize the volume of removed material that must be placed into controlled storage (i.e., remove no more material than is necessary to clean the surface), and
- perform the removal operation as quickly as possible.

These considerations and the techniques which are discussed apply equally to the decontamination of nonnuclear facilities.

CONCRETE SURFACE REMOVAL TECHNIQUES

Many techniques have been used for removing contaminated concrete surfaces in nuclear facilities. Table I provides a listing of common methods for reducing the contamination level, as well as two methods now under development. All of these techniques are aimed at keeping the contamination from spreading once it has been removed. The size of the facility to be cleaned plays an important part in determining the technique which physically can be used. (For example, an impactor mounted on a back hoe will not fit into a small room or an area with limited access.)

Sand blasting and flame spalling,[3] where intense heat is applied to concrete surfaces, remove only minimal surface depth and produce large quantities of small, contaminated particles. A large exhaust and air filtration system is needed with these methods.[4] These two techniques are also relatively slow if the contamination penetrated all but a thin surface layer.

The use of explosives has been shown to be an effective method for removing contaminated concrete surfaces and structures in relatively large, thick-walled facilities. Holes are drilled into and behind the surface to be removed and filled with the proper explosive.

During the dismantling of the Elk River nuclear reactor, five different types of explosives were used to remove concrete walls and floors. Careful attention was given to controlling airborne contamination, preventing the spread of contamination outside of the reactor cavity, and controlling the size and weight of the materials being removed.[5]

While the use of explosives is a fast method for removing concrete surfaces, significant quantities of dust are generated which must be controlled. The facility must also be strong enough to withstand the explosive forces.

Two surface removal methods are used more extensively than the rest. Those two methods are jack hammers and impactors. Jack hammers, which are powered by compressed air, are readily available and are easily operated by one man. They are used to chip off the surface material deep enough to remove the contamination. Because they are difficult to position on walls and ceilings, jack hammers are used primarily on floors. Impactors, which are similar in operation to a jack hammer but are much larger, have been used successfully in several decontamination projects. A pick or chisel point is driven into the concrete surface with high energy impacts at several times per second. The impactors are powered by either air or hydraulics and are held and

Table I. Comparison of Various Concrete Surface Removal Techniques.

Technique	Limitation	Type of Rubble Produced	Estimated Size of Air Filtration System Required	Estimated Relative Speed at Which a Unit of Surface Area Can Be Removed
Sand Blasting		Small particles	Large	Slow
Flame Spalling	Heat may cause undesirable chemical reactions	Small particles	Large	Slow
Explosives	Generates moderate quantities of dust which must be controlled	Medium-sized pieces and small particles	Large	Fast
Jack Hammer	Awkward to use on walls	Medium-sized pieces and small particles	Medium	Medium Fast
Impactor Powered by Air or Hydraulics	Limited to large accessible facilities	Medium-sized pieces and small particles	Medium	Fast
Scrubber	Awkward to use on walls	Small-sized pieces and small particles	Medium	Slow
Water Cannon				
Handheld Modified 458 Magnum Rifle	Gun powder combustion products are produced	Small pieces coated with glycerine and gun powder combustion products	Small	Slow (5-6 min/ft^2, (54-64 min/m^2)
Rapid Fire Model[a]	Limited to large accessible facilities	Small pieces coated with water	Small	Fast (6-10 sec/ft^2, (1.1-1.8 min/m^2)
Concrete Spaller with 38 Pound Air Drill to Make Holes		Medium-sized pieces and small particles	Small	Medium Fast (50-60 sec/ft^2, (9-11 min/m^2)

(a) The water cannon is presently being evaluated for surface removal. The stated performance is a best estimate.

positioned by linkages typical of those found on tractor-mounted back hoes and excavators.

These methods produce dust which is typically removed from the vicinity by pulling the dust-laden air into an exhaust duct. Filters remove and collect the dust for proper disposal.

Another tool used to chip away concrete surfaces is the scrubber, a handheld tool with a gang of carbide-tipped bits. The bits rapidly impact a surface, causing small concrete pieces to be spalled. A dust shield with a vacuum attachment is placed around the bit to remove airborne contamination. The scrubber is relatively slow and is difficult to use on walls and ceilings.[6]

Water spray is also used to keep dust from spreading as light coatings of water hold the dust to the bigger rubble. Care must be taken not to apply so much water that it flows off the rubble, spreading contamination.

WATER CANNON

The water cannon removes concrete surfaces by shooting very high pressure jets of liquid at the surface causing it to spall.[7] The advantages of this method are that no initial surface preparation is needed and the equipment does not contact the surface. The rubble which is removed by the water cannon is in small pieces and is coated with liquid. Because of the liquid, little or no dust is generated.

Two different versions of the water cannon have been developed which are applicable to concrete surface removal. One is a modified 458 magnum rifle which shoots solidified glycerine through a nozzle.[8] The second version uses stored compressed gas to drive a piston which forces water through a small diameter nozzle.[9]

A 458 magnum rifle has been modified by replacing the standard barrel with a shorter smooth-bored barrel. The end of the barrel is threaded to accept a nozzle which reduces the inside diameter from 0.45 in. (1.14 cm) to 0.17 in. (0.43 cm).

A 9 in. (23 cm) diameter shield in the shape of a funnel has been placed around the nozzle in order to protect the operator and to funnel the rubble into a vacuum hose mounted on the shield. The rubble consists of pieces which are 0.5 in. to 0.75 in. (1.3 cm to 1.9 cm) in diameter and small particles which are all coated with glycerine. The size of the rubble allows it to be easily transported to a collection bin by a vacuum system. The shield extends 1 in. (2.54 cm) beyond the end of the nozzle so that it can be placed against the surface and the correct nozzle-to-surface distance will always be achieved.

The gun fires projectiles made of solidified glycerine, 2 in. (5.08 cm) long and 0.45 in. (1.14 cm) in diameter. The glycerine projectiles are propelled by gun powder loaded into a conventional cartridge case. When the gun is fired, the glycerine accelerates down the barrel and is extruded through the nozzle emerging at a very high velocity.

Wax is placed in the cartridge case to hold the powder in place and when the gun is fired the wax helps to create a moving seal to keep the combustion gas from passing around the glycerine.

The modified 458 magnum rifle version of the water cannon has been extensively tested. Spall craters averaging between 3 and 4 in. (7 and 10 cm) in diameter and 0.75 in. (1.9 cm) deep in the center are typically obtained. This is shown in Figure 1. The shots are spaced about 3 in. (7.6 cm) apart in a triangular pattern. In a test conducted on nuclear reactor-grade concrete, 24 shots were required to remove 1 ft^2 (0.093 m^2) of surface. (See Figure 2.) This took 5 to 6 minutes. Size variations appear to be determined primarily by the type and distribution of the aggregate within the concrete. For example, if hard round river gravel aggregate is struck by the shot of glycerine "head-on", small spalls will generally result. The best spalls occur when the glycerine can work around and behind the embedded aggregate.

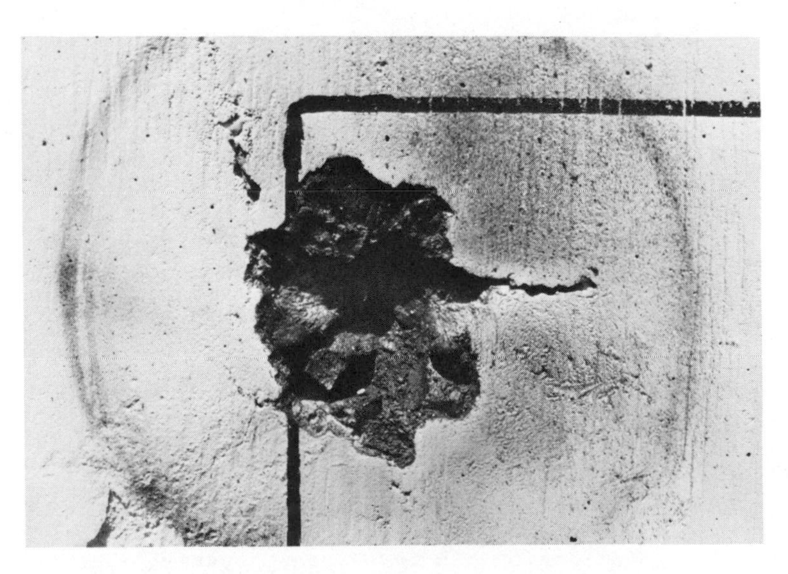

Figure 1. Single 458 magnum water cannon spall.

The 458 magnum water cannon is positioned and held by hand and can be operated as fast as a person can reload and position the gun. (See Figure 3.) This handheld spaller is suitable for cleaning small areas where large equipment would be impractical.

A second type of water cannon is also being investigated for spalling concrete surfaces. While the water cannon described above uses gun powder to drive glycerine through a nozzle, the second type uses compressed gas to drive a piston to impact a small quantity of water and force it through a nozzle. (See Figure 4.) It was initially developed to fracture rock for tunneling and mining applications, but has been found to work well for spalling surfaces.

The gas which drives the piston is compressed by a hydraulic impactor that will allow firing rates of up to 5 times per second. Water is added after each shot into a chamber in front of the piston.

The unit is mounted on a back hoe or excavator and is operated in a manner similar to a concrete or rock breaker. As a result, the unit is usable only in rooms that are large enough to accommodate the equipment.

Like the 458 magnum water cannon, a funnel-like shield will be placed around the nozzle to protect the operators and remove debris.

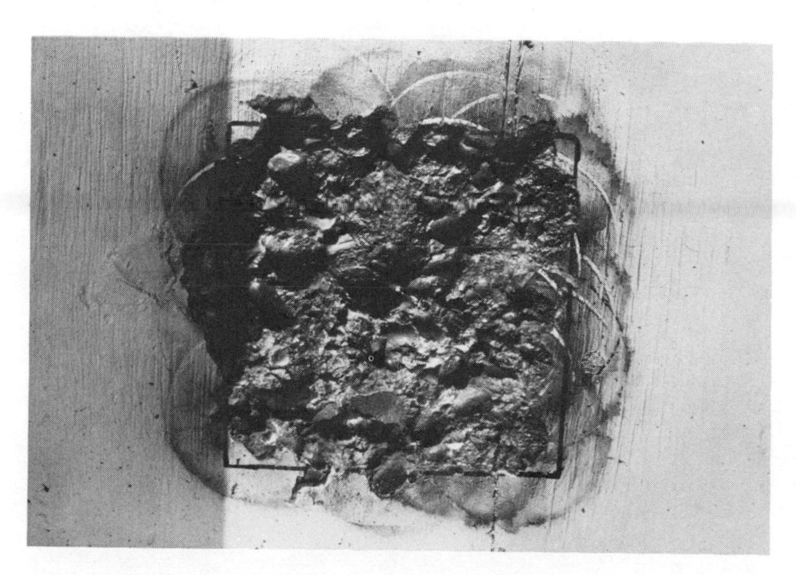

Figure 2. Test panel spalled by the 458 magnum water cannon (1 ft^2, 0.093 m^2).

The advantages of this rapid-fire water cannon over most other surface removal techniques are: 1) it is expected to have a removal rate of one square foot in 6 to 10 seconds (1.1 to 1.8 min/m^2), and 2) the water that is fired by the cannon coats the rubble pieces and particles, which helps to minimize the possibility of spreading contamination.

Figure 3. 458 magnum water cannon being fired.

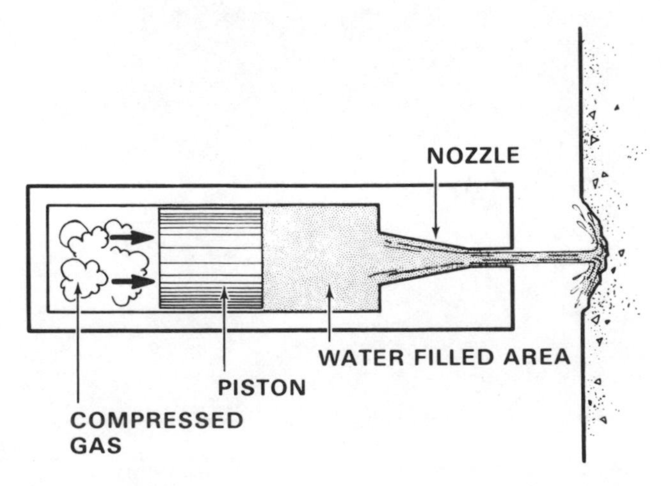

Figure 4. Schematic of a water cannon basic components.

CONCRETE SPALLER

The concrete spaller is a device which has been developed specifically for removing concrete surfaces by Pacific Northwest Laboratory (PNL). This device was developed to satisfy the need for a method of removing only the top layer of a contaminated surface. Also, the device was designed to be lightweight, easy to handle, and convenien·ly used on any contaminated surface without spreading the contamination.

The concrete spaller is similar in design to the commercially available Darda Rock and Concrete Spaller.[10] The Darda splitter, unlike the concrete spaller used to remove only the surface layer of concrete, is used to break large slabs, walls, and boulders into smaller pieces.

The concrete spaller consists of three basic parts: a hydraulic cylinder, a push rod, and a bit with expanding wedges. The hydraulic cylinder, which is attached at one end, activates a push rod, which is installed inside the bit. (See Figure 5.)

The bit is a piece of steel tubing, the inside diameter of which is tapered at one end. A circular wedge is machined into the tubing at the tapered end. The bit is split into 4 equally spaced segments parallel to its central axis.

Inside the tubing is placed a push rod with an outside diameter slightly smaller than the inside of the bit. This rod is also tapered at one end which matches the tapered end of the tubing.

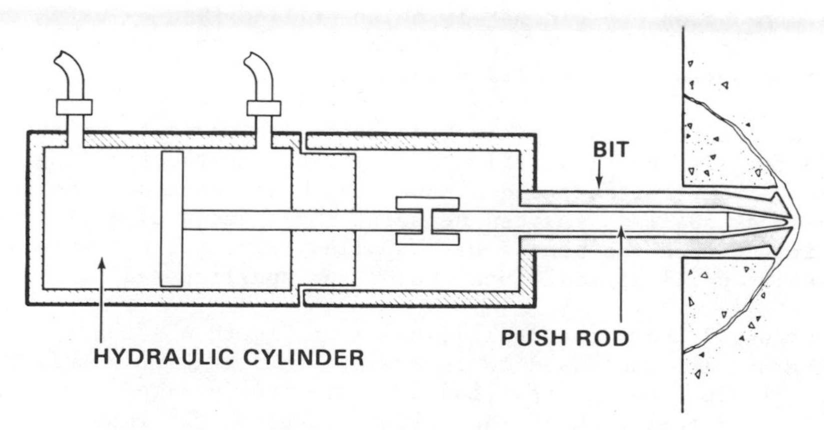

Figure 5. Schematic of a concrete spaller.

The concrete spaller is operated by inserting the action end of the bit into a pre-drilled hole, approximately 2 in. (5 cm) deep and 1 in. (2.5 cm) in diameter. The hydraulic cylinder, powered by a 10,000-psi (6895-N/cm^2) pump, is then activated, forcing the push rod toward the end of the bit. Two things cause the spalling to take place. First, the wedges are forced radially outward embedding into the walls of the hole. Second, when the tip of the push rod reaches the bottom of the drilled hole, it forces the wedges away from the bottom, causing the spalling to take place.

The initial hole drilling is the time-consuming portion of the use of the concrete spaller. Three different types of drills were tested to determine the most efficient: a compressed air powered drill (38-lb model), an electric core drill, and an electric rotary hammer. The electric core drill was discarded because it produced a fine dust and was also very slow. (Drilling a 2 in. (5 cm) deep hole required 90-120 seconds.) Although the electric rotary hammer was able to drill a 2 in. (5 cm) deep hole in 30 to 40 seconds, the compressed air powered drill was shown to be more efficient as it was able to drill the hole in 10 to 15 seconds. Both the electric rotary hammer and the compressed air powered drill produces small chips which are more easily handled by the air filtration system. To keep the drilling chips from contaminating the air, a vacuum attachment is placed around the drill bit to remove the chips generated during the drilling operation.

The holes were drilled in a triangular pattern. Tests show that the optimum space between the holes is 8 in. (20 cm). With 8-in. (20-cm) spacing, some areas of the surface were not removed due to variations in the distribution of the aggregate within the concrete. It is therefore necessary to redrill and spall the remaining surface. Closer spacing of holes is less efficient since more holes are ultimately being drilled than by using 8-in. (20-cm) spacing, even though additional holes to remove the random remaining surface are needed. (See Table II.)

The concrete spaller has been tested on various surfaces on the Hanford Reservation. Figure 6 illustrates the spalling which resulted from two tests and a panel which has been drilled and is ready to be spalled. As can be seen, small areas of surface were left intact when the panels were spalled. These areas were later removed by drilling additional holes and spalling again.

Figure 7 shows the drilled panel in Figure 6 after spalling and the spaller tool used to remove the surface. Note that the rubble produced by the spalling is conveniently sized for easy handling and that much of the surface layer of the rubble remains

Table II. Hole Spacing Versus Number of Holes Required.

Spacing Between Holes, in.	Average Number of Holes Required for 1 yd^2 (0.84 m^2) of Surface	% of Additional Holes Required Compared With 8-in. (20-cm) Spacing
4	93	299
5	59	155
6	41	77
7	30	30
8	23	--

intact, holding the contamination fixed. According to this testing, approximately 53 seconds were required to drill and spall 1 ft^2 (9.5 min/m^2) of surface area (or approximately 8 minutes to remove 1 yd^2, 0.84 m^2).

As with other surface removal techniques, if the spalled surface is still contaminated, it can be redrilled and spalled a second time.

Figure 6. Test panels spalled by concrete spaller (1 yd^2, 0.84 m^2).

Figure 7. Concrete spaller next to a spalled test panel.

FUTURE EQUIPMENT

Contamination removal equipment of the future will likely be remotely operated. The operators will possibly be stationed within view of the spalling operation, but far enough removed to reduce the amount of radiation exposure. In some circumstances the machine operation may be viewed by television and controlled from a remote location. Quite possibly, future machinery might be fully automated, so that it will operate unattended.

Included with the surface removal tools will be machinery to handle the contaminated rubble. Big pieces will be conveyed to a loading station for packing into disposal containers. Small chips and dust will be moved by a vacuum system to separators for removal and placement in storage containers.

The physical dimensions of future equipment will depend on the size of the facility. With facilities varying in size from very large canyon buildings with 60 to 80 ft (18 to 25 m) high walls to small control and equipment rooms, various specialized tools and rubble handling systems will be needed.

CONCLUSIONS

Several techniques are presently used for removing contaminated concrete surfaces. So far, none of them have been developed to the extent that they meet all three of these requirements: minimal radiation exposure, removal of only the contaminated portion of the surface, and performance of the removal in the least amount of time. The two techniques discussed in this paper, the water cannon and the concrete spaller, are felt to be promising developments toward equipment that will meet the needs of future decontamination projects.

ACKNOWLEDGMENTS

This work was performed for the U.S. Department of Energy, under Contract EY-76-C-06-1830.

The authors are grateful to M. K. Dunstan for editorial assistance and R. A. Keefe for word processing.

REFERENCES

1. L. E. Kusler, "Survey of Decontamination and Decommissioning Techniques," ARH-CD-984, Atlantic Richfield Hanford Company, Richland, Washington, May 1977.
2. P. R. Moore, in "Proceedings of the Conference on Decontamination and Decommissioning (D and D) of ERDA Facilities," p. 157 ERDA, Idaho Operations, Idaho Falls, Idaho, September, 1975.
3. N. G. W. Cook and V. R. Harvey, in "Proceedings 3rd International Society of Rock Mechanics Congress," Vol. 1, Part B, pp. 1599-1615, Denver, Colorado, 1974.
4. J. A. Ayres, Editor, "Decontamination of Nuclear Reactors and Equipment," p. 379, The Ronald Press Co., New York, 1970.
5. J. F. Nemac and K. G. Anderson, paper presented at the 1974 Annual Meeting American Nuclear Society, Philadelphia, Pennsylvania, June 1974.
6. B. S. Ureda, (1978), personal communication.
7. "Water Jet Technology," Hydromechanics Division, Exotech Incorporated, Gaithersburg, Maryland.
8. Exotech Incorporated, "Apparatus for Producing a Pulse of Liquid for Machining Operations," U.S. Patent No. 3746256, U.S. Patent Office, Washington, D.C.
9. Exotech Incorporated, "Apparatus for Forming Pulsed Jets of Liquid," U.S. Patent No. 3905552, U.S. Patent Office, Washington, D.C.
10. "Darda Rock and Concrete Splitter," Publication Number RS-73, Emaco, Incorporated, Elmwood Park, New Jersey.

PREPARATION AND CHARACTERIZATION OF CLEAN MINERAL SURFACES

P. Somasundaran and Brij M. Moudgil

Henry Krumb School of Mines
Columbia University
New York, NY 10027

Surface chemical properties of minerals are known to be strongly influenced by the surface heterogeneities which arise from either natural variations in the chemical composition and crystal structure of the mineral or pretreatments given during sample preparation. Large discrepancies are often found in the values reported for the surface properties of the same mineral. These variations mostly occur due to different sample preparation techniques employed by investigators and sometimes, due to impurities present in the liquid or gas phase with which the mineral particles are contacted. To elucidate the mechanisms underlying the various interfacial processes it is essential to avoid the artifacts introduced by sample preparation techniques. An understanding of the effects of pretreatments on the surface properties of the minerals can be helpful in this regard in obtaining consistent and reproducible results. In this paper commonly used pretreatments for obtaining clean mineral surfaces and their implications in the study of interfacial phenomena are discussed. Methods for characterizing the clean mineral surface are also presented.

INTRODUCTION

Interfacial properties of minerals are essentially determined by their physical and chemical nature. Surface heterogeneities, natural or induced by sample preparation techniques, are known to have a significant influence over the surface potential developed when a mineral surface is contacted with the aqueous phase. In the absence of a direct measure of surface potential, the interfacial processes such as flotation and flocculation have been correlated with the zeta potential of the minerals. Also, mechanisms of adsorption of surfactants and polymers in flotation and flocculation respectively have been studied by measuring the variation in zeta potential of minerals upon adsorption. However, for natural minerals widely differing values of zeta potential have often been reported. For example isoelectric point (IEP) values ranging from pH 3 to 10 and from pH 2 to 9 have been reported in the literature for alumina and natural hematite respectively[1-7]. For rutile, IEP values from pH 4.6 to 5.9 in some cases for material of the same origin have been reported by various investigators[8-12]. As noted before most of these differences in electrokinetic measurements can be attributed to the different pretreatments of the mineral samples.

In the case of synthetic minerals surface chemical properties can also depend on the reagents, temperature and other conditions used for the preparation of the samples.

SOURCES OF SURFACE HETEROGENEITIES

The lack of agreement between results obtained by various investigators for surface properties of a mineral is mostly due to the mineralogical or chemical heterogeneity of the particles. Mineralogical heterogeneity can arise from incomplete liberation, whereas chemical heterogeneity of the particles is encountered because of the type of pretreatment such as acid leaching often used for cleaning the samples. Various pretreatments commonly used in obtaining a mineral sample of desired particle size and their implications in surface chemical properties are discussed below.

Mineralogical Heterogeneity

The mineralogical heterogeneity of the surface and subsurface regions of mineral particles can significantly affect the interfacial properties. These can arise from the natural variations in the crystal structure or from treatments during sample preparation. In the case of low grade ores gangue minerals can be disseminated in a matrix of value mineral and vice versa. It is possible to obtain different ratios of value and gangue minerals on the surface upon subjecting an ore particle to different treatments. With such particles the electrokinetic properties of the mineral will lie between the values for the individual minerals present in the surface region. It is to be noted that the response of such ore particles to adsorption and flotation processes need not be affected to the same degree as the electrokinetic behavior. This is so because for the purpose of adsorption, except for a difference in the total available surface area and hence the amount of surfactant adsorbed, the surface is essentially unchanged from that of the original substrate. Any correlation between electrokinetic behavior of the ore particles with other interfacial properties such as adsorption and flotation can lead to misleading results in such cases. Using low angle x-rays and Auger spectroscopic techniques, Kulkarni and Somasundaran[13] found hematite particles to be richer in silica and clay in the surface region than in the bulk, the proportion of the minerals in the surface region being dependent upon the pretreatment that the particles received. In their study they used a massive natural Minnesota hematite sample. Chemical and x-ray analysis of the bulk sample showed it to consist of 94 wt. % hematite and the rest essentially of quartz. Samples were prepared by roll crushing and sieving. The various size fractions and the treatments to which these samples were subjected are given below in Table I.

Results from the low angle x-ray probe analysis of the samples listed above are given in Table II. A comparison of these results shows a decrease in iron content upon washing suggesting removal of hematite during that operation. Slimes are observed to analyze higher in Al, Na and Mg as compared to the deslimed fraction. This is possibly due to the larger amounts of clay in the slime fraction.

Ultrasonic treatment appears to be removing both Si and clay from the surface region. The amount of Fe in the ultrasonically treated sample is the same as that of the unwashed sample. Washing as seen earlier removes hematite slimes only.

Table I. Treatments Received by Hematite Samples.[13]

No.	Sample	Treatment Received
1.	½-in. hematite	unwashed
2.	½-in. hematite	washed with distilled water several times
3.	325/400 mesh hematite	deslimed
4.	slimes	obtained during preparation of sample 3
5.	½-in. hematite	ultrasonically treated

Table II. Low Angle X-ray Probe Analysis of Various Hematite Samples.[13]

Sample	Elements*, at.%				
	Fe	Si	Al	Na	Mg
1. Unwashed	88.7	8.4	1.9	-	0.9
2. Washed	79.4	10.3	1.9	4.1	3.4
3. Deslimed	81.9	10.6	4.4	0.5	1.7
4. Slime	75.1	10.5	7.7	1.5	4.9
5. Ultrasonically treated	88.5	7.7	2.0	0.7	0.9

*Oxygen is not taken into account for calculations of the atomic percentages.

The above results suggest that simple physical treatments such as washing and ultrasonic cleaning of the mineral surface can cause marked alterations in the surface chemical composition which in turn can influence the interfacial properties.

The isoelectric point of the natural hematite was determined by Kulkarni and Somasundaran[13] to be at about pH 3.0 using the streaming potential and electrophoretic measurements. The titration experiments yielded a point of zero charge (PZC) of 7.1 and the maximum in floatability of hematite using oleate which is expected to occur at its point of zero charge, was obtained at pH 8. The above discrepency between IEP and PZC can also be explained on the basis of the surface mineralogical heterogeneity identified for the surface regions. The presence of silica on the hematite particle surface has evidently caused a shift in the IEP of the sample from that of pure hematite towards that of pure silica. PZC as estimated from flotation results are expected to be much closer to that of pure hematite, since the magnitude of process such as adsorption and flotation is mainly controlled by the surface area and the surface charge density of minerals. It has been reported that effective surface charge density of hematite can be higher than that of the same amount of silica[14]. Complex interfacial behavior of natural ore particles can, therefore, be explained on the basis of the mineralogical heterogeneity of the surface and the concentration of impurities in the surface region. While the electrokinetic properties, such as isoelectric point, will be an average of those for the individual minerals present on the surface, adsorption properties, being the result of processes at a molecular level, can be characteristic of an individual mineral, if other minerals are inert towards the adsorbents. Adsorption, for example, of alkylsulfate on natural hematite can be considered as electrostatic in nature, even above its isoelectric point, if the effect of surface segregation of silica on the zeta potential is recognized. Alternative mechanisms, proposed in the past, involving for example chemisorption, are unwarranted. These observations emphasize the need for monitoring of surface composition while studying interfacial properties of natural minerals, and for the recognition of the effect of the variation in chemical and mineralogical composition on these properties.

Mechanical Treatments

Mechanical treatments such as comminution and screening can result in deformed surface layer which may not show the crystalline properties of the bulk material. In fact, in a recent work[15] the density of the quartz being ground was found to decrease as a function of grinding time apparently due to the creation of a deep amorphous layer on the particles. In a typical case, as much as 14 wt % of the material of size 50 μm was found to be amorphous. It is well known that the presence of such deformed layers can yield a more reactive surface with increased solubility and surface hydration. Shifts of the isoelectric point on account of mechanical surface disruption have been reported by Smolik et al[16] and Jorgensen and Jensen[17].

Depending on the comminution method used, fracture of the particles could occur either at random or along the cleavage planes resulting in different ratios of the chemical constituents and therefore different reactivities. Characteristics of surfaces produced during comminution could therefore vary significantly depending on the cleavage behavior of the mineral. Smolik et al[16] obtained different values of isoelectric point for three aluminosilicates with the same bulk composition but with different aluminum to silica ratios in the cleavage planes. Their results are summarized in Table III.

Table III. Isoelectric Points of Aluminosilicates with Different Cleavage Planes.[16]

Mineral	Cleavage Planes	Al/Si ratio	IEP
Sillimanite	(010)	2	6.8
Andalusite	(110)	3	7.8
Kyanite	(100) and (010)	3	7.9

It is interesting to note that even though the three minerals are of identical bulk chemical composition, isoelectric point of andalusite, the mineral with a higher aluminum to silica ratio in the cleavage plane, does give an isoelectric point closer to that of alumina (9.4) than to that of silica (2.5).

As mentioned earlier, prolonged grinding can lead to polymorphic transitions and solid state reactions and thereby change the surface properties significantly. In the case of grinding of massicot and calcite, Lin and Somasundaran[15] found polymorphic transitions to alter its structure to that of litharge and aragonite respectively. The amounts of aragonite formed during grinding were determined using the X-ray diffraction analysis. Results obtained for the polymorphic transition of calcite into aragonite due to prolonged dry grinding are presented in Figure 1. Decomposition temperatures of samples after grinding for various intervals are also plotted in this figure

It should be noted that the change in heat of solution often observed for ground calcite and usually attributed to change in particle size, might at least partly be due to the formation of some aragonite in the samples.

Solid state reactions have been found to take place during grinding. Lin and Somasundaran[15] observed formation of galena during grinding of massicott or litharge with sulfur. In addition, even contamination with grinding media, as reported by Jamieson and Goldsmith[18] who found mullite contamination in their sample ground in mortar and pestle, is to be expected when samples are prepared by grinding. The observations suggest, first, the need for establishing standard conditions of preparation of samples, and, second, the importance of recognizing the fact that the properties of small samples prepared by grinding need not necessarily be representative of the bulk materials and may differ depending on the technique used.

Chemical Treatments

Leaching. To remove impurities that are originally present on surfaces as well as from those introduced during comminution and screening stages it is a common practice to subject minerals to leaching treatment. Various reagents are used for cleaning the surfaces by leaching and each reagent has its own effect on the surface properties. Commonly used leaching agents are: hydrochloric acid, nitric acid, chromic acid, hydrofluoric acid and sodium or potassium hydroxide. Even though these leaching agents might remove the natural and induced impurities, some of the acids are known to contaminate the surface with their specifically adsorbing anions.

In this regard the effect reported for leaching with different reagents on the isoelectric point of quartz is quite typical.

Gaudin and Fuerstenau[19] reported an isoelectric point of 3.7 for quartz treated with boiling hydrochloric acid whereas Li[20] obtained a value of 2 for quartz from the same source treated with boiling aqua regia, conc. HCl, 48% HF solution and 10^{-1}M NaOH, successively. Ohyama and Usui[21] and Zucker[22] obtained values between those obtained by Gaudin and Li. Fujii[23] obtained a value of about 4 for HF leached quartz. Kulkarni and Somasundaran[24] have recently reported the results of their study on the effects of various pretreatments on the electrokinetic properties of quartz. The pretreatments included desliming, ultrasonic irradiation, leaching with nitric acid and leaching with hydrofluoric acid followed by cold or warm sodium hydroxide solutions. For HNO_3 treated quartz they reported an isoelectric point of 2. Hydrofluoric acid treatment was found to be severe and was considered to dissolve various surface silanol species and cause distortion of the surface. Contamination of the surface with silicon-fluoride complexes was also suspected. Use of warm sodium hydroxide solution after treating the quartz with hydrofluoric acid resulted in an increase in isoelectric point of quartz to about 6.5. The resultant surface however, was found to be highly unstable as indicated by the continuous shift in zeta potential towards more negative values during subsequent aging in water.

It should be pointed out that both F^- and Na^+ are highly surface active towards quartz[25,26]. Meyer[25] has reported that even if HF is used elsewhere in the laboratory building a significant level of fluoride can be detected on the quartz surface. Removal of F^- or Na^+ by washing with deionized water or even HCl solutions is extremely difficult. HF treatment has also been reported to produce a decrease in the hydration of the silica surface. All these changes in the surface can affect the electrokinetic behavior of the mineral particles.

Leaching with acids often leads also to preferential dissolution of ions from the surface resulting in a surface which is significantly different from the bulk chemical composition. Smolik et al[16] obtained a lowering in the isoelectric point of alumino-silicates on leaching them with acid. It was determined that silica content of the surface was higher after the leaching treatment indicating preferential dissolution of aluminum from the surface.

Buchanan and Openheim[27] reported a similar phenomenon for leaching of kaolinite with HCl. They observed preferential leaching of aluminum when kaolinite was treated with HCl solution for one hour. These investigators also determined the amount of aluminum ions leaching out when kaolinite was washed with doubly distilled water (pH 6.0). Clearly such preferential leaching can produce differences in the interfacial characteristics of kaolinite. Similar behavior is expected of other clay minerals also.

In the case of apatite (a salt type mineral) Somasundaran[28] obtained increased negative zeta potential upon acid leaching. Furthermore, considerably more negatively charged surface was obtained with hydrochloric acid leaching than with nitric acid treatment, possibly due to the specific adsorption of the chloride ions. Any preferential leaching of calcium ions would also result in a more negatively charged surface. The zeta potential of acid leached apatite are shown in Figure 2. It might be noted that the isoelectric point obtained for nitric acid leached apatite is in general agreement with the reported experimental and theoretical values[29].

In addition to the above, prolonged acid leaching leads to surface roughness which is well known to cause variations in experimental results especially contact angle studies[30].

Aging of Mineral Samples. Leaching of mineral samples results in the removal of impurities, but it is clear that it often leaves the surface which is not always representative of the bulk material. The leached surface, to an extent can be restored to its original structure by aging the sample in aqueous solutions. The aging time required to attain equilibrium, can often be very long depending upon the mineral as well as the solution properties such as pH. During the conditioning of nitric acid leached apatite in water, Somasundaran[28] observed that the equilibrium was not attained even for several days. The aging of the apatite samples was carried out at different pH values for various times. In all cases, the pH of those solutions with an initial pH value greater than the point of zero charge decreased and the pH of those with an initial pH value lower than it increased. Preferential hydrolysis of calcium ions in the basic pH range and of phosphate ions in the acidic range are the major reactions which govern the above pH changes. It was also observed that equilibrium was attained sooner when the conditioning was done at a pH value closer to the point of zero charge than away from it.

Time required for the equilibrium depends also on the pretreatment the mineral surface has undergone. Kulkarni and Somasundaran[24] have reported a sharp initial decrease in the zeta potential of freshly deslimed quartz sample followed by a slow increase before a constant value was attained. The equilibrium value was lower than the initial value. The nitric acid treated quartz did not exhibit any such change in the zeta potential. In contrast to the above, quartz sample leached with hydrofluoric acid and subsequently treated with warm sodium hydroxide solution was observed to exhibit a significant dependence of zeta potential on aging time. Ionic strength of the solution in which aging is carried out has also been reported to influence the electrokinetic properties. Kulkarni and Somasundaran[24] found that whereas the zeta potential continuously decreased upon aging a given quartz sample at 10^{-2}N ionic strength and at pH 4, at 10^{-3}N ionic strength the zeta potential was observed

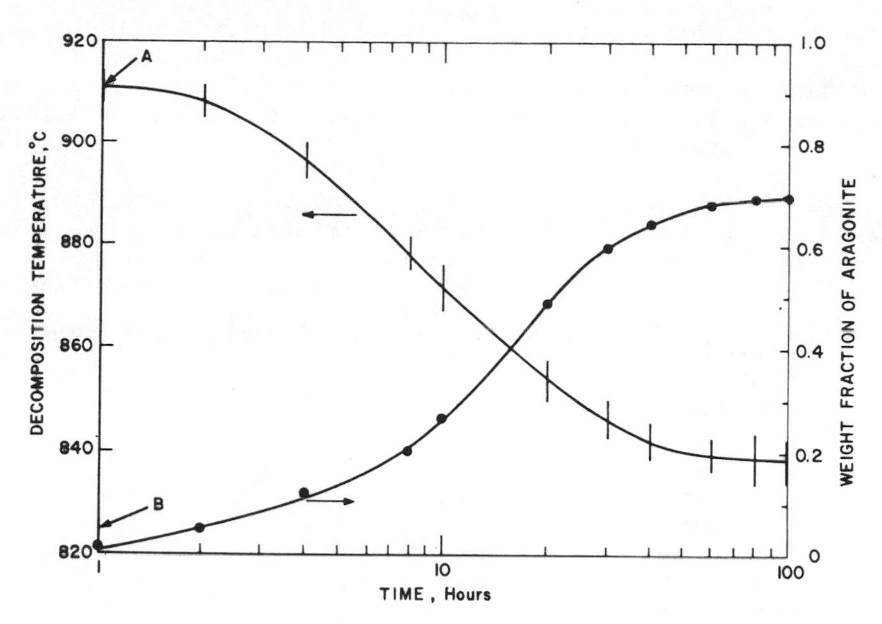

Figure 1. The decomposition temperature of ground calcite samples and the fraction of aragonite formed as a function of grinding time. (A and B indicate the decomposition temperature of pure calcite and of pure aragonite respectively.) (From reference 15.)

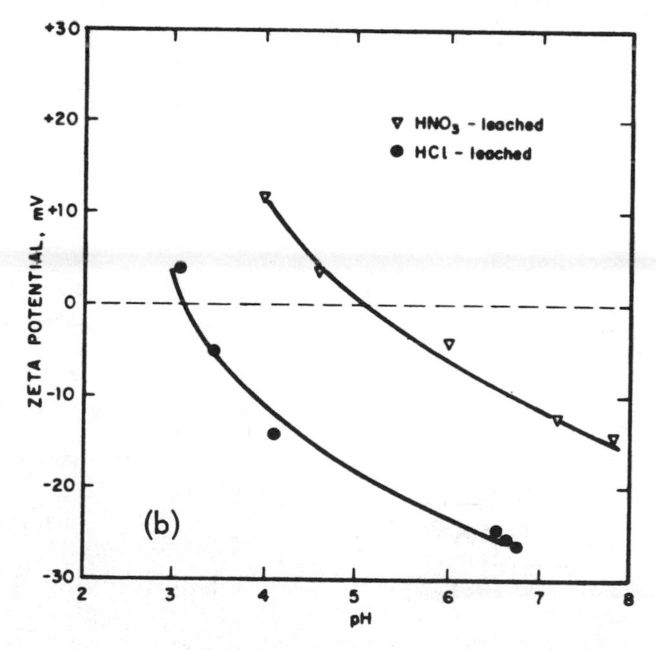

Figure 2. Zeta potential of HNO_3-leached and HCl-leached apatite as a function of pH. (From reference 28.)

to increase with the aging time. These changes in the electrokinetic behavior of quartz upon aging have been explained in terms of mechanisms which include the formation of silicic acid at the interface and its subsequent pH dependent dissolution and interfacial ion exchange reactions.

Clay minerals also have been reported to undergo interesting changes upon aging[28]. Originally the measured property may be representative of one kind of structure but on aging in water it will respond to the exposure of other structures by swelling of the clay mineral. This behavior can be attributed to the fact that these minerals have layers combining sheets of different structures. The surface properties in such cases would depend on the structure exposed to the environment.

OTHER SOURCES OF IMPURITIES

Structural Impurities

The effect of structural impurities on the surface properties can be significant. This effect is clearly illustrated by Park[5] in his review on the electrokinetic properties of oxides and complex oxide minerals. Selected data on the effect of structural impurities in Al_2O_3 and Fe_2O_3 substrates is given in Table IV. It is to be noted that levels of the structural impurities present on the surface can be different from those in the bulk by varying amounts depending upon the conditions of preparation and cleaning and such differences will indeed have effect on the surface properties.

Table IV. Effect of Structural Impurities on the Isoelectric Point (IEP) of Minerals.[5]

Substrate	Impurity	Impurity level mole % in substrate (approx.)	Change in IEP
Al_2O_3 (hydrous)	SO_4^{2-}	7.4	≥ -0.5
Fe_2O_3 (hydrous)	SO_4^{2-}	1.4	-1.6
Fe_2O_3 (anhydrous)	SO_4^{2-}	0.3	-0.2 to -0.6
Fe_2O_3 (hydrous)	Cl^-	0.5	≤ 1.4

Table V. Effect of Adsorbed Impurities on the Isoelectric Point (IEP) of Minerals.

Substrate	Impurity	Impurity level moles/liter	Change in IEP	Reference
Al_2O_3	$H_2PO_4^-$	$\sim 5 \times 10^{-6}$	-4.0	5
Al_2O_3	SO_4^{2-}	$\sim 3 \times 10^{-6}$	-2.9	5
TiO_2	SO_4^{2-}	$\sim 10^{-3}$	-0.7	5
TiO_2	Ca^{2+}	$\sim 10^{-4}$	+0.3	5
α-TiO_2	Sodium oleate	$\sim 10^{-3}$	-3.7	5
SiO_2	Dodecyl ammonium chloride	$\sim 3 \times 10^{-4}$	+4.0	31
SiO_2	Octadecyl ammonium chloride	$\sim 8 \times 10^{-6}$	+4.0	31

Contamination

Adsorbed Impurities. Both organic and inorganic contaminants can affect the surface and mechanical properties of mineral samples. Effect of adsorbed impurities on the isoelectric point of some selected minerals is shown in Table V. It can be seen from the table above that in the case of ionic impurities the effect of impurity increases with increasing valency and in the case of alkyl impurities the effect increases with increasing length of the hydrocarbon chain.

It has also been reported in the literature that adsorption of organic polar molecules can significantly change the mechanical properties of the materials. For example, adsorption of fatty acids, amines and alcohols on the metals has been observed to produce reductions in mechanical properties of several materials[32]. An increase in mechanical properties, on the other hand has also been reported[32,33,34]. This increase has been attributed to the locking of the surface dislocations by the adsorbed molecules.

Impurities can also significantly affect the properties of semi-conducting materials. Mular[35] for example has reported that the presence of aluminum ions can make zinc oxide more of an n-type conductor rendering the surface more negative and making it less hydrophobic in the presence of sodium oleate, whereas, acceptors such as copper ions have the opposite effect.

The sources of the adsorbing impurities can be improper handling of the material during preparation and cleaning stages. In the case of synthetic mineral samples the chemicals used during the preparation stage can alter the surface properties in a significant manner. Youssef[36] for example observed that addition of polymer during preparation of alumina caused significant changes in both porosity and adsorption properties. The surface area and total pore volume for alumina sample prepared in the absence of polymer was determined using nitrogen adsorption to be 412 m^2/g and 0.620 ml/g respectively. However, when alumina was precipitated in presence of 10% polymer the surface area and total pore volume were determined to be 460 m^2/g and 1.860 ml/g respectively. The addition of polymer was also observed to increase the adsorption capacity. The considerable increase in the total pore volume upon addition of polymer is not accompanied by a proportional increase in the total surface area. This has been attributed to pore widening in the presence of the polymer. It is expected that the amount of polymer added, polymer structure and molecular weight of the polymer added can cause changes in surface properties of the synthetic mineral sample.

Another major source of contamination is the hydrocarbons in the air. Kaye[37] has determined using low angle laser light scattering (LALLS) technique that organic matter constitutes the major class of particulates in water clarified by conventional methods such as distillation, filtration partitioning and ion exchange methods. Particle count in samples where air was totally excluded was determined to be 1,000 particles/ml which increased drastically after a few hours storage in sealed containers. Removal of these particles is essential for trace analysis studies and for accurate molecular weight and particle size determination by the light scattering technique. A considerable reduction in particle count was obtained by distilling deionized water in a quartz still modified to permit collection of water with very little agitation of the distillate surface. It is suspected that organic vapors are condensed on exposed water surface and then dispersed into submicron size droplets with even the slightest agitation of the surface. It is obvious that the presence of the hydrocarbon droplets in the aqueous phase can seriously affect the interfacial properties.

Drying. In addition to surface heterogeneities due to acid leaching and mechanical treatments, processes such as oven drying can impart surface impurities for example from the walls of the oven. Heating also causes non-stoichiometry of some oxide surfaces which can alter the electrochemical properties of the surface. Parks[5] has reported a shift in the isoelectric point of titania after heating. In the case of certain other minerals such as alumina, isoelectric point undergoes an acid shift on heating possibly due to dehydration of the surface. Excessive heating can also lead to structural changes thus altering the electrochemical behavior of the particular mineral.

Storage. Organic contaminants are often present in the environment and can alter the electrochemical properties of the surface during storage of samples. White and Pudvin[38] observed an increase in contact angle from <8° to 21 to 30° when highly oxidized aluminum (Al_2O_3) surface was exposed to the laboratory air for 2 hours.

Furlong and Parfitt[10] observed a significant change in the electrokinetic behavior of rutile upon exposing it to air for six months as shown in Figure 3.

For prolonged storage of mineral samples in aqueous environment, it is essential to select a suitable container because leaching of the container material itself can become a major source of contamination.

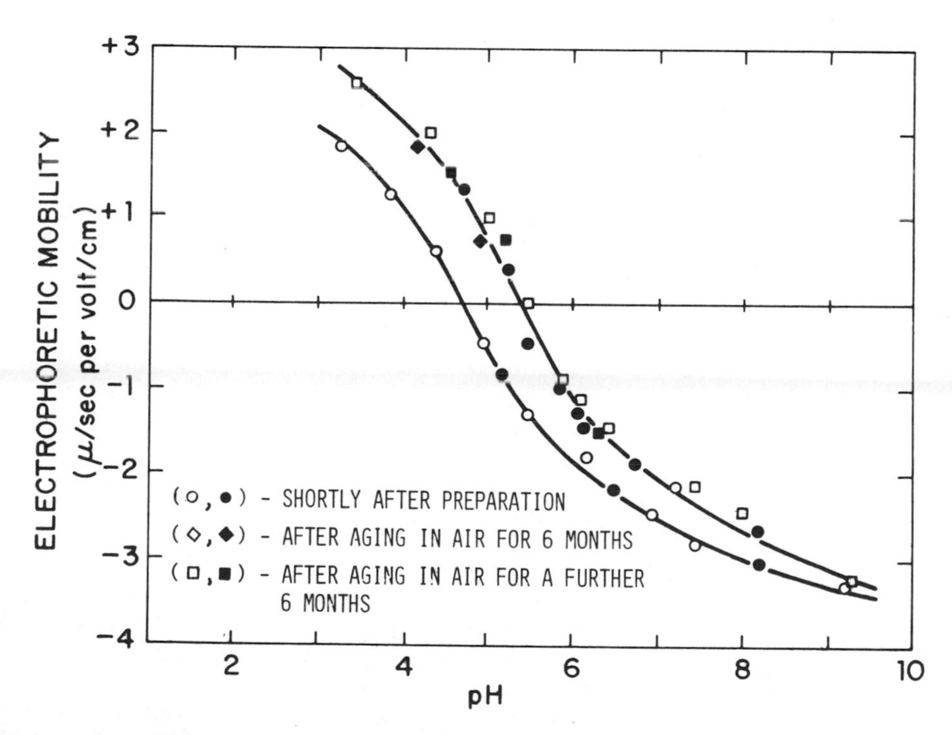

Figure 3. Effect of aging on the electrophoretic mobility of rutile as a function of pH. (Open symbols- decreasing pH; solid symbols- increasing pH.) (From reference 10.)

CHARACTERIZATION OF MINERALS

In basic research on the surface properties of minerals and processes dependent on such properties, it is important to have capabilities both to obtain reproducible and clean mineral surfaces and to characterize the surface. Techniques that exist for determining the major physical and chemical properties are discussed below.

Physical Properties

Physical properties that are relevant in mineral science include surface area, size analysis, porosity and particle shape. For surface area determination, B.E.T. adsorption, steric acid adsorption and permeability (e.g. Fisher subsieve analyzer) are among the commonly measured parameters[39,40]. It might be noted here that steric acid adsorption technique is limited to polar solids but may be more appropriate for the measurement of interfacial area accessible to surfactant molecules. The 'Quantasorb' continuous flow sorption system[41] (based on the B.E.T. surface area measurement principle) manufactured by Quantachrome Corporation has been reported to perform satisfactorily down to about 0.0028 m^2/gm. Size analysis of the mineral particles using sieves can be carried out down to about 25 microns (625 mesh). For single minerals 'Sedimentation Balance,' based on Stokes' law of settling can be used in the finer size range, while for heterogeneous fines produced, 'Coulter Counter' is better suited particularly in the range of 1-50 microns. In the colloidal range down to 0.05 microns, LADAL pipette centrifuge has been reported to perform satisfactorily[42].

Morphological Characterization

Morphology of particles can be important in determining their behavior in a number of interfacial processes. In this regard scanning electron microscopy[43] has proven to be a powerful tool for obtaining information on the topology.

Mineralogical Analysis

Bulk composition of mineral particles can be determined by wet analysis, and x-ray fluorescence. Surface chemical composition, as mentioned earlier is however more relevant for minerals research as segregation of minerals near the surface region can often produce considerable deviation of the surface composition from that of the bulk[13]. Techniques such as AUGER[44], ESCA[45] (Electron Spectroscopy for Chemical Analysis), ISS-SIME [46](Ion Scattering Spectroscopy in combination with Secondary Ion Mass Spectroscopy), SAM[47] (Scanning Auger Microscopy), are powerful tools that have become available towards this purpose. Recent advances in infrared spectroscopy also offer new methods for the characterization of surfaces of materials and the species adsorbed on them. Developments that are more commonly used include ATR (Attenuated Total Reflection Spectroscopy), and reflection adsorption spectroscopy. A review on characterization of solid surfaces[47] describes the current status of these techniques.

Interfacial Properties

The two most commonly studied interfacial properties in mineral processing research are zeta potential and wettability. For zeta potential measurements, streaming potential technique or electrophoretic mobility technique are usually employed. Commercial instruments that are available for measurement of electrophoretic mobility are Riddick Zeta Meter (Zeta Meter, Inc.), Laser Zee Meter (Pen Kem Co.), Electrophoresis and Mass Transport Analysis (Micromeretics), Zeta Reader (Komlin-Sanderson) and Rank Zeta Meter (Rank Bros., England). Electrophoretic mobility technique can be used only for colloidal particles of 0.1 to 2 microns. For measurement of zeta potential of coarse particles, streaming potential technique is used. A recently developed automated cell[48,49] can be employed to measure streaming potential as a function of different system variables such as temperature, pH, concentration of reagent, aging time, etc. The techniques that are used in minerals research to determine the wettability of the minerals are--(i) contact angle measurements using sessile drop[30] or captive bubble method[50], (ii) microflotation using modified Hallimond cell[51], (iii) bubble pick-up[52] method and (iv) suction potential measurements[53].

Among all the techniques listed above for determining wettability (or lack of it), contact angle measurements provide information of limited practical importance since it requires polished mineral surfaces which are rarely used in actual processing systems.

CONCLUDING REMARKS

It is obvious from the above discussion that the type of treatments used for the preparation, cleaning and storing of mineral samples can strongly influence the nature of the surface. To a large extent these factors probably can also explain the wide discrepancies among, for example, the electrokinetic values reported in the past. No single method of preparation and cleaning a mineral sample, however, can be suggested. The satisfactory method will depend upon the purpose and scope of the proposed study. However, some general guidelines to ensure the preparation of clean mineral surfaces are suggested on the basis of results obtained in the past. Grinding should probably be done in containers made up of a material that is harder than the sample itself. For removing contaminations at that stage mild leaching with dilute nitric acid or perchloric acid can be considered to be safe. Use of nitric acid as the leaching agent also minimizes the contamination of the mineral due to leaching itself, since nitrate ions do not adsorb specifically on the mineral surfaces. Treatment by hydrochloric acid and hydrofluoric acid seems unnecessarily severe and should be avoided.

Most of the minerals appear to reach equilibrium in short times when contacted with aqueous media, however recent studies[24,28,54] have shown that often equilibrium is not attained in such short times. Aging of the mineral samples in aqueous medium, has a significant effect on the surface chemical properties and should be a major factor in the designing of the experiments. To produce a clean reproducible mineral surface in a reasonable time, it has been suggested[28] that the conditioning or aging of the sample be done at a pH value closer to the point of zero charge. At this pH preferential dissolution of ions is expected to be minimum. In the case of minerals exhibiting ion exchange properties, it would be necessary to establish the conditions for minimum preferential dissolution by conducting chemical analysis of solutions that have been equilibrated with samples of the mineral.

It should be recognized that mineralogical heterogeneity, present either as a result of natural variations in the crystal structure or introduced due to treatments such as washing and leaching during preparation of the mineral surface, can significantly affect the interfacial properties. It is therefore, essential to monitor the surface properties, particularly the chemical composition, during investigations related to the interfacial properties of minerals.

ACKNOWLEDGMENT

Support of the National Science Foundation (Grant: ENG-76-80139) is gratefully acknowledged.

REFERENCES

1. H.J. Modi and D.W. Fuerstenau, J. Phys. Chem., 61, 640 (1957).
2. P. Somasundaran and R.D. Kulkarni, J. Colloid Interface Sci., 45 (3), 591 (1973).
3. P.G. Johansen and A.S. Buchanan, Australian J. Chem., 10, 398 (1957).
4. J. Schuylenborgh and A.M.H. Sanger, Rec. Trav. Chim., 68, 999 (1949).
5. G.A. Parks, Chem. Rev., 65, 177 (1965).
6. B. Dobias, J. Spurny and E. Frendlova, Collection Czech. Chem. Commun., 24, 3668 (1957).
7. A.S. Joy and D. Watson in "Proc. VIth Int. Mineral Process. Congress," p. 355, Pergamon Press, Oxford, 1965.
8. G.D. Parfitt, J. Ramsbotham and C.H. Rochester, J. Colloid Interface Sci., 41, 437 (1972).
9. R.H. Ottewill and J.M. Tiffany, J. Oil Colour Chem. Assoc. 50, 844 (1967).
10. D.N. Furlong and G.D. Parfitt, J. Colloid Interface Sci., 65, 548 (1978).
11. T.W. Healy, G.R. Weise, D.E. Yates and B.V. Kavanagh, J. Colloid Interface Sci., 42, 647 (1973).
12. R.O. James and T.W. Healy, J. Colloid Interface Sci., 40, 53 (1972).
13. R.D. Kulkarni and P. Somasundaran, Powder Technol., 14, 279 (1976).
14. A. Breeuwsma, "Adsorption of Ions on Hematite (α-Fe_2O_3), A Colloid Chemistry Study," Thesis, Agricultural Univ., Wageningen, p. 65, 1973.
15. I.J. Lin and P. Somasundaran, Powder Technol., 6, 171 (1972).
16. T. Smolik, J. Harman and D.W. Fuerstenau, Trans. AIME, 235, 367 (1966).
17. S.S. Jorgensen and A.T. Jensen, J. Phys. Chem., 71, 745 (1967).
18. J.C. Jamieson and J.R. Goldsmith, Amer. Mineralogist, 45, 818 (1960).
19. A.M. Gaudin and D.W. Fuerstenau, Trans. AIME, 208, 1365 (1957).
20. H.C. Li, Ph.D. Thesis, M.I.T., Cambridge, Mass., 1958.
21. Work of Ohyama and Usui cited in A.S. Joy, R.M. Manser, K. Lloyd and D. Watson, Trans. IMM London, 75, C81 (1966).
22. G.L. Zucker, D.Sc. Thesis, Columbia University, New York, 1959.
23. Y. Fujii, Unpublished Work, Henry Krumb School of Mines, Columbia University, New York, 1965.
24. R.D. Kulkarni and P. Somasundaran, Inter. J. Mineral Process., 4, 89 (1974).

25. D.E. Meyer, "Surface Impurities on SiO_2 and Their Effect on Water Adsorption Isotherm," paper presented at 42nd National Colloid Symposium, Chicago, June 1968.
26. F.M. Fowkes and T.E. Burgess in "Clean Surfaces: Their Preparation and Characterization for Interfacial Studies," G. Goldfinger, Editor, p. 351, Marcel Dekker, New York, 1970.
27. A.S. Buchanan and R.C. Openheim, Aust. J. Chem., 21, 2367 (1968).
28. P. Somasundaran, in "Clean Surfaces: Their Preparation and Characterization for Interfacial Studies," G. Goldfinger, Editor, p. 285, Marcel Dekker, New York, 1970.
29. G.A. Parks, Advan. Chem. Series, 67, 121 (1967).
30. A.M. Gaudin, "Flotation," Second Edition, McGraw Hill, New York, 1957.
31. P. Somasundaran and D.W. Fuerstenau, J. Phys. Chem., 70, 90 (1966).
32. V.I. Likhtman, P.A. Rebinder and G.V. Karpenko," Effect of a Surface Active Medium on the Deformation of Metals," Her Majesty's Stationary Office, London, 1958.
33. E.S. Machlin, in "Strengthening Mechanisms in Solids," American Society for Metals, Cleveland, p. 375, 1962.
34. A.R.C. Westwood, Phil. Mag., 7, 633 (1963).
35. A.L. Mular, Trans. AIME, 235, 204 (1965).
36. A.M. Yousseff, J. Colloid Interface Sci., 54, 447 (1976).
37. W. Kaye, J. Colloid Interface Sci., 46, 543 (1974).
38. M.L. White and J.F. Pudvin cited in "Environmental Control in Electronic Manufacturing," P.W. Morrison, Editor, p. 175, Van Nostrand Reinhold Co., New York, 1973.
39. K.S.W. Sing in "Characterization of Powder Surfaces," G.D. Parfitt and K.S.W. Sing, Editors, p. 1, Acad. Press, London, 1976.
40. A.W. Adamson, "Physical Chemistry of Surfaces," Interscience Publ., New York, 1960.
41. "Quantasorb Surface Area Analyzer," Bulletin, Quantachrome Corp., New York, 1977.
42. R. Hogg, Pennsylvania State University, Personal Communication, 1978.
43. O. Johan and A.V. Samudra in "Characterization of Solid Surfaces," P.F. Kane and G.R. Larrabee, Editors, p. 107, Plenum Press, New York, 1974.
44. D.F. Stein, J. Vac. Sci. Technol., 12, 1 (1975).
45. T.A. Carlson, Phys. Today, 25, 1 (1972).
46. J.A. Leys and J.T. Mackinney, paper presented at the 9th Annual Congress of Microbeam Analysis Soc. of America, July 22-24, Ottawa, Canada, 1974.
47. "Characterization of Solid Surfaces," P.F. Kane and G. R. Larrabee, Editors, Plenum Press, New York, 1974.
48. P. Somasundaran and R.D. Kulkarni, J. Colloid Interface Sci., 45, 591 (1973).
49. P. Somasundaran, S. Ramachandran and R.D. Kulkarni, J. Electrochem. Soc. India, 26 (2), 7 (1977).

50. A.F. Taggart, T.C. Taylor, and C.R. Ince, Trans. AIME, 87, 285 (1930).
51. D.W. Fuerstenau, P.H. Metzger and G.D. Seele, Eng. and Mining J., 158 (3), 93 (1957).
52. R.N. Hargreaves and H.P. Schreiber, CIM Bulletin, June, 84 (1974).
53. C.C. Harris in "Second Symposium on Coal Preparation," Papers No. 3 and 9, The University of Leeds, Dept. of Mining, October 1957.
54. R.D. Kulkarni and P. Somasundaran in "Proc. Symp. on Oxide-Electrolyte Interfaces," R.S. Alwitt, Editor, p. 31, The Electrochemical Society, 1973.

THE PREPARATION OF CLEAN WATER SURFACES FOR FLUID MECHANICS

John C. Scott

Fluid Mechanics Research Institute
University of Essex
Colchester, CO4 3SQ, England

This paper describes the criteria used in the preparation and characterization of clean and controlled water surfaces for experimental fluid mechanics. Small levels of surface contamination are shown to be capable of changing the whole character of certain free surface flows. The basis on which a surface can be considered clean is discussed, and acceptable techniques for determining the presence or absence of surface-active contamination are outlined.

INTRODUCTION

(a) Surface Tension

Liquid surfaces differ from solid surfaces in being mobile, adapting readily to changing body forces such as gravity, and in being determined in their reaction to these forces not principally by elastic constraints but by surface tension. Surface tension arises from the imbalance felt by surface molecules as they find dynamic equilibria among the attractive force fields of their near neighbours, those near the surface feeling a net inwards attraction. Although it has such basic molecular origins, and is capable of more or less rigorous thermodynamic treatment, surface tension may also be usefully considered solely in terms of its phenomenological existence. An extensive range of striking physical phenomena - static and dynamic -

are attributable to surface tension and surface tension variations. Boys' stimulating monograph on soap bubbles,[1] and Isenberg's recent monograph[2] which carries the subject much further and deeper with the same high standard, are both monuments to the persistent and increasing interest in these phenomena.

(b) Clean Water

The demand for clean water surfaces arises as a requirement of several types of research, both on the basic fluid dynamics of clean surfaces and surfaces with controlled added contamination, and also on the detailed effects of particular materials on the surface tension. It is obviously not possible to infer accurately the effects of contamination unless the surface is known to be clean initially. We will concentrate here on the fluid mechanical aspects of clean-water research, whose demands are rather different from those of pure colloid and interface science. The requirements of fluid mechanics are to some extent the simpler, in that chemical purity is far less important than surface cleanliness, while the two are of equal importance for pure interfacial science. In fluid mechanics, the important factor is the ability to specify accurately the mechanical properties of the interface, and the exact manner in which a given surface is achieved, in terms of what chemicals are used and how, is of rather less interest. Mysels and Florence[3] have given an excellent review of the chemical purity aspects of the problem neglected here.

It is in terms of the surface tension of a surface, and the way this affects surface motions, that we talk of clean and contaminated water surfaces in the context of fluid mechanics. At a given temperature, a perfectly pure liquid with a clean surface has a surface tension which is a constant, characteristic of the liquid molecules. In these circumstances, small expansions and contractions of the surface will only influence the surface tension if they occur on a time-scale that is short enough to be of the order of the time taken for the equilibrium to be established. It is usually reasonably assumed that for a liquid such as water, a time-scale of 1 μs is sufficiently large for this equilibration to be considered instantaneous.

If a second component is added to the pure liquid then the surface tension will inevitably change, as the force fields of the dissimilar molecules settle into different equilibria. In most cases it is unlikely that both components will find it equally acceptable, from an energetic point of view, to be situated near the surface, and the detailed composition of the surface regions will therefore usually differ from that of the bulk liquid. The 'surface excess' or adsorption of the added component may be either positive or negative, and the resulting surface tension may be either less than or greater than that of the pure liquid. Many water-soluble materials, such as the inorganic salts for example, have only a small effect on the surface tension of water, and the large concentrations needed for small changes in surface excess ensure a rapid re-establishment of the equilibrium value following small expansions or contractions of the surface. Aqueous solutions of inorganic salts may therefore be found to have surface properties quite similar to those of pure liquids, and they can therefore often be considered to be 'clean'.

(c) Contaminated Surfaces

Contamination, from the point of view of surface cleanliness, comes from a class of materials, mainly organic, whose effect on surface tension is much more pronounced than that of inorganic salts. These materials exhibit large values of the surface excess and they result in major changes of the surface tension. This class of materials is called surface-active, and they are often called surfactants. The importance of these materials in the fluid mechanics of liquid surfaces is that it may no longer be reasonably assumed that surface expansions and contractions will be immediately followed by re-establishment of the equilibrium. Relative motions of the surface will therefore in general lead to surface tension gradients - surface stresses acting to oppose the original relative motion - and the nature of the fluid motion may be thereby fundamentally altered.

In the general case, the surface-active molecules at the surface are in dynamic equilibrium with those in the bulk of the liquid. If the number of surface-active molecules in the bulk is sufficiently great, and the relative changes in area are sufficiently small, then expansions and contractions will eventually be followed by re-establishment of the original equilibrium. This will occur on a time-scale determined by the diffusional interchange between surface and bulk and by whatever time that may be required for individual adsorption or desorption events to take place once the molecules are close to the surface. On shorter time-scales, the variation of surface tension with expansion or contraction will be perceived as a surface dilatational elasticity which will be, in general, complex. It is this elasticity which usually gives rise to the most important fluid mechanical differences between clean and contaminated water surfaces. Mysels and Florence[3] prefer the term areal elasticity to dilatational elasticity, but the latter is more commonly used, and, I believe, it expresses adequately the concept involved.

SURFACE RHEOLOGY

(a) Surface Dilatational Elasticity

There is one particularly important case of contamination by surface-active materials in which a dilatational elasticity may be observed in all practical time-scales. This is the case of a material whose water-solubility is so low that a monomolecular layer (or monolayer) spread on the surface, perhaps from a liquid drop, is observed to persist on the surface without any appreciable desorption taking place. In this case, the observed complex dilatational elasticity may be wholly real. This case is quite common. and this fact, together with the considerable simplification that follows in theoretical analyses when the insoluble-film assumption is made, has had the effect that a large proportion of the theoretica discussion of the fluid mechanics of contaminated liquid surfaces has been concentrated in this area. Stable monolayers of this type on water usually consist of molecules that contain chemical groups of both hydrophobic and hydrophilic types, the net balance making the molecules "insoluble" - unacceptable in the bulk 'structure' of the water. Gaines[4] has given a detailed description of the surface-active properties of films of insoluble organic materials. However, even some normally water-soluble materials may behave in such an "insoluble" way when spread at the water surface.[5] Some water-soluble polymers may stay adsorbed at the surface simply because, although each surface-active monomer unit would normally become desorbed and pass by diffusion into the bulk of the water, it would require the simultaneous desorption of at least the great majority of the chain of monomer units in any macromolecule for that molecule to be allowed to leave the surface region.

(b) Surface Viscosity

Dilatational elasticity is not the only possible characteristic of surface-active contamination, although it is unquestionably the most general, requiring only that the material reduces the surface tension by an amount that increases as the film becomes more compressed. As there is a concentration of the active molecules in the surface region, then it is possible that the interactions between these molecules could become sufficiently great that relative motions in the surface become impeded by the interactions, in much the same way as the bulk properties of a two-dissimilar-component liquid will change as the proportion of the more viscous component becomes significant. Goodrich[6] has examined the possible ways in which the surface rheology might be affected in this way.

The most likely result of such intermolecular interactions is accepted to be a surface shear viscosity - a two-dimensional analogue of the shear flow viscosity of liquids in bulk. However, even this manifestation of surface rheology has proved rather elusive, because although its effects appear to be quite common in practice, its direct measurement is by no means simple. It is tempting to think of the surface as a form of separate 'phase', two-dimensional in nature, but this concept can be misleading, for the surface exists only in connection with its substrate, and simple shearing motions of the surface must inevitably be accompanied by flows of the more-or-less viscous substrate that are by no means simple. Following the work of Mannheimer and Schechter[7], however, the basis for the meaningful measurement of surface viscosity has been laid.

(c) Other Effects

As the surface concentration of the adsorbed material becomes great enough for molecular interactions to lead to surface viscous effects, other effects also become likely, including further contributions to the dilatational elasticity as the adsorbed molecules begin to feel each others force fields more strongly. Dilatational viscosity and shear elasticity are also possible. It may be shown, however, that whatever the origin of the rheological parameters of a surface, from an analytical point of view the surface may be completely described, for small strains, by only two complex quantities, representing the dilatational and shear properties of the surface film.[8]

THE MEASUREMENT OF SURFACE PROPERTIES

(a) Surface Tension

The surface tension of a static liquid surface may be assessed either by the force it exerts when the surface is given a known deformation or by the shape of the surface as it responds to a known force.[9] Techniques in the first category include: the Wilhelmy plate method, in which a solid object, such as a plate or a rod, is dipped into the liquid and the attractive force on the object is measured; the capillary rise method, which measures the hydrostatic force needed for equilibrium of a surface contained in and deformed by a circular bore capillary tube; the drop weight method, which measures the weight of a pendant drop of maximum size; and the ripple method, which measures indirectly the restoring forces exerted when the surface is distorted by a propagating train of surface waves, the surface tension being deduced from the observed wave velocity. The second category includes techniques that measure the shapes of surfaces distorted by known gravitational forces, such as the sessile drop method.

For our purposes, we need a technique that can be used on a flat surface of arbitrary size, such as might be found in the apparatus to be used in a fluid mechanical experiment. Of those described briefly above, only the Wilhelmy plate method is near ideal, with the ripple method as a possible alternative in some circumstances.

(b) The Wilhelmy Plate Method

This method, which seems to be rapidly gaining in popularity in both fluid mechanical and surface chemical applications, uses the fact that if a plate, made of material of high surface energy, is dipped into the liquid, and the liquid wets the plate, then the force tending to pull the plate into the liquid is in principle equal to the product of the surface tension with the perimeter of the line of contact between the liquid and the solid. The traditional problem with using this technique has been estimating that part of the plate perimeter associated with the vertical edges of the plate. Although theoretical estimates of this contribution are possible[10], it is convenient to use an expirical determination of the effect, with the use of two or more plates[9] identical except for their horizontal extent. Recent work by Padday[11,12] however, on the shapes and surface forces associated with rods suspended in the free surface, suggests that the substitution of a rod or cylinder for the flat plate may allow direct deduction of the surface tension from the force measurement.

However, the traditional method is still acceptable for fluid mechanical applications, and with suitable calibration, a precision of better than ±0.2 $mN.m^{-1}$ may be achieved without difficulty. There is considerable difficulty in specifying absolute values of surface tension to greater accuracy than this, and, in the case of water, Mysels and Florence[3] have demonstrated that there is considerable disagreement regarding the absolute value within even such relatively large limits, using experiments claiming precisions as much as a factor of ten better. For fluid mechanics, however, it is unlikely that the static value of surface tension should need to be known to a greater accuracy than this.

The Wilhelmy plate method is as suitable for contaminated surfaces as for clean surfaces, as long as the plate is adequately cleaned following each immersion, to avoid the build-up of dried surface-active material, which can drastically change the wetting properties of the plate and lead to erroneous results. This cleaning is usually done by first thoroughly rinsing and then burning off any remaining organic substances by heating the plate to red heat. The technique is suitable also for the continuous monitoring of a changing surface tension. As the surface tension force is usually measured with some form of electrical current balance[13] the automation of this to give a continuous-reading electrical analogue signal is possible. This property of the Wilhelmy plate technique has been used with success in the measurement of dilatational elasticity, as will be indicated later.

(c) The Ripple Method

The concept of increasing wave damping by the use of oily or surface-active materials is age-old, and is well known in the phrase "pouring oil on troubled waters'. The historical development of this concept has recently been outlined[14] and a detailed bibliography of the whole subject of waves on contaminated surfaces will be appearing shortly.[15] As a means of measuring surface tension, it is possible with a clean pure liquid to achieve a precision similar to that attainable with the Wilhelmy plate method,[16] but this requires much more complicated apparatus and a much more exacting experimental technique than does that method. A further disadvantage is that, for contaminated surfaces, straightforward ripple phase-velocity measurements are not sufficient for precise calculation of the surface tension, and the damping rate of the waves must also be measured simultaneously for correct inference of the surface tension. It should be mentioned, though, that an estimate of the dilatational elasticity also follows from the simultaneous measurement of velocity and damping rate.[17] Measurements using wave velocity alone have been found possible in cases where extreme precision is not required.[18]

(d) The Measurement of Surface Dilatational Elasticity

Surface dilatational elasticity is defined as $d\gamma/d(\ln A)$, where γ and A are the surface tension and the area of the surface, respectively. Its direct measurement therefore involves the measuring of the change of surface tension that results from an infinitesimal change in surface area. The variation of surface tension with the surface concentration is obviously not linear, as may be seen from Figure 1, and this means that, in general, the need to produce a measurable change in surface tension will lead to the deduced elasticity being some mean value taken over the range of areas considered.

If the surface tension vs. concentration curve is known with sufficient accuracy, than the dilatational elasticity at any concentration may be deduced from the slope of the curve at that point. Otherwise, a direct measurement is necessary, if possible with the surface in situ in the apparatus of the experiment. Some form of direct measurement is valuable in any case, as it provides a means of checking for the absence of unwanted contaminations.

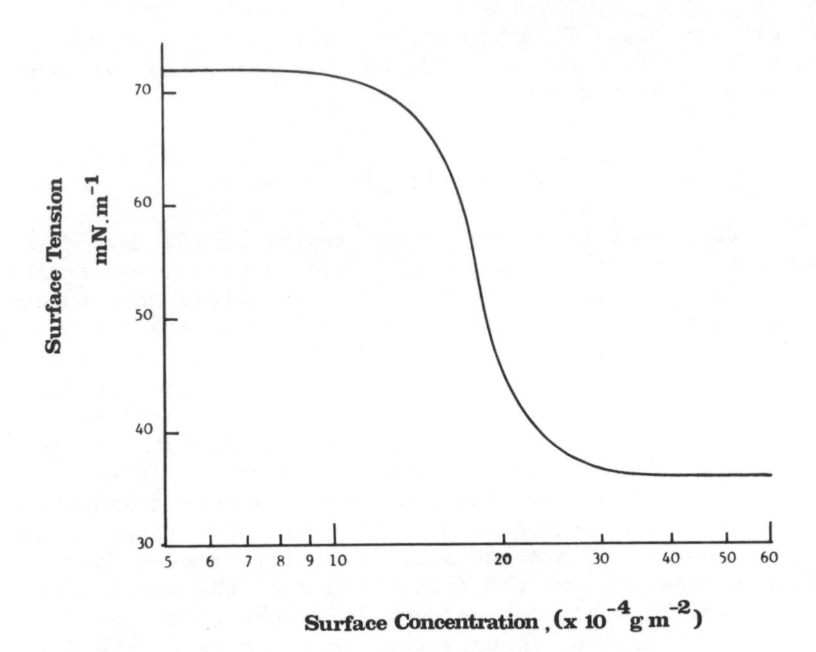

Figure 1. The general shape of the surface tension/surface concentration curve for typical adsorbed layers on water.

For basic surface chemical studies, the measurement may be made by varying the area of the Langmuir trough or tray used to contain the liquid surface, while monitoring continuously the surface tension using a current-balance type of Wilhelmy plate apparatus. This type of system or the ripple-measuring technique can be used for the study of the relaxation behaviour of the surface tension of a surfactant solution as well as allowing measurement of the elasticity of an insoluble surface film.[18] It is not, however, usually convenient for in situ measurements in an apparatus intended for other experiments.

As was mentioned earlier, it is possible, by simultaneous measurement of the phase velocity and damping rate of water waves, to deduce both the surface tension and the dilatational elasticity of a contaminated surface. This is principally true for dilute insoluble adsorbed monolayers, for which it may usually be assumed that any shear rheological properties of the surface are negligible. If high-frequency water waves (100 Hz or above) are used for this measurement, then the area of surface required need not be large, say 50mm x 100mm, and this may make such a procedure acceptable. The experimental technique required is quite sophisticated, however, and it may be considered better to avoid such time-consuming sensitive measurements if a less precise, more rapid method would suffice.

A much simpler method, which is suitable for use with a Wilhelmy plate surface tension measuring technique, and which may often be satisfactory for the requirements of fluid mechanics, uses a trapezoid device that encloses a limited part of the surface and allows the production of a small change in the enclosed area. One example of this method is illustrated in Figure 2. The trapezoid device is lowered, in its open position, into the surface, and the trapezoid is slowly closed, isolating a region of the surface some 100mm square. The surface tension in this region, assumed to be the same as that of the surface as a whole, is then measured using a Wilhelmy plate apparatus. While keeping the plate in position in the surface, the trapezoid is slightly compressed to a pre-set fixed position with a smaller surface area, determined by above-surface mechanical stops, and the surface tension of this reduced area is then measured. The change in surface tension for the known imposed change in area allows inference of the dilatational elasticity.

One point to be watched when any solid object is immersed in a monolayer-covered water surface is that a hydrophobic surface, (such as Perspex or PTFE) can draw on to its own surface a monolayer of the surface-active material as it passes through the surface of the water (Figure 3). This has the effect of reducing the surface concentration of the monolayer on the observed part of the surface,

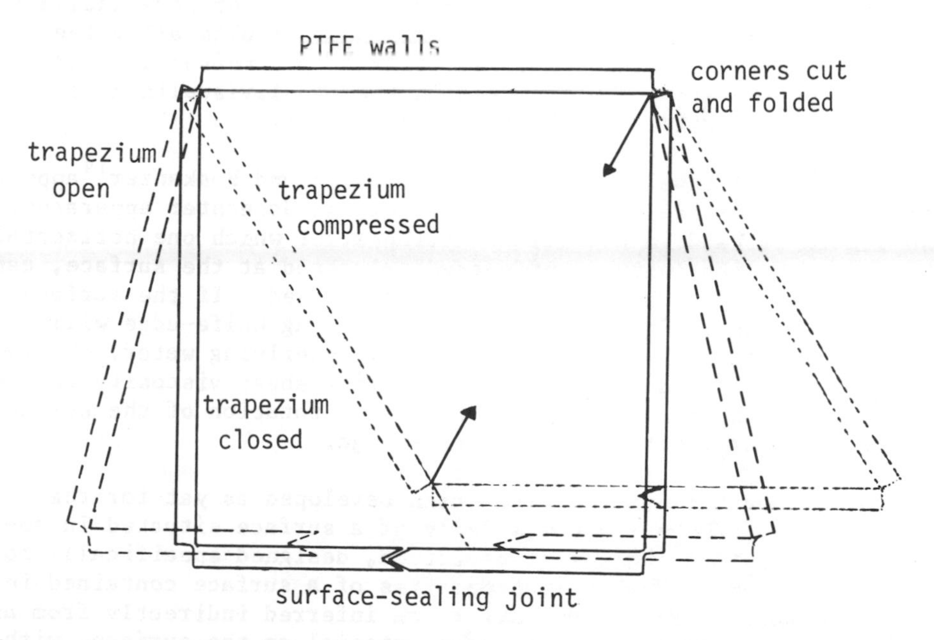

Figure 2. Trapezoid device for measuring dilatational elasticity.

and can lead to spurious results. The problem can be avoided by allowing the trapezoid just to touch the surface without appreciable immersion.

(e) The Measurement of Surface Shear Viscosity

As was mentioned earlier, the simple intuitive picture of the liquid surface as a form of separate two-dimensional "phase" is a misleading one. Any motion of the surface film, although it is determined by the rheological properties of the film, is also determined by the flow of the underlying liquid, the two contributions being inseparable. The whole fluid mechanical system must be subjected to rigorous theoretical analysis before the influence of the surface properties on the flow can be reliably deduced.

This has been done so far for two types of measurement apparatus the viscous traction annular canal apparatus of Mannheimer and Schechter[7]; and the rotating ring/fixed wall apparatus of Goodrich et al.[19]. In the former apparatus, sketched in Figure 4, the liquid is contained in a circular dish, and an annular canal, consisting of two concentric parallel-sided rings, is lowered centrally into the liquid. The liquid in the canal is forced into motion by the rotation of the circular dish, and the resulting motion of the surface is determined by the rotation of the dish, the viscosity of the liquid, the presence of the canal walls, and the shear resistance of the surface film. Mannheimer and Schechter[7] have analyzed the fluid mechanics of the system, and their results allow the deduction of the surface shear viscosity from a measurement of the velocity of the liquid surface within the canal, visualized using small buoyant particles resting on the surface.

The device developed by Goodrich, Allen and Poskanzer[19] appears to be more sensitive than the Mannheimer and Schechter apparatus. The liquid is held in a vertical cylinder, of which one horizontal thin-ring section (with a knife-edge), located at the surface, can rotate, leaving the rest of the apparatus fixed. If the surface possesses a shear viscosity, then the rotating knife-edge will induce a rotation of the surface and the underlying water, observable at the centre of the cylinder. The surface shear viscosity may be deduced from observations of the period of rotation of the surface relative to the rotation of the knife-edge.

No method has unfortunately been developed as yet for the measurement of the surface viscosity of a surface situated in apparatus other than those just described, designed specifically for that measurement. The shear properties of a surface contained in a fluid mechanical experiment have to be inferred indirectly from an auxiliary experiment using the same material on the surface, with a film of the same concentration or the same surface tension, whichever is known for the experimental conditions in use.

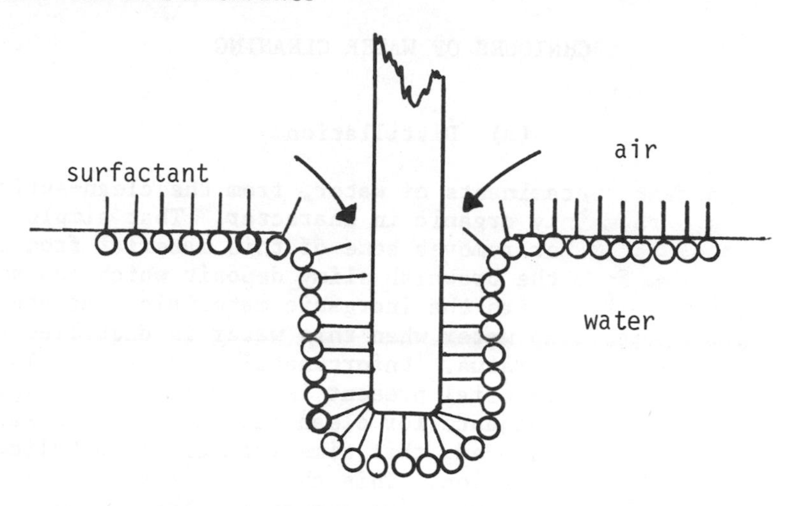

Figure 3. The removal of adsorbed surface-active material from a surface when a solid is lowered into the liquid.

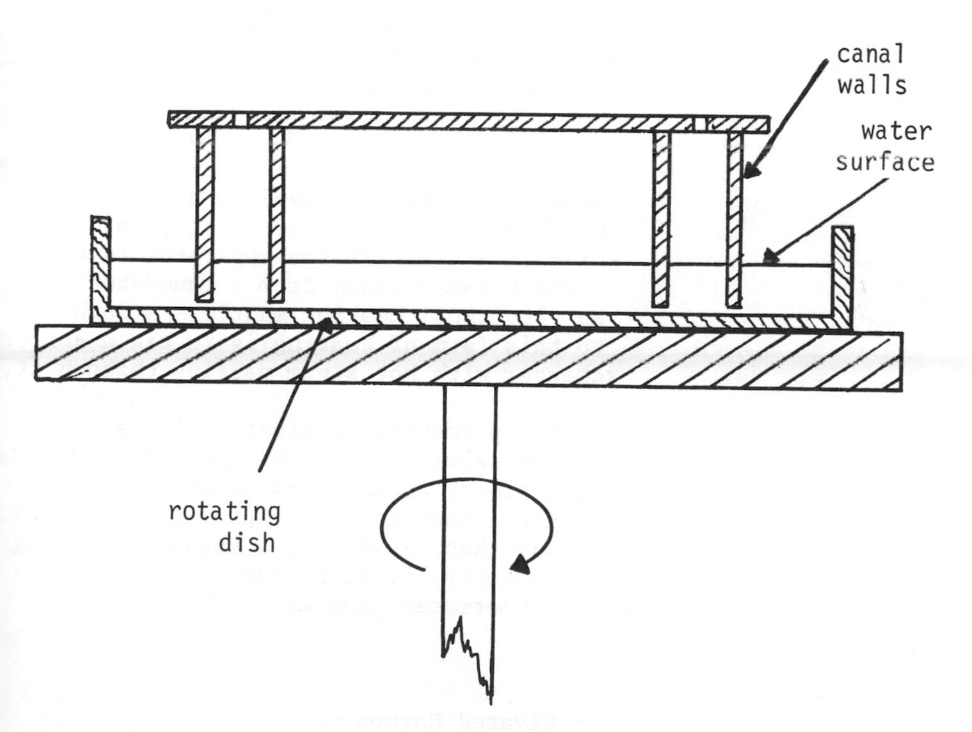

Figure 4. The principle of the Mannheimer and Schechter viscous traction annular canal apparatus.[7]

TECHNIQUES OF WATER CLEANING

(a) Distillation

The important contaminants of water, from the clean-surface point of view, are mainly organic in character. That simple distillation of the water removes some of this material from the water can be seen from the brownish slimy deposit which can sometimes be observed mixed with the inorganic materials that are deposited from "hard" tap water when this water is distilled from a continuously running apparatus. Unfortunately, however, a large proportion of the organic material present is volatile, and simple distillation is rarely adequate for clean surface work. A second distillation is usually needed, this time from dilute alkaline potassium permanganate solution. This chemical is effective in destroying most of the contaminating compounds, but amines and similar materials are unaffected, and pass into the distillate. These may often be removed by a third, acidic stage of distillation. It is found, however, that in batch distillation from potassium permanganate solution, the residual contamination appears to be distilled over within the first 10% of the operation, leaving the rest of the distillate of adequate quality for the majority of fluid mechanical experiments. Mysels and Florence[3] suggest the use of an electrically heated 'dry ring' in the condenser system to prevent the passage of non-volatile impurities through a continuous surface liquid film.

The arrangement used by the present author involves simple distillation, from a continuous 3kW still, followed by a 2kW permanganate stage, using 10ℓ batches, from which the first 1ℓ of distillate is discarded. All silica-glass/silica apparatus is used for the second distillation, which takes place from a round-bottomed flask heated externally by an electrical heating mantle. A much more efficient alternative to this form of heating is to use a coiled glass-tube internal heat exchanger with hot steam passing through it.

The limitation placed by the permanganate distillation step on the size of a fluid mechanical experiment should be noted. A 10ℓ total capacity batch still will yield, at most, 8ℓ of useful water. The ease with which surfaces become accidentally contaminated enough to require complete renewal means that, even with a means of storing considerable quantities of clean water, it is unwise to plan experiments using volumes significantly greater than about twice this volume, i.e. 16ℓ.

(b) Activated Carbon

While ion-exchangers are usually found to be completely unsatisfactory, as they tend to add more contamination than they remove (and this material may be harder to remove than was the original material), the activated carbon treatment has sometimes been used, either as a replacement for the permanganate distillation step described above, or as a further stage before use of the water. One problem in its use is knowing when the carbon has become too saturated to be of any further value, and Mysels and Florence[3] feel that, in any case, this process is inadvisable as a final purification step.

One advantage of activated carbon treatment is that it may be used with aqueous solutions of inorganic salts as well as for pure water, and this is an important asset for fluid mechanical work.

(c) Bubble Cleaning

The two methods described above attack the contamination using attributes of the contaminating materials that are to some extent removed from the aspect which it is actually hoped to remove - the property of becoming adsorbed on to a water/air surface. One way of attacking the contamination using precisely this same attribute is to generate a large water/air surface in the water, in the form of a dense cloud of bubbles at the base of a long vertical column of water. Surface-active materials will tend to become adsorbed on to this surface, and will tend to be carried upwards as the bubbles rise to the free surface. The dynamic nature of the adsorption/desorption process has the effect that it cannot be assumed that all material that becomes adsorbed will then automatically rise with the bubbles until they break. Nevertheless, a definite cleaning action is produced.

Work on this type of system has been reported by Anderson and Quinn,[20] who observed the effects of traces of contamination on the coalescence properties of rising gas bubbles in water. It is bubble coalescence, in fact, that limits the efficiency of this type of cleaning technique. When gas is blown through a fine porous glass membrane into water, the bubbles are initially very small indeed, representing a very high surface area per unit volume of gas. Within a very short distance, however, in pure water the bubbles coalesce to form a much smaller number of quite large bubbles, which may be in the region of 5mm diameter almost completely independent of the original small size on leaving the glass membrane. However, Krantz[21] has successfully used this type of procedure as a final step in pure water cleaning, and he has been able to deduce that considerable improvements may be produced in the contamination levels of water that is already very clean.

From the point of view of fluid mechanics, the bubble-cleaning technique has the major asset that it is suitable for any aqueous solutions. It is often desirable in certain experiments to use a liquid which, although fundamentally aqueous from its surface chemical aspect, differs in some other property, such as density or bulk-liquid viscosity. The bubble-cleaning method is just as suitable in these cases. Indeed, marked improvements in the performance of the technique may be observed compared with the results using pure water.[13] This improvement follows from the effect of the dissolved material on the coalescence properties of the bubbles in the tube, coalescence being significantly retarded. Bubble coalescence is by no means completely understood for aqueous solutions, but it is obvious that it is very sensitive to several of the characteristics changed by some dissolved materials, such as viscosity and ionic content.[22,23] The present author has found, with solutions of materials as dissimilar as sucrose and sodium chloride, sufficient reduction in the mean bubble size compared with water alon to allow a very efficient liquid cleaning action. This has, in fact, been found to be sufficiently efficacious for both sucrose and salt solutions - whether made with tap water or distilled water - that both have been produced with an adequate quality for clean-surface experiments.

(d) Surface Scraping

One method which is suitable for water that is already contained in some forms of fluid mechanical apparatus, such as trays and channels, uses the concentration of the surface into a relatively small area and the subsequent removal of the film from this area using suction. For use in a parallel-sided wave tank, for example, a scraper blade - consisting of a flexible rectangle of clean thin Perspex, braced against both sides of the channel - is inserted into the surface as close as possible to one end of the tank, to a depth of 10 - 20mm, and steadily moved along the tank to the other end, accumulating surface contamination on its forwards side. At some point close to the far end, the motion is halted, and the surface is sucked clean with a glass tube attached to a water-jet vacuum pump.

This method is obviously unsuitable for pure surface-chemical investigations, as, among other things, it does not adequately remove the contamination from the containing walls of the channel, and the Perspex 'scraper' must always miss a small area at the beginning of its travel. However, this technique has been found by some investigators to give surfaces on large volumes of tap water which, if not perfectly clean, will at least yield some reproducible results for a useful period before dissolved or dispersed contamination re-asserts itself at the surface. Several successive scraping operations, with a thorough rinsing of the scraper blade after each, are found to give a greater improvement than a single operation. While not completely acceptable from the clean-surface point of view, it appears that this technique may be all that can reasonably be done to clean the surface in many large-scale experiments that use volumes of water in excess of 30ℓ - 40ℓ. The properties of the surfaces that result from such techniques are presently being examined as part of a wider investigation, by the present author, of the surface properties of commonly used water surfaces. A more satisfactory variation of the surface scraping technique is possible if the water can be arranged to fill a channel or tray to the brim. If the channel material is hydrophobic, and water is replenished from below the surface, then the channel may be filled to near-overflowing and the surface layers can be scraped off completely in one movement, either using a hydrophobic bar running across the channel and touching the brim or using a tangential stream of clean air.

(e) Talc Cleaning

A related technique, that has been given some acceptance by surface chemistry research workers, is known as the talc cleaning method.[24] A nominally clean surface is covered with a light dusting of talc powder that has been heated to red heat to burn off organic materials. This talc covering forms a rigid surface layer that traps surface-active material, so that when the talc is removed from the surface using suction through a clean glass capillary, the talc and contamination are removed together. When combined with the use of a tangential stream of clean air, to concentrate the contaminated surface into a relatively small part of the surface, a talc covering gives an excellent visualization technique which clearly differentiates between the original contaminated surface and the freshly created surface.[3] Without this concentration of the surface, the method can be rather wasteful of the clean water, and this may be important when large areas of shallow water are involved. The technique can, however, be used on surfaces of awkward shape, such as would be completely impractical with the scraping method just described.

(f) Laser Burning

One novel technique has recently been proposed by Askar'yan et al.,[25] who observed that high-power laser beams focussed on to a water surface have the effect of selectively heating the surface and near-surface layers to high temperatures, at which selective evaporation of the surface material may take place. Ignition of layers such as of kerosene, and the forcible ejection of contaminating surface material to heights 200 - 300mm above the surface, were among the effects noted using a 50 W c.w. CO_2 laser, and a pulsed (1 MW, 1 J, pulse rate 1 - 2 Hz) CO_2 laser, both operating with 10 μm wavelength radiation. Although the authors suggest possible uses for this technique for the cleaning of liquids, it is unlikely that it will achieve wide acceptance in either surface chemistry or fluid mechanics, except perhaps as a means of rapidly removing the bulk of quite thick layers of volatile materials from small water surfaces

CHARACTERIZATION OF CLEAN SURFACES

(a) Surface Rheology

In theory, a water surface could be classed as "clean" using the techniques for measuring surface tension, surface dilatational elasticity and surface viscosity described above. A clean surface should have zero values of dilatational elasticity and of surface viscosity, and the surface tension should be close to the accepted value for the temperature observed. The dilatational elasticity measure is probably the most sensitive of the three, but it is not the most sensitive test we have at our disposal.

The most sensitive tests of surface cleanliness are in fact basically fluid mechanical in character. This fact serves to emphasize the crucial importance of surface effects in fluid mechanics, an importance which is not always fully realised. Surface contamination effects may be justifiably ignored in some free-surface flow experiments, in the same way that it is often possible in some flows of real liquids to proceed as if they were inviscid, but it is in general unwise to assume that contamination effects are negligible without a full understanding of how they might be affecting a given flow.

(b) Wave Damping

One good example of the subtle influence of surface films is in the damping of water waves, mentioned above as a means of measuring the dilatational elasticity of a surface. The effect of a surface film is to impede shear flows near the surface, influencing the flow through a viscous boundary layer similar to that found near a solid wall.[26] Two assumptions have commonly been made about this effect regarding wave damping: that the greatest wave damping should occur for an inextensible film (infinite dilatational elasticity); and that the contamination principally affects short-wavelength waves, or ripples. In fact, the damping can be much greater with a film of intermediate elasticity than with an inextensible film, and the potential effect of a surface film becomes greater for the larger wavelengths (waves not normally influenced significantly by surface tension), in terms of the change in damping rate between clean and contaminated surfaces.[17] Ripple damping was used as a cleanliness criterion even in the 19th Century, by Lord Rayleigh[27] and by Agnes Pockels,[28] and, as is shown by Lucassen-Reynders and Lucassen,[17] it can give a sensitive indication of the dilatational elasticity of the surface. However, without the detailed measurements of wave velocity and damping needed for this deduction, the observation of ripples can give no more than a rough-and-ready indication of the presence of significant contamination.

(c) Bubble Observation

One of the most sensitive generally accepted indicators is in fact quite a simple and widely applicable test: the persistence of gas bubbles at the free surface. When a bubble has arrived from below the surface, perhaps as a result of violent shaking of the liquid, it will break only when the liquid film has drained sufficiently for the film to be unstable. The process of draining this film is very similar to that involved in the coalescence of two bubbles, mentioned above in the section on bubble cleaning techniques, and it is similarly influenced by impurities in the water, especially surface-active impurities.[29] It is generally accepted that if a bubble persists at the surface for as long as 0.5s following its arrival at the surface, then the water is considered to be contaminated.[30]

For fluid mechanical experiments using surfaces created inside the bulk of the water, such as experiments using trapped or freely rising bubbles, measurements of the cleanliness at the surface are of limited value, although if the contamination present is distributed uniformly throughout the bulk, continuous measurement at the surface might indicate the gradual eventual contamination of all free surfaces.

A very sensitive test of surface-active contamination in bulk water is provided using another observation using bubbles: the measurement of the velocity of ascent of small bubbles of known size. The flow past a rising bubble concentrates adsorbed surface-active material towards the rear, lower part of the bubble, and the surface there becomes locally rigid, with the effect that a viscous boundary layer is formed adjacent to it. If the adsorbed film were to extend over the whole bubble, giving a completely immobile surface, then the bubble would rise as if it were a solid buoyant body. It has long been observed that small bubbles do tend usually to rise in this way, and this effect can be observed to some extent even in scrupulously cleaned water.[31] Bachhuber and Sanford[32] have recently reported experimental results indicating rise velocities close to those expected from completely free-surface bubbles, and it is significant that it was necessary in those experiments to use a technique which minimized the opportunity for contamination of the bubbles. This phenomenon therefore can provide a sensitive means of asessing the dissolved surfactant content of water in experiments where the surface is not easily accessible, if some means of generating bubbles of known size can be incorporated into the apparatus.

(d) The Krantz Surface Flow Test

A two-dimensional analogue of the rising bubble technique has been developed by Krantz and his co-workers[33] for general clean-surface assessment, and this technique is seen to give both a sensitive means of examining the efficacy of all surface-cleaning techniques and a form of quasiquantitative test of contamination level. In this technique, the water is arranged to flow along a wide, shallow channel of constant depth (Figure 5). A U-shaped barrier is inserted into the flow, just touching the surface, and surface-active material builds up against this barrier. The extent of the surface film thus produced is an indication of the level of contamination. It can be measured by observing the behaviour of small buoyant particles, which tend to circulate throughout the film and show the location of the leading edge. The cumulative effect of the viscous drag of the water beneath the film produces a surface tension variation along the film, which balances the drag force. This test gives a precisely measurable quantity - either the length of the stagnant film or the total surface stress - related to the concentration of contaminant. However, the deduction of the concentration from this test is not a simple matter, except in the case where one - and only one - (known) contaminant is present. Nevertheless, the system does allow a sensitive comparison of one water sample with another, and an important advantage for fluid mechanics is that the observation is closely related to the sort of surface property that is of interest.

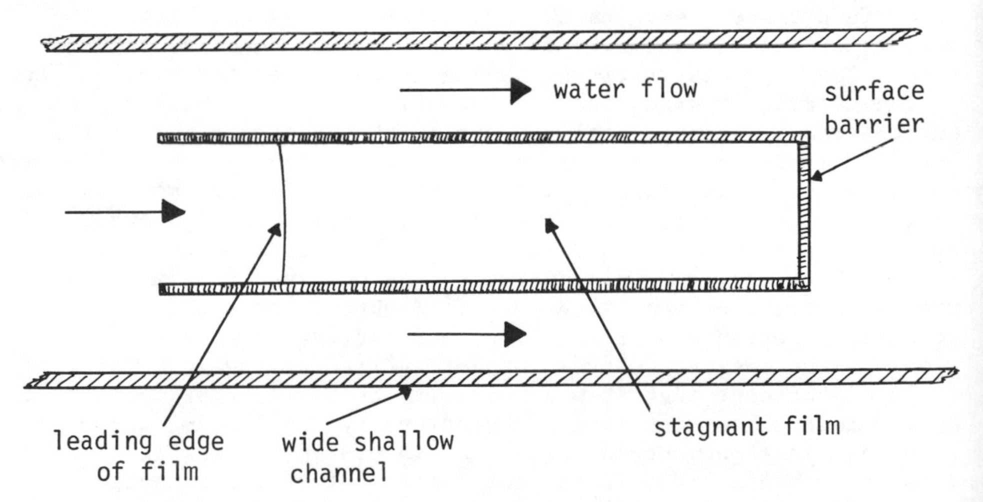

Figure 5. The principle of the Krantz surface flow test of surface purity

(e) The Talc Test

A useful simple test, suitable for in situ assessment of surface cleanliness, was described by Langmuir.[34] If a small amount of clean talc is sprinkled on a clean surface, then it may be moved to any part of the surface simply by blowing. If the surface is contaminated, the motion of the talc will be impeded by the surface film, and, after blowing, the talc will tend to return to its original location. Blowing vertically downwards towards the film will spread the talc outwards and make a clean region at the centre which will disappear when the blowing stops.

MATERIALS FOR CLEAN-SURFACE FLUID MECHANICS

Once we have prepared our clean water, it must be maintained in its clean state until this state is changed as part of the experiment. Severe restrictions are imposed on the constructional materials that may be used, by the need to keep surface-active matter away from the water.

Silica glass, or Pyrex, is acceptable for all but the purest of surface chemical research, and if storage bottles and their glass stoppers are thoroughly cleaned using hot chromic acid (a strong infusion of potassium dichromate in concentrated sulphuric acid), the storage of clean water presents few problems. For constructional purposes, however, glass is unfortunately a severely limited material, as the substances usually used to form water-tight joints between glass sheets are naturally surface-active in order to do their job, and they are therefore unacceptable.

The use of plastics is generally frowned upon, except in the case of polytetrafluoroethylene (PTFE, Teflon), which is sufficiently inactive for almost all purposes. The making of joints is as much of a problem as it is with glass, however, although PTFE may be milled from block material if the cost of this is justified.

Perspex (Lucite, polymethylmethacrylate) is not fully accredited, because of the fear that the lower molecular weight fractions and plasticizers used in sheet fabrication will be leached out into the water in sufficient quantities to become adsorbed at the water/air surface. The present author has found, however, that provided the Perspex is thoroughly cleaned with detergent and a soft sponge (to prevent surface damage), and given a copious rinse, finally with clean water, then it is left dry and clean. Clean water has been kept in a clean condition (observations with the bubble persistence test) in such apparatus for periods of many hours.

Perspex has the major advantage that it is easy to machine, and it can be jointed using solvent cements which, when set, leave a

Perspex surface of a quality similar to the original sheet. The precaution should be taken of allowing several days to elapse between cementing and any serious experiments, so that organic solvents can evaporate away.

Polyethylene may be a possibility in some cases where an apparatus constructed of an unacceptable material such as aluminium or iron can be completely covered by a continuous skin of the polymer.

Metals are generally unacceptable unless a) they have no reaction at all with water and air; and b) they may be treated with chromic acid or heated to red heat without oxidation. Platinum is acceptable, and is often used in the Wilhelmy plate method for the plate material, and stainless steel has been used in some applications.

Fresh surfaces of specially purified paraffin wax are acceptable coatings for unacceptable materials, but they are not readily cleaned, and complete re-coating may be necessary for each experiment. Rubbery materials, cements and glues must be avoided, as must any contact with silicone oils and greases. The machining of Perspex and PTFE should be done without lubricants. Contact with clothes and skin must be avoided, and the wearing of disposable polyethylene gloves is recommended during washing and handling operations. Water surfaces, and indeed all surfaces, should be protected as far as possible from contamination by atomosphere-borne contamination.

ACKNOWLEDGEMENT

The author is indebted to the Natural Environment Research Council for the provision of his Research Fellowship at the University of Essex.

REFERENCES

1. C.V.Boys,"Soap Bubbles: Their Colours and the Forces which Mould Them", S.P.C.K., London, 1890; enlarged edition, Dover, New York, 1959.
2. C. Isenberg, "The Science of Soap Films and Soap Bubbles", Tieto Ltd., Clevedon, U.K., 1978, ISBN 0-905028-02-3.
3. K.J. Mysels and A.T. Florence, in "Clean Surfaces", G. Goldfinge Editor, pp. 227-268, Marcel Dekker, New York, 1970.
4. G.L. Gaines, Jr., "Insoluble Monolayers at Liquid-Gas Interfaces", Wiley-Interscience, New York, 1966.
5. J.C. Scott and R.W.B. Stephens, J. Acoust. Soc. Amer., 52, 871 (1972).

6. F.C. Goodrich, J. Phys. Chem., 66 1858 (1962).
7. R.J. Mannheimer and R.S. Schechter, J. Colloid Interface Sci., 32, 195, 225 (1970).
8. T.B. Benjamin, in "Cavitation in Real Liquids", R. Davies, Editor, pp. 164 - 180, Elsevier, Amsterdam, 1964.
9. J.F. Padday, in "Surface and Colloid Science", E. Matijevic, Editor, Vol.1, pp. 101-149, Wiley-Interscience, New York, 1969.
10. F.P. Pike and J.C. Bonnet, J. Colloid Interface Sci., 34, 597 (1970).
11. J.F. Padday and A. Pitt, J. Colloid Interface Sci., 38, 323 (1972).
12. J.F. Padday, R. Pashley, and A. Pitt, J. Chem. Soc. Faraday Trans. I, 71, 1919 (1975).
13. J.C. Scott, J. Fluid Mech., 69, 339 (1975).
14. J.C. Scott, History of Technology, 3, 163 (1978).
15 J.C. Scott, "Oil on Troubled Waters: A Bibliography on the Effects of Surface-Active Films on Surface-Wave Motions", Multi-Science, London, 1979, ISBN 0-906522-00-5.
16. J. Lucassen and R.S. Hansen, J. Colloid Interface Sci., 22, 32 (1966).
17. E.H. Lucassen-Reynders and J. Lucassen, Adv. Colloid Interface Sci., 2, 347 (1969).
18. J. Lucassen and M. van den Tempel, J. Colloid Interface Sci., 41, 491 (1972).
19. F.C. Goodrich, L.A. Allen, and A. Poskanzer, J. Colloid Interface Sci., 52, 201 (1975).
20. J.L. Anderson and J.A. Quinn, Chem. Eng. Sci., 25, 373 (1970)
21. W.B. Krantz, personal communication, 1975.
22. V.G. Gleim, I.K. Shelomov, and B.R. Shidlovskii, J. Appl. Chem. USSR, 32, 1069 (1959).
23. S.A. Zieminski and R.C. Whittemore, Chem. Eng. Sci., 26, 509 (1971).
24. J.T. Davies and R.W. Vose, Proc Roy. Soc., 286A, 218 (1965).
25. G.A. Askar'yan, E.K. Karlova, R.P. Petrov, and V.B. Studenov, J.E.T.P. Lett., 18 389 (1973).
26. H. Lamb, "Hydrodynamics", Article 304, 2nd Edition, Cambridge University Press, Cambridge, 1895.
27. Lord Rayleigh, Phil. Mag., 30, 386 (1890).
28. A. Pockels, Nature, 43, 437 (1891).
29. J.D. Robinson and S. Hartland, Tenside, 9, 301 (1972).
30. J.A. Kitchener and C.F. Cooper, Quart.Rev.Chem.Soc., 13, 71 (1959).
31. J.F. Harper, Adv. Appl. Mech., 12, 59 (1972).
32. C. Bachhuber and C. Sanford, J. Appl. Phys., 45, 2567 (1974).
33. Anon., Anal. Chem., 46 799A (1974).
34. I. Langmuir, J. Am. Chem. Soc., 39, 1848 (1917).

TECHNIQUES TO OBTAIN ATOMICALLY CLEAN SURFACES

J. Verhoeven

FOM-Institute for Atomic and Molecular Physics
Kruislaan 407, Amsterdam/Wgm.
The Netherlands

In this paper techniques for preparing atomically clean surfaces are reviewed. The advantages and disadvantages of several methods are discussed. A combination of techniques is needed to obtain proper results in many cases. Bulk impurities play an important role.

Heat treatments are widely used but are not satisfactory in general. Combination with gas-solid surface reactions improves the results. Noble gas ion bombardment has turned out to be a successful technique. However ion implantation and surface structure damage can be caused. This can be prevented by sputtering at higher temperature or annealing afterwards.

INTRODUCTION

Contamination and cleaning of surfaces play an important role in many fields of surface science, like catalysis, oxidation, alloys, semi-conductors, etc. The surface of a solid can be considered as a discontinuity of the bulk. The crystal structure as well as the energy band structure are disturbed. As bonds between the bulk atoms are broken, several sites for binding other atoms on the surface are available. The new bindings can have a physical as well as a chemical character; the latter mostly being stronger. Moreover, a molecule arriving at the surface can dissociate. The character of the binding is determined by the chemical composition of the surface, the crystal structure of the surface and the type of adsorbing atom or molecule. The main source of surface contamination is the environment. However, impurities from the bulk diffusing to the surface cannot be neglected. So it is necessary to have a good vacuum and the remove intrinsic impurities from the bulk. Many techniques to remove surface contaminants are known. The use of different techniques or combinations can lead to the same results. On the other hand one technique can fail to give reproducible results, mostly as a consequence of different circumstances of the vacuum environment. It is therefore very difficult to give standard methods for the cleaning of a particular surface. This paper is a review of various techniques: some typical examples are given of applications and results.

PRETREATMENT

Before installation in an ultra-high vacuum system ($P<10^{-8}$ Pa) a pretreatment of the sample is required.

Rinsing in alcohol, trichlor ethylene or acetone will remove hydrocarbon contaminants. Freon should be avoided as a degreasing fluid, as it can cause fluorine and chlorine contamination. This explains the fluorine and chlorine peaks in a residual gas spectrum of an ultra-high vacuum system, which originate from freon contamination of the ion source of the residual gas analyser.

Very often some kind of polishing is performed. Mechanical, chemical, as well as electro-polishing will cause a smooth surface but can introduce contaminants also. For electro polishing chemical liquids of different composition are needed for different materials. Turkenburg[1] found that after electropolishing of his Ni(110) surface with a commercial polishing liquid, carbon and oxygen contamination could be observed.

THERMAL TREATMENT

Thermal treatment is one of the oldest and most widely used techniques of cleaning a surface. A well known method is ohmic heating. The sample can be spotwelded onto high purity molybdenum or tungsten rods. Tantalum rods cannot be purified. Leads need to be thick compared to the target. Another method is electron bombardment at the back of the sample. A simple filament can be used as electron source. One has to be careful not to use filament materials containing large amounts of impurities. Tungsten for example emits potassium and sodium ions during heating. Using electron bombardment it is possible to heat thicker samples. The sample has to be mounted in an insulated holder which prevents heat loss. Heating also can be performed by radiation. As a heat radiation source a small (halogen) lamp inside the vacuum system can be used. Temperatures of 1100°C are obtained[2]. This system works well for insulators.

In figure 1 an example is given of a target holder. The target is kept in position by three small sapphire insulating rods, which

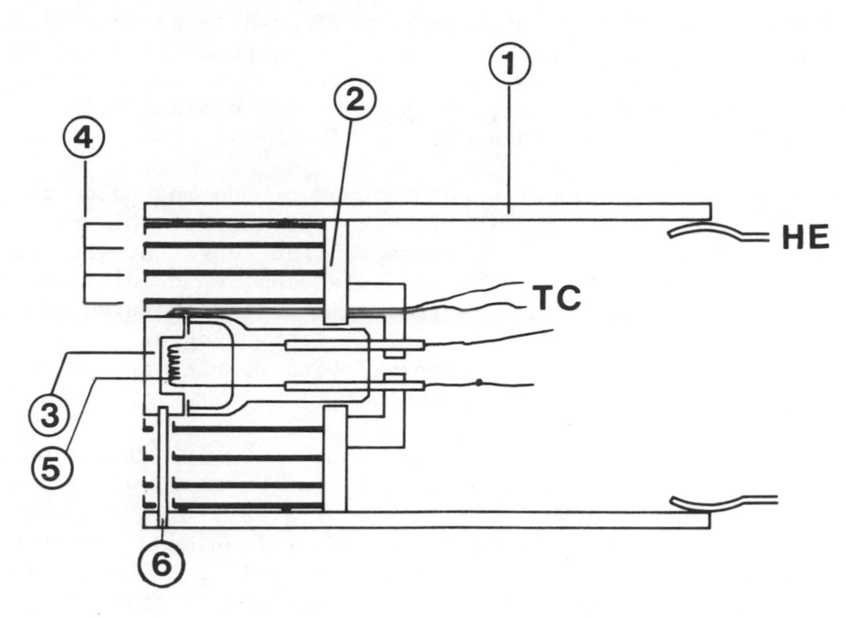

Figure 1. Example of a target holder. 1) Molybdenum cylinder; 2) Al_2O_3 holder; 3) target; 4) radiations screens; 5) filament; 6) sapphire rods; TC) thermocouple; HE) heat exchanger.

are fixed on a molybdenum cylinder. The sapphire rods have a rather good heat conductivity at lower temperatures, so by cooling the molybdenum cylinder the target can be cooled. The target can be baked either by radiation or by electron bombardment. For this purpose a lamp is used of which the glass bulb is removed. This provides a filament which can be installed easily. Radiation screens are mounted around the target. The filament and the screens are mounted on a Al_2O_3 holder. The temperature is measured by a thermocouple.

An advanced technique is pulsed laser bombardment[3,4,5]. This technique is more a thermal treatment than a photo-desorption treatment. The advantage is that as one uses high power and "short" pulses (40 ms), only the surface is heated, which prevents bulk impurities to diffuse to the surface. Another advantage is that the vacuum conditions are not degraded, as the heat source is on the outside of the vacuum system. Disadvantages are the possibility of local melting and of surface disorder produced by rapid quenching, which is observed e.g. after complete removal of oxygen and sulfur contamination from nickel surfaces by laser bombardment at 100 MW/cm^2 [4,5]. In the case of insulator- or semi-conductor surfaces this method is very useful.

Temperature measurements can be performed using an optical pyrometer or a thermocouple. In the lower temperature range an optical pyrometer cannot be used. Farnsworth warns for the use of chromel-alumel thermocouples, which can cause contamination by surface diffusion during bake-out.

Contaminants are usually not removed by heating below the melting point, unless the material is very refractory. This method is satisfactory for carbon-free polycrystalline tungsten. The main impurities in tungsten are sodium and potassium, which diffuse from the bulk to the surface and desorb at higher temperature. If the sample is in the form of a single crystal, heating at too high temperatures may produce thermal etching which results in facetting parallel to undesired crystal planes.

The main problem of heating is, that not only desorption takes place, but also diffusion from the bulk to the surface. It can therefore be useful to heat the sample at a very high temperature during a short time to keep diffusion low (flashing). For thin samples bulk impurities can be removed by heating at high temperatures during a long time. This is not possible for thick samples as the heating time would be too long[6], so flashing is preferred. Another solution is heating by pulsed laser radiation, as only the region near the surface is heated. In some cases it is possible to remove a contaminant or to anneal at such low temperature that diffusion from the bulk is negligible. Sickafus[7] reports for Ni(110) that surface carbon is removed at 650°C, but that heating at this temperature increases sulfur coverage. This sulfur contamination is removed by argon ion bombardment. But the subsequent annealing must take place below 650°C to prevent recontamination with sulfur.

One also has to take into account that impurities diffused to the surface due to heat treatment can diffuse back or undergo a reversible segregation to the surface at higher temperature. Mróz et al.[8] observed an increasing amount of sulfur on their Ni(111) and Ni(110) surfaces, reaching its maximum at 800 - 900°C. Above 1000°C the amount of sulfur decreased considerably, which they attributed to reversible segregation of sulfur of the nickel surface. So the surface of a thick nickel sample can be cleaned by heating during several minutes at 1100°C. Schouten et al.[9] observed diffusion of carbon from several nickel surfaces into the bulk. They have given a theoretical description of diffusion from surfaces into the bulk of solids.

GAS-SOLID SURFACE REACTIONS

One can also use gas-solid surface reactions to remove surface contaminants. For example it can be useful to oxidize the surface contaminant at lower temperatures. The contaminant oxide can be removed more easily in many cases by heating. The oxygen used for this purpose should be of a high purity and the gas inlet system is to be cleaned thoroughly.

Mróz et al.[8] report the behaviour of carbon and sulfur on (111) and (100) nickel faces on heating in a vacuum and in oxygen ($\sim 10^{-6}$ torr), which is investigated by means of Auger Electron Spectroscopy (AES). Their sample of 0.5 mm thickness was spot-welded to leads of molybdenum. In vacuum carbon disappeared from the surface on heating up to 500°C for Ni(111) and up to 800°C for Ni(100). Sulfur diffused onto the surface and could be removed by prolonged heating of the sample under investigation in oxygen (10^{-7} - 10^{-6} torr, 800°C) and also the amount of sulfur in the bulk was reduced. So a thin nickel sample can be cleaned by heating in an oxygen atmosphere, which also results in a reduced bulk concentration.

Lapujoulade and Neil[10] used a repetitive sequence of heating their nickel sample at 900°C to diffuse carbon to the surface, followed by heating at 1000 K in 10^{-8} torr oxygen to remove the carbon from the surface. Their cleaning procedure took ~70 hours. They do not mention how they removed sulphur, which is an important contamination for nickel.

For thick samples one can also combine gas-solid surface reactions with temperature flashes instead of long baking periods. As far as known to the author no results have been reported of the combination of gas-solid surface reactions and heating by lasers. Instead of oxidation of surface contaminants, also reduction can be used. For example carbon can be removed by heating in a hydrogen atmosphere giving methane. A disadvantage is that it is badly pumped by getter ion pumps and not pumped at all by titanium sublimation pumps.

SPUTTER ION CLEANING

Ion sputtering is a widely used technique to remove adsorbed atoms together with one or more surface layers. The most simple method is to create a discharge between an electrode and the sample after filling the vacuum system with a noble gas up to a certain pressure. If we use the simple discharge method, there is a chance that during pressure or voltage adjusting, a discharge will take place between the electrode and another part of the system.

A more advanced method is to use an ion gun, which basically consists of an ion source and lens, which also acts as an extractor[18]. The hot cathode ion source in principle consists of a cathode filament and an anode grid, like a Bayard Alpert vacuum gauge. The ions formed within the grid are extracted by the lens system. The advantages of this source are that the ion yield depends on the gas pressure within the source and that the ion energy can be chosen by varying the extraction voltage. The only limiting factor is discharge in the source or elsewhere in the gun.

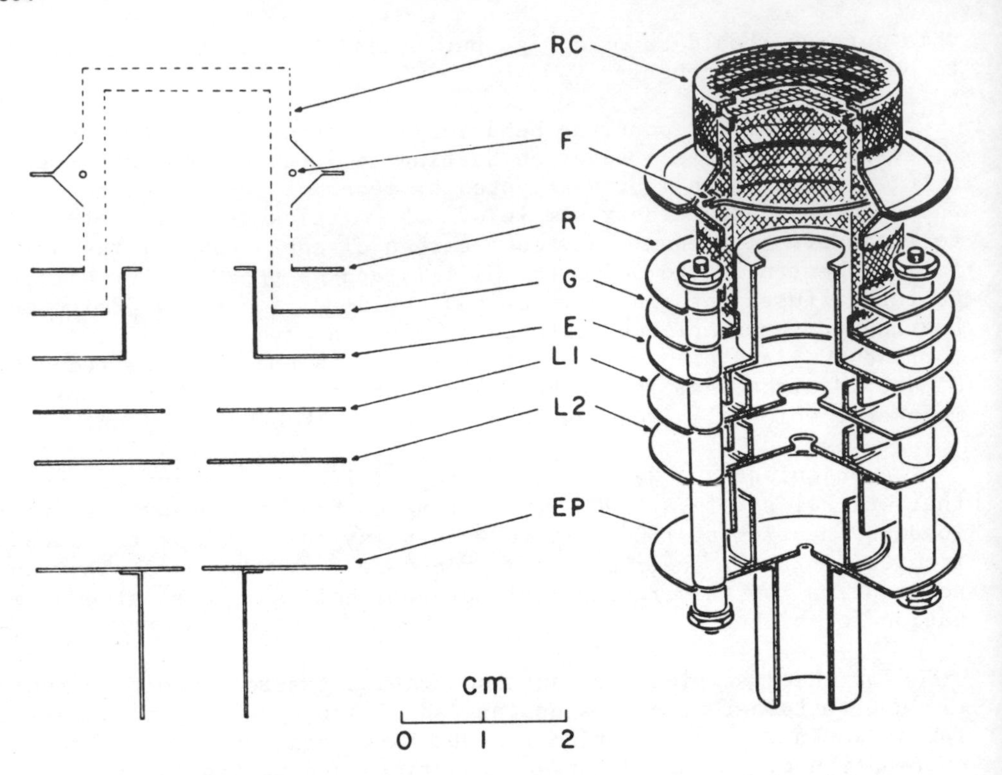

Figure 2. Example of an ion gun of the hot cathodetype. (a) Schematic arrangement of ion source electrodes; (b) Cut-away drawing of source showing details of elements and mounting structure; RC - electron repeller cap; F - filament; R - electron repeller; G - grid; E - beam forming electrode; L1 - first aperture lens plate; L2 - second aperture lens plate; EP - source exit plate, (taken from Ref. 41).

Figure 2 shows an example of an ion gun in the hot cathode type as designed by Kornelsen[41]. This gun is specially designed for operation at pressures $\leq 10^{-5}$ torr. It produces ion beams at voltages 100 V to 10 kV and currents up to 1×10^{-7} A.

One can also use an ion gun with a discharge source. Basically one has to deal with the Paschen curve (fig. 3), giving the discharge voltage as a function of the product of the distance between the parallel electrodes and the pressure for various gases. This means that, as the source dimensions are fixed, the discharge voltage depends on the gas pressure in the source. There is also a minimum in the Paschen curve which means that taking into account the source dimensions, there is a minimum voltage at which discharge can occur. So depending on the source dimensions it can be

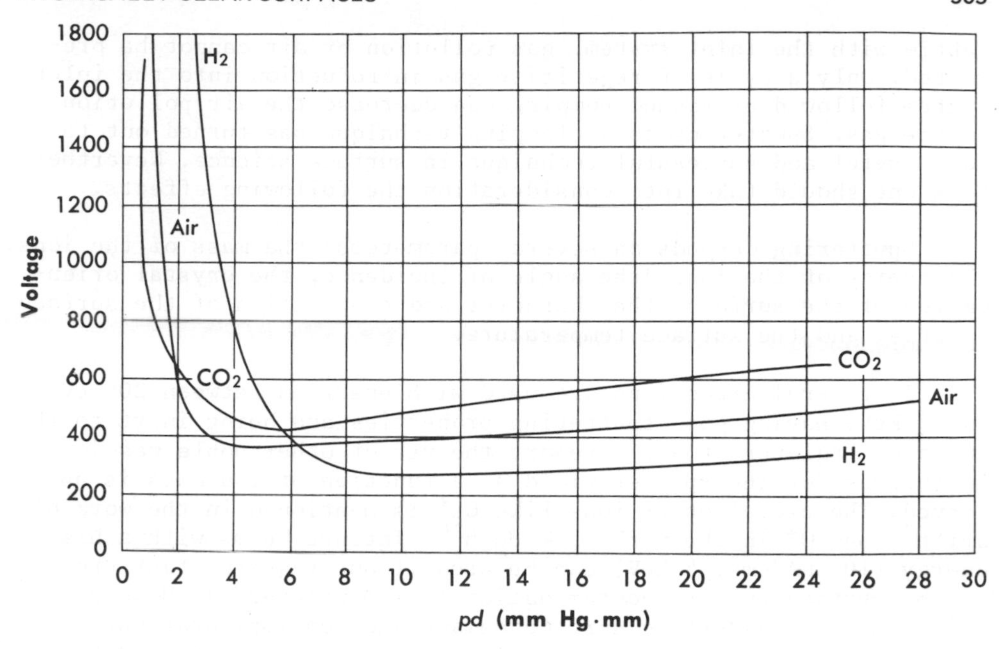

Figure 3. Spark breakdown voltage for plane-parallel electrodes (temperature 20°C).

difficult to obtain ions of low energy. This can be improved by a combination of a magnetic field and an electric field, like a Penning discharge source or by the choice of special field geometries like the saddle field source[19] or the hollow cathode source[20].

The advantage of the discharge source is that the source itself often produces a focused beam with an energy equal to the discharge voltage, so one does not need a lens. In general the ion yield with respect to the pressure in the source is higher than for a hot cathode ion source.

Finally it should be mentioned that a uniform beam intensity can be very important[21] to obtain spatially uniform sputtering. One can use the gun by filling the whole experimental chamber with the gas to be ionized up to the required working pressure of the ion source. A refinement is the installation of the ion source or the whole gun in a differentially pumped part of the vacuum system. In that case gas should be admitted directly into the ion source.

The gas admitted into the ion source needs to be pure, to prevent contamination of the surface with undesirable active gas ions. This means that not only purified gas has to be used, but that also the gas inlet system has to be bakeable. If one connects the gas bottle with the inlet system, gas pollution by air cannot be prevented. Only a cycle of repetitive gas introduction into the inlet system followed by vacuum pumping can decrease the air pollution in the gas. Sputtering as a cleaning technique has turned out to be a useful and successful technique in surface science. Nevertheless one should take into consideration the following effects.

Sputtering depends on several parameters: the mass of the ions, the energy of the ions, the angle of incidence, the crystal orientation of the surface, the character and composition of the surface material and the surface temperature.

In general argon ions are used with energies between 200 eV and 2 keV, having good sputtering properties and being inert to the surface. Taglauer et al.[22] report the use of other noble gas ions. An increase of the sputter yield as a function of ion mass is observed. The use of other ions like Cs^+ is mentioned in the work of Smith[23] and O^+ in the work of McHugh[24]. Intense beams with a high energy (100 μA/cm^2, 5 keV) can be used if one requires fast erosion of the surface and its contamination (> 1 monolayer/s). However, this causes considerable surface damage and ion implantation. A rearrangement of the surface crystal structure can be obtained by annealing at temperatures higher than during sputtering. However, this in turn can cause diffusion of bulk impurities to the surface, as we have seen in the previous section, about sulfur impurities in nickel. Use of a beam of low energy and lower intensity (1 μA/cm^2, 500 eV), will cause less structure damage. The erosion time will be in the order of one monolayer per 10 minutes[23]. Staib however observed that even an argon ion beam of 500 eV and 0.2 μA can cause considerable damage to a mica surface[25].

It is well known that secondary, tertiary and higher-order collisions play a dominant role in determining the consequences on an ion-solid interaction[21,22]. However, direct momentum transfer causes motion of the atoms in the sample lattice, referred as "knock on" or "knock in" process. This process depends on the ion energy. Defects of the bulk caused by ion impact can cause enhanced diffusion.

Another important parameter concerning the sputtering cross section is the angle of incidence relative to the surface[21,22,26].

Taglauer et al.[22] observed a maximum sputter yield for various noble gas ions on Ni(111) and Ni(110) at an impact angle of 30°. At certain orientations of the surface with respect to the ion beam, enhanced implantation due to channeling can take place[21,26].

Concerning the character and composition of the surface, preferential sputtering plays an important role, since the sputter yields of various elements differ substantially. This will lead to the formation of conical structures around the species with a low sputtering rate, which tend to protect the underlying bulk material [21]. The sputter yield of elements from pure materials also differ from that in multi-component systems, as observed by Ogar et al. in their work on Cu - Au alloys[27].

Ion bombardment of an electrically insulating solid will create electric fields on the surface and in the solid. A surface of a good conductivity has a higher sputter yield[23]. It is also observed that the sputter yield can be greatly enhanced in the region exposed to an electron beam during ion bombardment. The technique of exposing a surface to an electron beam during sputtering is also used in Secondary Ion Mass Spectrometry (SIMS). Electric fields in the solid can cause migration of mobile ionized species. This is demonstrated to be a severe problem in SIMS analysis of SiO_2 implanted with Na^+ ions[28].

The temperature of the sample is a parameter which can influence some of the processes by sputter etching as mentioned before. The sputter yield of pure metals tends to be independent of the temperature. However, in the sputter yield ratio of alloys, changes are observed[29]. By the choice of a high sample temperature during ion bombardment, ion embedding can be decreased.

Titanium presents the special problem that during annealing Cl and S segregate and cover the surface, although the bulk concentration is only about 0.1 ppm. A minimum contamination ($<$ 0.1 monolayer) by Ar, Cl and S is obtained by bombarding with Ar^+ ions at about 525°C sample temperature[23]. The choice of a low temperature during ion bombardment can prevent migration or diffusion problems as mentioned before[30].

It can be concluded that ion bombardment is useful to clean a solid surface, but damage due to bombardment is difficult to avoid. Also changes in surface composition play an important role, mainly in the investigation of alloy surfaces. It can therefore be advised to use an ion energy as low as possible. Also the annealing temperature should be as low as possible to avoid diffusion from the bulk.

ELECTRON STIMULATED DESORPTION AND PHOTODESORPTION

Electron Stimulated Desorption (ESD) is well known either as a surface analysis technique or as an effect which disturbs other surface analysis methods in which electron beams are used. However, this technique is also useful to remove weakly bound contaminants on a surface. One can use any electron gun with an intensity $> 10\ \mu A/cm^2$. The lowest energy threshold for ESD is about 15 eV[31]. The ESD effect is reviewed extensively by Madey and Yates[32]. They state that direct energy and momentum transfer will not break chemisorption bonds, although physisorption bonds may be affected. The

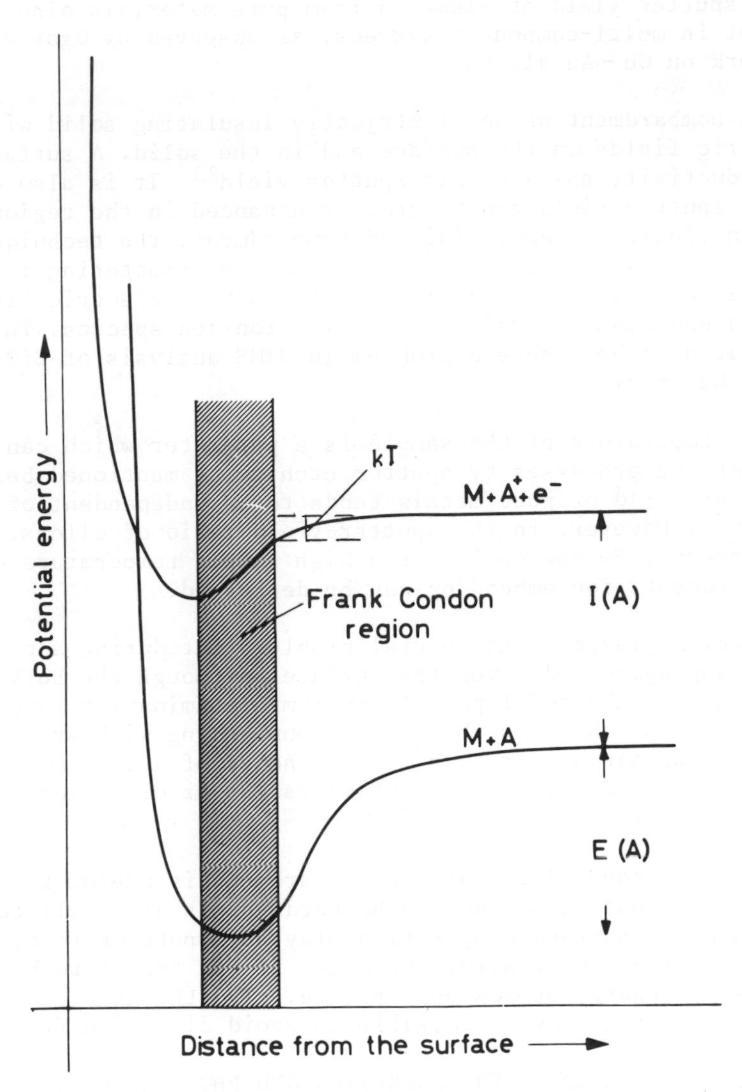

Figure 4. Schematic potential curves for interaction between a surface M and an atom A, and between M and the ion A^+. A possible electronic transition resulting in electron stimulated desorption of A^+ as indicated in the Franck-Condon region. E(A) is the binding energy of A to M and I(A) the ionization potential of A.

ideas of the theory of ESD have a foundation in the Franck-Condon principle, which states that during an electronic transition in a molecule, the nuclear separation and relative velocity are essentially unchanged. The potential curve in fig. 4 which represents the interaction energy between the metal substrate M and the adsorbate A, is different from the curve which belongs to the system $M + A^+ + e^-$ or $M + A^* + e^-$. So taking into account the Franck-Condon principle, the before mentioned ionized or excited systems are probably not in the equilibrium part of the potential curve just after ionization or excitation. It can happen that under these circumstances thermal energy is enough to desorb an ion or a neutral.

The cross section for this process is not only dependent on the character of the adsorbate but also on the state which an adsorbate occupies at the surface. The cross section for a more weakly bound state can be smaller than for a more strongly bound state [33]. Not only desorption can take place under electron beam impact, but also dissociation and formation of new compounds[34]. The effect of ESD is even observed within a radius of 1 cm around the electron beam impact spot[35].

In our work on nickel (110) surfaces we found that a combination of electron bombardment (500 eV, 140 μA) and heating is useful to remove carbon[14]. Using a quadrupole mass spectrometer, the appearance of a C^+ peak could be observed between 300 and 400°C.

Photodesorption is limited by the low photon yield and the difficulty of obtaining high fluxes, in the high energy ranges (> 10 eV) where the cross section increases. Another problem is the appreciable heating caused by intense radiation. So it has often not been possible to discriminate between photodesorption and photo-induced thermodesorption. Menzel et al. concluded that ESD and photodesorption of neutral CO on W proceed via the same excited state[36]. Franchy and Menzel[37] mention in their work the importance of photoelectrons and secondary electrons during photodesorption.

Further it should be taken into account that photodesorption from semiconductors involves desorption of chemical species by photogenerated carriers[38]. If the bandgap is illuminated, photo holes are drawn towards the surface by the barrier field. If the hole recombination lifetime is greater than the transition time through the barrier, some holes can reach the surface and neutralize the negatively charged adsorbate ions, which can be thermally released from the surface.

Electron impact on a solid surface not only causes desorption but can also give rise to enhanced adsorption[11,12,13,14,15] (ESA).

It is observed that enhanced oxidation can occur during oxygen exposure when a nickel surface is bombarded by electrons[13,14]. The presence of CO at a nickel surface during electron bombardment can cause a considerable carbide contamination[16]. Even the presence of electrons from a vacuum gauge can influence surface contamination. Bootsma and coworkers[17] discovered enhanced sticking probability of methane to several nickel surfaces, due to electrons from the vacuum gauge. It is believed that ionisation or excitation of the residual gas plays a minor role in this effect[43].

This also implies that the use of electron beams for surface analyses should be done in a very good vacuum ($< 10^{-7}$ Pa). So it can be concluded that the electron bombardment, eventually combined with baking, can be used to remove very specific contaminating species. However possible ESA effects can influence the surface composition in the reverse way if no attention is payed to a proper vacuum.

CLEAVAGE IN ULTRA HIGH VACUUM

Cleavage is used in some applications with single crystals of a certain class of materials. However, results are rarely mentioned in literature.

The topography of the resulting surface depends on the method of cleaving. The technique of a wedge shaped knife produced a surface containing many irregularities[39]. On the other hand there is an indication that one can cleave more easily along a plane which contains irregularities. An improved surface has been produced by fracturing a long (111) oriented silicon specimen when applying a bending force[39].

Menyhard[42] reports investigations on cleaved surfaces of polycrystalline tungsten wires. He investigated the enrichment of contaminations or alloy elements on the grain boundaries, which may influence very much the mechanical behaviour of polycrystalline metals. The tungsten wires were cleaved in ultra-high vacuum. After some slight deformation the wires could be cleaved parallel to their axis. The cleavage runs along the grain boundaries which are also parallel to the axis of the wire. Cleaving did turn out not always to produce a reproducible surface[42].

The group of Rivière[40] has studied pure iron extensively. The polycrystalline iron samples are cut in the vacuum with a strong knife. Another method they use is to pull apart iron bars in vacuum. For this purpose one needs special equipment for pulling and placing the sample in the holder afterwards.

CONCLUSIONS

It can be concluded that all surface treatments have their specific advantages and disadvantages. The use of a combination of techniques is inevitable. In the literature one often reports different combinatings of cleaning methods for one material, giving the same result. One should be very careful not to change the composition or to damage the surface to be cleaned.

Even the techniques used to monitor the cleanliness of the surface can cause damage or change of composition or even enhanced adsorption.

REFERENCES

1. W. Turkenburg, Ph.D. Thesis, University of Amsterdam (1976).
2. J.F. van der Veen, R.G. Smeenk, R.M. Tromp and F.W. Saris, Surface Sci. 79, 212 (1974).
3. H.E. Farnsworth, DAGV-Symposium, Moderne Verfahren der Oberflächenanalyse (1972).
4. S.M. Bedair and H.P. Smith, J.Appl.Phys. 40, 4776 (1969).
5. R. Gautier, P. Pinard and F. Devaine, Vide 24, 140, 109 (1969).
6. L. Bakker, Ph.D. Thesis, Technical University Delft (1972).
7. E.N. Sickafus, Surface Sci. 19, 181 (1970).
8. S. Mróz, C. Koziol and J. Kołaczkiewicz, Vacuum 26, 61 (1975).
9. F.C. Schouten, D.L.J. Gijzeman and G.A. Bootsma, Nederl.Tijdschr.v.Vacuüm Techn. 15, 46 (1977).
   F.C. Schouten, E. te Brake, O.L.J. Gijzeman and G.A. Bootsma, Surface Sci., to be published.
10. J. Lupujoulade and K. Neil, J.Chem.Phys. 57, 3535 (1972).
11. Y. Margoninsky, D. Segal and R.E. Kirby, Surface Sci. 51, 488 (1975).
12. Y. Margoninsky, J.Appl.Phys. 47, 3868 (1976).
13. J. Verhoeven and J. Los, Surface Sci. 58, 566 (1976).
14. J. Verhoeven, in: "Proceedings 7th Intern.Vacuum Congres and the 3rd Conference on Solid Surfaces" pp. 915, Vienna (1977).
15. J.P. Coad, H.E. Bishop and J.C. Rivière, Surface Sci. 21, 253 (1970).
16. J. Verhoeven and J. Los, to be published.
17. G.A. Bootsma, personal communications.
18. Mahboob Khan and Juergen M. Schroeer, Rev.Sci.Instrum. 42, 1348 (1971).
19. J. Franks and A.M. Chandler, Vacuum 24, 489 (1974).
20. A.M. Chandler and R.K. Fitch, Vacuum 24, 483 (1974).
21. J.W. Coburn, J.Vac.Sci.Technol. 13, 1037 (1976).
22. E. Taglauer, U. Beitat, G. Marin and W. Heiland, Nuclear Materials 63, 193 (1976).
23. T. Smith, Surface Sci. 27, 45 (1971).
24. J.A. McHugh, Radiat.Eff. 18, 211 (1973).
25. P. Staib, Radiat.Eff. 18, 217 (1973).

26. J. Kistemaker and H.E. Roosendaal, Jpn.J.Appl.Phys.Suppl. 2, Pt.Z. 571 (1974).
27. W.T. Ogar, N.T. Olson and H.P. Smith, J.Appl.Phys. 40, 4997 (1969).
28. H.L. Hughes, R.D. Baxter and B.R. Phillips, IEEE Trans.Nucl. Sci. NS-19, 6, 256 (1972).
29. H. Shimizu, M. Ono and K. Nahayama, J.App.Phys. 46, 460 (1975).
30. C.G. Pantano, D.B. Dove and G.Y. Onoda, J.Vac.Sci.Technol. 13, 414 (1976).
31. S. Sayyid and E.M. Williams, Surface Sci. 62, 431 (1977).
32. T.E. Madey and J.T. Yates, J.Vac.Sci.Technol. 8, 525 (1971).
33. T.E. Madey, Surface Sci. 36, 281 (1973).
34. J. Anderson and P.J. Estrup, Surface Sci. 9, 463 (1968).
35. J.P. Coad, M. Gettings and J.C. Rivières, internal report, AERE, Harwell, Oxfordshire, England.
36. D. Menzel, P. Kronauer and W. Jeland, Ber.Bunsenges.Phys.Chem. 75, 1074 (1971).
37. R. Franchy and D. Menzel, in: "Proc.7th Intern.Vac.Congr.and 3rd Intern.Conf.Solid Surf.", pp. 1209, Vienna (1977).
38. S. Baydyavoy, W.R. Bottoms and P. Mark, Surface Sci. 28, 517 (1971).
39. G.W. Gobeli and F.G. Allen, J.Phys.Chem.Solids 14, 23 (1960).
40. J.C. Rivière, personal communication.
41. E.V. Kornelsen, J.Vac.Sci.Technol. 13, 716 (1976).
42. M. Menyhárd, Nederl.Tijdschr.v.Vacuümtechniek 16, 220 (1978).
42'. M. Menyhárd, personal communication.
43. W. Ranke, J.Phys.D 11, L37 (1978).

ABOUT THE CONTRIBUTORS

Here are included brief biographical sketches of only those authors who have contributed to this volume. Biosketches of contributors to Volume 2 are included in that volume.

P. B. Adams is presently Research Supervisor of Surface Analysis at Corning Glass Works, Corning, New York, where he has been employed since 1951. He received his B.S. degree in 1951 from Hobart College. He is a Fellow of the American Ceramic Society and a recipient of the H. V. Churchill Award for Service to ASTM Committee E2 on Emission Spectroscopy. He has about 30 publications in the fields of glass analysis and chemical properties of glass surfaces.

R. N. Anderson is Professor of Metallurgy in the Department of Materials Engineering at San Jose State University. He received his Ph.D. from Stanford University. He was with the U.S. Navy as research engineer and at Stanford as Associate Professor before coming to San Jose State University.

W. W. Balwanz is President of Mattox, Inc. He has worked for the Federal Government since 1935, continuing as a consultant at the Naval Research Laboratory since his retirement from there in 1974. He received his MSEE degree from the University of Maryland in 1948. His work for the past 30 years has been mainly on plasmas, their properties and their interaction with naval systems.

Bharat Bhushan is Senior Engineering Scientist at the R & D Division of Mechanical Technology Inc., Latham, New York. He received his M.S. in mechanical engineering in 1971 from MIT, and M.S. in mechanics in 1973 and Ph.D. in mechanical engineering in 1976 from the University of Colorado. He is interested in areas of material interface phenomena, material behavior and product development. He is the author of numerous technical papers.

Robert N. Bolster is with the Surface Chemistry Branch of the Naval Research Laboratory, Washington, D.C., which he joined in 1961. He received his B.S. degree in Chemistry from the University of Virginia in 1958. He has been engaged in research on synthetic lubricants, corrosion inhibition, and cleaning and salvage techniques, and has publications and patents in these fields. He has served as a consultant to the Navy and other government agencies on the recovery of electronic equipment contaminated by fire and flooding.

Harry Bonham is presently affiliated with the Electronic Systems Group, Rockwell International, Dallas, Texas. He received his M.S. degree in Electrical Engineering from the University of Missouri in 1971. He has studied hybrid interconnection techniques for 12 years (with Bendix and Rockwell), and has developed lead frame wire and beam lead bonding techniques. At present, he is developing an automatic intraconnection method for assembly of hybrid circuits.

Robert Bouwman is affiliated with Shell Research B.V., Amsterdam, The Netherlands. He received his doctorate degree based on his work dealing with surface and adsorption studies of metals and alloys from the University of Leiden. Since joining Shell Research in 1971, he has been concerned with studies of the solid/gas interface using a variety of sophisticated surface analytical techniques. He has published on the characterization of alloy surfaces, the quantification of electron spectroscopic data, work functions, and surface cleaning.

Dale A. Brandreth is Senior Research Engineer at E. I. du Pont de Nemours & Co, Freon® Products Laboratory, Chestnut Run, Wilmington. He received his Ph.D. in chemical engineering in 1965 from the University of Toronto. Before joining du Pont in 1966, he had worked in both industry and academia. His technical interests are applications of mathematics in chemical engineering, thermodynamics, surface chemistry and contamination control.

Charles O. Brewer is Supervisor of Weld Development and Fabrication, Mound Facility, Monsanto Research Corp., Miamisburg, Ohio. He received his B.S. degree in 1961 from Morehead State College, Morehead, Kentucky.

J. W. Costerton is Professor of Microbiology at the University of Calgary, Calgary, Canada. He obtained his Ph.D. in Microbiology from the University of Western Ontario. He has published numerous papers and reviews on the ultrastructure and molecular architecture of the bacterial cell wall. Recently he has turned his attention to those bacterical cell wall structures that are involved in adhesion to surfaces, and his studies form the basis of a popular article in the January, 1978 issue of Scientific American.

J. A. Cross is manager of the Wolfson Electrostatic Advisory Unit at the University of Southampton, England. She graduated in Physics from Manchester University in 1968 and obtained her

Ph.D. in high vacuum physics from London University. Since that time she has been at Southampton University and has performed fundamental research into adhesion of charged dusts.

E. C. Ethridge is presently with the Department of Materials Science and Engineering, University of Florida, Gainesville. He received his B.S., M.S., and Ph.D. degrees in Materials Science and Engineering from the same university where he is now working. He is the coauthor of a dozen scientific publications and a book in the areas of glass surfaces and biomaterials.

R. G. Fekula is an Associate Member of the Technical Staff in the Film and Hybrid Technology Laboratory at Bell Telephone Laboratories in Allentown, Pennsylvania. He has a B.S. degree in Electrical Engineering from Lafayette College.

W. J. Flood is a Member of the Integrated Circuit Packaging Department at Bell Laboratories in Allentown, Pennsylvania. He attended the U.S. Naval Academy and California State College at Long Beach. He is concerned with the design and process development of nonhermetic packaging techniques.

Fernando Galembeck is currently in the faculty of the Institute of Chemistry of the University of Sao Paulo, Sao Paulo, Brazil, where he is working in colloid and surface chemistry. He obtained his Ph.D. in Chemistry in Sao Paulo, and did postdoctoral research in Denver, Colorado, and Davis, California from 1972 to 1974. He has authored papers on the chemistry of metal carbonyls, protein-protein interactions and on polymer surface modifications.

Sergio Galembeck is a B.Sc. student at the Institute of Chemistry of the University of Sao Paulo, Sao Paulo, Brazil.

C. G. Geesey is Assistant Professor of Biochemistry and Microbiology at the University of Victoria, B.C., Canada. He obtained his Ph.D. in Microbiology from Oregon State University. He has published a number of papers on the physiology and distribution of aquatic bacteria. During his postdoctoral fellowship, he supervised work on the pristine alpine Marmot Basin study which is being reported as an exhaustive monograph and is the basis of a popular article in the January, 1978 issue of Scientific American.

C. C. Ghizoni is a Senior Researcher at INPE (Brazilian Aerospace Research Institution.) He is a Ph.D. Electronics Engineer who graduated from ITA (São José dos Campos, Brazil.) He is interested in the transducing properties of solids.

Michael Halter is a Research Engineer with Battelle Pacific Northwest Laboratories, Richland, Washington. A graduate of the University of Missouri in Columbia, he specializes in mechanical design and analysis. He holds a Ph.D. degree.

M. S. Hassan is currently an Engineer at American Micro Devices in Sunnyvale, California. He received his Ph.D. in 1977 from Stanford University.

Larry L. Hench is Professor of Materials Science and Engineering and Head of the Ceramics Division of the University of Florida, Gainesville, where he also serves as Director of the Biomedical Engineering Program of Distinction for the State University System. He received his B.S. and Ph.D. degrees in ceramic engineering from the Ohio State University. The author of over 120 scientific publications and editor/author of eight books, his research concerns characterization and physical properties of ceramics, glasses, glass-ceramics, and medical and dental applications of specially designed surface active bioglasses and bioglass-ceramics.

Anton A. Holscher is affiliated with Shell Research B.V., Amsterdam, The Netherlands. His doctoral thesis was in the area of field emission and field ionization microscopy, the work he started in 1961 upon joining the Shell Laboratories. Since 1970, he is leader of a research group engaged in the structural characterization of solids and surfaces using a variety of optical, spectroscopic, and thermal techniques. He has coauthored various papers on microscopy, work function measurements, Auger and photoelectron spectroscopy, and ellipsometry.

Rulon E. Johnson, Jr. is a Research Associate in the Central Research Department, E. I. du Pont de Nemours & Company, Wilmington. He has been with du Pont since 1957. He received his Ph.D. in Physical Chemistry in 1957 from Stanford University. He is past Secretary-Treasurer, Colloid and Surface Science Division of ACS; past member, Advisory Board, J. Colloid and Interface Science; member, Advisory Board, Journal of Adhesion. His technical interests are surface chemistry, wettability, particle adhesion, and dispersions.

R. G. Jungst is affiliated with the Sandia Laboratories, Albuquerque. He received his Ph.D. in Inorganic Chemistry from the University of Illinois. He is working in the area of corrosion chemistry, and explosive synthesis, characterization and stability.

Hisazo Kawakatsu is currently the Chief of the Electron and Ion-Beam Section at the Electrotechnical Laboratory, Tokyo, which he had joined in 1954. He received his B.Sc. degree in 1953 from Kyoto University and Eng.D. degree in 1962 from Osaka University, Osaka, Japan. At Electrotechnical Laboratory, he has worked in the field of electron and ion optics.

William G. Kenyon is an organic chemist with DuPont's Petrochemicals Department based in Wilmington, DE, where he is responsible for fluorosolvent cleaning technical development programs. He received his Ph.D. in Organic Chemistry in 1963 from Duke University and joined du Pont in 1965 after two years of postdoctoral research. His present interest is electronic assembly cleaning and contamination control, and he is Chairman of the IPC Cleaning and Contamination Control Committee. He has authored a number of publications dealing with use of solvents in cleaning.

Y. Kharlamov holds a Chair in Engineering at the Voroshilovgrad Machine Construction Institute, Voroshilovgrad, U.S.S.R. He was a visiting scholar at Stanford University during 1977-1978.

J. A. Koutsky is Professor of Chemical Engineering at the University of Wisconsin, Madison. He received his Ph.D. in Polymer Science and Engineering in 1966 from Case-Western Reserve University. His interests have been in the solid state structure and morphology of polymers; surface properties of polymers and its relation to adhesion, fracture strength and durability of bonded polymer systems.

Howard H. Manko has his own consulting service in Teaneck, New Jersey. Before establishing his firm, he has had a number of responsible industrial appointments. He holds an MBA in Marketing, an M.S. in Metallurgical Engineering, a P.E. diploma, and a B.S. in Chemical Engineering (Technion, 1952). He is the author of Solders and Soldering, the section on "Joining" in the Printed Circuits Handbook, Effective Technical Speeches and Sessions, as well as over 40 technical papers. He has been holding a series of annual Solder Technology Seminars since 1964 at various universities in the United States, and these seminars have also been presented in various countries overseas. He is presently the U.S.A. Technical Expert on the International Electrostandards Commission W.G. 3 Committee, and received the President's Award from the IPC for his work in solderability testing. Among his many R & D activities are new soldering and brazing materials and solutions to solder joining problems.

T. M. Massis is affiliated with the Sandia Laboratories, Albuquerque. He received his B.S. in Chemistry from the University of Albuquerque. He has extensive experience in the areas of explosive stability, materials compatibility, thermal analysis and gas chromatographic techniques.

Toshiaki Matsunaga is Professor in the Department of Fuel Chemistry, Mining College, Akita University, Japan. From 1962 until March, 1978, he worked at the Chemical Research Institue of Nonaqueous Solutions at Tohoku University, Sendai, Japan. He earned his B.Eng. from Tokyo University and Ph.D. from Tohoku University. His fields of interest include surface chemistry (especially surface energetics of solids) and coal gasification.

Donald M. Mattox is Supervisor of the Surface Metallurgy Division at Sandia Laboratories, Albuquerque. He has been active in the fields of research, development, and technology of thin films for many years. At present his principal interests are in coatings for solar energy applications, fusion reactor application, and for wear and erosion resistance.

L. C. M. Miranda is Professor of Physics at the State University of Campinas, Brazil. He is a Ph.D. (Oxford) physicist, and is the author or coauthor of many papers on solid state physics and molecular physics, mainly theoretical.

Kashmiri Lal Mittal is presently employed at the IBM Corportation in Hopewell Junction, New York. He received his B.Sc. in 1964 from Panjab University, M.Sc. (first class first) in Chemistry in 1966 from Indian Institute of Technology, New Delhi, and Ph.D. in Colloid Chemistry in 1970 from the University of Southern California. During his college days, based on his scholastic records, he was recipient of many scholarships. In the last few years, he has organized and chaired a number of very successful international symposia and in addition to this two-volume set, he has edited seven more volumes as follows: Adsorption at Interfaces, and Colloidal Dispersions and Micellar Behavior (1975); Micellization, Solubilization and Microemulsions, Volumes 1 & 2 (1977); Adhesion Measurement of Thin Films, Thick Films, and Bulk Coatings (1978); and Solution Chemistry of Surfactants, Volumes 1 & 2 (July, 1979). In addition to these volumes he has published about 40 papers in the areas of surface and colloid chemistry, adhesion, polymers, etc. He has given or is scheduled to give many invited talks on the multifarious facets of surface science, particularly adhesion, on the invitation of various societies and organizations in many countries, and is always a sought-after speaker. He is a member of many professional and honorary societies, is a Fellow of the American Institute of Chemists, is listed in American Men and Women of Science and Who's Who in the East. Recently he has been appointed a member of the editorial boards of a number of scientific and technical journals. He started the highly acclaimed short course on adhesion in the United States in 1976.

Brij M. Moudgil is currently at the Henry Krumb School of Mines, Columbia University on leave from Occidental Research Corporation. He received his B.E. degree from the Indian Institute of Science, Bangalore, India, and M.S. degree from Columbia University.

Gloria B. Munier is a member of the Electrochemical and Contamination Research Department at Bell Laboratories, Holmdel, New Jersey. She received her B.S. degree in Chemistry and Zoology from Douglas College, and joined Bell Laboratories in 1953. Her work involves investigations of contamination problems and research on the interactions of materials and the environment.

W. E. J. Neal currently holds the posts of Senior Lecturer and Senior Tutor in Physics at Aston University, Birmingham, England. He was awarded a B.Sc. Hons. Physics degree (London, 1949) and a doctorate in 1953. He is a Fellow of the Institute of Physics and a chartered engineer. His research interests include transport properties in gases, electrical superconducting and optical properties of thin films, and solar energy applications.

Howard G. Patton is a Project Mechanical Engineer in the Laser Fusion Program at the Lawrence Livermore Laboratory. He graduated from the San Francisco State University and has been at LLL since 1959. Since joining LLL, he has worked on a number of projects including nuclear weapons research and magnetic confinement fusion reactor research. He is Chairman of the Education Committee of the American Vacuum Society and lectures nationally on the subject of vacuum engineering. He is the author or coauthor of a number of research papers.

Steve I. Petvai is currently employed at the IBM Corporation, Hopewell Junction, New York. He received his M.S.M.E. degree in 1953 from the Technical University of Budapest, Hungary. He is a certified manufacturing engineer and a registered professional engineer. He has received Invention Award of IBM and has contributed articles to professional journals.

P. V. Plunkett is presently with the Sandia Laboratories, Albuquerque, New Mexico. Before his present position, he was employed by Rockwell International since 1973. He received his M.S. degree in Materials Science in 1978 from the University of Texas at Arlington. He has developed resistor-conductor metallization processes for a variety of materials and has extensively worked in all areas of hybrid processing.

Linda A. Psota is a member of the Electrochemical and Contamination Research Department at Bell Telephone Laboratories, Holmdel, New Jersey. She received her B.A. degree in Chemistry from Hunter College in 1976 and joined Bell Laboratories in April, 1977. Her work involves investigations of contamination problems and research on the interactions of materials and the environment.

R. K. Quinn is affiliated with the Sandia Laboratories, Albuquerque. He received his Ph.D. in Physical Chemistry from the University of Texas. He has done research in solid state chemistry, interfacial electrochemistry, and surface properties of reactive materials.

Barbara T. Reagor is a member of the Electrochemical and Contamination Research Department at Bell Laboratories, Holmdel, New Jersey. She joined Bell Laboratories in 1970 and is currently working on a Ph.D. degree in Inorganic Chemistry at Seton Hall University. Her work involves investigation of contamination problems and research on the interactions of materials and the environment.

A. S. Rehal is currently carrying out his postgraduate research in a collaborative programme between the University of Aston, Birmingham, and Alcan International. He obtained an honours degree in Physics in 1975 from the University of Aston.

D. L. Rehrig is a Planning Engineer in the Semiconductor Materials Engineering Department at the Western Electric Allentown Works. He received an M.S. in Metallurgy and Materials Science in 1975 from Lehigh University. He has been responsible for projects including plasma cleaning, electronic packaging and electroplating. He is President of the Allentown Reading Branch of the American Electroplaters' Society.

C. A. Ribeiro is Assistant Professor at the State University of Campinas, Brazil. He is a physicist with a D.Sc. degree from Grenoble, France. He is interested in electron microscopical methods in the study of solid materials.

R. N. Roberts is affiliated with the Sandia Labortories, Albuquerque. He received his M.S. in Physical Chemistry from Columbia. He has worked at Los Alamos Scientific Laboratory and Lockheed Propulsion. He is presently working in the areas of crystal chemistry and optics.

Blanche Russiello retired from Bell Telephone Laboratories in June, 1978 and is now residing in the Bronx, New York. She received her B.A. degree in Zoology and Natural Science in 1936 from Hunter College and joined Bell Laboratories in 1942.

Dale R. Schaeffer is Senior Research Engineer at the Mound Facility, Monsanto Research Corporation, Miamisburg, Ohio. He received his B.S. in Metallurgical Engineering in 1967 from the University of Missouri, Rolla, and his M.S. also in Metallurgical Engineering in 1973 from Ohio State University.

Randolph H. Schnitzel is currently employed at the IBM Corporation, Hopewell Junction, New York. He received his B.S. and M.S. degrees in Metallurgical Engineering from Michigan College of Mining and Technology, and Columbia University, respectively. He has published a number of papers and holds patents in semiconductor and metallurgy areas.

John C. Scott is Research Fellow in the Fluid Mechanics Research Institute, University of Essex. He received his Ph.D. degree from Imperial College, London, in 1971. Although his research background is in the field of acoustics--he is editor of the journal Acoustics Letters--his main interest at present is in a wide range of effects associated with surface-active films on water surfaces. He was recently awarded a Science Research Council Advanced Fellowship to enable him to continue this work.

Keizo Shimizu is affiliated with the Electrotechnical Laboratory, Tokyo, which he had joined in 1965. He received his B.Eng., M.Eng. and Eng.D. degrees from Tohoku University, Sendai, Japan in 1963, 1965, and 1977 respectively. At Electrotechnical Laboratory, he has worked on the fundamental study of ion-beam applications.

J. Douglas Sinclair is a member of the Electrochemical and Contamination Research Department at Bell Telephone Laboratories, Holmdel, New Jersey. He received his Ph.D. degree in Inorganic Chemistry from the University of Wisconsin in 1972 and joined the Bell Laboratories the same year. His work involves investigations of contamination problems and research on the interaction of materials and the environment.

P. Somasundaran is Professor of Mineral Engineering in the Henry Krumb School of Mines, Columbia University. He received his Ph.D. from the University of California, Berkeley, in 1964. He has authored more than 100 scientific publications and is the author/editor of two books. His research interests include premicellar aggregation and other solution chemistry aspects of surfactants, flotation, flocculation, enhanced oil recovery, and chemomechanical effects on grinding and biosurfaces.

Irving F. Stowers is currently a Project Engineer and Associate Group Leader of the Laser Program Solid State Mechanical Engineering Group at the Lawrence Livermore Laboratory. The

Laser Program is currently designing the Nova Laser System for the Department of Energy. He received his D.Sc. degree in Mechanical Engineering from MIT in 1974. Since joining LLL in 1976, he has worked on and published several articles concerning causes of laser damage to optics, the maintenance of damage free laser components, and techniques for reducing contaminants on laser components and optics.

Robert Sullivan is a mechanical specialist with Battelle Pacific Northwest Laboratories, Richland, Washington, where he has worked for 11 years. His R & D field is prototype mechanical design and instrumentation.

Jan B. van Mechelen joined Shell Laboratories in Amsterdam, The Netherlands, in 1965. He studied technical physics at the Technical College of Dordrecht, The Netherlands. He is coauthor of a number of publications concerned with the quantification of Auger spectroscopic data and the surface cleaning of solids.

Helion Vargas is Professor of Physics at the State University of Campinas, Brazil, where his group works on spectroscopic studies of solids and on the genetic improvement of corn using ESR spectroscopy. He obtained his Ph.D. from Grenoble, France.

J. Verhoeven is affiliated with the FOM Institute for Atomic and Molecular Physics in Amsterdam, The Netherlands, where he is the leader of the Vacuum Department. At present, he is studying the effect of electron bombardment on the adsorption or desorption of various gases from a nickel surface.

John R. Vig is currently employed in the Electronics Technology and Devices Laboratory at Fort Monmouth, New Jersey, where he is engaged in studies aimed at improving the stability of high precision quartz crystal resonators. He is currently leader of the Quartz Crystal Task Force. From 1969-1972 he served in the U.S. Army, and developed and received a patent on a superconductive high Q turntable filter. He received his Ph.D. degree from Rutgers - The State University, New Brunswick, in 1969. He holds ten patents, one of which is on a UV/ozone surface cleaning method.

R. E. Whan is presently Supervisor of the Explosive Materials Division at Sandia Laboratories, Albuquerque. She received her Ph.D. in Physical Chemistry from the University of New Mexico. She has done research in semiconductor physics, radiation effects, and electron microscopy.

W. J. Whitfield is currently Division Supervisor, Isotopes Source and Applications Division, Sandia Laboratories, Albuquerque. He received his D.Sc. degree in 1970 from Hardin-Simmons University. Before coming to Sandia Laboratories, he was at the Naval Research Laboratory. He has been very active in a number of societies and was appointed to National Academy of Science Committee on Airborne Bacterial Control in Hospital Operating Rooms in 1974. He has contributed a great deal in the area of contamination control and has been the recipient of a number of awards which include Individual Scientific Technical Achievement Award by the American Association for Contamination Control, 1969, and Distinguished Service Award, New Mexico Hardin-Simmons University Alumni Association in 1973.

SUBJECT INDEX

Pages 1-524 appear in Volume 1
Pages 525-1042 appear in Volume 2

†Ionograph is a trademark of Alpha Metals, Inc.

†Meseran is a trademark of ERA Systems, Inc.

---

†Omega Meter is a trademark of Kenco Chemical Company

*ESCA - Electron Spectroscopy for Chemical Analysis
ISS - Ion Scattering Spectrometry
SIMS - Secondary Ion Mass Spectroscopy